AF333954

DATA MINING IN BIOMEDICINE

DATA MINING IN BIOMEDICINE

Edited by

PANOS M. PARDALOS
University of Florida, Gainesville, FL

VLADIMIR L. BOGINSKI
Florida State University, Tallahassee, FL

ALKIS VAZACOPOULOS
Dash Optimization, Englewood Cliffs, NJ

 Springer

Library of Congress Control Number: 2006938889

ISBN-10: 0-387-69318-1   e-ISBN-10: 0-387-69319-X
ISBN-13: 978-0-387-69318-7   e-ISBN-13: 978-0-387-69319-4

Printed on acid-free paper.

AMS Subject Classifications: 65K05, 90C90, 62-07

Contents

Part II Data Mining Techniques in Disease Diagnosis

Part III Data Mining Studies in Genomics and Proteomics

Preface

Data mining techniques are applied in a great variety of practical problems nowadays. With the overwhelming growth of the amounts of data arising in diverse areas, the development of appropriate methods for extracting useful information from this data becomes a crucial task.

Biomedicine has always been one of the most important areas where information obtained from massive datasets can assist medical researchers and practitioners in understanding the structure of human genome, exploring the dynamics of human brain, disease diagnosis and treatment, drug discovery, etc. Data mining techniques play an essential role in analyzing and integrating these datasets, as well as in discovering biological processes underlying this data.

This volume presents a collection of chapters covering various aspects of data mining problems in biomedicine. The topics include new approaches for the analysis of biomedical data, applications of data mining techniques to diverse problems in medical practice, and comprehensive reviews of recent trends in this exciting research area.

A significant part of the book is devoted to applications of data mining techniques in genomics. The success of the Human Genome Project has provided the data on the DNA sequences of the human genome. New tools for analyzing this data have been recently developed, including the widely used DNA microarrays. A number of chapters present novel approaches to microarray data analysis with applications in disease diagnosis based on gene expression profiling.

Analyzing protein structure and protein fold prediction is another interesting research field addressed in this volume. The methods discussed here include global optimization models and topological methods that proved to be applicable in practice.

One more exciting research area discussed in this book deals with data mining techniques for studying human brain dynamics. Recent advances in this field are associated with the extensive use of electroencephalographic (EEG) recordings, which can be treated as a quantitative representation of the brain function. The analysis of EEG data combines different methodologies, including statistical preprocessing and hypothesis testing, chaos theory, classification models, and network-based techniques.

Moreover, several chapters present new promising methodological frameworks for addressing data mining problems in biomedicine, including Logical Analysis of Data, Sparse Component Analysis, and Entropy Minimization.

We believe that this book will be of interest to both theoreticians and practitioners working in diverse fields of biomedical research. It can also be helpful to graduate students and young researchers looking for new exciting directions in their work.

We would like to take the opportunity to thank the authors of the chapters for their valuable contributions, and Springer staff for their assistance in producing this book.

Gainesville, Florida, USA

Panos M. Pardalos
Vladimir Boginski
Alkis Vazacopoulos

List of Contributors

G. Addonisio
Dipartimento di Statistica, Probabilità e Statistiche Applicate
Università di Roma "La Sapienza",
Italy

S. Alexe
RUTCOR - Rutgers University
Center for Operations Research
Piscataway, NJ, USA
salexe@rutcor.rutgers.edu

H. Bakardjian
Brain Science Institute, RIKEN,
Wako-shi, Japan
hova@bsp.brain.riken.go.jp

C. Bauco
Unita Operativa Geriatria
Ospedale "G. De Bosis", Cassino,
Italy

H. Bensmail
University of Tennessee
Statistics Department
Knoxville, TN 37996-0532, USA
bensmail@utk.edu

V. Boginski
Industrial Engineering Department
Florida State University
Tallahassee, FL 32310, USA
boginski@eng.fsu.edu

M.W. Brauner
Department of Radiology, Fédération
MARTHA, UFR Bobigny,
Université Paris 13 et Hôpital
Avicenne AP-HP, 125, route de
Stalingrad, 93009 Bobigny Cedex,
France
michel.brauner@wanadoo.fr

N. Brauner
Laboratoire Leibniz-IMAG, 46 av.
Felix Viallet, 38031 GRENOBLE
Cedex, France
Nadia.Brauner@imag.fr

S. Busygin
Industrial and Systems Engineering
Department
University of Florida
303 Weil Hall, Gainesville, FL 32611,
USA
busygin@ufl.edu

Y. Cao
1026 Springfield Drive, Campbell,
CA 95008, USA
contact@biosieve.com

P.R. Carney
Department of Pediatrics
University of Florida, USA
carnepr@peds.ufl.edu

G. Ceci
Department of Mathematics, Second
University of Naples
via Vivaldi 43, I-81100 Caserta, Italy

W. Chaovalitwongse
Department of Industrial and
Systems Engineering
Rutgers, The State University of
New Jersey
wchaoval@rci.rutgers.edu

D. Chen
Department of Preventive Medicine
and Biometrics
Uniformed Services University of the
Health Sciences
4301 Jones Bridge Road, Bethesda,
MD 20814, USA
dchen@usuhs.mil

L. Chen
Computer Science Department
University of Northern British
Columbia
Prince George, BC, Canada V2N
4Z9
chenl@unbc.ca

X. Chen
Electrical Engineering and Computer
Science Department
The University of Kansas
1520 West 15th Street, Lawrence,
KS 66045, USA
xwchen@eecs.ku.edu

S. Chen
Center of Neuroproteomics and
Biomarkers Research & Computer
and
Information Science and Engineering
Department
University of Florida, Gainesville,
FL, USA
suchen@cise.ufl.edu

X. Cheng
Department of Computer Science
The George Washington University
801 22nd St. NW, Washington, DC
20052, USA
cheng@gwu.edu

M.L. Chiusano
Department of Genetics, General
and Molecular Biology
University of Naples "Federico II",
via Mezzocannone 8, 80134
Naples, Italy
chiusano@unina.it

A. Cichocki
Brain Science Institute, RIKEN,
Wako-shi, Japan
cia@bsp.brain.riken.go.jp

G. Colonna
Department of Biochemistry and
Biophysics
Second University of Naples
via Costantinopoli 16, I-80138
Naples, Italy
colonna@unina2.it

T. Conway
Department of Botany and Microbi-
ology
University of Oklahoma
Norman, OK, USA
tconway@ou.edu

S. Costantini
Department of Biochemistry and
Biophysics
Second University of Naples
via Costantinopoli 16, I-80138
Naples, Italy
susan.costantini@unina2.it

L.K. Dance
Department of Neuroscience,
University of Florida
Malcolm Randall Department of
Veteran's Affairs Medical Center
Gainesville, FL, USA
lkdance@mbi.ufl.edu

M. D'Apuzzo
Department of Mathematics, Second
University of Naples
via Vivaldi 43, I-81100 Caserta, Italy
marco.dapuzzo@unina2.it

N. Denslow
Center of Neuroproteomics and
Biomarkers Research
University of Florida, Gainesville,
FL 32610, USA
ndenslow@ufl.edu

M. Deshpande
Department of Computer Science
and Engineering
University of Minnesota
4-192 EE/CSci Building, 200 Union
Street SE
Minneapolis, MN 55455
deshpand@cs.umn.edu

L. Di Cioccio
Area Dipartimentale Geriatrica
ASL Frosinone, Frosinone, Italy

D. Di Serafino
Department of Mathematics, Second
University of Naples
via Vivaldi 43, I-81100 Caserta, Italy
daniela.diserafino@unina2.it

D.-Z. Du
Computer Science Department
University of Texas at Dallas
Richardson, TX 75083, USA
dzdu@utdallas.edu

A. Facchiano
Institute of Food Science, CNR, via
Roma 52 A/C, I-83100 Avellino,
Italy
angelo.facchiano@isa.cnr.it

T. Faraggiana
Dipartimento di Medicina Sperimen-
tale e Patologia
Università di Roma "La Sapienza",
Italy

L. Frusciante
Department of Soil, Plant and
Environmental Sciences (DISSPA)
University of Naples "Federico II",
via Università 100, 80055
Portici, Italy
fruscian@cds.unina.it

J.-B. Gao
Department of Electrical and
Computer Engineering
University of Florida, Gainesville,
FL 32611, USA
gao@ece.ufl.edu

P.G. Georgiev
ECECS Department
University of Cincinnati
Cincinnati, Ohio 45221-0030
USA
pgeorgie@ececs.uc.edu

C. Giannakakis
Dipartimento di Medicina Sperimen-
tale e Patologia
Università di Roma "La Sapienza",
Italy

P.L. Hammer
RUTCOR - Rutgers University
Center for Operations Research
640 Bartholomew Rd., Piscataway
NJ, 08854-8003 USA
hammer@rutcor.rutgers.edu

A. Haoudi
Eastern Virginia Medical School
Department of Microbiology and
Molecular Cell Biology
Norfolk, VA 23507, USA
haoudia@evms.edu

W.E. Haskins
Center of Neuroproteomics and
Biomarkers Research
University of Florida, Gainesville,
FL 32610, USA

R.L. Hayes
Center for Traumatic Brain Injury
Studies
University of Florida, Gainesville,
FL 32610, USA
hayes@ufbi.ufl.edu

C. Huang
Department of Computer Science
and Engineering
University of Minnesota
Minneapolis, MN 55455, USA
huang@cs.umn.edu

L.D. Iasemidis
The Harrington Department of
Bioengineering
Arizona State University, Tempe, AZ
85287 USA
Leon.Iasemidis@asu.edu

G. Karypis
Department of Computer Science
and Engineering
University of Minnesota
4-192 EE/CSci Building, 200 Union
Street SE
Minneapolis, MN 55455
karypis@cs.umn.edu

H. Kim
Computer and Information Science
and Engineering Department
University of Florida, Gainesville,
Florida 32611, USA
hykim@cise.ufl.edu

B. Krishnamoorthy
Department of Mathematics
Washington State University
kbala@wsu.edu

M. Kuramochi
Department of Computer Science
and Engineering
University of Minnesota
4-192 EE/CSci Building, 200 Union
Street SE
Minneapolis, MN 55455
kuram@cs.umn.edu

Y. Li
Department of Computer Science
and Engineering
University of Minnesota
Minneapolis, MN 55455, USA
yili@cs.umn.edu

Z. Liu
Bioinformatics Cell, TATRC
110 North Market Street, Frederick,
MD 21703, USA
liu@bioanalysis.org

I. Lozina
RUTCOR - Rutgers University
Center for Operations Research
640 Bartholomew Rd., Piscataway
NJ, 08854-8003 USA
ilozina@rutcor.rutgers.edu

M. Mammadov
Centre for Informatics and Applied
Optimization
University of Ballarat
Victoria, 3353, Australia
m.mammadov@ballarat.edu.au

C.N. Meneses
Department of Industrial and
Systems Engineering
University of Florida
303 Weil Hall, Gainesville, FL,
32611, USA
claudio@ufl.edu

A. Mucherino
Department of Mathematics, Second
University of Naples
via Vivaldi 43, I-81100 Caserta, Italy
antonio.mucherino@unina2.it

S.P. Nair
Department of Biomedical Engineer-
ing, University of Florida
Malcolm Randall Department of
Veteran's Affairs Medical Center
Gainesville, FL, USA
spnair@mbi.ufl.edu

K. Narayanan
The Harrington Department of
Bioengineering
Arizona State University, Tempe, AZ
85287 USA
Narayanan.Krishnamurthi@asu.edu

W.M. Norman
Department of Pediatrics
University of Florida
Gainesville, FL, USA
normanw@mail.vetmed.ufl.edu

A. Okafor
Department of Industrial and
Systems Engineering
University of Florida
Gainesville, FL, 32611, USA
aokafor@ufl.edu

C.A.S. Oliveira
School of Industrial Engineering and
Management
Oklahoma State University
Stillwater, OK, USA
coliv@okstate.edu

A. Onetti Muda
Dipartimento di Medicina Sperimen-
tale e Patologia
Università di Roma "La Sapienza",
Italy

A.K. Ottens
Center of Neuroproteomics and
Biomarkers Research
University of Florida, Gainesville,
FL 32610, USA
aottens@mbi.ufl.edu

P.M. Pardalos
Industrial and Systems Engineering
Department
University of Florida
303 Weil Hall, Gainesville, FL 32611,
USA
pardalos@ufl.edu

G. Patrizi
Dipartimento di Statistica, Proba-
bilità e Statistiche Applicate
Università di Roma "La Sapienza",
Italy
g.patrizi@caspur.it

Gr. Patrizi
Dipartimento di Scienze Chirurgiche
Università di Roma "La Sapienza",
Italy

J. Pei
School of Computing Science
Simon Fraser University
Burnaby, BC Canada V5A 1S6
jpei@cs.sfu.ca

A. Prasad
Department of Physics and Astro-
physics
University of Delhi
New Delhi, India 110009

O.A. Prokopyev
Department of Industrial Engineering
University of Pittsburgh
Pittsburgh, PA, 15261, USA
prokopyev@engr.pitt.edu

S. Provan
Department of Statistics and
Operations Research
University of North Carolina
scott_provan@unc.edu

M. Ragle
Department of Industrial and
Systems Engineering
University of Florida
Gainesville, FL, 32611, USA
raglem@ufl.edu

A. Rubinov
Centre for Informatics and Applied
Optimization
University of Ballarat
Victoria, 3353, Australia
a.rubinov@ballarat.edu.au

S. Sabesan
Department of Electrical Engineering
Arizona State University, Tempe, AZ
85287 USA
shivkumar.sabesan@asu.edu

J.C. Sackellares
Department of Neuroscience
University of Florida
Gainesville, FL, USA
sackellares@mbi.ufl.edu

B. Santosa
School of Industrial Engineering,
University of Oklahoma
Norman, OK, USA
bsantosa@gmail.com

O.J. Semmes
Eastern Virginia Medical School
Department of Microbiology and
Molecular Cell Biology
Norfolk, VA 23507, USA
semmesoj@evms.edu

D.-S. Shiau
Department of Neuroscience,
University of Florida
Malcolm Randall Department of
Veteran's Affairs Medical Center
Gainesville, FL, USA
shiau@mbi.ufl.edu

A. Spanias
Department of Electrical Engineering
Arizona State University, Tempe, AZ
85287 USA
spanias@asu.edu

W. Suharitdamrong
Department of Industrial and
Systems Engineering
University of Florida
Gainesville, FL, 32611, USA
wichais@ufl.edu

F. Theis
Institute of Biophysics
University of Regensburg
D-93040 Regensburg, Germany
fabian@theis.name

G. Toraldo
Department of Agricultural Engineering
University of Naples "Federico II",
via Università 100, 80055
Portici, Italy
toraldo@unina.it

T. Trafalis
School of Industrial Engineering,
University of Oklahoma
Norman, OK, USA
ttrafalis@ou.edu

A. Tropsha
School of Pharmacy
University of North Carolina
alex_tropsha@unc.edu

K. Tsakalis
Department of Electrical Engineering
Arizona State University, Tempe, AZ
85287 USA
tsakalis@asu.edu

W. Tung
National Center for Atmospheric
Research
P.O. BOX 3000, Boulder, CO
80307-3000, USA
wwtung@ucar.edu

D. Valeyre
Department of Pneumology,
Fédération MARTHA, UFR
Bobigny,
Université Paris 13 et Hôpital
Avicenne AP-HP, 125, route de
Stalingrad, 93009 Bobigny Cedex,
France

K.K.W. Wang
Center of Neuroproteomics and
Biomarkers Research
University of Florida, Gainesville,
FL 32610, USA
kwang1@ufl.edu

W. Wu
Computer Science Department
University of Texas at Dallas
Richardson, TX 75083, USA
weiliwu@utdallas.edu

M.C.K. Yang
Department of Statistics
University of Florida
Gainesville, FL, USA
yang@stat.ufl.edu

V.A. Yatsenko
Institute of Space Research
40 Glushkov Ave, Kyiv 02022,
Ukraine
vitaliy_yatsenko@yahoo.com

J. Yearwood
Centre for Informatics and Applied
Optimization
University of Ballarat
Victoria, 3353, Australia
j.yearwood@ballarat.edu.au

J. Zeng
Computer Science Department
University of Northern British
Columbia
Prince George, BC, Canada V2N
4Z9
zeng@unbc.ca

Z. Zhang
Department of Biomedical Engineer-
ing
University of North Carolina
Chapel Hill, North Carolina 27599,
USA
zhangz@email.unc.edu

## Part I

Recent Methodological Developments for Data Mining Problems in Biomedicine

Pattern-Based Discriminants in the Logical Analysis of Data

Sorin Alexe and Peter L. Hammer *

RUTCOR - Rutgers University Center for Operations Research,
Piscataway, NJ, USA

Summary. Based on the concept of patterns, fundamental for the Logical Analysis of Data (LAD), we define a numerical score associated to every observation in a dataset, and show that its use allows the classification of most of the observations left unclassified by LAD. The accuracy of this extended LAD classification is compared on several publicly available benchmark datasets to that of the original LAD classification, and to that of the classifications provided by the most frequently used statistical and data mining methods.

Key words: Data mining, machine learning, classification, rule-based inductive learning, discriminants.

1 Introduction

The development of large databases in various sectors of the economy as well as in numerous areas of science and technology led to the creation of powerful instruments for their analysis. Besides the classical domain of statistics, entire new disciplines including data mining and machine learning appeared, having as aim the development of data analysis tools. Among the many new sophisticated data analysis methods we mention in particular decision trees [10,11], artificial neural networks [21], nearest neighborhood methods [7], and support vector machines [13,20,22].

The Logical Analysis of Data (LAD) is a combinatorics and optimization-based method [12,16] to extract knowledge from a dataset, consisting of "positive" and "negative" observations, represented as n-dimensional real vectors. A specific feature of LAD is the extraction of large collections of *patterns*, or *rules*, specific for either positive or negative observations in the dataset. One of the typical uses of LAD is the pattern-based classification of new observations, i.e., of real-valued vectors not included in the original dataset. The basic

---

* Corresponding author. Phone: 732-445-4812, Fax: 732-445-5472, e-mail: `hammer@rutcor.rutgers.edu`.

assumption of *LAD* is that new observations satisfying some of the *positive patterns* (i.e., patterns characteristic for positive observations), but none of the *negative patterns* (i.e., patterns characteristic for negative observations), are positive, while those satisfying some negative patterns, but no positive ones, are negative. On the other hand, new observations satisfying some of the positive as well as some of the negative patterns are not classified; similarly, those observations which do not satisfy any of the detected patterns are also left unclassified.

The main objective of this chapter is to propose an extension of the original *LAD* classification system, capable of classifying all those observations which satisfy some of the patterns in the collection, even if the satisfied patterns include both positive and negative ones. The proposed extension of the *LAD* classification system relies on the determination of a numerical score which balances the relative importance of the positive patterns satisfied by a new observation, compared to that of the negative patterns satisfied by it. The performance of the proposed extension of the *LAD* classification scheme is demonstrated on several publicly available benchmark datasets, and its accuracy - evaluated by computational experiments - is shown to compare positively with that of the best, frequently used statistical and data mining classification methods.

The first section of the chapter describes the basic elements on which the derivation of the scoring systems relies, including the description of the way in which observations are represented, the definition of the patterns considered in *LAD* and *LAD*'s pattern classification system, discussion of the accuracy measures used in this chapter, and of the benchmark datasets used in computational experiments. The second section of the chapter discusses a new pattern-based representation system of observations, introduces two numerical scores associated to observations, and shows that the classification systems based on the signs of these two scores are equivalent. In the third section, it is shown that the score-based extension of the *LAD* classification system increases the accuracy of classification, and compares favorably with most frequently used statistical and data mining classification systems.

2 Methods and Data

2.1 Basic Elements of *LAD*

A. Discrete Space Representation of Observations

The datasets considered in a large variety of applications consist of finite sets of vectors (sometimes called data points, or observations, or cases) defined in the n dimensional real space, along with "positive" or "negative" outcomes associated to them, signifying the "quality" of the corresponding vector. Because of the inherent difficulties of working in real space, it is frequently assumed that each dimension i of the real space is subdivided by *cutpoints* into intervals

$$I_i^0 = \left(-\infty, c_i^1\right], \;\; I_i^1 = \left(c_i^1, c_i^2\right], \;\; ..., \;\; I_i^{k_i} = \left(c_i^{k_i}, +\infty\right).$$

In this way, each observation $(\xi_1, \xi_2, ..., \xi_n)$ in R^n is transformed into a vector $(\eta_1, \eta_2, ..., \eta_n)$ in a discrete space D^n, by defining $\eta_i = h$ if the corresponding ξ_i is in I_i^h $(h = 0, 1, ..., k_i)$. This type of transformation was studied in [8] where it was shown that finding a minimum set of cutpoints which separate the positive observations from the negative ones (i.e., which have the property that none of the n-dimensional intervals defined by them will contain simultaneously positive and negative observations) is NP hard; an extension of the set-covering model introduced in [9,10] for cutpoint identification provides however efficient heuristics for finding reasonably small sets of cutpoints. Within this study we shall not be concerned with the identification of cutpoints, and shall simply assume the dataset to be given in a discrete space.

Given a dataset $\Omega = \left\{\omega^1, \omega^2, ..., \omega^m\right\}$ consisting of m *positive* and *negative* observations represented as points in D^n, we shall associate to every observation ω^i an *outcome* $\omega_i^0 \in \{0, 1\}$, defined to be 1 if the observation is positive, and 0 if it is negative.

B. Patterns

The set of points $y = (y_1, y_2, ..., y_n)$ in D^n satisfying a system of constraints

$$\omega_i \le \alpha_i \;\; (i \in I) \;\; \text{and} \;\; \omega_j \ge \beta_j \;\; (j \in J) \tag{1}$$

will be called a *cylinder*. The *positive (negative) homogeneity* of a cylinder is the proportion of positive (negative) points among all the points $\omega = (\omega_1, \omega_2, ..., \omega_n) \in \Omega$ contained in the cylinder. Clearly, the sum of the positive and negative homogeneities of any non-empty cylinder is 100%. The number of variables used in (1) for the definition of the cylinder is its *degree*. It can be assumed that (1) does not contain redundant constraints. It can also be noted that if a variable appears bounded from above, as well as from below, it is counted for the degree calculation only once.

A *positive pure pattern* is a cylinder containing some of the positive but none of the negative points of Ω. *Negative pure patterns* are defined in a similar way.

In many practical problems, the concepts of positive or negative pure patterns may turn out to be too restrictive, since their numbers as well as their coverages may be too limited for a meaningful analysis. One way of overcoming this difficulty is offered by the use of positive and negative (not necessarily pure) patterns. A cylinder is called a *positive (negative) pattern* if its positive (negative) homogeneity exceeds a given threshold (set usually close to 100%, and lowered in the case of less "clean" datasets to 90% or even to 80%).

Beside the degree and homogeneity, *prevalence* is another important characteristic parameter associated with patterns. The *positive (negative) prevalence* of a pattern is the proportion of positive (negative) observations in Ω which is covered by the pattern, i.e., which satisfies the pattern-defining system of constraints (1). Positive (negative) patterns having a low positive (negative) prevalence have a low significance for classification. On the other hand, many of the benchmark datasets frequently quoted in the literature contains several patterns with extremely large prevalences, going in some cases up to 75%-90%, or even to 100%.

We shall illustrate the concept of pure patterns on the case of the Wisconsin Breast Cancer (**bcw**) dataset [15]. This dataset includes records of 239 positive (malignant breast tumor) patients and 444 negative (benign breast tumor) cases, each of the patients being described as a 9 dimensional vector with positive integer coordinates ranging from 1 to 10. The 9 discrete valued variables considered in this dataset are the following: 1. *Clump Thickness*, 2. *Uniformity of Cell Size*, 3. *Uniformity of Cell Shape*, 4. *Marginal Adhesion*, 5. *Single Epithelial Cell Size*, 6. *Bare Nuclei*, 7. *Bland Chromatin*, 8. *Normal Nucleoli*, 9. *Mitoses*.

Example

A typical positive pure pattern is defined by the constraints "Uniformity of Cell Size $\geq$ 5, and Bland Chromatin $\geq$ 5". This pattern is of degree 2, and it covers 129 of the 239 positive cases, and none of the 444 negative ones; clearly, its positive and negative prevalences are of 54% and 0%, respectively, while its positive and negative homogeneities are of 100% and 0%, respectively. The constraints "Uniformity of Cell Size $\leq$ 3, Bare Nuclei $\leq$ 2, and Normal Nucleoli $\leq$ 3" define a negative pure pattern of degree 3, which covers 402 of the 444 negative cases, and none of the 239 positive ones; the positive and negative prevalences of this pattern are 0% and 90.5%, respectively, and its positive and negative homogeneities are 0% and 100%, respectively.

C. Pattern-Based Classification

Patterns represent the key concept on which the LAD method is based, and several efficient algorithms have been developed [1,5,15] for the generation of large collections of patterns. In this chapter, we shall concentrate on LAD models consisting of *prime patterns* (i.e., inclusion wise maximal patterns), since a series of recent case studies showed that, on one hand large classes of these patterns can be efficiently enumerated [5], and on the other hand, they can be used for constructing high accuracy LAD classification models.

Although the problem of finding all patterns corresponding to a dataset is NP hard, efficient ways for finding certain types of patterns are known. For example, a polynomial method has been given in [5] for enumerating all the *bounded degree prime patterns*, i.e., those prime patterns for which $|I|+|J| \leq \delta$, for some small value of δ (e.g., 2, 3, or 4). For instance, it has been shown in [5] that in all the benchmark problems considered in that study, the set of all prime patterns of degree 3 or less can be generated in 2-4 seconds, while those of degree 4 or less can be still generated in a few minutes.

Let $\mathcal{C}$ be a collection of positive patterns $P_1, P_2, ..., P_q$, and negative patterns $N_1, N_2, ..., N_r$, and let us assume that every point $\omega^i \in \Omega$ is *covered* by at least one of the patterns in $\mathcal{C}$, i.e., its coordinates satisfy all the defining conditions of at least one of the patterns in $\mathcal{C}$. The existence of a collection with this property can be easily obtained, e.g., by associating with every point $\omega^i \in \Omega$ a set of conditions of type (1) covering ω^i, and no other point in Ω, and then "relaxing" this pattern in order to increase its coverage as much as possible. In fact, efficient heuristics [2,17] can be found for constructing such systems which cover the entire set Ω and contain only patterns of relatively large coverage.

Example (continued)

*For the **bcw** dataset, a collection $\mathcal{C}$ consisting of 13 positive and 13 negative patterns is given in Table 1. Each pattern is shown as a row of the table, e.g., the pattern P1 is defined by the conditions "Uniformity of Cell Size $\geq$ 5, and Bland Chromatin $\geq$ 5", it covers 129 of the 239 positive points, and none of the 444 negative points in the dataset.*

A point $\omega \in D^n$ is classified by LAD as positive if it is covered by at least one of the patterns $P_1, P_2, ..., P_q$, but is not covered by any of the patterns $N_1, N_2, ..., N_r$. Similarly, it is classified by LAD as negative if it is covered by at least one of the patterns $N_1, N_2, ..., N_r$, but is not covered by any of the patterns $P_1, P_2, ..., P_q$. If the point is covered both by some positive and some negative patterns in $\mathcal{C}$, as well as the case when the point is not covered by any of the patterns in $\mathcal{C}$, LAD leaves the point unclassified.

Example (continued)

In order to illustrate the LAD classification system, we shall first introduce a random partition of the dataset into a "training set" and a "test set", develop then, using only information in the training set, a set of positive and negative patterns covering all the observations in the training set, and finally using these patterns, classify the observations in the test set. In this example, we have randomly generated a training set consisting of 50% of the positive and 50% of the negative observations in the original dataset. The model built on this training set consists of the 12 positive and 11 negative patterns shown in Table 2. The collection of positive and negative patterns described in Table 2 provides the following LAD classification of the 119 positive and 222 negative points in the test set. Table 3 shows that 325 (=108+217) of the 341 observations in the test set, i.e., 95.3%, are correctly classified.

Table 1. Patterns for the *bcw* Dataset (LAD Model)

Pattern	\<Pattern Descriptions\> Clump Thickness	Uniformity of Cell Size	Uniformity of Cell Shape	Marginal Adhesion	Single Epithelial Cell Size	Bare Nuclei	Bland Chromatin	Normal Nucleoli	Mitoses	# positive points covered	# negative points covered
P1		≥ 5					≥ 5			129	0
P2						≥ 9	≥ 4			121	0
P3			≥ 3	≥ 6						110	0
P4			≥ 5	≥ 4				≥ 3		106	0
P5	≥ 9									83	0
P6	≥ 7					≥ 3		≤ 7		80	0
P7								> 8		75	0
P8			≥ 4		≤ 3	≥ 5				41	0
P9			≥ 3	≤ 4					≥ 2	39	0
P10					≤ 6		≥ 4	$=4,5$ or 6		33	0
P11	$=4$ or 5			≥ 4				≥ 4		33	0
P12					$=2$	≥ 3	≥ 4			30	0
P13				$=2$ or 3		≥ 5		≤ 6		27	0
N1		≤ 3				≤ 2		≤ 3		0	402
N2	≤ 8				≤ 3	≤ 2				0	400
N3	≤ 6	≤ 2				≤ 4				0	396
N4	≤ 7			≤ 2	≤ 2					0	369
N5	≤ 5		≤ 2		$=2, 3$ or 4					0	342
N6				≤ 3		≤ 1	$=3$ or 4			0	108
N7	$=3$ or 4		$=2, 3$ or 4	≤ 4						0	32
N8	≤ 6.5							$=2$		0	29
N9	$=6$ or 7		≤ 3				≤ 3			0	12
N10	$=6$				≤ 3		≤ 3			0	12
N11	≤ 5			≤ 1		$=5,6,7,8$ or 9				0	8
N12		≤ 5				≤ 6		$=6,7$ or 8		0	7
N13	$=5,6,7,$ or 8		≤ 4	$=5$ or 6						0	5

Table 2. Patterns for the *bcw* Dataset (LAD Model)

Pattern	Clump Thickness	Uniformity of Cell Size	Uniformity of Cell Shape	Marginal Adhesion	Single Epithelial Cell Size	Bare Nuclei	Bland Chromatin	Normal Nucleoli	Mitoses	Training Set # positive points covered	Training Set # negative points covered	Test Set # positive points covered	Test Set # negative points covered
P1			≥ 5				≥ 5			62	0	63	0
P2						≥ 9	≥ 4			60	0	61	0
P3			≥ 5			≥ 5		≥ 3		59	0	60	1
P4		≥ 5			≥ 4			≥ 3		58	0	67	1
P5		≥ 2		≥ 6						58	0	55	1
P6	≥ 6				≥ 4			≥ 3		42	0	33	0
P7	≥ 7		≥ 3					≤ 7		42	0	40	2
P8						≥ 5		≥ 4	≤ 1	29	0	33	2
P9				≥ 4				≥ 4	≤ 1	28	0	31	1
P10		$=2,3,4,5,6,7$ or 8				≥ 4	$=4, 5$ or 6			24	0	19	1
P11				$=2$ or 3		≥ 4	≥ 2			21	0	21	1
P12						≤ 9	≥ 2		≥ 3	19	0	17	0
N1		≤ 3				≤ 2		≤ 3		0	199	0	203
N2	≤ 6	≤ 2				≤ 4				0	192	0	204
N3	≤ 8			≤ 2	≤ 2					0	185	1	185
N4	≤ 7			≤ 2		≤ 1				0	182	1	176
N5		≤ 2			$=2,3,4,$ or 5			≤ 2		0	174	1	180
N6	≤ 4			≤ 3		≤ 4				0	159	0	171
N7	$=3$ or 4		$=2,3$ or 4	≤ 4						0	17	0	15
N8	≤ 7							$=2$	≤ 2	0	14	0	14
N9	> 5		≤ 3				≤ 3			0	10	2	4
N10	$=5$						≤ 3	≤ 8		0	9	1	3
N11		≤ 4				≤ 7	$=5$ or 6			0	6	5	1

Table 3. LAD Classification for the *bcw* Dataset

	Classified as		Unclassified		Total
	Positive	Negative	Mixed Coverings	Not Covered	
Positive Observations	108	1	9	1	119
Negative Observations	4	217	1	0	222

2.2 Accuracy of Classification

A *classifier C* is a function which associates a binary value $C(x)$ to the vectors x in D^n. Given an arbitrary finite subset S of Ω along with the outcome ω^0 associated to every element $\omega \in S$, *the accuracy of the classifier C on the set S* is defined in [14] as the frequency of correct classifications in S, i.e.,

$$acc_S(C) = \frac{\left|\{\omega \in S \,|\, C(\omega) = \omega^0\}\right|}{|S|}. \tag{2}$$

We shall introduce two changes in this formula in order to address problems related on the one hand to the fact that in frequent applications the number of positive observations can be disproportionately larger or smaller than that of negative observations, and on the other hand to the fact that the function $C(x)$ constructed by *LAD* does not necessarily classify every observation as *positive* or *negative*, but may leave some (usually few) of the observations *unclassified*.

Let us first deal with the problems caused by the possibly disproportionate sizes of the subsets of positive and of negative points in the dataset. For this, let us associate to a classifier C a *classification matrix* $M_C = \begin{pmatrix} t^+ & f^- \\ f^+ & t^- \end{pmatrix}$, whose entries are defined in the following way: t^+ and t^- represent respectively the number of positive and negative, points in S correctly classified by C, while f^- and f^+ represent the number of positive observations in S which are classified by C as negative, respectively the number of negative observations in S classified by C as positive. Then the proportions of correct classifications among all observations classified as positive, respectively as negative will be equal to $\frac{t^+}{t^+ + f^-}$ and $\frac{t^-}{f^+ + t^-}$ respectively.

Let C be a classifier and let $\mu(C) = a + \frac{bt^+}{t^+ + f^-} + \frac{ct^-}{f^+ + t^-}$ be a measure of its accuracy, which takes into account the possibly unequal sizes of the subsets of positive and negative observations. It is natural to demand that if a classifier does not make any errors, than the corresponding value of μ should be 1; since in this case $\frac{t^+}{t^+ + f^-} = \frac{t^-}{f^+ + t^-} = 1$, it follows that $a + b + c = 1$. Similarly, if a classifier misclassifies every observation, then the corresponding value of μ should be 0; since in this case $\frac{t^+}{t^+ + f^-} = \frac{t^-}{f^+ + t^-} = 0$ it follows that $a = 0$. Replacing now b by a parameter λ which takes values between 0 and 1, and replacing c by $1 - \lambda$, we can see that all the accuracy measures which satisfy the above described conditions are of the form

$$\mu_\lambda(C) = \lambda \frac{t^+}{t^+ + f^-} + (1 - \lambda) \frac{t^-}{f^+ + t^-}. \qquad (3)$$

This accuracy measure will be used in the evaluation of the classification tools to be introduced in this chapter. In the absence of any additional information, we shall fix the value of the parameter λ to $1/2$. However, in problems in which the major objective is the avoidance of false positives (false negatives) the parameter λ can be fixed to a higher (lower) value, i.e., closer to 1 (respectively to 0).

Turning now to the second problem mentioned at the beginning of this section, we shall modify the proposed accuracy measure by allowing the classifier not to classify every observation as positive or as negative, but to leave some of them unclassified. In such cases, we shall replace the classification matrix M_C by

$$\tilde{M}_C = \begin{pmatrix} t^+ & f^- & u^+ \\ f^+ & t^- & u^- \end{pmatrix},$$

where u^+ and u^- represent the number of *unclassified* positive and negative points in S, respectively, while the numbers t^+, t^-, f^+ and f^- are the same as in the definition of M_C.

In order to calculate an expression of accuracy in this case, we shall assume that an unclassified observation has the same chance of being correct as of being incorrect. Clearly, $\frac{t^+ + 0.5u^+}{t^+ + f^- + u^+}$ and $\frac{t^- + 0.5u^-}{t^- + f^+ + u^-}$ represent respectively the proportions of good classifications among all observations classified as positive, respectively as negative. It is easy to see that a measure of classification accuracy $\tilde{\mu}(C)$ can be obtained analogously to the previous case. The resulting expression of the classification accuracy

$$\tilde{\mu}(C) = \lambda \frac{t^+ + 0.5u^+}{t^+ + f^- + u^+} + (1 - \lambda) \frac{t^- + 0.5u^-}{t^- + f^+ + u^-} \qquad (4)$$

includes again a parameter λ which takes values between 0 and 1. Clearly, if C classifies all the observations in S as positive or negative (i.e., $u^+ = u^- = 0$), the formulas (2) and (3) give the same results.

Throughout this chapter, we shall evaluate the accuracy of the various classification schemes using (4), with $\lambda = 0.5$.

Given a dataset Ω, the set of all prime positive patterns having a limited degree (say, $\leq \delta^+$), a sufficiently large positive prevalence (say, $\geq \pi^+$), and a sufficiently large positive homogeneity (say, $\geq \chi^+$) is called the $(\delta^+, \pi^+, \chi^+)$-*positive pandect of* Ω. The $(\delta^-, \pi^-, \chi^-)$-*negative pandect of* Ω is defined in a similar way. The values $\delta^+, \pi^+, \chi^+, \delta^-, \pi^-, \chi^-$ are called the *control parameters* defining a pandect.

Since the generation of every conceivable prime pattern is computationally intractable, in every classification problem we determine first a set of control parameters and generate all the prime patterns of the corresponding pandect. The determination of control parameters for which the associated pandect offers a high accuracy classification of new points is carried out through an experimental procedure in which various sets of control parameters are produced, for each such set of values the corresponding pandect is generated, and its accuracy statistically estimated then through k-folding experiments. In these experiments, the dataset is randomly divided into k approximately equal subsets, and k classification experiments are carried out then out; in each of these experiments the observations appearing in k-1 of the subsets form the training set, while the remaining subset is taken as the test set.

The experimental procedure described above identifies the values of the control parameters which gave the highest average accuracy during the k-folding test process. The pandect defined by these control parameter values is selected then as the proposed classification system.

2.3 Computational Experiments

The concepts discussed in this chapter have applied to four frequently analyzed benchmark datasets, which are publicly available at the repository `http://www1.ics.uci.edu/~mlearn/` `MLRepository.html` of the University of California at Irvine. The four datasets are: *Wisconsin Breast Cancer* (**bcw**), *BUPA Liver Disorders* (**bld**), *Congressional Voting Records* (**vot**) and *StatLog Heart Disease* (**hea**). Table 4 presents some basic information about these datasets.

Table 4. Basic Information on Benchmark Datasets

	Dataset	Dataset ID	Number of Attributes		Number of Observations	Proportion of Positive Observations
			Binary	Numerical		
1	Congressional voting records	**vot**	16	0	435	61.38%
2	Wisconsin breast cancer	**bcw**	0	9	683	34.99%
3	StatLog heart disease	**hea**	3	10	297	46.13%
4	BUPA liver disorder	**bld**	0	6	345	57.97%

In order to make possible the execution of a large number (tens of thousands) of computational experiments, we have selected the following control parameters. First we have selected $\delta^+ = \delta^- = 3$ as the upper bound on the degrees of all the positive and negative patterns generated; clearly, the generation of higher degree patterns would have required excessive computer time. The following values of the other control parameters have been considered $\pi^+ = 5\%, 10\%, \ldots, 50\%$; $\pi^- = 5\%, 10\%, \ldots, 50\%$; $\chi^+ = 80\%, 85\%, \ldots, 100\%$; $\chi^- = 0\%, 5\%, \ldots, 20\%$. For each of these values, the generated pandects were sufficiently "rich" to cover almost completely the corresponding datasets. For the validation process, we have set $k = 10$. Examining the 25,000 experiments (i.e., 10-folding cross-validations for each of the $10 \times 10 \times 5 \times 5 = 2,500$ parameter value combinations) for each of the four datasets, we found that the optimal accuracies of LAD classification schemes were detected for the following values of the control parameters (Table 5).

Table 5. Optimal Control Parameter Values for Classification by LAD among 2,500 Combinations/Dataset

| Dataset | Prevalence (%) | | Homogeneity (%) | | Average Accuracy (%) |
	Positive	Negative	Positive	Negative	(on 10 folds)
vot	35	45	100	5	**94.50**
bcw	25	45	100	0	**95.93**
hea	35	50	95	10	**81.67**
bld	10	30	100	20	**70.45**

3 Extended *LAD* Classification

3.1 Pattern-Space Representation

The existence of a collection C of positive and negative patterns $P_1, P_2, ..., P_q$, respectively $N_1, N_2, ..., N_r$, leads naturally to the idea of a *pattern-space representation* of every point x in D^n, i.e., the representation of every point y in D^n as a 0-1 vector $\tilde{y} = (P_1(y), ..., P_q(y); N_1(y), ..., N_r(y))$, whose entries equal 1 if the vector x satisfies the defining conditions of the corresponding pattern, and 0 otherwise. We shall denote by Π the set of binary $(q + r)$-vectors associated to the points y in D^n.

The pattern-space representation of points in D^n has been shown to be extremely informative [2,4]; in particular, the application of clustering techniques in Π has lead to the identification of previously unknown groups of ovarian cancer [3], and breast cancer patients.

It is also interesting to remark that the sets of positive and negative observations in Ω are *linearly separable* in Π. Indeed, for any set of positive coefficient $\alpha_1, \alpha_2, ..., \alpha_q, \beta_1, \beta_2, ..., \beta_r$, the expression

$$\Delta(y) = \sum_{i=1}^{q} \alpha_i P_i(y) - \sum_{j=1}^{r} \beta_j N_j(y) \tag{5}$$

takes a positive value for the positive observations in Ω, and a negative value for the negative observations in Ω.

The idea of considering weighted sum of classifiers is known in the literature [11]. It should be remarked however, that the patterns appearing in the weighted sum (18) are not classifiers. Indeed, if a positive or negative pattern takes the value 1 in an observation, that is an indication that the observation is positive respectively negative. However, the fact that a pattern takes the value 0 in an observation does not lead to any conclusion.

3.2 Compass Index and Score

Using now the pattern space representation of the data, we shall define the vector $\uparrow = (1, 1, ..., 1, 0, 0, ..., 0)$ whose first q components are 1 and last r components are 0, as the *positive pole* of Π, while its "complement" $\downarrow = (0, 0, ..., 0, 1, 1, ..., 1)$ will be called the *negative pole* of Π. Clearly, the definition of poles does not assume the actual existence of points in D^n whose image in Π is $\uparrow$ or $\downarrow$.

It is natural to expect the outcome of a point to be positive or negative according to the class of the "closest" pole. In order to measure closeness, we shall define the *positive* and *negative attractions* $a_\uparrow(y)$ and $a_\downarrow(y)$ of a point x in D^n, as $corr(\uparrow, \tilde{y})$ and $corr(\downarrow, \tilde{y})$, respectively; here $corr$ denotes the Pearson correlation coefficient. Clearly, the equality $corr(\uparrow, \tilde{y}) = -corr(\downarrow, \tilde{y})$ holds, and therefore, we shall focus only on the first of these two measures.

The numerical value of $a_\uparrow(y)$ is called the *compass score* of y in D^n, while the *compass index* $c(y)$ is defined as

$$c(y) = \begin{cases} 1 \ if \ a_\uparrow(y) > 0 \\ 0 \ if \ a_\uparrow(y) < 0 \\ \frac{1}{2} \ if \ a_\uparrow(Y) = 0 \end{cases}.$$

3.3 Balance Index and Score

An other natural way of evaluating the positive or the negative nature of a point y in D^n can be based on the sign of a discriminant $\Delta(y)$ defined as in (4). In order specify (18), we shall take here all the coefficients $\alpha_i = 1/q$, and all the coefficients $\beta_j = 1/r$. A simplified version of the balance score using a limited number of patterns was used efficiently for risk stratification among cardiac patients [6,19].

The *balance score* of a point x in D^n is defined then as

$$\Delta^*(y) = \frac{1}{q} \sum_{i=1}^{q} P_i(y) - \frac{1}{r} \sum_{j=1}^{r} N_j(y), \tag{6}$$

while the *balance index* $b(y)$ of y is defined as

$$b(y) = \begin{cases} 1 \ if \ \Delta^*(y) > 0 \\ 0 \ if \ \Delta^*(y) < 0 \\ \frac{1}{2} \ if \ \Delta^*(y) = 0 \end{cases}.$$

3.4 Extended *LAD* Classification by Compass or Balance Index

The compass index can be used for classifying the observations in D^n by saying that an observation y in D^n is *positive* if $c(y)=1$, *negative* $c(y)=0$, and *unclassified* is $c(y) = 1/2$. Similarly, the balance index can also be used for classification by using the corresponding values of $b(y)$.

Theorem 1. *For any observation, the classification provided by the balance index is identical to that provided by the compass index, i.e., $c(y) = b(y)$, for every y in D^n.*

Proof. Clearly, the condition $c(y) = b(y)$, is equivalent to the following two relations: $a_\uparrow(y) > 0$ iff $\Delta^*(y) > 0$, and $a_\downarrow(y) > 0$ iff $\Delta^*(y) < 0$. Let us introduce now some notations. First, let

$E[\tilde{y}] = \frac{1}{q+r}\left(\sum_{i=1}^{q} P_i(y) + \sum_{j=1}^{r} N_j(y)\right)$ denote the average value of the components of the vector $\tilde{y}$, and

$E[\uparrow] = \frac{q}{q+r}$ denote the average value of the components of the vector $\uparrow$.

With these notations, obviously the positive compass of an observation y can be expressed as:

$$a_\uparrow(y) = corr(\uparrow, \tilde{y}) = \frac{\sum_{k=1}^{q+r}(\tilde{y}_k - E[\tilde{y}])(\uparrow_k - E[\uparrow])}{\sqrt{\sum_{k=1}^{q+r}(\tilde{y}_k - E[\tilde{y}])^2}\sqrt{\sum_{k=1}^{q+r}(\uparrow_k - E[\uparrow])^2}}.$$

Simple algebraic manipulations applied to the numerator of the above fraction show that

$$\sum_{k=1}^{q+r}(\tilde{y}_k - E[\tilde{y}])(\uparrow_k - E[\uparrow]) = \sum_{k=1}^{q}(\tilde{y}_k - E[\tilde{y}])(\uparrow_k - E[\uparrow]) +$$

$$\sum_{k=1}^{r}(\tilde{y}_{q+k} - E[\tilde{y}])(\uparrow_{q+k} - E[\uparrow])$$

Since $\uparrow_k = 1$, $\uparrow_{q+k} = 0$, $1 - E[\uparrow] = \frac{r}{q+r}$, the above expression equals to:

$$\frac{r}{q+r}\sum_{k=1}^{q}(\tilde{y}_k - E[\tilde{y}]) - \frac{q}{q+r}\sum_{k=1}^{r}(\tilde{y}_{q+k} - E[\tilde{y}]) =$$

$$\frac{r}{q+r}\sum_{k=1}^{q}\tilde{y}_k - \frac{r}{q+r}qE[\tilde{y}] - \frac{q}{q+r}\sum_{k=1}^{r}\tilde{y}_{q+k} + \frac{q}{q+r}rE[\tilde{y}] =$$

$$\frac{r}{q+r}\sum_{k=1}^{q}\tilde{y}_k - \frac{q}{q+r}\sum_{k=1}^{r}\tilde{y}_{q+k} = \frac{qr}{q+r}\left(\frac{1}{q}\sum_{k=1}^{q}\tilde{y}_k - \frac{1}{r}\sum_{k=1}^{r}\tilde{y}_{q+k}\right) =$$

$$\frac{qr}{q+r}\left(\sum_{k=1}^{q}\alpha_k\tilde{y}_k - \sum_{k=1}^{r}\beta_k\tilde{y}_{p+k}\right) = \frac{qr}{q+r}\Delta^*\left(y\right).$$

In conclusion, $a_{\uparrow}\left(y\right) = \frac{qr}{q+r}\sqrt{\sum_{k=1}^{q+r}\left(\tilde{y}_k - E\left[\tilde{y}\right]\right)^2}\sqrt{\sum_{k=1}^{q+r}\left(\uparrow_k - E\left[\uparrow\right]\right)^2}\Delta^*\left(y\right)$

showing that – in view of the positivity of $\frac{qr}{q+r}$ – the sign of $a_{\uparrow}\left(y\right)$ is the same as that of $\Delta^*\left(y\right)$, thus completing the proof.

4 Accuracy of Extended *LAD* (*e-LAD*) Classification

In order to evaluate the accuracy of the extended *LAD* (*e-LAD*) classification system, we have conducted a series of computational experiments to compare it with that of *LAD*, of the Fisher discriminant, and of several frequently used classification techniques in data mining. In view of the equivalence of the classification systems given by the compass index and the balance index, and the easiness of calculating the balance index, the discussions in the following sections of this chapter will assume that *e-LAD* is based on the balance index.

4.1 Computational Evaluation of *e-LAD*

Given a dataset and a collection of positive an negative patterns covering all (or at least "almost all") given data, the classification scheme used by *LAD* declares a new observation to be *positive* (*negative*) if it is covered by at least one positive (negative) pattern in the collection, and it is not covered by any of the negative (positive) patterns in the collection; an observation which is covered both by positive and negative patterns, or by none of the patterns in the collection is declared *unclassified*.

We propose here an extension of the above *LAD* classification scheme, using the balance or the compass index for defining the class of a new observation. This classification leaves unchanged the classification of those observations which are classified as positive or negative by *LAD*, but provides classifications also of those observations which are covered both by positive and negative patterns. Since Theorem 1 shows that classification using the compass index is identical to the one using the balance index, we shall only describe below one of these two classification systems. In view of the ease of calculating the balance index, we shall focus the discussion on the system which uses this index.

In order to find those values of the control parameters which provide the highest accuracy for the proposed classification system, we re-examined the 25,000 experiments described above for each of the four datasets, and found the results presented in Table 6.

Table 6. *e-LAD* Classification: Optimal Control Parameter Values among 2,500 Combinations/Dataset

Dataset	Prevalence (%)		Homogeneity (%)		Average Accuracy (%)
	Positive	Negative	Positive	Negative	(on 10 folds)
vot	5	50	100	5	**95.77**
bcw	50	20	95	0	**97.78**
hea	35	15	95	5	**84.62**
bld	10	35	100	20	**70.44**

4.2 Comparison between *LAD* and *e-LAD*

Comparing the accuracy of the *e-LAD* classification scheme with the accuracy given by *LAD*, it can be seen (Table 7) that both its *maximum* and its *mean* show an increase; in the 25,000 experiments the maximum increases on the average by 2.5%, and the mean by 3.5%. It is also important to remark that the *standard deviation* of the *e-LAD* classification accuracies shows an average decrease of 55%. In conclusion the proposed classification scheme increases clearly both the accuracy and the robustness of the results.

Table 7. Accuracy of *LAD* and *e-LAD* Classification

	Accuracy					
	Maximum		Mean		Standard Deviation	
	LAD	e-LAD	LAD	e-LAD	LAD	e-LAD
vot	94.50%	95.77%	93.07%	94.46%	0.72%	0.53%
bcw	95.93%	97.78%	94.86%	97.57%	0.73%	0.07%
hea	81.67%	84.62%	76.79%	83.13%	3.78%	0.56%
bld	70.45%	72.74%	62.95%	64.08%	4.74%	5.76%
Average	**85.64%**	**87.73%**	**81.92%**	**84.81%**	**2.49%**	**1.73%**
e-LAD/LAD	1.02		1.04			

The observations which remained unclassified by *LAD* are classified remarkably well by the *e-LAD*, as shown in Table 8. This table examines the set of those observations which remained unclassified by *LAD*, showing what percentages of these are correctly classified, incorrectly classified, or remain unclassified by *e-LAD*.

It can be seen that a remarkable proportion (99%) of the observations which could not be classified by *LAD* are classified by *e-LAD*, and that on the average, 4 out of 5 of these observations are classified correctly.

All those observations which are classified by *LAD* as positive or negative remain classified in the same way by *e-LAD*. On the other hand, almost all the observations which remain unclassified by *LAD* are classified by *e-LAD*, and the vast majority of them is classified correctly. Taking into account the fact

Table 8. Accuracy of *e-LAD* on the Set of Observations Left Unclassified by *LAD*

Dataset	Correct	Incorrect	Unclassified
vot	84.3%	12.0%	3.7%
bcw	85.7%	14.3%	0.0%
hea	74.4%	25.6%	0.0%
bld	68.8%	30.7%	0.5%
average	78.3%	20.7%	1.1%

that the proportion of observations not classified by *LAD* can be substantial (between 8.21% and 97.3% for the benchmark datasets considered above), the improvement of the *e-LAD* based classification over the simple *LAD* based classification can be considerable. Moreover, the accuracy provided by *e-LAD* turns out to be consistently superior to the accuracy given by *LAD*.

4.3 *e-LAD* vs. Other Classifiers

Having noticed the enhancement of classification provided by *e-LAD* over *LAD*, in the following sections we shall examine the performance of *e-LAD* compared to that of some frequently used statistical and data mining classification methods.

The accuracy of any *LAD* or *e-LAD* classification scheme depends on the collection of positive and negative patterns used by it. In its turn, this collection is entirely determined by the set of control parameter values. Clearly, the best combination of control parameter values among the 2,500 combinations examined is not necessarily optimal. By additional fine-tuning experiments, these control parameter values (with the exception of δ, which was kept equal to 3 during all the experiments) have been further improved. The best values found for the four benchmark datasets examined in this chapter, along with the corresponding average accuracies on 10-folding cross validation experiments with the *e-LAD* classification system are reported in Table 9.

Table 9. Classification by *e-LAD* using Enhanced Control Parameter Values

Dataset	Prevalence (%)		Homogeneity (%)		Average Accuracy (%)
	Positive	Negative	Positive	Negative	(on 10 folds)
vot	15	15	100	3	**97.44**
bcw	10	10	100	0	**98.16**
hea	10	10	87	17	**85.59**
bld	10	10	87	17	**72.91**

4.4 *e-LAD* vs. Fisher Discriminant

One of the most frequently used classification methods is the Fisher (linear) discriminant analysis. Since the balance score can be viewed as a linear discriminant in pattern space, we have compared the accuracy given by *e-LAD* with that obtained by the Fisher discriminant classifier [18] applied to the original attribute space. The Fisher discriminant was calculated using the S-Plus 6.1 software (Insightful Corp., 2002), and the results are presented in Table 10.

Table 10. Accuracy of *e-LAD* and of Fisher Discriminant Classifiers

Experiment	vot		bcw		hea		bld	
	Fisher	e-LAD	Fisher	e-LAD	Fisher	e-LAD	Fisher	e-LAD
Fold 1	91.11%	94.23%	97.83%	100.00%	85.34%	85.34%	83.21%	79.64%
Fold 2	100.00%	100.00%	96.69%	98.86%	85.34%	82.93%	67.86%	67.86%
Fold 3	94.95%	96.88%	100.00%	98.86%	75.96%	84.13%	69.29%	64.29%
Fold 4	93.03%	94.95%	100.00%	98.86%	96.88%	86.78%	64.29%	77.86%
Fold 5	96.15%	98.08%	94.52%	98.86%	82.93%	96.51%	53.57%	66.07%
Fold 6	96.15%	98.08%	95.55%	96.59%	76.68%	86.78%	56.43%	67.50%
Fold 7	100.00%	100.00%	94.52%	97.73%	86.06%	82.93%	68.21%	70.00%
Fold 8	96.15%	98.08%	93.38%	96.59%	81.49%	86.06%	69.29%	83.21%
Fold 9	96.88%	100.00%	91.30%	97.83%	74.52%	82.57%	74.29%	80.71%
Fold 10	94.89%	94.13%	91.15%	97.40%	81.88%	81.88%	58.95%	71.97%
Average	95.93%	97.44%	95.49%	98.16%	82.71%	85.59%	66.54%	72.91%
Standard Deviation	2.75%	2.33%	3.17%	1.11%	6.47%	4.24%	8.76%	6.85%
e-LAD/Fisher Accuracy	1.02		1.03		1.03		1.10	

It can be seen that the improvements of accuracy by using *e-LAD* on the four benchmark problems are 2%, 3%, 3% and 10%, respectively (i.e., an average increase of 4.5%) over that of the Fisher discriminant classifier. It can also be seen that in all four problems the standard deviation of the results obtained by *e-LAD* are substantially lower than those given by the Fisher discriminant classifier, indicating the higher robustness of *e-LAD*.

4.5 *e-LAD* vs. "Best" Data Mining Classifiers

Since many classification techniques are currently available, it is important to position *e-LAD* within the realm of classification schemes. In order to evaluate the relative performance of *e-LAD*, we have compared its accuracy with that of the best known classifiers, using the estimations reported in [15] for 33 data mining algorithms (including classification trees and rules, statistical techniques, and artificial neural networks). It was noticed in [15] that the best accuracy among the compared 33 methods varied from dataset to dataset. In the case of the **bcw** dataset, the best accuracy was obtained by neural networks (*LVQ*), in the case of the **vot** dataset it was obtained by classification trees (*QL0*), in the case of the **hea** dataset it was obtained by linear discriminant analysis (*Fisher*), and for the **bld** dataset it was obtained by classification trees (*OCM*).

Since there is no universally best method among the 33 reported in [15] (which include the Fisher discriminant analysis and 32 other methods), for each of the 4 datasets examined in our chapter (**bcw**, **hea**, **vot** and **bld**) we have selected one of the 33 methods which gave the best performance for that specific dataset. The performance of *e-LAD* was then compared with that of the 4 methods selected in this way. Clearly, the positive results of the comparison with these 4 methods imply that the comparison with other 29 algorithms examined in [15] is also positive.

The results of the comparison are presented in Table 11, and show that in three of the four cases the performance of *e-LAD* exceeds that of the best of the other methods (by an average of 1.5%), and underperforms (by 0.3%) the best of the other methods in only one case. On the average, the performance of *e-LAD* is 4.5% above the average performance of the four selected methods.

Table 11. Comparison of *e-LAD* with Best Performing Dataset-Specific Classifiers [15].

Dataset	Classification Accuracy Results Reported in [7]					*e-LAD*
	Neural Networks (LVQ)	Classification Trees (QL0)	Fisher Discriminant	Classification Trees (OCM)	Average of 4 Methods	
vot	**95.0%**	96.4%	95.4%	94.2%	95.6%	97.4%
bcw	97.2%	**96.9%**	96.1%	95.9%	96.7%	98.2%
hea	65.9%	84.8%	**85.9%**	77.8%	78.9%	85.6%
bld	67.1%	69.4%	67.4%	**72.1%**	68.0%	72.9%
Average	**81.3%**	**86.9%**	**86.2%**	**85.0%**	84.8%	**88.5%**
e-LAD/Other Classifiers	1.089	1.019	1.027	1.041	1.045	

5 Conclusions

Classification by e-LAD vs. classification by LAD

The *e-LAD* classification significantly increases the number of observations classified in the dataset and the accuracy of these classifications. Those observations which are classified by *LAD* are classified in the same way by *e-LAD*. On average, the accuracy of *e-LAD* classification among the observations left unclassified by *LAD* is 78.3%. The average percentage of observations left unclassified by *e-LAD* is 1.1%. The average increase of accuracy over 25,000 tests on the datasets examined in this chapter is 2.9%.

The stability of classification by *e-LAD* is substantially higher than that of *LAD* for various choices of control parameters and training/test samplings, illustrated in the computational experiments by a 55% reduction of the standard deviation.

Classification by e-LAD vs. other classification methods

The accuracy of classification by *e-LAD* exceeds that given by the Fisher discriminant by more than 4% for the benchmark datasets.

The accuracy of classification by *e-LAD* exceeds that of other data mining classification methods. In comparison with the 33 classification methods reported in [15], the average improvement for the four benchmark datasets is 4.5%.

Relative Positions of LAD and e-LAD Classification Methods in the Ranking of [15]

Lim, Loh and Shih in the Appendix of [15] rank the accuracies of 33 methods on a series of datasets, which include the four datasets examined in this chapter. Table 12 compares the accuracies of *LAD* and *e-LAD* with those reported in [15], indicating the ranks of *LAD* and *e-LAD* within the collection of methods analyzed by Lim, Loh and Shih.

Table 12. Positioning of *LAD* and *e-LAD* within the Hierarchy of Methods Ranked by Accuracy in [15 - Appendix].

	vot	bcw	hea	bld
Rank of LAD	31	15	10	6
Rank of e-LAD	5	1	6	1

Acknowledgements

This work was supported in part by NSF Grant IIS-0312953 and ONR Grant N00014-92-J-1375.

References

1. G. Alexe and P.L. Hammer. Spanned Patterns for the Logical Analysis of Data, RUTCOR Research Report RRR 15-2002, *Discrete Applied Mathematics* (in print).
2. G. Alexe, S. Alexe, and P.L. Hammer. Pattern-Based Clustering and Attribute Analysis, *RUTCOR Research Report*, RRR 10-2003, *Soft Computing* (in print).
3. G. Alexe, S. Alexe, P. L. Hammer, L. Liotta, E. Petricoin, and M. Reiss. Logical Analysis of the Proteomic Ovarian Cancer Dataset. *Proteomics*, 4: 766-783, 2004.
4. G. Alexe, S. Alexe, P.L. Hammer, and B. Vizvari. Pattern-Based Feature Selection in Genomics and Proteomics. *RUTCOR Research Report*, RRR 7-2003.
5. S. Alexe and P.L. Hammer. Accelerated Algorithm for Pattern Detection in Logical Analysis of Data. RUTCOR Research Report RRR 59-2001, *Discrete Applied Mathematics* (in print).
6. S. Alexe, E. Blackstone, P.L. Hammer, H. Ishwaran, M.S. Lauer, and C.E.P. Snader. Coronary Risk Prediction by Logical Analysis of Data. *Annals of Operations Research*, 119: 15-42, 2003.
7. M. Berthold and D. Hand, editors. *Intelligent Data Analysis - An introduction*. Springer, 1999.
8. E. Boros, P.L. Hammer, T. Ibaraki, and A. Kogan. Logical Analysis of Numerical Data. *Mathematical Programming*, 79: 163-190, 1997.
9. E. Boros, P.L. Hammer, T. Ibaraki, A. Kogan, E. Mayoraz, and I. Muchnik. An Implementation of Logical Analysis of Data. *IEEE Transactions on Knowledge and Data Engineering*, 12(2): 292-306, 2000.
10. L. Breimann, J.H. Friedman, R.A. Olshen, and C.J. Stone, *Classification and Regression Trees*, Wadsworth International Group, 1984.
11. W. Buntine. Learning Classification Trees. *Statistics and Computing*, 2: 63-73, 1992.
12. Y. Crama, P.L. Hammer, and T. Ibaraki. Cause-Effect Relationships and Partially Defined Boolean Functions. *Annals of Operations Research*, 16: 299-326, 1988.
13. N. Cristianini and J. Shawe-Taylor. *An Introduction to Support Vector Machines*, Cambridge University Press, 2000.
14. R. Kohavi. *Wrappers for Performance Enhancement and Oblivious Decision Graphs*. PhD Thesis, Stanford University, 1995.
15. T.S. Lim, W.Y. Loh, and Y.S. Shin. A Comparison of Prediction Accuracy, Complexity, and Training Time of Thirty-three Old and New Classification Algorithms. *Machine Learning*, 40: 203-229, 2000. (Appendix: www.stat.wisc.edu/p/stat/ftp/pub/loh/treeprogs/quest1.7/appendix.pdf)
16. P.L. Hammer. Partially Defined Boolean Functions and Cause-Effect Relationships. *International Conference on Multi-Attribute Decision Making Via OR-Based Expert Systems*, University of Passau, Passau, Germany, 1986.
17. P.L. Hammer, A. Kogan, B. Simeone, and S. Szedmak. Pareto-Optimal Patterns in Logical Analysis of Data. *Discrete Applied Mathematics*, 144: 79-102, 2004.

18. J.D. Jobson. *Applied Multivariate Data Analysis*. Springer-Verlag, 1991.

19. M.S. Lauer, S. Alexe, C.E.P. Snader, E.H. Blackstone, H. Ishwaran, and P.L. Hammer. Use of the "Logical Analysis of Data" Method for Assessing Long-Term Mortality Risk After Exercise Electrocardiography. *Circulation*, 106: 685-690, 2002.

20. S. Schölkopf, C.J.C. Burges, and A.J. Smola. *Advances in Kernel Methods: Support Vector Learning*. MIT Press, Cambridge, MA, 1999.

21. S.H. Simon. *Neural Networks: A Comprehensive Foundation*. Prentice Hall, 1998.

22. V. Vapnik. *Statistical Learning Theory*. Wiley-Interscience, New York, 1998.

Exploring Microarray Data with Correspondence Analysis

Stanislav Busygin and Panos M. Pardalos

Industrial and Systems Engineering Department
University of Florida
303 Weil Hall, Gainesville, FL 32611
{busygin,pardalos}@ufl.edu

Summary. Due to the rapid development of DNA microarray chips it has become possible to discover and predict genetic patterns relevant for various diseases on the basis of exploration of massive data sets provided by DNA microarray probes. A number of data mining techniques have been used for such exploration to achieve the desirable results. However, high dimensionality and uncertain accuracy of microarray datasets remain the major obstacles in revealing the most crucial genetic factors determining a particular disease. This chapter describes a microarray data processing technique based on the correspondence analysis that helps to handle this issue.

1 Introduction

The importance of data analysis in life sciences is steadily increasing. Up to recently, biology was a descriptive science providing relatively small amount of numerical data. However, nowadays it has become one of the main applications of data mining techniques operating on massive data sets. This transformation can be particularly attributed to two recent advances which are complementary to each other. First, the Human Genome Project and some other genome-sequencing undertakings have been successfully accomplished. They have provided the DNA sequences of the human genome and the genomes of a number of other species having various biological characteristics. Second, revolutionary new tools able to monitor quantitative data on the genome-wide scale have appeared. Among them, there are the *DNA microarrays* widely used at the present time. These devices measure gene expression levels of thousands of genes simultaneously, allowing researchers to observe how the genes act in different types of cells and under various conditions.

As a consequence of this progress, the traditional approach of studying one particular gene per experiment has been changed. Now it is possible to investigate not only how a gene behaves itself, but also how it *interacts* with other genes and which gene expression patterns are formed. It is natual to expect that on the basis of microarray data, the genes characterizing certain medical phenomena (such as diseases) can be detected and classified. Especially, such a study is crucial for understanding *genetic diseases* caused by a mutation in a gene or a set of genes. They make the mutant genes inappropriately expressed or even not expressed at all. For example, it is known that cancer can be caused by inactivation, deletion or, on the contrary, by constitutive activity of *p53 tumor suppressor gene*. Furthermore, some genetic diseases have subtypes that are indistinguishable clinically but differ from each other in the underlying genetic mechanism. Most likely, it would imply that these subtypes require different methods of treatment. However, unless a sophisticated diagnostic technique is available, it would be impossible to properly make the right choice. One illustrative example of such a situation considered in this chapter is discriminating *acute lymphoblastic leukemia* (ALL) versus *acute myeloid leukemia* (AML).

However, the analysis of microarray data is not an easy task. High dimensionality of the data, poor accuracy of microarray probes, and practical difficulties with taking the probes (the procedure might be very painful for alive patients while the gene expression levels rapidly degrade in dead tissue) hinder the success of microarray technology. Hence, the microarray datasets must be processed by a sophisticated data mining technique applicable in the case of high-dimensional data and still able to refine particular data values known to be critically inaccurate.

Generally, data mining techniques may be divided into three major classes that sometimes overlap: statistical analysis, clustering, and dimensionality reduction (projection methods). The statistical analysis for microarray data usually consists in calculating *fold change* of particular genes across different groups of samples and applying classical statistic tests such as t-test, ANOVA, Wilcoxon test, etc. These techniques are appropriate when a proper separation of samples into classes is known, the number of outliers in each class is insignificant, and the data may be assumed to have certain statistical properties (e.g., normal distribution). While the normality assumption is believed to be feasible for microarray data [4], the other conditions are harder to guarantee, taking into account the issues with accuracy of microarray data mentioned above. Furthermore, the statistical analysis cannot reveal more general patterns in the data rather than up- or downregulation of single genes.

Clustering techniques can be divided into *supervised* and *unsupervised* learning. Supervised learning techniques are also called sometimes *classification* methods . They take predetermined classes of objects as input and aim at deriving characteristics (features) common for samples of a class and discriminating them against samples of other classes. Examples of supervised learning techniques include linear discriminant analysis, classification and regression trees, support vector machines, etc. Clearly, supervised learning requires a set of *training samples* whose separation into different classes is known beforehand. On the contrary, unsupervised clustering techniques do not require such a training set; they build classes (clusters) of samples starting from scratch.

Examples of clustering techniques are hierarchical clustering, k-means clustering, self-organizing maps (SOM), etc. Common drawbacks of these methods are significant dependence of the results on initialization of the clustering and the absence of a clear mathematical criterion to judge quality of the results. (i.e., a universal objective function whose optimal value would signify best clustering in all instances does not exist). Furthermore, it has been proved in [8] that there is no clustering algorithm simultaneously satisfying three simple properties one might expect to be required: *scale-invariance* (i.e. multiplying all distances by the same positive number should not change the result), *richness* (all partitions should be achievable), and *consistency* (decreasing the distances within the clusters with increasing the distances between the clusters should not change the result).

The dimensionality reduction methods do not aim at delivering strict categorization of data into classes or separation of relevant versus non-relevant features. They rather produce a low-dimensional projection of an originally high-dimensional data set. As soon as such a projection is presented in the form of biplot or 3D diagram, there is the opportunity for a researcher of the data domain to eyeball the picture and gain an understanding of the crucial data patterns. Clearly, there are at least two advantages comparing to strict categorization of the data. First, when the data dimensionality is small, the human eye becomes an analytic tool of remarkable power able to grasp complex data patters undetectable by any statistical methods generally aimed at simple linear relations. Second, optimization of projecting high-dimensional spaces onto low-dimensional subspaces is nicely supported by extensive theoretical background of linear algebra. The core of projection methods is *singular value decomposition* (SVD), which can provide the subspace of any desirable dimension preserving the maximum possible similarity between the original data set and its projection onto the subspace. Another important point here is that the SVD procedure is computationally efficient (i.e. it can be performed in a short polynomial time, especially if only few dominating singular vectors are sought). This compares favorably to many iterative clustering procedures. For instance, the convergence of SOM cannot even be guaranteed without gradually decreasing the learning rate parameter with each iteration. Hence the projection techniques are also attractive from the computational complexity viewpoint. Lastly, the projection techniques do not depend on any parameters that should be specified by the user before the algorithm is applied. This potentially makes them more appealing for biological researchers typically not familiar in detail with the data mining algorithms. We refer the reader to [5] for detailed introduction of relevant algebra and SVD algorithms.

In this chapter we describe one specific dimensionality reduction technique called *correspondence analysis* (CA), and consider its application to microarray data. The effectiveness of CA is illustrated by discovering AML- and ALL-relevant genes from a well-known microarray dataset published by Golub at al [6].

2 Correspondence Analysis

2.1 Basic Algorithm

Like other dimensionality reduction techniques, correspondence analysis is an *exploratory data analysis* technique providing a view of the data set as a whole. The main advantage of correspondence analysis over other dimensionality reduction techniques is that it allows for simultaneous observation of data samples (usually given by columns of the data matrix) and data points (correspondingly, represented by rows of the data matrix) in *one* low-dimensional space. This becomes possible due to the bidirectional nature of correspondence analysis, investigating not only relations within the set of samples and the set of data points, but also cross-relations between elements of these two sets. The only restriction of this technique is that all data values must be nonnegative.

Thus, correspondence analysis maps all samples and data points of a data set onto one low-dimensional space, which can be visualized as a biplot (2-D) or 3-D diagram. Each axis of this diagram tends to reveal a profound characterization of the data set, and samples/data points having high similarity with respect to this characterization have similar coordinates on it. Like in other dimensionality reduction techniques, the construction of the low-dimensional space is performed by means of singular value decomposition (SVD). However, in case of the correspondence analysis, SVD is not applied directly to the data matrix, but is used after its specific *correspondence* matrix is constructed. We refer the reader to the existing literature (e.g., [7]) to review theoretical background of the method and related algebraic proofs. Here we describe the algorithm of correspondence analysis and its generalization in case when some data entries are missing or cannot be trusted.

A data set is normally given as a rectangular matrix $A = (a_{ij})_{m \times n}$ of n samples (columns) and m data points (rows). In the case of microarray data, rows represent genes and the value a_{ij} shows the expression level of gene i in sample j. For the sake of simplicity, we assume further on that $m > n$. However, this does not restrict the generality of the discussed technique, since both the columns and the rows of the data matrix are to be treated in a unified way, and it is always possible to work with the transposed matrix without making any changes in the algorithm. So, to perform correspondence analysis, we first construct the *correspondence* matrix $P = (p_{ij})_{m \times n}$ by computing

$$p_{ij} = a_{ij}/a_{++}, \tag{1}$$

where

$$a_{++} = \sum_{i=1}^{m} \sum_{j=1}^{n} a_{ij} \tag{2}$$

is called the *grand total* of A. The correspondence matrix is somewhat analogous to a two-dimensional probability distribution table, whose entries sum up to 1. We also compute *masses* of rows and columns (having the analogy with the marginal probability densities):

$$r_i = a_{i+}/a_{++}, \tag{3}$$

$$c_i = a_{+j}/a_{++}, \tag{4}$$

where

$$a_{i+} = \sum_{j=1}^{n} a_{ij}, \tag{5}$$

$$a_{+j} = \sum_{i=1}^{m} a_{ij}. \tag{6}$$

Then the matrix $S = (s_{ij})_{m \times n}$, to which SVD is applied, is formed:

$$s_{ij} = (p_{ij} - r_i c_j)/\sqrt{r_i c_j}. \tag{7}$$

The SVD of $S = U \Lambda V^T$ represents the provided matrix as the product of three specific matrices. Columns of the matrix $U = (u_{ij})_{m \times n}$ are orthonormal vectors spanning the columns of S, columns of the matrix $V = (v_{ij})_{n \times n}$ are also orthonormal vectors but they span the rows of S, and finally $\Lambda = \mathrm{diag}(\lambda_1, \lambda_2, \ldots, \lambda_n)$ is a diagonal matrix of nonnegative *singular values* of S having a nondecreasing order: $\lambda_1 \geq \lambda_2 \geq \ldots \geq \lambda_n \geq 0$. It can be shown algebraically that the optimal low-dimensional subspace to project the columns of S onto, with the minimum possible information loss, is formed by a desired number of first columns U. Similarly, the optimal low-dimensional subspace for plotting the rows of S is formed by the same number of first columns of V. Furthermore, due to specific properties of the matrix S, the columns and rows of the original data set matrix A may be represented in one low-dimensional space of dimensionality $K < n$ as follows:

$$f_{ik} = \lambda_k u_{ik}/\sqrt{r_i}, \ \ k = 1, 2, \ldots, K, \tag{8}$$

gives the k-th coordinate of row i, and

$$g_{jk} = \lambda_k v_{jk}/\sqrt{c_j}, \ \ k = 1, 2, \ldots, K, \tag{9}$$

gives the k-th coordinate of column j in the new space. Obviously, we select $K = 2$ if we want to obtain a biplot and $K = 3$ if we want to obtain a 3-D diagram of the analyzed data set.

2.2 Treatment of missing values

Correspondence analysis allows for an easy and natural treatment of missing data values. We just need to look at the procedure backward and asnwer the question: if f and g were the positions of rows and columns on the low-dimensional plot, what value of a data entry a_{ij} would minimize the information loss incurred due to the dimensionality reduction with respect to the constructed low-dimensional representation? It is necessary to mention here that the correspondence analysis algorithm constructing the low-diminsional space actually solves the following least-squares problem [7]:

$$\min \sum_{i=1}^{m} \sum_{j=1}^{n} (a_{ij} - \hat{a}_{ij})^2 / (a_{i+} a_{+j}), \tag{10}$$

where $\hat{A} = (\hat{a}_{ij})_{m \times n}$ is the sought low-dimensional approximation of the data that can be expressed as

$$\hat{a}_{ij} = (a_{i+} a_{+j} / a_{++}) \left(1 + \sum_{k=1}^{K} f_{ik} g_{jk} / \sqrt{\lambda_k} \right). \tag{11}$$

So, the relation (11) gives the best guess for the data entry a_{ij} provided that we already have the low-dimensional coordinates f and g, and the singular values λ. From here we can infer an iterative *E-M algorithm* performing simultaneously construction of the low-dimensional plot of the data and approximation of missing data entries (the latter is called *imputing* the values) [7].

1. Make some initial guesses for the missing data entries.
2. Perform the K-dimensional correspondence analysis as specified by the formulas (1)-(9) (the M-step, or *maximization* step of the E-M algorithm).
3. Obtain new estimations for the imputing data entries by (11) (the E-step, or *expectation* step of the algorithm).
4. If the new estimations are close enough to the previous estimations, STOP. Otherwise repeat from Step 2 with the new estimations.

The initial guesses for the imputing data entries for Step 1 of the algorithm should be made such that $p_{ij} = r_i c_j$ for these entries [7]. This condition is equivalent to the equalities

$$a_{ij} = a_{i+} a_{+j} / a_{++} \tag{12}$$

for the missing data entries (i, j). To find the a_{ij} values satisfying (12), we employ a simple iterative algorithm:

1. Initialize all missing data entries with 0.
2. Compute a_{++} and all a_{i+}, $i = 1, 2, \ldots, m$ and a_{+j}, $j = 1, 2, \ldots, n$ by (5).
3. Compute new values a_{ij} for the missing data entries by (12). If all the new values are close enough to the previous values, STOP. Otherwise repeat from Step 2.

The E-M algorithm is known to converge properly in "well-behaved" situations (for example, no row or column should be entirely missing). This condition is plausible for most microarray experiments.

3 Test Framework

We applied correspondence analysis to a well-researched microarray data set containing samples from patients diagnosed with ALL and AML diseases [6]. It has been the subject of a variety of research papers, e.g. [1, 2, 10, 11]. This data set was also used in the CAMDA data contest [3]. Our primary concern was to try to designate genes whose expression levels significantly correlate with one of the diseases. It is natural to assume that those genes may be responsible to the development of particular conditions causing one of the considered leukemia variations. The paper [6] pursues a similar goal, but the authors used a statistical analysis algorithm followed by SOM clustering. The data set was divided into two parts – the training set (27 ALL, 11 AML samples) and the test set (20 ALL, 14 AML samples) – as the authors employed a learning algorithm. We considered it without any division since correspondence analysis does not require training. Hence there were 72 samples, 47 of which are ALL and 25 are AML. All the samples were obtained with Affymetrix GeneChip^{TM} microarray technology and contained 7129 data points. First 59 data points were miscelaneous Affymetrix control values and were removed from the data set, the rest 7070 data points were human genes. Affymetrix GeneChip^{TM} data values represent the difference between perfect match (PM) and mismatch (MM) probes that is expected to be significant. Usually when such a value is below 20, or even negative, it is not considered reliable. Hence we regarded all the data entries that are below 20 in the data set to be missing. Furthermore, genes having more than half of the missing entries were removed from the data set since the imputing can also not be reliable in this case. The residual 4902 data points were used in the analysis.

4 Computational Results

Fig. 1 shows the biplot obtained. It becomes immediately clear from the visual inspection that the first principal axis (horizontal) discriminates ALL samples from AML samples. Now, we may regard the genes having most positive coordinates on this axis signifying the ALL condition, while those genes having most negative coordinates there signify the AML condition. Similarly to [6], we listed top 25 ALL genes and top 25 AML genes with respect to our analysis. They are presented in Tables 1 and 2.

To validate the obtained top gene sets, we tried to estimate their relevance by observing the references made to these genes in MEDLINE articles. Each article in the MEDLINE database is categorized by the Medical Subject Headings (MeSH) [13]. We employed the same approach as the High-Density Array Pattern Interpreter (HAPI) of the University of California, San Diego [9, 12], and simply observed into which MeSH categories the articles mentioning the found genes fall predominantly. The HAPI web service reports the number of terms from each MeSH category matching genes from a provided

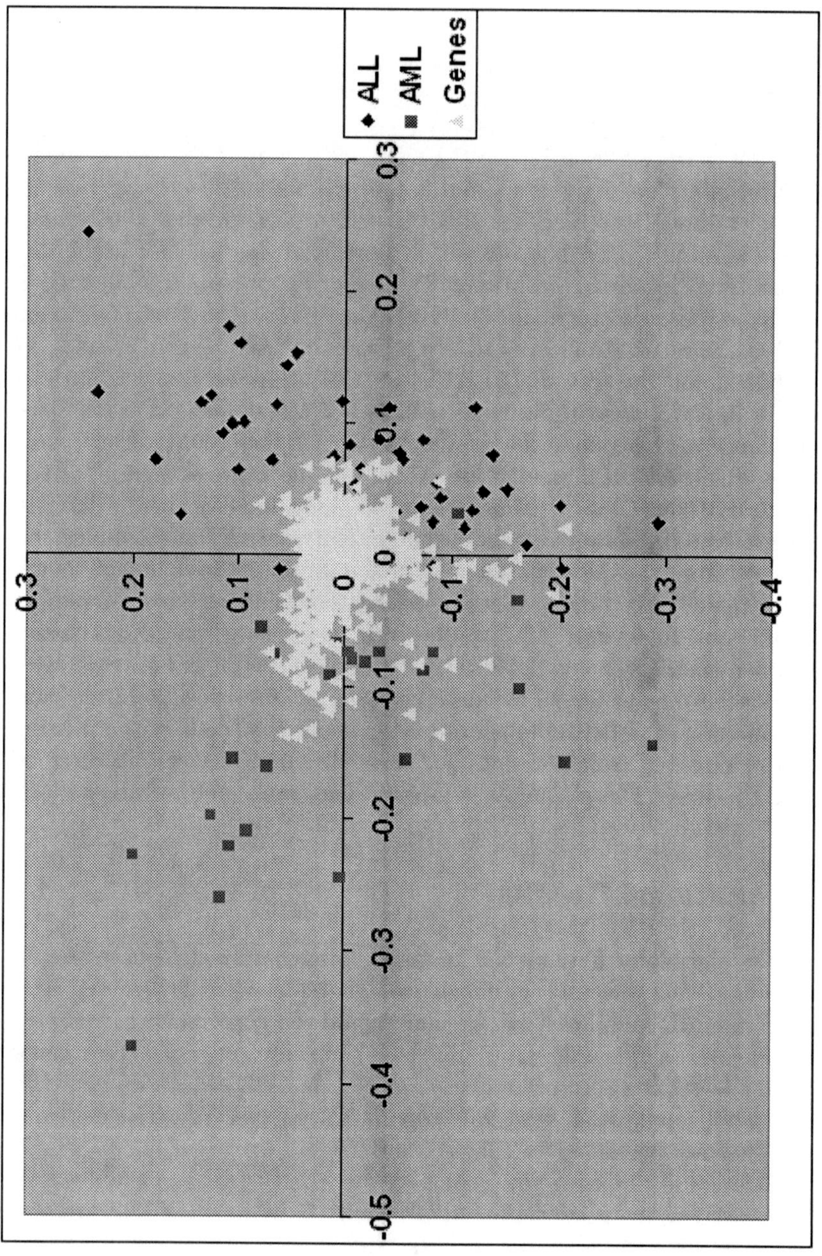

Fig. 1. Correspondence analysis biplot for the ALL vs. AML dataset

Table 1. 25 Top ALL Genes

#	Name	Description
1	M89957	IGB Immunoglobulin-associated beta (B29)
2	K01911	NPY Neuropeptide Y
3	AF009426	Clone 22 mRNA, alternative splice variant alpha-1
4	D13666 s	Osteoblast specific factor 2 (OSF-2os)
5	M83233	TCF12 Transcription factor 12 (HTF4, helix-loop-helix transcription factors 4)
6	D87074	KIAA0237 gene
7	X82240 rna1	TCL1 gene (T cell leukemia) extracted from H.sapiens mRNA for Tcell leukemia/lymphoma 1
8	S50223	HKR-T1
9	X53586 rna1	Integrin alpha 6 (or alpha E) protein gene extracted from Human mRNA for integrin alpha 6
10	D88270	GB DEF = (lambda) DNA for immunoglobin light chain
11	M38690	CD9 CD9 antigen
12	L33930 s	CD24 signal transducer mRNA and 3' region
13	U05259 rna1	MB-1 gene
14	U36922	GB DEF = Fork head domain protein (FKHR) mRNA, 3' end
15	D21262	KIAA0035 gene, partial cds
16	M94250	MDK Midkine (neurite growth-promoting factor 2)
17	M11722	Terminal transferase mRNA
18	M54992	CD72 CD72 antigen
19	D25304	KIAA0006 gene, partial cds
20	U31384	G protein gamma-11 subunit
21	X97267 rna1 s	LPAP gene
22	M29551	Serine/threonine protein phosphatase 2B catalytic subunit, beta isoform
23	M92934	CTGF Connective tissue growth factor
24	X84373	Nuclear factor RIP140
25	X17025	Homolog of yeast IPP isomerase

list. Furthermore, such a report is stored online, so the matchings found for our ALL and AML genes are available for future references [14, 15].

The report for the ALL genes shows most singnificant matching in such categories as "Cells" (37), "Cell Nucleus" (8), "Cells, Cultured" (10), "Hemic and Immune Systems" (16), "Immune System" (10), "Neoplasms" (12), "Neoplasms by Histologic Type" (8), "Hormones, Hormone Substitutes, and Hormone Antagonists" (8), "Enzymes, Coenzymes, and Enzyme Inhibitors" (30), "Enzymes" (30), "Hydrolases" (14), "Esterases" (10), "Transferases" (8), "Amino Acids, Peptides, and Proteins" (104), "Proteins" (90), "DNA-Binding Proteins" (8), "Glycoproteins" (12), "Membrane Glycoproteins" (8), "Membrane Proteins" (24), "Membrane Glycoproteins" (8), "Receptors, Cell Surface" (12), "Receptors, Immunologic" (12), "Transcription Factors" (12), "Nucleic Acids, Nucleotides, and Nucleosides" (14), "Nucleic Acids" (10), "Immunologic and Biological Factors" (78), "Biological Factors" (26), "Biolog-

Table 2. 25 Top AML Genes

#	Name	Description
1	U60644	HU-K4 mRNA
2	U16306	CSPG2 Chondroitin sulfate proteoglycan 2 (versican)
3	M69203 s	SCYA4 Small inducible cytokine A4 (homologous to mouse Mip-1b)
4	M33195	Fc-epsilon-receptor gamma-chain mRNA
5	M21119 s	LYZ Lysozyme
6	D88422	Cystatin A
7	M27891	CST3 Cystatin C (amyloid angiopathy and cerebral hemorrhage)
8	M57731 s	GRO2 GRO2 oncogene
9	M31166	PTX3 Pentaxin-related gene, rapidly induced by IL-1 beta
10	D83920	FCN1 Ficolin (collagen/fibrinogen domain-containing) 1
11	X97748 s	GB DEF = PTX3 gene promotor region
12	M23178 s	Macrophage inflammatory protein 1-alpha precursor
13	M92357	B94 protein
14	HG2981-HT3127 s	Epican, Alt. Splice 11
15	X04500	IL1B Interleukin 1, beta
16	M57710	LGALS3 Lectin, galactoside-binding, soluble, 3 (galectin 3)
17	U02020	Pre-B cell enhancing factor (PBEF) mRNA
18	Y00787 s	Interleukin-8 precursor
19	M28130 rna1 s	Interleukin 8 (IL8) gene
20	K01396	PI Protease inhibitor 1 (anti-elastase), alpha-1-antitrypsin
21	D38583	Calgizzarin
22	J04130 s	SCYA4 Small inducible cytokine A4 (homologous to mouse Mip-1b)
23	X62320	GRN Granulin
24	J03909	Gamma-interferon-inducible protein IP-30 precursor
25	M14660	GB DEF = ISG-54K gene (interferon stimulated gene) encoding a 54 kDA protein, exon 2

ical Markers" (18), "Antigens, Differentiation" (18), "Antigens, CD" (12), "Cytokines" (14), "Receptors, Immunologic" (12), "Investigative Techniques" (20), "Genetic Techniques" (9), "Biological Phenomena, Cell Phenomena, and Immunity" (7), "Genetics" (60), "Genes" (7), "Genetics, Biochemical" (43), "Molecular Sequence Data" (32), "Base Sequence" (13), "Sequence Homology" (13), "Physical Sciences" (11), "Chemistry" (11), and "Chemistry, Physical" (9).

The most significant matchings for the AML genes are in the categories "Nervous System" (13), "Cells" (102), "Blood Cells" (18), "Leukocytes" (15), "Cells, Cultured" (23), "Cell Line" (13), "Cytoplasm" (13), "Hemic and Immune Systems" (51), "Blood" (18), "Vertebrates" (18), "Algae and Fungi" (14), "Fungi" (14), "Organic Chemicals" (14), "Heterocyclic Compounds" (18), "Enzymes, Coenzymes, and Enzyme Inhibitors" (40), "Enzymes" (38), "Hydrolases" (24), "Carbohydrates and Hypoglycemic Agents" (46), "Carbohydrates" (46), "Polysaccharides" (36), "Glycosaminoglycans" (18), "Proteoglycans" (14), "Lipids and Antilipemic Agents" (16), "Lipids" (16), "Amino Acids, Peptides, and Proteins" (384), "Proteins" (370), "Blood Proteins" (48), "Acute-Phase Proteins" (14), "Contractile Proteins" (14), "Muscle Proteins" (14), "Cytoskeletal Proteins" (28), "Microtubule Proteins" (14), "Globulins" (14), "Serum Globulins" (14), "Glycoproteins" (62), "Membrane Glycoproteins" (26), "Proteoglycans" (14), "Membrane Proteins" (72), "Membrane Glycoproteins" (26), "Receptors, Cell Surface" (28), "Receptors, Immunologic" (20), "Nerve Tissue Proteins" (24), "Scleroproteins" (14), "Extracellular Matrix Proteins" (14), "Nucleic Acids, Nucleotides, and Nucleosides" (64), "Nucleic Acids" (34), "DNA" (20), "Nucleotides" (26), "Immunologic and Biological Factors" (262), "Biological Factors" (116), "Biological Markers" (26), "Antigens, Differentiation" (26), "Antigens, CD" (18), "Chemotactic Factors" (16), "Growth Substances" (32), "Interleukins" (18), "Toxins" (18), "Immunologic Factors" (146), "Antibodies" (16), "Antigens" (42), "Antigens, Surface" (38), "Antigens, Differentiation" (26), "Antigens, CD" (18), "Cytokines" (68), "Growth Substances" (20), "Interleukins" (18), "Monokines" (22), "Receptors, Immunologic" (20), "Specialty Chemicals and Products" (14), "Chemical Actions and Uses" (16), "Diagnosis" (17), "Laboratory Techniques and Procedures" (14), "Immunologic Tests" (13), "Investigative Techniques" (63), "Genetic Techniques" (18), "Immunologic Techniques" (16), "Immunohistochemistry" (13), "Technology, Medical" (20), "Histological Techniques" (15), "Histocytochemistry" (15), "Immunohistochemistry" (13), "Biological Phenomena, Cell Phenomena, and Immunity" (33), "Cell Physiology" (18), "Genetics" (160), "Genes" (21), "Genetics, Biochemical" (97), "Gene Expression" (15), "Gene Expression Regulation" (17), "Molecular Sequence Data" (63), "Base Sequence" (35), "Sequence Homology" (20), "Sequence Homology, Nucleic Acid" (14), "Biochemical Phenomena, Metabolism, and Nutrition" (100), "Biochemical Phenomena" (88), "Molecular Sequence Data" (61), "Base Sequence" (33), "Physiological Processes" (15), "Growth and Embryonic Development" (13), "Physical Sciences" (25).

Obviously, such literature scoring can only give an indicative measure of the quality of the obtained results. Furthermore, it should be noted that the data set contains only leukemia samples and no control samples, so it provides no information about the normal state of the gene expressions in absence of the diseases. Hence, the data analysis can only discover genes differentiating the sample classes. However, the HAPI scoring suggests that correspondence analysis enhanced by the missing data imputing feature uncovered genes highly relevant to the leukemia conditions. Moreover, the obtained numbers of matchings in the relevant MeSH categories compares favorably to the matchings of 25 ALL and AML genes reported by Golub at al. [16, 17].

5 Conclusions

Correspondence analysis is able to deliver informative projections of high-dimensional microarray data onto low-dimensional spaces. Such results in the form of pictures can be obtained in absence of any prior information about classification of samples and/or data points of the data set. In contrast to many other data mining techniques, correspondence analysis is computationally efficient, does not involve any parameters that must be tuned before the algorithm is executed, and successfully handles missing/inaccurate data values as long as their number is moderate. Furthermore, the method proves to be useful in uncovering hidden relations between groups of samples and data points (genes), possibly outperforming in efficiency more complicated statistical analysis techniques. The obtained lists on genes discriminating ALL and AML conditions may be useful for oncology researchers, providing further insights about the roles of particular human genes in the development of the acute leukemia cases.

References

1. A. Ben-Dor, L. Bruhn, I. Nachman, M. Schummer, and Z. Yakhini. Tissue classification with gene expression profiles. *Journal of Computational Biology*, 7:559–584, 2000.
2. A. Ben-Dor, N. Friedman, and Z. Yakhini. Class discovery in gene expression data. *Proceedings of the Fifth Annual International Conference on Computational Molecular Biology (RECOMB)*, 2001.
3. CAMDA 2001 Conference Contest Datasets.
   `http://www.camda.duke.edu/camda01/datasets/`.
4. P. J. Giles and D. Kipling. Normality of oligonucleotide microarray data and implications for parametric statistical analyses. *Bioinformatics*, 19:2254–2262, 2003.
5. G. H. Golub and C. F. Van Loan. *Matrix Computations, 3rd ed. (Johns Hopkins Series in the Mathematical Sciences)*. Baltimore, MD: John Hopkins, 1996.
6. T. R. Golub, D. K. Slonim, P. Tamayo, C. Huard, M. Gaasenbeek, J. P. Mesirov, H. Coller, M. L. Loh, J. R. Downing, M. A. Caligiuri, C. D. Bloomfield, and E. S. Lander. Molecular classification of cancer: class discovery and class prediction by gene expression monitoring. *Science*, 286:531–537, 1999.
7. M. J. Greenacre. *Theory and Applications of Correspondence Analysis*. Academic Press, 1984.
8. J. Kleinberg. The impossibility theorem for clustering. *Proceedings of the NIPS 2002 Conference*, 2002.
9. D. R. Masys, J. B. Welsh, J. L. Fink, M. Gribskov, I. Klacansky, and J. Corbeil. Use of keyword hierarchies to interpret gene expression patterns. *Bioinformatics*, 17:319–326, 2001.
10. J. Weston, S. Mukherjee, O. Chapelle, M. Pontil, T. Poggio, V. Vapnik. Feature selection for SVMs. *Proceedings of the NIPS 2000 Conference*, 2001.
11. E. P. Xing and R. M. Karp. CLIFF: Clustering of high-dimensional microarray data via iterative feature filtering using normalized cuts. *Bioinformatics Discovery Note*, 1:1–9, 2001.
12. High-Density Array Pattern Interpreter (HAPI).
   `http://array.ucsd.edu/hapi/`
13. National Library of Medicine – MeSH.
   `http://www.nlm.nih.gov/mesh/meshhome.html`
14. Hierarchy of keywords from literature associated with the top 25 ALL genes reported by correspondence analysis.
   `http://132.239.155.52/HAPI/ALL25_453.HTML`
15. Hierarchy of keywords from literature associated with the top 25 AML genes reported by correspondence analysis.
   `http://132.239.155.52/HAPI/AML25_502.HTML`
16. Hierarchy of keywords from literature associated with the top 25 ALL genes reported by Golub at al.
   `http://132.239.155.52/HAPI/goluball_911.HTML`
17. Hierarchy of keywords from literature associated with the top 25 AML genes reported by Golub at al.
   `http://132.239.155.52/HAPI/golubaml_161.HTML`

An Ensemble Method of Discovering Sample Classes Using Gene Expression Profiling

Dechang Chen[1], Zhe Zhang[2], Zhenqiu Liu[3], and Xiuzhen Cheng[4]

[1] Department of Preventive Medicine and Biometrics
Uniformed Services University of the Health Sciences
4301 Jones Bridge Road, Bethesda, MD 20814, USA
`dchen@usuhs.mil`
[2] Department of Biomedical Engineering
University of North Carolina
Chapel Hill, North Carolina 27599, USA
`zhangz@email.unc.edu`
[3] Bioinformatics Cell, TATRC
110 North Market Street, Frederick, MD 21703, USA
`liu@bioanalysis.org`
[4] Department of Computer Science
The George Washington University
801 22nd St. NW, Washington, DC 20052, USA
`cheng@gwu.edu`

Summary. Cluster methods have been successfully applied in gene expression data analysis to address tumor classification. Central to cluster analysis is the notion of dissimilarity between the individual samples. In clustering microarray data, dissimilarity measures are often subjective and predefined prior to the use of clustering techniques. In this chapter, we present an ensemble method to define the dissimilarity measure through combining assignments of observations from a sequence of data partitions produced by multiple clusterings. This dissimilarity measure is then subjective and data dependent. We present our algorithm of hierarchical clustering based on this dissimilarity. Experiments on gene expression data are used to illustrate the application of the ensemble method to discovering sample classes.

Key words: Cluster analysis, Dissimilarity measure, Gene expression

1 Introduction

Microarrays provide a very effective approach to interrogate hundreds or thousands of genes simultaneously. Such high throughput capability poses great challenges in terms of analyzing the data and transforming the data into useful information. As an exploratory data analysis tool, clustering has become a useful technique for identifying different and previously unknown cell types [3, 4, 9, 10].

Among many clustering methods applied to cluster samples, hierarchical clustering and its variations have received a special attention [2, 4, 6, 8]. This is mainly because a hierarchical method can produce a dendrogram, which provides a useful graphical summary of the data. However, dendrograms depend on the measures of dissimilarity for each pair of observations. A dissimilarity measure is amplified graphically by means of a dendrogram. In microarray data analysis, dissimilarity measures are commonly based on Pearson correlation. Such measures are restrictive, since Pearson correlation coefficient only describes the linear relationship between the observations on two variables. Therefore, it is difficult to see how one might be capable of making valid biological inferences. In this chapter, we present an ensemble method to define the dissimilarity measure. This method derives dissimilarity by combining assignments of observations from a sequence of data partitions produced by multiple clusterings. Thus, the dissimilarity measure is subjective and data dependent. We then present our algorithm of hierarchical clustering based on this dissimilarity. Experimental results show that the ensemble method is efficient in discovering sample classes using gene expression profiles.

2 Methods

Assume that there are $k(\geq 2)$ distinct tumor tissue classes for the problem under consideration, and that there are p genes (inputs) and n tumor mRNA samples (observations). Suppose x_{li} is the measurement of the expression level of the lth gene from the ith sample for $l = 1, \ldots, p$ and $i = 1, \ldots, n$. Let $\mathbf{G} = (x_{li})_{p \times n}$ denote the corresponding gene expression matrix. Note that the columns and rows of the expression matrix $\mathbf{G}$ correspond to samples and genes, respectively. The ith column may be written as a vector $\mathbf{x}_i = (x_{1i}, x_{2i}, \ldots, x_{pi})^T$, where T represents the transpose operation. We consider clustering the data $\{\mathbf{x}_1, \mathbf{x}_2, \cdots, \mathbf{x}_n\}$. Clustering may be viewed as a process of finding natural groupings of the observations $\mathbf{x}_i$. A key issue related to this type of groupings is how one measures the dissimilarity between data points. Most clustering algorithms presume a measure of dissimilarity. For example, K-means clustering uses Euclidean distance as a dissimilarity measure and Hierarchical clustering often uses correlation based dissimilarity measures. In this section, we briefly review K-means method and linkage methods, special cases of Hierarchical clustering techniques. Then we present an ensemble clustering algorithm using K-means and linkage methods.

K-means

Cluster analysis aims at partitioning the observations into clusters (or groups) so that observations within the same cluster are more closely related to each other than those assigned to different clusters. Partitioning is one of the major clustering approaches. A partitioning method constructs a partition of the data into clusters that optimizes the chosen partitioning criterion. The K-means is one of the most popular partitioning methods. This method uses Euclidean distance as the dissimilarity measure. It starts with a given assignment and proceeds to assign an observation to the cluster whose mean is closest. The process is repeated until the assignments do not change.

Linkage Methods

Commonly used linkage methods include single linkage (SL), complete linkage (CL), and average linkage (AL). They are special cases of agglomerative clustering techniques and follow the same procedure: beginning with the individual observations, at each intermediate step two least dissimilar clusters are merged into a single cluster, producing one less cluster at the next higher level [7]. The difference among the linkage methods lies in the dissimilarity measures between two clusters, which are used to merge clusters. SL, CL, and AL define the dissimilarity between two clusters to be the minimum distance between the two clusters, the maximum distance between the two clusters, and the average distance between the two clusters, respectively. Specifically, suppose we have n observations $\mathbf{x}_1, \mathbf{x}_2, \ldots, \mathbf{x}_n$. Let $d(\mathbf{x}_i, \mathbf{x}_j)$ denote the predefined dissimilarity between $\mathbf{x}_i$ and $\mathbf{x}_j$. Given two clusters G_1 and G_2, containing n_1 and n_2 observations, respectively, the dissimilarity between G_1 and G_2 defined by SL, CL, and AL are

$$d_{SL}(G_1, G_2) = \min_{\mathbf{x}_i \in G_1, \mathbf{x}_j \in G_2} d(\mathbf{x}_i, \mathbf{x}_j),$$

$$d_{CL}(G_1, G_2) = \max_{\mathbf{x}_i \in G_1, \mathbf{x}_j \in G_2} d(\mathbf{x}_i, \mathbf{x}_j),$$

$$d_{AL}(G_1, G_2) = \frac{1}{n_1 n_2} \sum_{\mathbf{x}_i \in G_1, \mathbf{x}_j \in G_2} d(\mathbf{x}_i, \mathbf{x}_j).$$

Hierarchical Clustering by Ensemble Procedure

Learning Dissimilarity from Data

The clustering procedure partitions the data into clusters so that observations in one cluster are more like one another than like observations in other clusters. This procedure requires a measure of "closeness" or "similarity" between two observations. Such a measure can be made by using any metric $d(\cdot, \cdot)$, which, for any observations $\mathbf{x}_i$, $\mathbf{x}_j$, and $\mathbf{x}_k$, satisfies the following four properties: (i) (nonnegativity) $d(\mathbf{x}_i, \mathbf{x}_j) \geq 0$; (ii) (reflexivity) $d(\mathbf{x}_i, \mathbf{x}_j) = 0$ if and only if $\mathbf{x}_i = \mathbf{x}_j$; (iii) (symmetry) $d(\mathbf{x}_i, \mathbf{x}_j) = d(\mathbf{x}_j, \mathbf{x}_i)$; (iv) (triangle inequality) $d(\mathbf{x}_i, \mathbf{x}_j) + d(\mathbf{x}_j, \mathbf{x}_k) \geq d(\mathbf{x}_i, \mathbf{x}_k)$. The smaller the measure $d(\mathbf{x}_i, \mathbf{x}_j)$ is, the more similar the two observations. As an example, the Euclidean distance $d(\mathbf{x}_i, \mathbf{x}_j) = (\sum_{l=1}^{p}(x_{li} - x_{lj})^2)^{1/2}$ is a metric.

Clustering methods also accept other measures of "closeness" that may not meet the reflexivity or triangle inequality. In general, one can introduce a dissimilarity function $dis(\mathbf{x}_i, \mathbf{x}_j)$ to measure 'closeness" or "similarity" between $\mathbf{x}_i$ and $\mathbf{x}_j$. A dissimilarity function is a function that satisfies nonnegativity and symmetry. For example, $dis(\mathbf{x}_i, \mathbf{x}_j) = 1 - \cos(\mathbf{x}_i, \mathbf{x}_j)$ is a dissimilarity function, where $\cos(\mathbf{x}_i, \mathbf{x}_j)$ refers to the cosine of the angle between $\mathbf{x}_i$ and $\mathbf{x}_j$. (It is easy to show that neither reflexivity nor the triangle inequality is satisfied by this measure.)

The notion of dissimilarity is central to cluster analysis. Different choices of dissimilarity functions can lead to quite different results. Prior knowledge is often helpful in selecting an appropriate dissimilarity measure for a given problem. However, it is possible to learn a dissimilarity function from the data. We describe such a procedure as follows.

Partitioning methods are usually not stable in the sense that the final results often depend on initial assignments. However, if two observations are assigned to the same cluster by a high percentage of the times of use of the same partitioning method, it is then very likely that these two observations come from a common "hidden" group. This heuristic implies that the "actual" dissimilarity between two observations may be derived by combining the various clustering results from repeated use of the same partitioning technique. Here we formalize this combining process using K-means partitioning method.

For the data $\{\mathbf{x}_1, \mathbf{x}_2, \cdots, \mathbf{x}_n\}$, we can select K centroids and then run K-means technique to partition the data into K clusters. It is known that the final assignment usually depends on the initial reallocation. Now we run K-means N times. Each time a number K is randomly picked from a given interval $[K_1, K_2]$. By doing this, we may end up with N possibly different final assignments. Given observations (samples) $\mathbf{x}_i$ and $\mathbf{x}_j$, let p_{ij} denote the probability that they are not placed into the same cluster by the final assignment of a run of K-means clustering. This probability p_{ij} can be estimated by using the results of repeated K-means clustering method. Define $\delta_m(i, j) = 1$ if the mth use of the K-means algorithm does not assign samples $\mathbf{x}_i$ and $\mathbf{x}_j$ into the same cluster; and $\delta_m(i, j) = 0$ otherwise. Then $\delta_1(i, j)$, $\delta_2(i, j)$, $\cdots$, $\delta_N(i, j)$ are iid Bernoulli(p_{ij}). It is well known that the best unbiased estimator of p_{ij} is $\sum_{m=1}^{N} \delta_m(i, j)/N$. This estimate will be used as the dissimilarity measure between $\mathbf{x}_i$ and $\mathbf{x}_j$, i.e.,

$$dis(\mathbf{x}_i, \mathbf{x}_j) = \frac{\sum_{m=1}^{N} \delta_m(i, j)}{N}. \tag{1}$$

A smaller value of $dis(\mathbf{x}_i, \mathbf{x}_j)$ is expected to imply a bigger chance that samples $\mathbf{x}_i$ and $\mathbf{x}_j$ come from the same "hidden" group.

We now state our ensemble method using K-means technique and linkage methods. Here the word ensemble refers to the sequence of K-means procedures involved in the method.

Algorithm

1. Given N, K_1, and K_2, run the K-means clustering method N times with each K randomly chosen from $[K_1, K_2]$.
2. Construct the pairwise dissimilarity measure $dis(\mathbf{x}_i, \mathbf{x}_j)$ between the observations by using the equation (1).
3. Cluster the n samples (observations) by applying a linkage method and the dissimilarity measure $dis(\mathbf{x}_i, \mathbf{x}_j)$ learnt in Step 2.

3 Results and Discussion

Datasets

We considered two gene expression datasets: COLON [1] and OVARIAN [11]. The COLON dataset consists of expression profiles of 2000 genes from 22 normal tissues and 40 tumor samples. The OVARIAN dataset contains expression profiles of 7129 genes from 5 normal tissues, 27 epithelial ovarian tumor samples, and 4 malignant epithelial ovarian cell lines.

Standardization and Gene Selection

Following Dudoit and Fridlyand [3], we standardized the gene expression data so that the samples have mean 0 and variance 1 across genes. This simple standardization procedure achieves a location and scale normalization of different arrays.

Each gene dataset considered here contains several thousand genes. Genes showing almost constant expression levels across samples are not expected to be useful for clustering. Thus, in our analysis, we used $p = 100$ genes which correspond to 100 largest variances of the gene expression levels across the samples.

Parameter Setting

To run our algorithm, we chose parameters as follows. The choice of N depends on the rate at which dis in (1) converges to p_{ij}. A large number should be chosen for N. For obtaining quick results, we set $N = 1000$. Since the number of true clusters for each dataset is at least 2, we set $K_1 = 2$. A large K_2 is required to guarantee that $\delta_1(i, j)$, $\delta_2(i, j)$, $\cdots$, $\delta_N(i, j)$ are iid Bernoulli(p_{ij}). However, in practice, the algorithm will not work for very large values of K_2. For example, $K_2 > n$ is clearly impractical. In our examples, we used $K_2 = 30$.

Cluster Analysis

For both gene expression datasets, we used single linkage in our algorithm. Cluster analysis for COLON and OVARIAN is shown in Fig. 1 and Fig. 2, respectively.

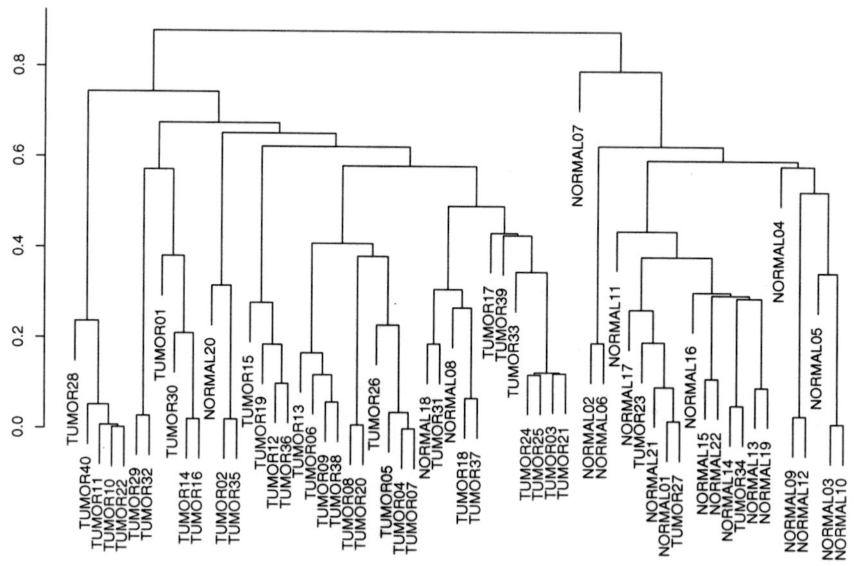

Fig. 1. Dendrogram from clustering of COLON data.

The dendrogram for COLON clearly shows that all the samples are clustered into two classes with 6 misclassified samples (NORMAL08, NORMAL18, NORMAL20, TUMOR23, TUMOR27, and TUMOR34). Alon et al. [1] conducted a two-way clustering analysis based on 20 genes with the most significant difference between tumor and normal tissues. They found two clusters with 8 misclassified samples: 5 tumor samples clustered with normal tissues and 3 normal tissues clustered with tumor samples.

From the dendrogram for OVARIAN, our algorithm shows that all the samples are clustered into three classes: the class for normal tissues, the class for malignant epithelial ovarian cell lines, and the class for tumor samples. The figure indicates that only one sample BCELL is misclassified. Clearly,

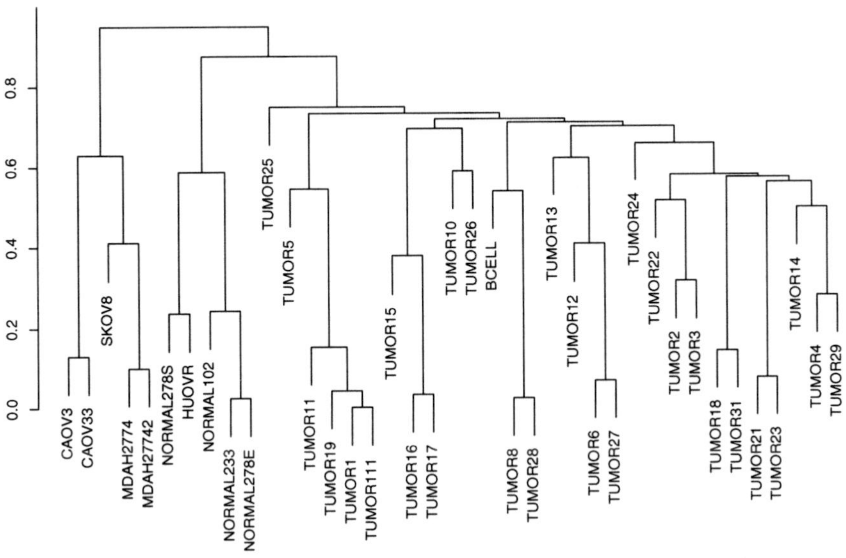

Fig. 2. Dendrogram from clustering of OVARIAN data.

our result is better than that of Welsh et al. [11], where the samples are mixed together.

The above clustering analysis shows that our algorithm could be employed to effectively discover sample classes using gene expression profiling. From the above two examples, we have found that the dendrograms do not show any significant change when increasing N or p or both, where p is the number of genes used in the analysis. This robustness will be further studied both theoretically and experimentally. SL was used in our algorithm for the above two gene expression datasets. The effect of linkage methods other than SL will be investigated in our future work. Our algorithm uses K-means to define dissimilarity. In our future work, we will also investigate the effect of replacing K-means by other partitioning methods. The ensemble method of clustering presented here was discussed in [5] from the perspective of evidence accumulation.

Acknowledgments

D. Chen was supported by the USUHS grant R087RH-01 and the National Science Foundation grant CCR-0311252.

References

1. U. Alon, N. Barkai, D. A. Notterman, K. Gish, S. Ybarra, D. Mack, and A. J. Levine. Broad patterns of gene expression revealed by clustering analysis of tumor and normal colon tissues probed by oligonucleotide arrays. *Proceedings of the National Academy of Sciences*, 96: 6745-6750, 1999.

2. M. Bittner, P. Meltzer, Y. Chen, Y. Jiang, E. Seftor, M. Hendrix, M. Radmacher, R. Simon, X. Yakhini, A. Ben-Dor, E. Dougherty, E. Wang, F. Marincola, C. Gooden, J. Lueders, A. Glatfelter, P. Pollock, E. Gillanders, D. Leja, K. Dietrich, C. Beaudry, M. Berens, D. Alberts, V. Sondak, N. Hayward, and J. Trent. Molecular classification of cutaneous malignant melanoma by gene expression profiling. *Nature*, 406: 536-540, 2000.

3. S. Dudoit and J. Fridlyand. A prediction based resampling method to estimate the number of clusters in a dataset. *Genome Biology*, 3(7): 0036.1-0036.21, 2002.

4. M. B. Eisen, P. T. Spellman, P. O. Brown, and D. Botstein. Cluster analysis and display of genome-wide expression patterns. *Proceedings of the National Academy of Sciences*, 95: 14863-14868, 1998.

5. A. Fred and A. K. Jain. Data clustering using evidence accumulation. *Proceedings of the 16th International Conference on Pattern Recognition, ICPR 2002*, 276-280, 2002.

6. D. R. Goldstein, D. Ghosh, and E. M. Conlon. Statistical issues in the clustering of gene expression data. *Statistica Sinica*, 12: 219-240, 2002.

7. T. Hastie, R. Tibshirani, and J. Friedman. *The Elements of Statistical Learning*. Springer-Verlag, New York, 2001.

8. W. Li, M. Fan, and M. Xiong. SamCluster: An integrated scheme for automatic discovery of sample classes using gene expression profile. *Bioinformatics*, 19(7): 811-817, 2003.

9. P. Tamayo, D. Slonim, J. Mesirov, Q. Zhu, S. Kitareewan, E. Dmitrovsky, E. S. Lander, and T. R. Golub. Interpreting patterns of gene expression with self-organizing maps: methods and application to hematopoietic differentiation. *Proceedings of the National Academy of Sciences*, 96: 2907-2912, 1999.

10. S. Tavazoie, J. Hughes, M. Campbell, R. Cho, and G. Church. Systematic determination of genetic network architecture. *Nature Genetics*, 22: 281-285, 1999.

11. J. B. Welsh, P. P. Zarrinkar, L. M. Sapinoso, S. G. Kern, C. A. Behling, R. A. Burger, B. J. Monk, and G. M. Hampton. Analysis of gene expression in normal and neoplastic ovarian tissue samples identifies candidate molecular markers of epithelial ovarian cancer. *Proceedings of the National Academy of Sciences*, 98: 1176-1181, 2001.

CpG Island Identification with Higher Order and Variable Order Markov Models

Zhenqiu Liu[1], Dechang Chen[2], and Xue-wen Chen[3]

[1] Bioinformatics Cell, TATRC
110 North Market Street, Frederick, MD 21703, USA
`liu@bioanalysis.org`
[2] Department of Preventive Medicine and Biometrics
Uniformed Services University of the Health Sciences
4301 Jones Bridge Road, Bethesda, MD 20814, USA
`dchen@usuhs.mil`
[3] Electrical Engineering and Computer Science Department
The University of Kansas
1520 West 15th Street, Lawrence, KS 66045, USA
`xwchen@eecs.ku.edu`

Summary. Identifying the location and function of human genes in a long sequence of genome is difficult due to lack of sufficient information about genes. Experimental evidence has suggested that there exists strong correlation between CpG islands and genes immediately following them. Much research has been done to identify CpG islands in a DNA sequence using various models. In this chapter, we introduce two alternative models based on high order and variable order Markov chains. Compared with the popular models such as the first order Markov chain, HMM, and HMT, these two models are much easier to compute and have higher identification accuracies. One unsolved problem with the Markov model is that there is no way to decide the exact boundary point between CpG and non-CpG islands. In this chapter, we provide a novel tool to decide the boundary points using the sequential probability test. Sequential data from GeneBank are used for the experiments in this chapter.

Key words: DNA sequences, CpG islands, Markov models, Probability Suffix Trees (PST), sequential probability ratio test (SPRT)

1 Introduction

Genome is defined by combining the word "gene" with the suffix "ome" for mass. Most genomes are made of DNA that contains the biochemical codes for the inheritance. A portion of the DNA that is associated with a specific trait or function is defined as a gene. A gene is responsible to initiate the biochemical reactions known as gene expression. DNA is a macromolecule made up of unbranched, linear chains of monomeric subunits called nucleotides. These subunits are comprised of a nitrogenous base, a sugar, and a phosphate group generally denoted dNTP for deoxyribonucleotide triphosphate. The nitrogen-base can be adenine (A), cytosine (C), guanine (G), or thyamine (T). The list of the bases in a DNA strand is called a DNA sequence.

Recently, DNA sequences of the human genome have been fully determined. And identifying probable locations of genes in a long DNA sequence has become one objective of the field of bioinformatics. In human genome the dinucleotide CG is called CpG. Because there are not an adequate number of marked genes available in human genome, scientists have designed methods to detect the CpG islands instead of identifying the probable location of genes directly. When the CpG occurs, the C nucleotide is chemically modified by methylation. The chances of this methyl-C mutating to a T become higher. Therefore, the CpG dinucleotides are usually rarer in the genome than that expected from the independent C and G. Based on the biological principle, methylation process is suppressed at the starting regions of many genes in a genome. As a result, these regions have CpG dinucleotide much higher than elsewhere. These regions are called CpG islands. About 56% of the human genes are associated with a region of a CpG island. The goal of distinguishing the CpG islands is to locate the probable gene in a long DNA sequence indirectly.

There are many methods available for identifying the CpG island. The first order Markov model, hidden Markov model (HMM), and Hidden Markov Tree (HMT) are the popular tools. Because of complexity of the real life sequence, the short memory assumption of the first order Markov chain usually is not satisfied. The HMM model, on the other hand, is a Markov chain with outputs. It is more complex and can be slow for large problems. It has been proven that HMMs can not be trained in polynomial time in the alphabet size [5]. Besides, the algorithm of HMM can only be guaranteed to converge to a local minimum.

In this chapter we introduce two alternative models to identify CpG islands: higher order Markov chains and variable order Markov chains. The first order Markov chain has only 4 states A, C, G, T. The number of states in the forth/fifth order chains is not very large (for a forth order chain, there are $4^4 = 256$ states). Therefore, certain higher order Markov chains could be used in CpG island identification. To overcome the drawback that the size of a Markov chain grows exponentially with order, variable order Markov chains can be used. In the variable order Markov chain, the order of the Markov chain is not fixed [3]. The model is based on the probability suffix trees and can also be explained by probability suffix automata [5].

This chapter is organized as follows. In Section 2, we introduce higher order Markov chains. In Section 3, we provide some experimental results from real life DNA sequences. Conclusions and remarks are given in Section 4.

2 Higher Order Markov Chains

In this section, we introduce higher order Markov chains and how to control the model complexity. A variable order Markov chain algorithm is also given.

2.1 Markov Models

Assume a discrete time random variable X_t takes values in the finite set $\{1, \ldots, m\}$. The first order Markov hypothesis says that the present value at time t is entirely explained by the value at time $t-1$, so that we can write

$$P(X_t = i_0 | X_0 = i_t, \ldots, X_{t-1} = i_1) = p_{i_1 i_0}(t),$$

where $i_t, \ldots, i_0 \in \{1, \ldots, m\}$. Assuming the chain is time invariant, we have a *homogeneous* Markov chain:

$$P = \begin{pmatrix} p_{11} & p_{12} & \cdots \\ p_{21} & p_{22} & \cdots \\ \vdots & \vdots & \ddots \end{pmatrix}.$$

Often we need to solve two problems with Markov chains. The first is prediction, and the second is estimation [2]. The prediction problem is defined as follows. Given a system being in state $X_t = i$ at time t, what is the probability distribution over the possible states X_{t+k} at time $t+k$? The answer is obtained by using the transition probabilities and the partition technique. For example,

$$P_2(X_{t+2} | X_t) = \sum_{X_{t+1}} P_1(X_{t+1} | X_t) P_1(X_{t+2} | X_{t+1}),$$

where $P_k(X'|X)$ is the k-step transition probability.

For the estimation problem, we need to estimate the initial state distribution $P_0(X_0)$ and the transition probability $P_1(X'|X)$. The answers can be obtained using the empirical estimates. Given L observed sequences of different lengths

$$X_0^{(1)} \to X_1^{(1)} \to X_2^{(1)} \to \ldots \to X_{n_1}^{(1)},$$

$$\ldots$$

$$X_0^{(L)} \to X_1^{(L)} \to X_2^{(L)} \to \ldots \to X_{n_L}^{(L)},$$

the maximum likelihood estimates (observed fractions) are

$$\hat{P}_0(X_0 = i) = \frac{1}{L} \sum_{l=1}^{L} \delta(X_0^{(l)}, i),$$

where $\delta(x,y) = 1$ if $x = y$ and zero otherwise. The transition probabilities are obtained as observed fractions out of a specific state. The joint estimate over successive states is

$$\hat{P}_{X,X'}(X = i, X' = j) = \frac{1}{\sum_{l=1}^{L} n_l} \sum_{l=0}^{L} \sum_{t=0}^{n_l - 1} \delta(X_t^{(l)}, i)\delta(X_{t+1}^{(l)}, j),$$

and the transition probability estimates are

$$\hat{P}_1(X' = j | X = i) = \frac{P_{X,X'}(X = i, X' = j)}{\sum_k P_{X,X'}(X = i, X' = k)}.$$

In a situation where the present depends on the last l observations, we have an lth order Markov chain whose probabilities are

$$P(X_t = i_0 | X_0 = i_t, \ldots, X_{t-1} = i_1)$$
$$= P(X_t = i_0 | X_{t-l} = i_l, \ldots, X_{t-1} = i_1) = p_{i_l \ldots i_0}.$$

For instance, if we set $l = 2$ and $m = 3$, the corresponding transition matrix is

$$P = \begin{pmatrix}
p_{111} & 0 & 0 & p_{112} & 0 & 0 & p_{113} & 0 & 0 \\
p_{211} & 0 & 0 & p_{212} & 0 & 0 & p_{213} & 0 & 0 \\
p_{311} & 0 & 0 & p_{312} & 0 & 0 & p_{313} & 0 & 0 \\
0 & p_{121} & 0 & 0 & p_{122} & 0 & 0 & p_{123} & 0 \\
0 & p_{221} & 0 & 0 & p_{222} & 0 & 0 & p_{223} & 0 \\
0 & p_{321} & 0 & 0 & p_{322} & 0 & 0 & p_{323} & 0 \\
0 & 0 & p_{131} & 0 & 0 & p_{132} & 0 & 0 & p_{133} \\
0 & 0 & p_{231} & 0 & 0 & p_{232} & 0 & 0 & p_{233} \\
0 & 0 & p_{331} & 0 & 0 & p_{332} & 0 & 0 & p_{333}
\end{pmatrix}.$$

When the order is greater than 1, the transition matrix P contains several elements corresponding to transitions that can not occur. For instance, it is impossible to go from the row defined by $X_{t-2} = 1$ and $X_{t-1} = 2$ to the column defined by $X_{t-1} = 1$ and $X_t = 1$. The probability of this transition is then 0. We can rewrite P in a more compact form excluding zeros. For example, the reduced form of the matrix corresponding to $l = 2$ and $m = 3$ is

$$Q = \begin{pmatrix}
p_{111} & p_{112} & p_{113} \\
p_{211} & p_{212} & p_{213} \\
p_{311} & p_{312} & p_{313} \\
p_{121} & p_{122} & p_{123} \\
p_{221} & p_{222} & p_{223} \\
p_{321} & p_{322} & p_{323} \\
p_{131} & p_{132} & p_{133} \\
p_{231} & p_{232} & p_{233} \\
p_{331} & p_{332} & p_{333}
\end{pmatrix}.$$

Each possible combination of l successive observations of the random variable X is called the state of the model. The number of states is equal to m^l (=9 in our example). Whatever the order is, there are $m - 1$ independent probabilities in each row of the matrix P and Q, the last one depending on the others since each row is a probability distribution summing to one. The total number of independent (free) parameters is $p = m^l(m - 1)$. Given a set of observations, these parameters can be computed as follows. Let $n_{i_l\dots i_0}$ denote the number of transitions of the type

$$X_{t-l} = i_l, \dots, X_{t-1} = i_1, X_t = i_0$$

in the data. The maximum likelihood estimate of the corresponding transition probability is

$$\hat{p}_{i_l\dots i_0} = \frac{n_{i_l\dots i_0}}{n_{i_l\dots i_1}},$$

where

$$n_{i_l\dots i_1} = \sum_{i_0=1}^{m} n_{i_l\dots i_0}.$$

Given the sequences, the order of a Markov chain can be determined using the minimum description length (MDL) [4]. MDL is defined to be

$$MDL = -2LL + p\log(n),$$

where

$$LL = \log(L) = \sum_{i_l,\dots,i_0=1}^{m} n_{i_l\dots i_0} \log p_{i_l\dots i_0}$$

is the log-likelihood of the model, p is the number of independent parameters, and n is the number of components in the likelihood. Another popular criterion used to determine the order of a Markov chain is the Akaike's Information Criterion (AIC):

$$AIC = -2LL + 2p.$$

The optimal order of a Markov chain can be determined by minimizing either MDL or AIC.

2.2 Variable Order Markov Chains

Variable Markov chains are based on Probability Suffix Trees (PST). A probability suffix tree describes a set of conditional distributions for the given sequences. Let Σ be the set of alphabet of 4 nucleic acids (A, C, T, or G) for DNA sequences. Let s be a substring in the sequences. For $\sigma \in \Sigma$, we define $P(\sigma|s)$ to be the probability of a symbol σ given a suffix s. Let the length of substring s be $|s| = l$. Let N_s be the number of occurrences of string s in the sequences and $N_{\sigma s}$ be the number of occurrences of string σs in the sequences.

We have the following estimate of $p(\sigma|s)$: $\hat{P}(\sigma|s) = N_{\sigma s}/N_s$. The probability of a sequence calculated by PST is the product of the probabilities of each letter given its longest suffix in the PST model.

A *complete* PST is a PST with full leaves. A complete PST of depth L is equivalent to a higher order Markov chain of order L. A noncomplete PST is equivalent to a variable order Markov chain and the order depends on the suffix. Hence, PST is a general model, including both higher order and variable order Markov chains as special cases. In the following, we first build a PST and then extract the leaf nodes of PST to form our variable Markov model.

The Algorithm for PST and Variable Order Markov Chains

Let us define a substring of s without first letter as $suf(s) = s_2 s_3, \ldots, s_l$. $suf(s)$ is the parent node of string s in PST. The algorithm for building the PST is based on a simple search method. First we initialize the PST with a single root node and an empty string. Then we gradually calculate the conditional probability $P(\sigma|s)$ of each symbol given a substring in the substring space, starting at the single letter substrings. We then check if it is large enough and if it is significantly different from the conditional probability of the same letter given the $suf(s)$, which means that we check if the current node's distributions are significantly different from its parent's distributions. If the difference is significant, the substring and all necessary nodes on its path are added to our PST. After the PST is established, we can extract the variable Markov model based on the leaf nodes and their transitional probabilities of the PST.

Let L be the maximum order of the variable Markov chain, α the minimum conditional probability in order for a node to be included in the PST model, P_{min} the minimum probability of a string in the sequences, and ε the minimum difference of the distribution between the current node and its parents node. For a string of length n, the complexity of building a PST takes $O(Ln^2)$ time by the method of Apostolico and Bejerano [1]. The procedure of their algorithm for building PST and variable Markov chains is given as follows.

1. Initialize the tree T with a single empty node e and substring set $S \leftarrow \{\sigma | \sigma \in \Sigma \text{ and } P(\sigma) > P_{min}\}$.
2. Build PST. While $S \neq \emptyset$, we pick up $s \in S$ and perform the following steps:
   - Remove s from S.
   - For a symbol $\sigma \in \Sigma$, if
$$P(\sigma|s) \geq \alpha$$
   and
$$\frac{|P(\sigma|s) - P(\sigma|suf(s))|}{P(\sigma|suf(s))} \geq \varepsilon,$$
   or

$$\frac{|P(\sigma|s) - P(\sigma|suf(s))|}{P(\sigma|s)} \geq \varepsilon,$$

we add the node corresponding to s and all the nodes on the path from s to the root to our PST.

- If $s < L$ and $P(\gamma s) \geq P_{min}$ for $\gamma \in \Sigma$, then add γs to S.

3. Extract the leaf nodes and related probabilities to form a transitional probability matrix of the variable Markov model.

3 Discrimination with Markov Models and SPRT

A Markov model is fully defined by its states and state transition matrix. The first order Markov chain for DNA sequences has the structure shown in Figure 1. In this section, we introduce how to use Markov models to identify

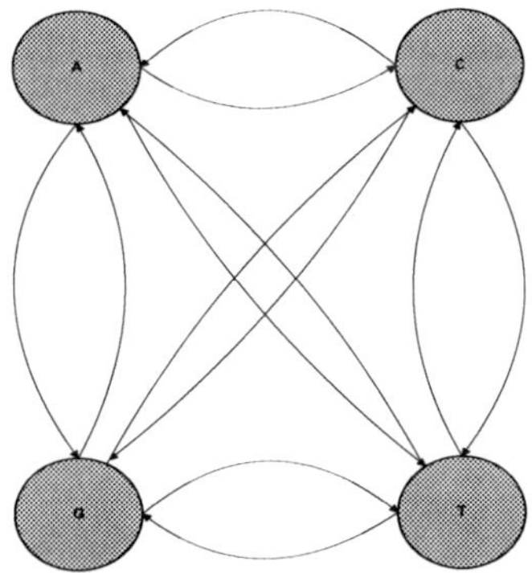

Fig. 1. The structure of the first order chain for a DNA sequence.

the CpG islands, with an aid of a sequential probability ratio test (SPRT).

3.1 SPRT

Assume a sequence of observations $\{x_i\}$ $(i = 1, 2, \ldots)$. The decision for discriminating two simple hypotheses $H_0 : \theta = \theta_0$ and $H_1 : \theta = \theta_1$ $(\theta_0 \neq \theta_1)$ is: accept H_0 if $Z_n \leq b$ and reject H_0 if $Z_n \geq a$. Here, θ represents the set of parameters of the model. For a Markov model, θ is the transition probability matrix. The random variable Z_n is the natural logarithm of the probability ratio at stage n:

$$Z_n = \ln \frac{f(\mathbf{x}_n|\theta_1)}{f(\mathbf{x}_n|\theta_0)} \quad \text{with} \quad \mathbf{x}_n = x_1 x_2 \ldots x_n.$$

Without any loss of generality, Z_n can be set to be zero when $f(\mathbf{x}_n|\theta_1) = f(\mathbf{x}_n|\theta_0) = 0$.

The numbers a and b are two stopping bounds. They are estimated by the following Wald approximations:

$$b \simeq \ln \frac{\beta}{1-\alpha} \quad \text{and} \quad a \simeq \ln \frac{1-\beta}{\alpha},$$

where α is the probability that H_1 is chosen when H_0 is true and β is the probability that H_0 is accepted when H_1 is true. It has been a standard practice to employ the above equations to approximate the stopping bounds in all practical applications. The continuation region $b \leq Z_n \leq a$ is called the *critical inequality* of the SPRT at length n.

3.2 SPRT for Markov Model

To use Markov models in identifying the CpG islands for DNA sequences, we need to train two models separately: one for the CpG island, the other for the non-CpG island. For simplicity, we shall represent CpG and non-CpG regions by '+' and '-', respectively. Let the transitional probability for the CpG island be a_{ij}^+ and a_{ij}^- for the non-CpG island. Given a test DNA sequence $\mathbf{x}$ of length n, we can discriminate the CpG islands from the non-CpG island using the following log-likelihood ratio:

$$R_n(\mathbf{x}) = Z_n = \ln \frac{P(\mathbf{x}|model+)}{P(\mathbf{x}|model-)} = \sum_{i=1}^{n} \ln \frac{a_{ij}^+}{a_{ij}^-} = \sum_{i=1}^{n} \ln a_{ij}^+ - \sum_{i=1}^{n} \ln a_{ij}^-.$$

If $R_n(\mathbf{x}) > C$, the sequence is the CpG island, where $C = \ln \frac{1-\beta}{\alpha}$ for given α and β. For instance, if $\alpha = 0.1$ and $\beta = 0.1$, we have $C = \ln(0.9/0.1) = 2.1972$. In our models, we may use different order Markov models for CpG islands and non-CpG islands, which is much more flexible in applications.

4 Computational Results

All the datasets in our experiments are from GeneBank. GeneBank is a repository for DNA sequences maintained by NIH. In our training and testing datasets, the locations of CpG islands are marked. We first extracted the marked CpG regions from the training DNA sequences and built two models, one for CpG island and one for non-CpG island. Figures 2 and 3 show the monomer density of CpG island and non-CpG island, respectively. The figures indicate that CpG island has higher nucleotide percentages of C+G, while the non-CpG island does not.

To test the efficiency of our models, we first split our testing sequences into fixed length windows. The step for the window to move forward was set

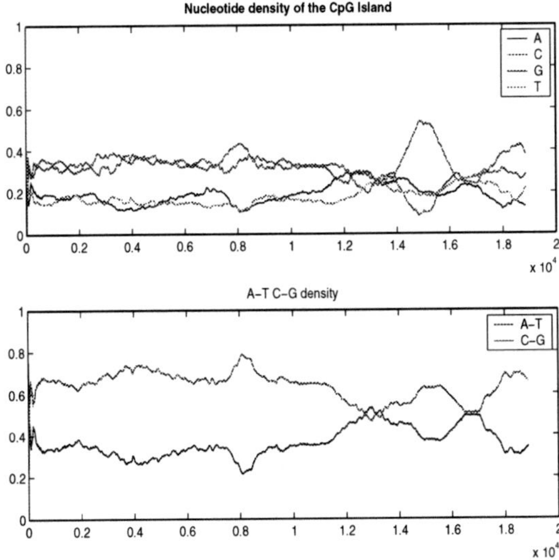

Fig. 2. The nucleotide density of CpG islands.

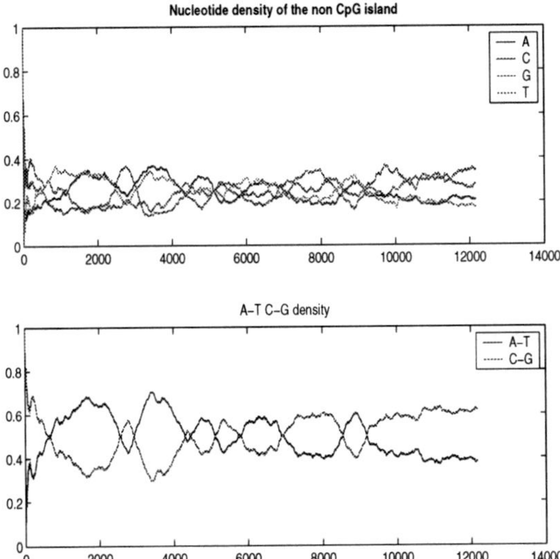

Fig. 3. The nucleotide density of non-CpG islands.

to be 1. The decision was then made using the likelihood of each subsequence given the two models. With $\alpha = 0.01$ and $\beta = 0.01$, the decision boundary for SPRT is $C = 4.6$. The computational results were also compared with those from the corresponding HMMs.

Our first test sequence is a human creatine kinase (CK) pseudogene, with an insert of 3 tandem Alu repeats (HSCKBPS). The length of the sequence is 2980. We split it into subsequences with 120 symbols. The testing results are given in Table 1. The numbers in Table 1 represent the relative position of the nucleotides with respect to the starting location of the DNA sequences used for testing. Table 1 indicates that the third order Markov model (MC3) has the best performance, since its boundary prediction is most close to the true boundary of the sequence. It is interesting to note that the hidden Markov model lead to a new region that does not exist in the sequence. The reason for this might be the noise in the data sequence.

Table 1. Performances of different order Markov chains on HSCKBPS

True Islands	MC2	MC3	VMC	HMM
581-921	569-900	569-917	545-897	570-894
937-1244	915-1237	919-1236	919-1189	914-1218
2015-2232	2048-2104	2040-2112	2075-2201	2090-2110
				2401-2437

Our second test sequence is a human alpha-1 collagen type II gene, exons 1, 2 and 3 (HSCOLII). This sequence has 6723 symbols. We split it into subsequences with 120 symbols. The testing results for the HSCOLII sequence are given in Table 2. The table shows that for this sequence the second order Markov and the variable Markov model have the best performance. All the models also predicted a region that does not exist.

Table 2. Performances of different order Markov chains on HSCOLII

True Islands	MC2	MC3	VMC	HMM
570-1109	154-1042	154-1047	160-1124	161-1057
1567-1953	1494-1728	1494-1726	1508-1838	1492-1726
2168-2406	2129-2299	2129-2291	2138-2373	2123-2299
3102-3348	3031-3349	3039-3277	3042-3373	3039-3373
	3384-3540	3398-3541	3398-3728	3399-3581

Our last test sequence is a human collagen alpha-1-IV and alpha-2-IV genes, exons 1-3 (HSCOLAA). This sequence has 2184 symbols. We split it into subsequences with 100 symbols. The testing results for the HSCOLAA sequence are given in Table 3. The results in Table 3 indicate that the variable Markov model has the best performance.

Table 3. Performances of different order Markov chains on HSCOLAA

True Islands	MC2	MC3	VMC	HMM
49-877	15-858	1-862	46-887	7-833
953-1538	908-1460	908-1445	919-1489	916-1444
1765-2100	1910-2015	1859-2011	1769-2007	1858-2007

5 Conclusions

In this chapter, we have presented CpG island identification with higher order and variable order Markov models. With these simple models, we have identified the CpG islands from 3 DNA sequences. We showed that different order Markov chains may have different prediction accuracies and it is possible for us to get more accurate identification with higher order or variable order Markov models. Sequential probability ratio test was used to eliminate the false alarm and find more accurate boundaries of CpG islands.

Acknowledgements

D. Chen was supported by the USUHS grant R087RH-01 and the National Science Foundation grant CCR-0311252.

References

1. A. Apostolico and G. Bejerano. Optimal amnesic probabilistic automata or how to learn and classify proteins in linear time and space. *Journal of Computational Biology*, 7(3):381-393, 2000.
2. Y. Bengio. Markovian models for sequential data. *Neural Computing Surveys*, 2:129-162, 1999.
3. P. Laird and R. Saul. Discrete sequence prediction and its applications. *Machine Learning*, 15:43-68, 1994.
4. J. Rissanen. Hypothesis selection and testing by the MDL principle. *The Computer Journal*, 42(4): 260-269, 1999.
5. D. Ron, Y. Singer, and N. Tishby. The power of amnesia: learning probabilistic automata with variable memory length. *Machine Learning*, 25: 117-142, 1996.

Data Mining Algorithms for Virtual Screening of Bioactive Compounds

Mukund Deshpande, Michihiro Kuramochi, and George Karypis

Department of Computer Science and Engineering,
University of Minnesota,
4-192 EE/CSci Building, 200 Union Street SE,
Minneapolis, MN 55455
{deshpand,kuram,karypis}@cs.umn.edu

Summary. In this chapter we study the problem of classifying chemical compound datasets. We present a sub-structure-based classification algorithm that decouples the sub-structure discovery process from the classification model construction and uses frequent subgraph discovery algorithms to find all topological and geometric sub-structures present in the dataset. The advantage of this approach is that during classification model construction, all relevant sub-structures are available allowing the classifier to intelligently select the most discriminating ones. The computational scalability is ensured by the use of highly efficient frequent subgraph discovery algorithms coupled with aggressive feature selection. Experimental evaluation on eight different classification problems shows that our approach is computationally scalable and on the average, outperforms existing schemes by 10% to 35%.

Key words: Classification, Chemical Compounds, Virtual Screening, Graphs, SVM.

1 Introduction

Discovering new drugs is an expensive and challenging process. Any new drug should not only produce the desired response to the disease but should do so with minimal side effects and be superior to existing drugs. One of the key steps in the drug design process is the identification of the chemical compounds (*hit* compounds) that display the desired and reproducible behavior against the specific biomolecular target [47] and represents a significant hurdle in the early stages of drug discovery. The 1990s saw the widespread adoption of high-throughput screening (HTS) and ultra HTS [11, 32], which use highly automated techniques to conduct the biological assays and can be used to screen a large number of compounds. Although the number of compounds that can be evaluated by these methods is very large, these numbers are small in comparison to the millions of drug-like compounds that exist or can be synthesized by combinatorial chemistry methods. Moreover, in most cases it is hard to find all desirable properties in a single compound and medicinal chemists are interested in not just identifying the hits but studying what part of the chemical compound leads to the desirable behavior, so that new compounds can be rationally synthesized (*lead* development).

Computational techniques that build models to correctly assign chemical compounds to various classes of interest can address these limitations, have many applications in pharmaceutical research, and are used extensively to replace or supplement HTS-based approaches. These techniques are designed to computationally search large compound databases to select a limited number of candidate molecules for testing in order to identify novel chemical entities that have the desired biological activity. The combination of HTS with these *virtual screening* methods allows a move away from purely random-based testing, toward more meaningful and directed iterative rapid-feedback searches of subsets and focused libraries. However, the challenge in developing practical virtual screening methods is to develop chemical compound classification algorithms that can be applied fast enough to rapidly evaluate potentially millions of compounds while achieving sufficient accuracy to successfully identify a subset of compounds that is significantly enriched in hits.

In recent years two classes of techniques have been developed for solving the chemical compound classification problem. The first class builds a classification model using a set of physico-chemical properties derived from the compounds structure, called quantitative structure-activity relationships (QSAR) [33, 34, 2], whereas the second class operates directly on the structure of the chemical compound and try to automatically identify a small number of chemical sub-structures that can be used to discriminate between the different classes [12, 76, 36, 44, 21]. A number of comparative studies [69, 39] have shown that techniques based on the automatic discovery of chemical sub-structures are superior to those based on QSAR properties and require limited user intervention and domain knowledge. However, despite their success, a key limitation of these techniques is that they rely on heuristic search methods to discover these sub-structures. Even though such approaches reduce the inherently high computational complexity associated with these schemes, they may lead to sub-optimal classifiers in cases in which the heuristic search failed to uncover sub-structures that are critical for the classification task.

In this chapter we present a sub-structure-based classifier that overcomes the limitations associated with existing algorithms. One of the key ideas of this approach is to decouple the sub-structure discovery process from the classification model construction step and use frequent subgraph discovery algorithms to find all chemical sub-structures that occur a sufficiently large number of times. Once the complete set of these sub-structures has been identified, the algorithm then proceeds to build a classification model based on them. The advantage of such an approach is that during classification model construction, all relevant sub-structures are available allowing the classifier to intelligently select the most discriminating ones. To ensure that such an approach is computationally scalable, we use recently developed [42, 44] highly efficient frequent subgraph discovery algorithms coupled with aggressive feature selection to reduce both the amount of time required to build as well as to apply the classification model. In addition, we present a sub-structure discovery algorithm that finds a set of sub-structures whose geometry is conserved, further improving the classification performance of the algorithm.

We experimentally evaluated the performance of these algorithms on eight different problems derived from three publicly available datasets and compared their performance against that of traditional QSAR-based classifiers and existing sub-structure classifiers based on SUBDUE [17] and SubdueCL [29]. Our results show that these algorithms, on the average, outperform QSAR-based schemes by 35% and SUBDUE-based schemes by 10%.

The rest of the chapter is organized as follows. Section 2 provides some background information related to chemical compounds, their activity, and their representation. Section 3 provides a survey on the related research in this area. Section 2 provides the details of the chemical compound classification approach. Section 5 experimentally evaluates its performance and compares it against other approaches. Finally, Section 6 provides outlines directions of future research and provides some concluding remarks.

2 Background

A chemical compound consists of different atoms being held together via bonds adopting a well-defined geometric configuration. Figure 2(a) represents the chemical compound Flucytosine from the DTP AIDS repository [24] it consists of a central aromatic ring and other elements like N, O and F. The representation shown in the figure is a typical graphical representation that most chemists work with.

There are many different ways to represent such chemical compounds. The simplest representation is the molecular formula that lists the various atoms making up the compound; the molecular formula for Flucytosine is `C4H4FN3O`. However this representation is woefully inadequate to capture the structure of the chemical compound. It was recognized early on that it was possible for two chemical compounds to have identical molecular formula but completely different chemical properties [28]. A more sophisticated representation can be achieved using the SMILES [72] representation, it not only represents the atoms but also represents the bonds between different atoms. The SMILES representation for Flucytosine is `Nc1nc(O)ncc1F`. Though SMILES representation is compact it is not guaranteed to be unique, furthermore the representation is quite restrictive to work with [41].

The activity of a compound largely depends on its chemical structure and the arrangement of different atoms in 3D space. As a result, effective

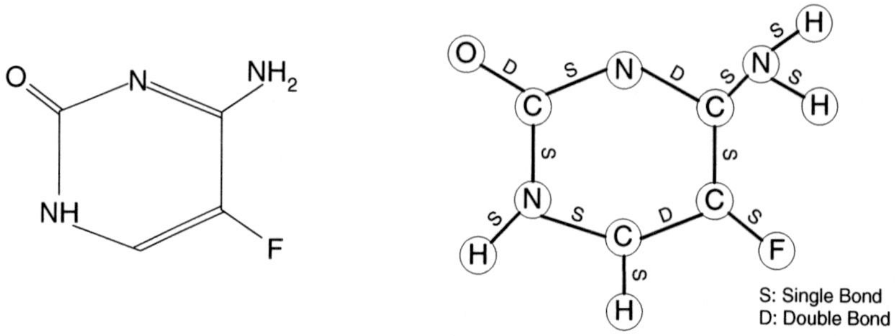

(a) NSC 103025 Flucytosine   (b) Graph Representation

Fig. 1. Chemical and Graphical representation of Flucytosine

classification algorithms must be able to directly take into account the structural nature of these datasets. In this chapter we represent each compound as undirected graphs. The vertices of these graphs correspond to the various atoms, and the edges correspond to the bonds between the atoms. Each of the vertices and edges has a label associated with it. The labels on the vertices correspond to the type of atoms and the labels on the edges correspond to the type of bonds. As an example, Figure 2(b) shows the representation of Flucytosine in terms of this graph model. We will refer to this representation as the *topological graph* representation of a chemical compound. Note that such representations are quite commonly used by many chemical modeling software and are referred as the *connection table* for the chemical compound [47].

In addition, since chemical compounds have a physical three-dimensional structure, each vertex of the graph has a 3D-coordinate indicating the position of the corresponding atom in 3D space. However, there are two key issues that need to be considered when working with the compound's 3D structure. First, the number of experimentally determined molecular geometries is limited (about 270,000 X-ray structures in the Cambridge Crystallographic Database compared to 15 millions known compounds). As a result, the 3D geometry of a compound needs to be computationally determined, which may introduce certain amount of error. To address this problem, we use the Corina [27] software package to compute the 3D coordinates for all the chemical compounds in our datasets. Corina is a rule- and data-based system that has been experimentally shown to predict the 3D structure of compounds with high-accuracy. Second, each compound can have multiple low-energy conformations (*i.e.*, multiple 3D structures) that need to be taken into account in order to achieve the highest possible classification performance. However, due to time constraints, in this study we do not take into account these multiple conformations but instead use the single low-energy conformation that is returned by Corina's default settings. However, as discussed in Section 4.1, the presented approach for extracting geometric sub-structures can be easily extended to cases in which multiple conformations are considered as well. Nevertheless, despite this simplification, as our experiments in Section 5 will show, incorporating 3D structure information leads to measurable improvements in the overall classification performance. We will refer to this representation as the *geometric graph* representation of a chemical compound.

The meaning of the various classes in the input dataset is application dependent. In some applications, the classes will capture the extent to which a particular compound is toxic, whereas in other applications they may capture the extent to which a compound can inhibit (or enhance) a particular factor and/or active site. In most applications each of the compounds is assigned to only one of two classes, that are commonly referred to as the *positive* and *negative* class. The positive class corresponds to compounds that exhibit the property in question, whereas the compounds of the negative class do not. Throughout this chapter we will be restricting ourselves to only two classes, though all the techniques described here can be easily extended to multi-class as well as multi-label classification problems.

Another important aspect of modeling chemical compounds is the naming of single and double bonds inside aromatic rings. Typically in an aromatic ring of a chemical compound, though the number of single and double bonds is fixed, the exact position of double and single bonds is not fixed, this is because of the phenomenon of resonance [28]. It is worth noting that the exact position of double and single bond in an aromatic ring does not affect the chemical properties of a chemical compound. To capture this uncertainty in the position of single and double bond we represent all the bonds making up the aromatic ring with a new bond type called the *aromatic bond*. Another aspect of the chemical compounds is that the number of hydrogen bonds connected to a particular carbon atom can usually be inferred from the bonds connecting that carbon atom [28], therefore in our representation we do not represent the hydrogen atoms that are connected to the carbon atoms, such hydrogen atoms are referred as non-polar hydrogen atoms. Note that the above transformations are widely used by many chemistry modeling tools and are usually referred to as *structure normalization* [47].

3 Related Research

Many approaches have been developed for building classification models for chemical compounds. These approaches can be grouped into two broad categories. The first contains methods that represent the chemical compounds using various descriptors and then apply various statistical or machine learning approaches to learn the classification models. The second category contains methods that automatically analyze the structure of the chemical compounds involved in the problem to identify a set of substructure-based rules, which are then used for classification. A survey of some of the key methods in both categories and a discussion on their relative advantages and disadvantages is provided in the remaining of this section.

Approaches based on Descriptors A number of different types of descriptors have been developed that are based on frequency, physicochemical properties, topological, and geometric substructures [74, 6]. The quality of these descriptors improves as we move from frequency-, to property-, to topology-, to geometry-based descriptors, and a number of studies have shown that topological descriptors are often superior to those based on simple physicochemical properties, and geometric descriptors outperform their topological counterparts [67, 5, 10]. The types of properties that are captured/measured by these descriptors are identified a priori in a dataset independent fashion and rely on extensive domain knowledge. Frequency descriptors are counts that measure basic characteristics of the compounds and include the number of individual atoms, bonds, degrees of connectivity, rings, *etc.* Physicochemical descriptors correspond to various molecular properties that can be computed directly from the compounds structure. This includes properties such as molecular weight, number of aromatic bonds, molecular connectivity index, $\log P$, total energy, dipole moment, solvent accessible surface area, molar refractivity, ionization potential, atomic electron densities, van der Waals volume, *etc* [13, 52, 5]. Topological descriptors are used to measure various aspects of the compounds two-dimensional structure, *i.e.*, the connectivity pattern of the compound's atoms, and include a wide-range of descriptors that are based on topological indices and 2D fragments. Topological indices are similar to physicochemical properties in the sense that they characterize some aspect of molecular data by a single value. These indices encode information about the shape, size, bonding and branching pattern [7, 31]. 2D fragment descriptors correspond to certain chemical substructures that are present in the chemical compound. This includes various atom-centered, bond-centered, ring-centered fragments [1], fragments based on atom-pairs [15], topological torsions [59], and fragments that are derived by performing a rule-based compound segmentation [8, 9, 48]. Geometric descriptors measure various aspects of the compounds 3D structure that has been either experimentally or computationally determined. These descriptors are usually based on pharmacophores [13]. Pharmacophores are based on the types of interaction observed to be important in ligand-protein binding interactions. Pharmacophore descriptors consist of three or four points separated by well-defined distance ranges and are derived by considering all combinations of three or four atoms over all conformations of a given molecule [67, 19, 4, 61, 30]. Note that information about the 2D fragments and the pharmacophores present in a compound are usually stored in the form of a fingerprint, which is fixed-length string of bits each representing the presence or absence of a particular descriptor.

The actual classification model is learned by transforming each chemical compound into a vector of numerical or binary values whose dimensions correspond to the various descriptors that are used. Within this representation, any classification technique capable of handling numerical or binary features can be used for the classification task. Early research on building these classification models focused primarily on regression-based techniques [13]. This work was pioneered by Hansch *et al.* [33, 34], which demonstrated that the biological activity of a chemical compound is a function of its physicochemical properties. This led to the development of the quantitative structure-activity relationship (QSAR) methods in which the statistical techniques (*i.e.*, classification model) enable this relationship to be expressed mathematically. However, besides regression-based approaches, other classification techniques have been used that are in general more powerful and lead to improved accuracies. This includes techniques based on principle component regression and partial least squares [75], neural networks [3, 51, 79, 23], recursive partitioning [16, 65, 2], phylogenetic-like trees [58, 70], binary QSAR [45, 26], linear discriminant analysis [60], and support vector machines [14].

Descriptor-based approaches are very popular in the pharmaceutical industry and are used extensively to solve various chemical compound classification problems. However, their key limitation stems from the fact that, to a large extent, the classification performance depends on the successful identification of the relevant descriptors that capture the structure-activity relationships for the particular classification problem.

Approaches based on Substructure Rules The pioneering work in this field was done by King *et al.* in the early 1990s [40, 39]. They applied an inductive logic programming (ILP) system [56], Golem [57], to study the behavior of 44 trimethoprin analogues and their observed inhibition of Escherichia coli dihydrofolate reductase and reported a considerable improvement in classification accuracy over the traditional QSAR-based models. In this approach the chemical compound is expressed using first order logic. Each atom is represented as a predicate consisting of atomID and the element, and a bond is represented as a predicate consisting of two atomIDs. Using this representation, an ILP system discovers rules (*i.e.*, conjunction of predicates) that are good for discriminating the different classes. Since these rules consist of predicates describing atoms and bonds, they essentially correspond to substructures that are present in the chemical compounds. Srinivasan *et al.* [69] present a detailed comparison of the features generated by ILP with the traditional QSAR properties used for classifying chemical compounds and show that for some applications features discovered by ILP approaches lead to a significant lift in the performance.

Though ILP-based approaches are quite powerful, the high computational complexity of the underlying rule-induction system limits the size of the dataset for which they can be applied. Furthermore, they tend to produce rules consisting of relatively small substructures (usually three to four atoms [18, 20]), limiting the size of structural constraints that are being discovered and hence affecting the classification performance. Another drawback of these approaches is that in order to reduce their computational complexity they employ various heuristics to prune the explored search-space [55], potentially missing substructures that are important for the classification task. One exception is the WARMR system [18, 20] that is specifically developed for chemical compounds and discovers all possible substructures above a certain frequency threshold. However, WARMR's computational complexity is very high and can only be used to discover substructures that occur with relatively high frequency.

One of the fundamental reasons limiting the scalability of ILP-based approaches is the first order logic-based representation that they use. This representation is much more powerful than what is needed to model chemical compounds and discover sub-structures. For this reason a number of researchers have explored the much simpler graph-based representation of the chemical compound's topology and transformed the problem of finding chemical sub-structures to that of finding subgraphs in this graph-based representation [12, 76, 36]. The best-known approach is the SUBDUE system [35, 17]. SUBDUE finds patterns which can effectively compress the original input data based on the minimum description length (MDL) principle, by substituting those patterns with a single vertex. To narrow the search-space and improve its computational efficiency, SUBDUE uses a heuristic beam search approach, which quite often results in failing to find subgraphs that are frequent. The SUBDUE system was also later extended to classify graphs and was referred as SubdueCL [29]. In SubdueCL instead of using minimum description length as a heuristic a measure similar to confidence of a subgraph is used as a heuristic. Finally, another heuristic-based scheme is MOLFEA [41] that takes advantage of the compound's SMILES string representation and identifies substructures corresponding to frequently occurring sub-sequences.

4 Classification Based on Frequent Subgraphs

The previous research on classifying chemical compounds (discussed in Section 3) has shown that techniques based on the automatic discovery of chemical sub-structures are superior to those based on QSAR properties and require limited user intervention and domain knowledge. However, despite their success, a key limitation of both the ILP- and the subgraph-based techniques, is that they rely on heuristic search methods to discover the sub-structures to be used for classification. As discussed in Section 3, even though such approaches reduce the inherently high computational complexity associated with these schemes, they may lead to sub-optimal classifiers in cases in which the heuristic search fails to uncover sub-structures that are critical for the classification task.

To overcome this problem, we developed a classification algorithm for chemical compounds that uses the graph-based representation and limits the number of sub-structures that are pruned a priori. The key idea of our approach is to decouple the sub-structure discovery process from the classification model construction step, and use frequent subgraph discovery algorithms

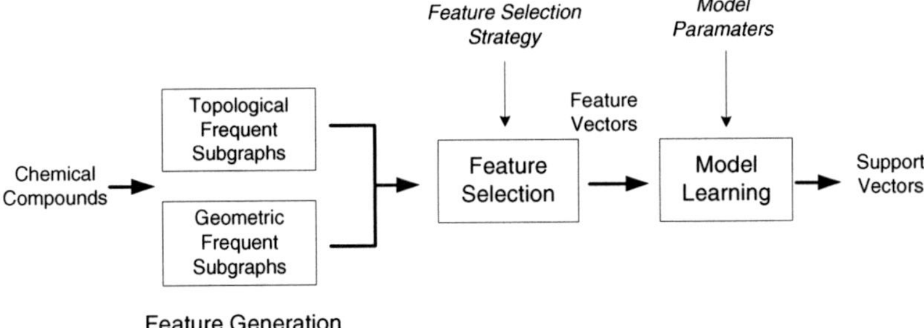

Fig. 2. Frequent Subgraph Based Classification Framework

to find all chemical sub-structures that occur a sufficiently large number of times. Once the complete set of such sub-structures has been identified, our algorithm then proceeds to build a classification model based on them. To a large extent, this approach is similar in spirit to the recently developed frequent-itemset-based classification algorithms [50, 49, 22] that have been shown to outperform traditional classifiers that rely on heuristic search methods to discover the classification rules.

The overall outline of our classification methodology is shown in Figure 4. It consists of three distinct steps: (i) feature generation, (ii) feature selection, and (iii) classification model construction. During the feature generation step, the chemical compounds are mined to discover the frequently occurring sub-structures that correspond to either topological or geometric subgraphs. These sub-structures are then used as the features by which the compounds are represented in the subsequent steps. During the second step, a small set of features is selected such that the selected features can correctly discriminate between the different classes present in the dataset. Finally, in the last step each chemical compound is represented using these set of features and a classification model is learned.

This methodology, by following the above three-step framework is designed to overcome the limitations of existing approaches. By using computationally efficient subgraph discovery algorithms to find all chemical substructures (topological or geometric) that occur a sufficiently large number of times in the compounds, they can discover substructures that are both specific to the particular classification problem being solved and at the same time involve arbitrarily complex substructures. By discovering the complete set of frequent subgraphs and decoupling the substructure discovery process from the feature generation step, they can proceed to select and synthesize the most discriminating descriptors for the particular classification problem that take into account all relevant information. Finally, by employing advanced machine learning techniques and utilizing alternate representations, they can account for the relationships between these features at different levels of granularity and complexity leading to high classification accuracy.

4.1 Feature Generation

Our classification algorithm finds sub-structures in a chemical compound database using two different methods. The first method uses the topological graph representation of each compound whereas the second method is based on the corresponding geometric graph representation (discussed in Section 2). In both of these methods, our algorithm uses the topological or geometric connected subgraphs that occur in at least $\sigma\%$ of the compounds to define the sub-structures.

There are two important restrictions on the type of the sub-structures that are discovered by our approach. The first has to do with the fact that we are only interested in sub-structures that are connected and is motivated by the fact that connectivity is a natural property of such patterns. The second has to do with the fact that we are only interested in frequent sub-structures (as determined by the value of σ) as this ensures that we do not discover spurious sub-structures that will in general not be statistically significant. Furthermore, this minimum support constraint also helps in making the problem of frequent subgraph discovery computationally tractable.

Frequent Topological Subgraphs

Developing frequent subgraph discovery algorithms is particularly challenging and computationally intensive as graph and/or subgraph isomorphisms play a key role throughout the computations. Despite that, in recent years, four different algorithms have been developed capable of finding all frequently occurring subgraphs with reasonable computational efficiency. These are the AGM algorithm developed by Inokuchi et al [36], the FSG algorithm developed by members of our group [42], the chemical sub-structure discovery algorithm developed by Borgelt and Berthold [12], and the gSpan algorithm developed by Yan and Han [76]. The enabling factors to the computational efficiency of these schemes have been (i) the development of efficient candidate subgraph generation schemes that reduce the number of times the same candidate subgraph is being generated, (ii) the use of efficient canonical labeling schemes to represent the various subgraphs; and (iii) the use of various techniques developed by the data-mining community to reduce the number of times subgraph isomorphism computations need to be performed.

In our classification algorithm we find the frequently occurring subgraphs using the FSG algorithm. FSG takes as input a database D of graphs and a minimum support σ, and finds all connected subgraphs that occur in at least $\sigma\%$ of the transactions. FSG, initially presented in [42], with subsequent improvements presented in [44], uses a breadth-first approach to discover the lattice of frequent subgraphs. It starts by enumerating small frequent graphs consisting of one and two edges and then proceeds to find larger subgraphs by joining previously discovered smaller frequent subgraphs. The size of these subgraphs is grown by adding one-edge-at-a-time. The lattice of frequent patterns is used to prune the set of candidate patterns and it only explicitly computes the frequency of the patterns which survive this downward closure pruning. Despite the inherent complexity of the problem, FSG employs a number of sophisticated techniques to achieve high computational performance. It uses a canonical labeling algorithm that fully makes use of edge and vertex labels for fast processing, and various vertex invariants to reduce the complexity of determining the canonical label of a graph. These canonical labels are then used to establish the identity and total order of the frequent and candidate subgraphs, a critical step of redundant candidate elimination and downward closure testing. It uses a sophisticated scheme for candidate generation [44] that minimizes the number of times each candidate subgraph gets generated and also dramatically reduces the generation of subgraphs that fail the downward closure test. Finally, for determining the actual frequency of each subgraph, FSG reduces the number of subgraph isomorphism operations by using TID-lists [25, 66, 78, 77] to keep track of the set of transactions that supported the frequent patterns discovered at the previous level of the lattice. For every candidate, FSG takes the intersection of TID-lists of its parents, and performs the subgraph isomorphism only on the transactions contained in the resulting TID-list. As the experiments presented in Section 5 show, FSG is able to scale to large datasets and low support values.

Frequent Geometric Subgraphs

Topological sub-structures capture the connectivity of atoms in the chemical compound but they ignore the 3D shape (3D arrangement of atoms) of the sub-structures. For certain classification problems the 3D shape of the sub-structure might be essential for determining the chemical activity of a compound. For instance, the geometric configuration of atoms in a sub-structure is crucial for its ability to bind to a particular target [47]. For this reason we developed an algorithm that find all frequent sub-structures whose topology as well as geometry is conserved.

There are two important aspects specific to the geometric subgraphs that need to be considered. First, since the coordinates of the vertices depend on a particular reference coordinate axes, we would like the discovered geometric subgraphs to be independent of these coordinate axes, *i.e.*, we are interested in geometric subgraphs whose occurrences are translation, and rotation invariant. This dramatically increases the overall complexity of the geometric subgraph discovery process, because we may need to consider all possible geometric configurations of a single pattern. Second, while determining if a geometric subgraph is contained in a bigger geometric graph we would like to allow some tolerance when we establish a match between coordinates, ensuring that slight deviations in coordinates between two identical topological subgraphs do not lead to the creation of two geometric subgraphs. The amount of tolerance (r) should be a user specified parameter. The task of discovering such r-tolerant frequent geometric subgraphs dramatically changes the nature of the problem. In traditional pattern discovery problems such as finding frequent itemsets, sequential patterns, and/or frequent topological graphs there is a clear definition of what a pattern is, given its set of supporting transactions. On the other hand, in the case of r-tolerant geometric subgraphs, there are many different geometric representations of the same pattern (all of which will be r-tolerant isomorphic to each other). The problem becomes not only that of finding a pattern and its support, but also finding the right representative for this pattern. The selection of the right representative can have a serious impact on correctly computing the support of the pattern. For example, given a set of subgraphs that are r-tolerant isomorphic to each other, the one that corresponds to an *outlier* will tend to have a lower support than the one corresponding to the *center*. These two aspects of geometric subgraphs makes the task of discovering the full fledged geometric subgraphs extremely hard [43].

To overcome this problem we developed a simpler, albeit less discriminatory, representation for geometric subgraphs. We use a property of a geometric graph called the ***average inter-atomic distance*** that is defined as the average Euclidean distance between all pairs of atoms in the molecule. Note that the average inter-atomic distance is computed between all pairs of atoms irrespective of whether a bonds connects the atoms or not. The average inter-atomic distance can be thought of as a geometric signature of a topological subgraph. The geometric subgraph consists of two components, a topological subgraph and an interval of average inter-atomic distance associated with it. A geometric graph contains this geometric subgraph if it contains the topological subgraph and the average inter-atomic distance of the embedding (of the topological subgraph) is within the interval associated with the geometric subgraph. Note that this geometric representation is also translation and rotation invariant, and the width of the interval determines the tolerance displayed by the geometric subgraph. We are interested in discovering such geometric subgraphs that occur above $\sigma\%$ of the transactions and the interval of average inter-atomic distance is bound by r.

Since a geometric subgraph contains a topological subgraph, for the geometric subgraph to be frequent the corresponding topological subgraph has to be frequent, as well. This allows us to take advantage of the existing approach to discover topological subgraphs. We modify the frequency counting stage of the FSG algorithm as follows. If a subgraph g is contained in a transaction t then all possible embeddings of g in t are found and the average inter-atomic distance for each of these embeddings is computed. As a result, at the end of the frequent subgraph discovery each topological subgraph has a list of average inter-atomic distances associated with it. Each one of the average inter-atomic distances corresponds to one of the embeddings *i.e.*, a geometric configuration of the topological subgraph. This algorithm can be easily extended to cases in which there are multiple 3D conformations associated with each chemical compound (as discussed in Section 2), by simply treating each distinct conformation as a different chemical compound.

The task of discovering geometric subgraphs now reduces to identifying those geometric configurations that are frequent enough, *i.e.*, identify intervals of average inter-atomic distances such that each interval contains the minimum number geometric configurations (it occurs in $\sigma\%$ of the transactions) and the width of the interval is smaller than the tolerance threshold (r). This task can be thought of as 1D clustering on the vector of average inter-atomic distances such that each cluster contains items above the minimum support and the spread of each cluster is bounded by the tolerance r. Note that not all items will belong to a valid cluster as some of them will be infrequent. In our experiments we set the value of r to be equal to half of the minimum distance between any two pairs of atoms in the compounds.

To find such clusters we perform agglomerative clustering on the vector of average inter-atomic distance values. The distance between any two average inter-atomic distance values is defined as the difference in their numeric values. To ensure that we get the largest possible clusters we use the maximum-link criterion function for deciding which two clusters should be merged [38]. The process of agglomeration is continued until the interval containing all the items in the cluster is below the tolerance threshold (r). When we reach a stage where further agglomeration would increase the spread of the cluster beyond the tolerance threshold, we check the number of items contained in the cluster. If the number of items is above the support threshold, then the interval associated with this cluster is considered as a geometric feature. Since we are clustering one-dimensional datasets, the clustering complexity is low. Some examples of the distribution of the average inter-atomic distance values and the associated clusters are shown in Figure 4.1. Note that the average inter-atomic distance values of the third example are uniformly spread and lead to no geometric subgraph.

Note that this algorithm for computing geometric subgraphs is approximate in nature for two reasons. First, the average inter-atomic distance may map two different geometric subgraphs to the same average inter-atomic distance value. Second, the clustering algorithm may not find the complete set of geometric subgraphs that satisfy the r tolerance. Nevertheless, as our experiments in Section 5 show the geometric subgraphs discovered by this approach improve the classification accuracy of the algorithm.

Additional Considerations

Even though FSG provides the general functionality required to find all frequently occurring sub-structures in chemical datasets, there are a number of issues that need to be addressed before it can be applied as a black-box tool for feature discovery in the context of classification . One issue deals with the

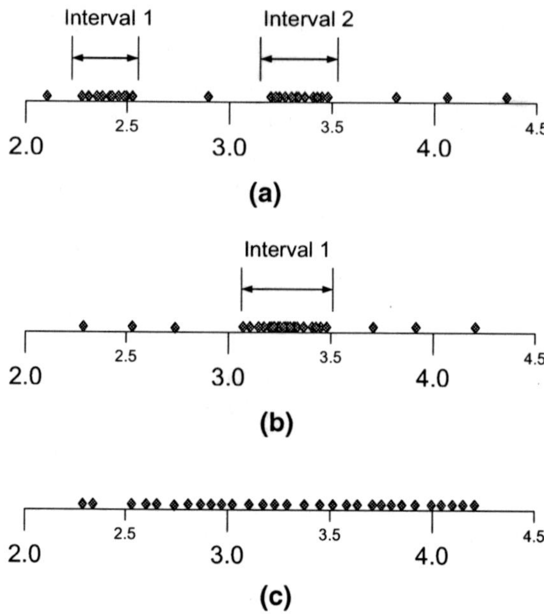

Fig. 3. Some examples of the one-dimensional clustering of average inter-atomic distance values.

selecting the right value for the σ, the support constraint used for discovering frequent sub-structures. The value of σ controls the number of subgraphs discovered by FSG. Choosing a good value of σ is especially important for the dataset containing classes of significantly different sizes. In such cases, in order to ensure that FSG is able to find features that are meaningful for all the classes, it must use a support that depends on the size of the smaller class.

For this reason we first partition the complete dataset, using the class label of the examples, into specific class specific datasets. We then run FSG on each of these *class datasets*. This partitioning of the dataset ensures that sufficient subgraphs are discovered for those class labels which occur rarely in the dataset. Next, we combine subgraphs discovered from each of the *class dataset*. After this step each subgraph has a vector that contains the frequency with which it occurs in each class.

4.2 Feature Selection

The frequent subgraph discovery algorithms described in Section 4.1 discovers all the sub-structures (topological or geometric) that occur above a certain support constraint (σ) in the dataset. Though the discovery algorithm is computationally efficient, the algorithm can generate a large number of features. A large number of features is detrimental for two reasons. First, it could increase the time required to build the model. But more importantly, a large number of features can increase the time required to classify a chemical compound, as we need to first identify which of the discovered features it contains before we can apply the classification model. Determining whether a compound contains a particular feature or not can be computationally expensive as it may require a subgraph isomorphism operation. This problem is especially critical in the drug discovery process where the classification model is learned on a small set of chemical compounds and it is then applied on large chemical compound libraries containing millions of compounds.

One way of solving this problem is to follow a heuristic subgraph discovery approach (similar in spirit to previously developed methods [17, 29]) in which during the subgraph discovery phase itself, the discriminatory ability of a particular subgraph is determined, and the discovery process is terminated as soon as a subgraph is generated that is less discriminatory than any of its subgraphs. By following this approach, the total number of features will be substantially reduced, achieving the desired objective. However, the limitation with such an approach is that it may fail to discover and use highly discriminatory subgraphs. This is because the discriminatory ability of a subgraph does not (in general) consistently increase as a function of its size, and subgraphs that appear to be poor discriminators may become very discriminatory by growing their size. For this reason, in order to develop an effective feature selection method, we use a scheme that first finds all frequent subgraphs and then selects among them a small set of discriminatory features. The advantage of this approach is that during feature selection all frequent subgraphs are considered irrespective of when they were generated and whether or not they contain less or more discriminatory subgraphs.

The feature selection scheme is based on the *sequential covering paradigm* used to learn rule sets [53]. To apply this algorithm we assume that each discovered sub-structure corresponds to a rule, with the class label of the sub-structure as the *target attribute*, such rules are referred to as *class-rules* in [50]. The sequential covering algorithm takes as input a set of examples and the features discovered from these examples, and iteratively applies the feature selection step. In this step the algorithm selects the feature that has the highest estimated accuracy. After selecting this feature all the examples containing this feature are eliminated and the feature is marked as selected. In the next iteration of the algorithm the same step is applied, but on a smaller set of examples. The algorithm continues in an iterative fashion until either all the features are selected or all the examples are eliminated.

In this chapter we use a computationally efficient implementation of sequential covering algorithm known as CBA [50], this algorithm proceeds by first sorting the features based on confidence and then applying the sequential covering algorithm on this sorted set of features. One of the advantages of this approach is that it requires minimal number of passes on the dataset, hence is very scalable. To obtain a better control over the number of selected features we use an extension of the sequential covering scheme known as *Classification based on Multiple Rules* (CMAR) [49]. In this scheme instead of removing the example after it is covered by the selected feature, the example is removed only if that example is covered by δ selected features. The number of selected rules increases as the value of δ increases, an increase in the number of features usually translates into an improvement in the accuracy as more features are used to classify a particular example. The value of δ is specified by the user and provides a means to the user to control the number of features used for classification .

4.3 Classification Model Construction

Given the frequent subgraphs discovered in the previous step, our algorithm treats each of these subgraphs as a feature and represents the chemical compound as a frequency vector. The ith entry of this vector is equal to the number of times (frequency) that feature occurs in the compound's graph. This mapping into the feature space of frequent subgraphs is performed both for the training and the test dataset. Note that the frequent subgraphs were identified by mining *only* the graphs of the chemical compounds in the training set. However, the mapping of the test set requires that we check each frequent subgraph against the graph of the test compound using subgraph isomorphism. Fortunately, the overall process can be substantially accelerated by taking into account the frequent subgraph lattice that is also generated by FSG. In this case, we traverse the lattice from top to bottom and only visit the child nodes of a subgraph if that subgraph is isomorphic to the chemical compound.

Once the feature vectors for each chemical compound have been built, any one of the existing classification algorithms can potentially be used for classification. However, the characteristics of the transformed dataset and the nature of the classification problem itself tends to limit the applicability of certain classes of classification algorithms. In particular, the transformed dataset will most likely be high dimensional, and second, it will be sparse, in the sense that each compound will have only a few of these features, and each feature will be present in only a few of the compounds. Moreover, in most cases the positive class will be much smaller than the negative class, making it unsuitable for classifiers that primarily focus on optimizing the overall classification accuracy.

In our study we built the classification models using support vector machines (SVM) [71], as they are well-suited for operating in such sparse and high-dimensional datasets. Furthermore, an additional advantage of SVM is that it allows us to directly control the cost associated with the misclassification of examples from the different classes [54]. This allows us to associate a higher cost for the misclassification of positive instances; thus, biasing the classifier to learn a model that tries to increase the true-positive rate, at the expense of increasing the false positive rate.

5 Experimental Evaluation

We experimentally evaluated the performance of our classification algorithm and compared it against that achieved by earlier approaches on a variety of chemical compound datasets. The datasets, experimental methodology, and results are described in subsequent sections.

5.1 Datasets

We used three different publicly available datasets to derive a total of eight different classification problems. The first dataset was initially used as a part of the Predictive Toxicology Evaluation Challenge [68] which was organized as a part of PKDD/ECML 2001 Conference.[1] It contains data published by the U.S. National Institute for Environmental Health Sciences, the data consists of bio-assays of different chemical compounds on rodents to study the carcinogenicity (cancer inducing) properties of the compounds [68]. The goal being to estimate the carcinogenicity of different compounds on humans. Each compound is evaluated on four kinds of laboratory animals (*male Mice, female Mice, male Rats, female Rats*), and is assigned four class labels each indicating the toxicity of the compound for that animal. There are four classification problems one corresponding to each of the rodents and will be referred as *P1*, *P2*, *P3*, and *P4*.

The second dataset is obtained from the National Cancer Institute's DTP AIDS Anti-viral Screen program [24, 41].[2] Each compound in the dataset is evaluated for evidence of anti-HIV activity. The screen utilizes a soluble formazan assay to measure protection of human CEM cells from HIV-1 infection [73]. Compounds able to provide at least 50% protection to the CEM cells were re-tested. Compounds that provided at least 50% protection on retest were listed as *moderately active* (CM, confirmed moderately active). Compounds that reproducibly provided 100% protection were listed as *confirmed active* (CA). Compounds neither active nor moderately active were listed as *confirmed inactive* (CI). We have formulated three classification problems on this dataset, in the first problem we consider only *confirmed active* (CA) and *moderately active* (CM) compounds and then build a classifier to separate these two compounds; this problem is referred as *H1*. For the second problem we combine *moderately active* (CM) and *confirmed active* (CA) compounds to form one set of *active* compounds, we then build a classifier to separate these *active* and *confirmed inactive* compounds; this problem is referred as *H2*. In the last problem we only use *confirmed active* (CA) and *confirmed inactive* compounds and build a classifier to categorize these two compounds; this problem is referred as *H3*.

The third dataset was obtained from the Center of Computational Drug Discovery's anthrax project at the University of Oxford [64]. The goal of

---

[1] http://www.informatik.uni-freiburg.de/~ml/ptc/.

[2] http://dtp.nci.nih.gov/docs/aids/aids_data.html.

this project was to discover small molecules that would bind with the heptameric protective antigen component of the anthrax toxin, and prevent it from spreading its toxic effects. A library of small sized chemical compounds was screened to identify a set of chemical compounds that could bind with the anthrax toxin. The screening was done by computing the binding free energy for each compound using numerical simulations. The screen identified a set of 12,376 compounds that could potentially bind to the anthrax toxin and a set of 22,460 compounds that were unlikely to bind to the chemical compound. The average number of vertices in this dataset is 25 and the average number of edges is also 25. We use this dataset to derive a two-class classification problem whose goal is to correctly predict whether or not a compound will bind the anthrax toxin or not. This classification problem is referred to as *A1*.

Table 1. The characteristics of the various datasets. N is the number of compounds in the database. $\bar{N}_A$ and $\bar{N}_B$ are the average number of atoms and bonds in each compound. $\bar{L}_A$ and $\bar{L}_B$ are the average number of atom- and bond-types in each dataset. $\max N_A/\min N_A$ and $\max N_B/\min N_B$ are the maximum/minimum number of atoms and bonds over all the compounds in each dataset.

	Toxic.	*Aids*	*Anthrax*	*Class Dist.* (% +ve class)	
N	417	42,687	34,836	**Toxicology**	
$\bar{N}_A$	25	46	25	P1: Male Mice	38.3%
$\bar{N}_B$	26	48	25	P2: Female Mice	40.9%
$\bar{L}_A$	40	82	25	P3: Male Rats	44.2%
$\bar{L}_B$	4	4	4	P4: Female Rats	34.4%
$\max N_A$	106	438	41	**AIDS**	
$\min N_A$	2	2	12	H1: CA/CM	28.1%
$\max N_B$	1	276	44	H2: (CA+CM)/CI	3.5%
$\min N_B$	85	1	12	H3: CA/CI	1.0%
				Anthrax	
				A1: active/inactive	35%

Some important characteristics of these datasets are summarized in Table 1. The right hand side of the table displays the class distribution for different classification problems, for each problem the table displays the percentage of positive class found in the dataset for that classification problem. Note that both the DTP-AIDS and the Anthrax datasets are quite large containing 42,687 and 34,836 compounds, respectively. Moreover, in the case of DTP-AIDS, each compound is also quite large having on an average 46 atoms and 48 bonds.

5.2 Experimental Methodology & Metrics

The classifications results were obtained by performing 5-way cross validation on the dataset, ensuring that the class distribution in each fold is identical to the original dataset. For the SVM classifier we used SVMLight library [37]. All the experiments were conducted on a 1500MHz Athlon MP processors having a 2GB of memory.

Since the size of the positive class is significantly smaller than the negative class, using *accuracy* to judge a classifier would be incorrect. To get a better understanding of the classifier performance for different cost settings we obtain the ROC curve [62] for each classifier. ROC curve plots the false positive rate (X-axis) versus the true positive rate (Y-axis) of a classier; it displays the performance of the classifier regardless of class distribution or error cost. Two classifiers are evaluated by comparing the area under their respective ROC curves, a larger area under ROC curve indicating better performance. The area under the ROC curve will be referred by the parameter A.

5.3 Results

Varying Minimum Support

The key parameter of the proposed frequent sub-structure-based classification algorithm is the choice of the minimum support (σ) used to discover the frequent sub-structures (either topological or geometric). To evaluate the sensitivity of the algorithm on this parameter we performed a set of experiments in which we varied σ from 10% to 20% in 5% increments. The results of these experiments are shown in the left sub-table of Table 2 for both topological and geometric sub-structures.

Table 2. Varying minimum support threshold (σ). "A" denotes the area under the ROC curve and "N_f" denotes the number of discovered frequent subgraphs.

D	$\sigma=10.0\%$				$\sigma = 15.0\%$				$\sigma = 20.0\%$			
	Topo.		*Geom.*		*Topo.*		*Geom.*		*Topo.*		*Geom.*	
	A	N_f	A	N_f	A	N_f	A	N_f	A	N_f	A	N_f
P1	66.0	1211	65.5	1317	66.0	513	64.1	478	64.4	254	60.2	268
P2	65.0	967	64.0	1165	65.1	380	63.3	395	64.2	217	63.1	235
P3	60.5	597	60.7	808	59.4	248	61.3	302	59.9	168	60.9	204
P4	54.3	275	55.4	394	56.2	173	57.4	240	57.3	84	58.3	104
H1	81.0	27034	82.1	29554	77.4	13531	79.2	8247	78.4	7479	79.5	7700
H2	70.1	1797	76.0	3739	63.6	307	62.2	953	59.0	139	58.1	493
H3	83.9	27019	89.5	30525	83.6	13557	88.8	11240	84.6	7482	87.7	7494
A1	78.2	476	79.0	492	78.2	484	77.6	332	77.1	312	76.1	193

$Dset$	Optimized σ					
	Topo.		*Geom.*		Per class	$Time_p$
	A	N_f	A	N_f	σ	(sec)
P1	65.5	24510	65.0	23612	3.0, 3.0	211
P2	67.3	7875	69.9	12673	3.0, 3.0	72
P3	62.6	7504	64.8	10857	3.0, 3.0	66
P4	63.4	25790	63.7	31402	3.0, 3.0	231
H1	81.0	27034	82.1	29554	10.0, 10.0	137
H2	76.5	18542	79.1	29024	10.0, 5.0	1016
H3	83.9	27019	89.5	30525	10.0, 10.0	392
A1	81.7	3054	82.6	3186	5.0, 3.0	145

From Table 2 we observe that as we increase σ, the classification performance for most datasets tends to degrade. However, in most cases this degradation is gradual and correlates well with the decrease on the number of sub-structures that were discovered by the frequent subgraph discovery algorithms. The only exception is the H2 problem for which the classification performance (as measured by ROC) degrades substantially as we increase the minimum support from 10% to 20%. Specifically, in the case of topological subgraphs, the performance drops from 70.1 down to 59.0, and in the case of geometric subgraphs it drops from 76.0 to 58.1.

These results suggest that lower values of support are in general better as they lead to better classification performance. However, as the support decreases, the number of discovered sub-structures and the amount of time required also increases. Thus, depending on the dataset, some experimentation may be required to select the proper values of support that balances these conflicting requirements (*i.e.*, low support but reasonable number of sub-structures).

In our study we performed such experimentation. For each dataset we kept on decreasing the value of support down to the point after which the number of features that were generated was too large to be efficiently processed by the SVM library. The resulting support values, number of features, and associated classification performance are shown in the right sub-table of Table 2 under the table header "Optimized σ". Note that for each problem two different support values are displayed corresponding to the supports that were used to mine the positive and negative class, respectively. Also, the last column shows the amount of time required by FSG to find the frequent subgraphs and provides a good indication of the computational complexity at the feature discovery phase of our classification algorithm.

Comparing the ROC values obtained in these experiments with those obtained for $\sigma = 10\%$, we can see that as before, the lower support values tend to improve the results, with measurable improvements for problems in which the number of discovered sub-structures increased substantially. In the rest of our experimental evaluation we will be using the frequent subgraphs that were generated using these values of support.

Varying Misclassification Costs

Since the number of positive examples is in general much smaller than the number of negative examples, we performed a set of experiments in which the misclassification cost associated with each positive example was increased to match the number of negative examples. That is, if n^+ and n^- is the number of positive and negative examples, respectively, the misclassification cost β was set equal to $(n^-/n^+ - 1)$ (so that $n^- = \beta n^+$). We refer to this value of β as the *"EqCost"* value. The classification performance achieved by our algorithm using either topological or geometric subgraphs for $\beta = 1.0$ and $\beta = EqCost$ is shown in Table 3. Note that the $\beta = 1.0$ results are the same with those presented in the right subtable of Table 2.

From the results in this table we can see that, in general, increasing the misclassification cost so that it balances the size of positive and negative class

Table 3. The area under the ROC curve obtained by varying the misclassification cost. "$\beta = 1.0$" indicates the experiments in which each positive and negative example had a weight of one, and "$\beta = EqCost$" indicates the experiments in which the misclassification cost of the positive examples was increased to match the number of negative examples.

Dataset	Topo		Geom	
	$\beta = 1.0$	$\beta = EqCost$	$\beta = 1.0$	$\beta = EqCost$
P1	65.5	65.3	65.0	66.7
P2	67.3	66.8	69.9	69.2
P3	62.6	62.6	64.8	64.6
P4	63.4	65.2	63.7	66.1
H1	81.0	79.2	82.1	81.1
H2	76.5	79.4	79.1	81.9
H3	83.9	90.8	89.5	94.0
A1	81.7	82.1	82.6	83.0

tends to improve the classification accuracy. When $\beta = EqCost$, the classification performance improves for four and five problems for the topological and geometric subgraphs, respectively. Moreover, in the cases in which the performance decreased, that decrease was quite small, whereas the improvements achieved for some problem instances (*e.g.*, P4, H1, and H2) was significant. In the rest of our experiments we will focus only on the results obtained by setting $\beta = EqCost$.

Feature Selection

We evaluated the performance of the feature selection scheme based on sequential covering (described in Section 4.2) by performing a set of experiments in which we varied the parameter δ that controls the number of times an example must be covered by a feature, before it is removed from the set of yet to be covered examples. Table 4 displays the results of these experiments. The results under the column labeled "Original" shows the performance of the classifier without any feature selection. These results are identical to those shown in Table 3 for $\beta = EqCost$ and are included here to make comparisons easier.

Two key observations can be made by studying the results in this table. First, as expected, the feature selection scheme is able to substantially reduce the number of features. In some cases the number of features that was selected decreased by almost two orders of magnitude. Also, as δ increases, the number of retained features increases; however, this increase is gradual. Second, the overall classification performance achieved by the feature selection scheme when $\delta \geq 5$ is quite comparable to that achieved with no feature selection. The actual performance depends on the problem instance and whether or not we use topological or geometric subgraphs. In particular, for the first four problems (P1, P2, P3, and P4) derived from the PTC dataset, the performance actually improves with feature selection. Such improvements are possible even in the context of SVM-based classifiers as models learned on lower dimensional spaces will tend to have better generalization ability [22]. Also note that for

Table 4. Results obtained using feature selection based on sequential rule covering. "δ" specifies the number of times each example needs to be covered before it is removed, "A" denotes the area under the ROC curve and "N_f" denotes the number of features that were used for classification.

Topological Features

Dataset.	Original		$\delta = 1$		$\delta = 5$		$\delta = 10$		$\delta = 15$	
	A	N_f	A	N_f	A	N_f	A	N_f	A	N_f
P1	65.3	24510	65.4	143	66.4	85	66.5	598	66.7	811
P2	66.8	7875	69.5	160	69.6	436	68.0	718	67.5	927
P3	62.6	7504	68.0	171	65.2	455	64.2	730	64.5	948
P4	65.2	25790	66.3	156	66.0	379	64.5	580	64.1	775
H1	79.2	27034	78.4	108	79.2	345	79.1	571	79.5	796
H2	79.4	18542	77.1	370	78.0	1197	78.5	1904	78.5	2460
H3	90.8	27019	88.4	111	89.6	377	90.0	638	90.5	869
A1	82.1	3054	80.6	620	81.4	1395	81.5	1798	81.8	2065

Geometric Features

Dataset.	Original		$\delta = 1$		$\delta = 5$		$\delta = 10$		$\delta = 15$	
	A	N_f	A	N_f	A	N_f	A	N_f	A	N_f
P1	66.7	23612	68.3	161	68.1	381	67.4	613	68.7	267
P2	69.2	12673	72.2	169	73.9	398	73.1	646	73.0	265
P3	64.6	10857	71.1	175	70.0	456	71.0	241	66.7	951
P4	66.1	31402	68.8	164	69.7	220	67.4	609	66.2	819
H1	81.1	29554	80.8	128	81.6	396	81.9	650	82.1	885
H2	81.9	29024	80.0	525	80.4	1523	80.6	2467	81.2	3249
H3	94.0	30525	91.3	177	92.2	496	93.1	831	93.2	1119
A1	83.0	3186	81.0	631	82.0	1411	82.4	1827	82.7	2106

some datasets the number of features decreases as δ increases. Even though this is counter-intuitive it can happen in the cases in which due to a higher value of δ, a feature that would have been skipped it is now included into the set. If this newly included feature has a relatively high support, it will contribute to the coverage of many other features. As a result, the desired level of coverage can be achieved without the inclusion of other lower-support features. Our analysis of the selected feature-sets showed that for the instances in which the number of features decreases as δ increases, the selected features have indeed higher average support.

Topological versus Geometric Subgraphs

The various results shown in Tables 2–4 also provide an indication on the relative performance of topological versus geometric subgraphs. In almost all cases, the classifier that is based on geometric subgraphs outperforms that based on topological subgraphs. For some problems, the performance advantage is marginal whereas for other problems, geometric subgraphs lead to measurable improvements in the area under the ROC curve. For example, if we consider the results shown in Table 3 for $\beta = EqCost$, we can see the geometric subgraphs lead to improvements that are at least 3% or higher for P2, P3, and H3, and the average improvement over all eight problems is 2.6%. As discussed in Section 4.1, these performance gains is due to the fact that conserved geometric structure is a better indicator of a chemical compounds activity than just its topology.

5.4 Comparison with Other Approaches

We compared the performance of our classification algorithm against the performance achieved by the QSAR-based approach and the approach that uses the SUBDUE system to discover a set of sub-structures.

Comparison with QSAR

As discussed in Section 3 there is a wide variety of QSAR properties each of which captures certain aspects of a compounds chemical activity. For our study, we have chosen a set of 18 QSAR properties that are good descriptors of the chemical activity of a compound and most of them have been previously used for classification purposes [2]. A brief description of these properties are shown in Table 5. We used two programs to compute these attributes; the geometric attributes like solvent accessible area, total accessible area/vol, total Van der Waal's accessible area/vol were computed using the programs SASA [46], the remaining attributes were computed using Hyperchem software.

Table 5. QSAR Properties.

Property	Dim.	Property	Dim.
Solvent accessible area	$\mathring{A}^2$	Moment of Inertia	*none*
Total accessible area	$\mathring{A}^2$	Total energy	*kcal/mol*
Total accessible volume	$\mathring{A}^3$	Bend energy	*kcal/mol*
Total Van der Waal's area	$\mathring{A}^2$	Hbond energy	*kcal/mol*
Total Van der Waal's volume	$\mathring{A}^3$	Stretch energy	*kcal/mol*
Dipole moment	*Debye*	Nonbond energy	*kcal/mol*
Dipole moment comp. (X, Y, Z)	*Debye*	Estatic energy	*kcal/mol*
Heat of formation	*Debye*	Torsion energy	*kcal/mol*
Multiplicity	*Kcal*	Quantum total charge	*eV*

We used two different algorithms to build classification models based on these QSAR properties. The first is the C4.5 decision tree algorithm [63] that has been shown to produce good models for chemical compound classification based on QSAR properties [2], and the second is the SVM algorithm that was used to build the classification models in our frequent sub-structure-based approach. Since the range of values of the different QSAR properties can be significantly different, we first scaled them to be in the range of $[0, 1]$ prior to building the SVM model. We found that this scaling resulted in some improvements in the overall classification results. Note that C4.5 is not affected by such scaling.

Table 6 shows the results obtained by the QSAR-based methods for the different datasets. The values shown for SVM correspond to the area under the ROC curve and can be directly compared with the corresponding values obtained by our approaches (Tables 2–4). Unfortunately, since C4.5 does not produce a ranking of the training set based on its likelihood of being in the positive class, it is quite hard to obtain the ROC curve. For this reason, the

Table 6. Performance of the QSAR-based Classifier.

Dataset	SVM	C4.5		Freq. Sub. Prec.	
	A	Precision	Recall	Topo	Geom
P1	60.2	0.4366	0.1419	0.6972	0.6348
P2	59.3	0.3603	0.0938	0.8913	0.8923
P3	55.0	0.6627	0.1275	0.7420	0.7427
P4	45.4	0.2045	0.0547	0.6750	0.8800
H1	64.5	0.5759	0.1375	0.7347	0.7316
H2	47.3	0.6282	0.4071	0.7960	0.7711
H3	61.7	0.5677	0.2722	0.7827	0.7630
A1	49.4	0.5564	0.3816	0.7676	0.7798

values shown for C4.5 correspond to the precision and recall of the positive class for the different datasets. Also, to make the comparisons between C4.5 and our approach easier, we also computed the precision of our classifier at the same value of recall as that achieved by C4.5. These results are shown under the columns labeled "*Freq. Sub. Prec.*" for both topological and geometric features and were obtained from the results shown in Table 3 for $\beta = EqCost$. Note that the QSAR results for both SVM and C4.5 were obtained using the same cost-sensitive learning approach.

Comparing both the SVM-based ROC results and the precision/recall values of C4.5 we can see that our approach substantially outperforms the QSAR-based classifier. In particular, our topological subgraph based algorithm does 35% better compared to SVM-based QSAR and 72% better in terms of the C4.5 precision at the same recall values. Similar results hold for the geometric subgraph based algorithm. These results are consistent with those observed by other researchers [69, 39] that showed that sub-structure based approaches outperform those based on QSAR properties.

Comparison with SUBDUE & SubdueCL

Finally, to evaluate the advantage of using the complete set of frequent sub-structures over existing schemes that are based on heuristic sub-structure discovery, we performed a series of experiments in which we used the SUB-DUE system to find the sub-structures and then used them for classification . Specifically, we performed two sets of experiments. In the first set, we obtain a set of sub-structures using the standard MDL-based heuristic sub-structure discovery approach of SUBDUE [35]. In the second set, we used the sub-structures discovered by the more recent SubdueCL algorithm [29] that guides the heuristic beam search using a scheme that measures how well a subgraph describes the positive examples in the dataset without describing the negative examples.

Even though there are a number of parameters controlling SUBDUE's heuristic search algorithm, the most critical among them are the width of the beam search, the maximum size of the discovered subgraph, and the total number of subgraphs to be discovered. In our experiments, we spent a considerable amount of time experimenting with these parameters to ensure that

Table 7. Performance of the SUBDUE and SubdueCL-based approaches.

Dataset	SUBDUE			SubdueCL		
	A	N_f	$Time_p$	A	N_f	$Time_p$
P1	61.9	1288	303sec	63.5	2103	301sec
P2	64.2	1374	310sec	63.3	2745	339sec
P3	57.4	1291	310sec	59.6	1772	301sec
P4	58.5	1248	310sec	60.8	2678	324sec
H1	74.2	1450	1,608sec	73.8	960	1002sec
H2	58.5	901	232,006sec	65.2	2999	476,426sec
H3	71.3	905	178,343sec	77.5	2151	440,416sec
A1	75.3	983	56,056sec	75.9	1094	31,177sec

SUBDUE was able to find a reasonable number of sub-structures. Specifically, we changed the width of the beam search from 4 to 50 and set the other two parameters to high numeric values. Note that in the case of the SubdueCL, in order to ensure that the subgraphs were discovered that described all the positive examples, the subgraph discovery process was repeated by increasing the value of beam-width at each iteration and removing the positive examples that were covered by subgraphs.

Table 7 shows the performance achieved by SUBDUE and SubdueCL on the eight different classification problems along with the number of subgraphs that it generated and the amount of time that it required to find these subgraphs. These results were obtained by using the subgraphs discovered by either SUBDUE or SubdueCL as features in an SVM-based classification model. Essentially, our SUBDUE and SubdueCL classifiers have the same structure as our frequent subgraph-based classifiers with the only difference being that the features now correspond to the subgraphs discovered by SUBDUE and SubdueCL. Moreover, to make the comparisons as fair as possible we used $\beta = EqCost$ as the misclassification cost. We also performed another set of experiments in which we used the rule-based classifier produced by SubdueCL. The results of this scheme was inferior to those produced by the SVM-based approach and we are not reporting them here.

Comparing SUBDUE against SubdueCL we can see that the latter achieves better classification performance, consistent with the observations made by other researchers [29]. Comparing the SUBDUE and SubdueCL-based results with those obtained by our approach (Tables 2–4) we can see that in almost all cases both our topological and geometric frequent subgraph-based algorithms lead to substantially better performance. This is true both in the cases in which we performed no feature selection as well as in the cases in which we used the sequential covering based feature selection scheme. In particular, comparing the SubdueCL results against the results shown in Table 4 without any feature selection we can see that on the average, our topological and geometric subgraph based algorithms do 9.3% and 12.2% better, respectively. Moreover, even after feature selection with $\delta = 15$ that result in a scheme that have comparable number of features as those used by SubdueCL, our algorithms are still better by 9.7% and 13.7%, respectively. Finally, if we compare the amount of time required by either SUBDUE or SubdueCL to that

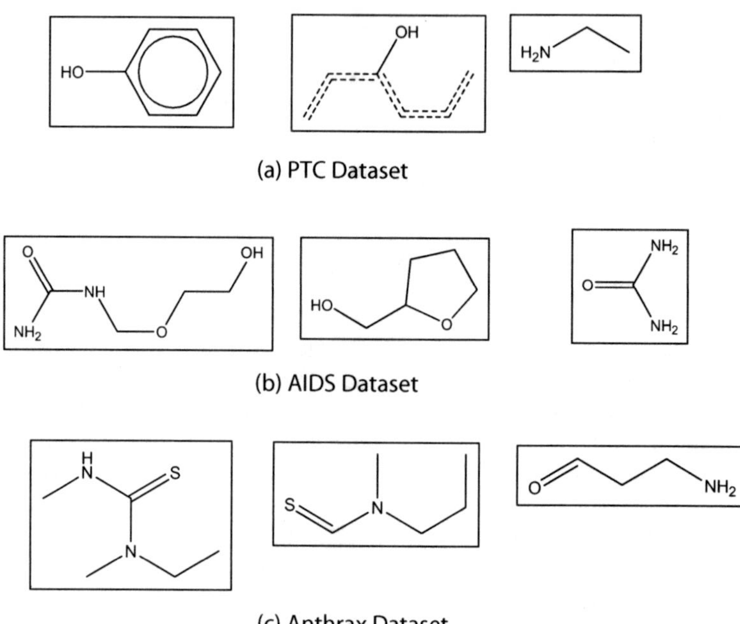

(a) PTC Dataset

(b) AIDS Dataset

(c) Anthrax Dataset

Fig. 4. The three most discriminating sub-structures for the PTC, AIDS, and Anthrax datasets.

required by the FSG algorithm to find all frequent subgraphs (last column of Table 2) we can see that despite the fact that we are finding the complete set of frequent subgraphs our approach requires substantially less time.

6 Conclusions and Directions for Future Research

In this chapter we presented a highly-effective algorithm for classifying chemical compounds based on frequent sub-structure discovery that can scale to large datasets. Our experimental evaluation showed that our algorithm leads to substantially better results than those obtained by existing QSAR- and sub-structure-based methods. Moreover, besides this improved classification performance, the sub-structure-based nature of this scheme provides to the chemists valuable information as to which sub-structures are most critical for the classification problem at hand. For example, Figure 4 shows the three most discriminating sub-structures for the PTC, DTP AIDS, and Anthrax datasets that were obtained by analyzing the decision hyperplane produced by the SVM classifier. A chemist can then use this information to understand the models and potentially use it to design better compounds.

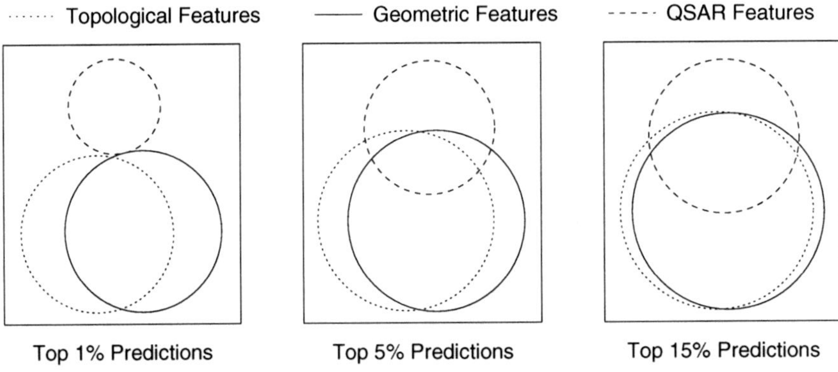

Fig. 5. Venn diagrams displaying the relation between the positive examples that were correctly classified by the three approaches at different cutoff values for the Anthrax dataset. The different cutoffs were obtained by looking at only the top 1%, 5%, and 15% of the ranked predictions. Each circle in the Venn diagram corresponds to one of the three classification schemes and the size of the circle indicates the number of positive examples correctly identified. The overlap between two circles indicates the number of common correct predictions.

The classification algorithms presented in this chapter can be improved along three different directions. First, as already discussed in Section 2 our current geometric graph representation utilizes a single conformation of the chemical compound and we believe the overall classification performance can be improved by using all possible low-energy conformations. Such conformations can be obtained from existing 3D coordinate prediction software and as discussed in Section 4.1 can be easily incorporated in our existing framework. Second, our current feature selection algorithms only focus on whether or not a particular sub-structure is contained in a compound and they do not take into account how these fragments are distributed over different parts of the molecule. Better feature selection algorithms can be developed by taking this information into account so that to ensure that the entire (or most of) molecule is covered by the selected features. Third, even though the proposed approaches significantly outperformed that based on QSAR, our analysis showed that there is a significant difference as to which compounds are correctly classified by the sub-structure- and QSAR-based approaches. For example, Figure 5 shows the overlap among the different correct predictions produced by the geometric, topological, and QSAR-based methods at different cutoff values for the Anthrax dataset. From these results we can see that there is a great agreement between the substructure-based approaches but there is a large difference among the compounds that are correctly predicted by the QSAR approach, especially at the top 1% and 5%. These results suggest that better results can be potentially obtained by combining the substructure- and QSAR-based approaches.

References

1. G. W. Adamson, J. Cowell, M. F. Lynch, A. H. McLure, W. G. Town, and A. M. Yapp. Strategic considerations in the design of a screening system for substructure searches of chemical structure file. *Journal of Chemical Documentation*, 13:153–157, 1973.
2. A. An and Y. Wang. Comparisons of classification methods for screening potential compounds. In *ICDM*, 2001.
3. T. A. Andrea and Hooshmand Kalayeh. Applications of neural networks in quantitative structure-activity relationships of dihydrofolate reductase inhibitors. *Journal of Medicinal Chemistry*, 34:2824–2836, 1991.
4. M. J. Ashton, M. C. Jaye, and J. S. Mason. New perspectives in lead generation ii: Evaluating molecular diversity. *Drug Discovery Today*, 1(2):71–78, 1996.
5. J. Bajorath. Integration of virtual and high throughput screening. *Nature Review Drug Discovery*, 1(11):822–894, 2002.
6. John M. Barnard, Geoffery M. Downs, and Peter Willet. Descriptor-based similarity measures for screening chemical databases. In H.J. Bohm and G. Schneider, editors, *Virtual Screening for Bioactive Molecules*, volume 10. Wiley-VCH, 2000.
7. S. C. Basak, V. R. Magnuson, J. G. Niemi, and R. R. Regal. Determining structural similarity of chemicals using graph theoretic indices. *Discrete Applied Mathematics*, 19:17–44, 1988.
8. Guy W. Bemis and Mark A. Murcko. The properties of known drugs. 1. molecular frameworks. *Journal of Medicinal Chemistry*, 39(15):2887–2893, 1996.
9. Guy W. Bemis and Mark A. Murcko. The properties of known drugs. 2. side chains. *Journal of Medicinal Chemistry*, 42(25):5095–5099, 1999.
10. K. H. Bleicher, Hans-Joachim Bohm, K. Muller, and A.I. Alanine. Hit and lead generation: Beyond high throughput screening. *Nature Review Drug Discover*, 2(5):369–378, 2003.
11. H.J. Bohm and G. Schneider. *Virtual Screening for Bioactive Molecules*, volume 10. Wiley-VCH, 2000.
12. Christian Borgelt and Michael R. Berthold. Mining molecular fragments: Finding relevant substructures of molecules. In *Proceedings of the ICDM*, 2002.
13. Gianpaolo Bravi, Emanuela Gancia ; Darren Green, V.S. Hann, and M. Mike. Modelling structure-activity relationship. In H.J. Bohm and G. Schneider, editors, *Virtual Screening for Bioactive Molecules*, volume 10. Wiley-VCH, 2000.
14. Evgeny Byvatov, Uli Fechner, Jens Sadowski, and Gisbert Schneider. Comparison of support vector machine and artificial neural network systems for drug/nondrug classification. *Journal of Chemical Information and Computer Science*, 43(6):1882–1889, 2003.
15. R. E. CarHart, D. H Smith, and R. Venkataraghavan. Atom pairs as molecular features in atructure-activity studies: Definition and applications. *Journal of Chemical Information and Computer Science*, 25(2):64–73, 1985.
16. Xin Chen, Andrew Rusinko, and Stanley S. Young. Recursive partitioning analysis of a large structure-activity data set using three-dimensional descriptors. *Journal of Chemical Information and Computer Science*, 38(6):1054–1062, 1998.
17. D. J. Cook and L. B. Holder. Graph-based data mining. *IEEE Intelligent Systems*, 15(2):32–41, 2000.

18. King Ross D., Ashwin Srinivasan, and L. Dehaspe. Warmr: A data mining tool for chemical data. *Journal of Computer Aided Molecular Design*, 15:173–181, 2001.

19. E. K. Davies. Molecular diversity and combinatorial chemistry: Libraries and drug discovery. *American Chemical Society*, 118(2):309–316, 1996.

20. L. Dehaspe, H. Toivonen, and R. D. King. Finding frequent substructures in chemical compounds. In R. Agrawal, P. Stolorz, and G. Piatetsky-Shapiro, editors, *4th International Conference on Knowledge Discovery and Data Mining*, pages 30–36. AAAI Press, 1998.

21. Mukund Deshpande and George Karypis. Automated approaches for classifying structure. In *Proceedings of the 2nd ACM SIGKDD Workshop on Data Mining in Bioinformatics*, 2002.

22. Mukund Deshpande and George Karypis. Using conjunction of attribute values for classification. In *Proceedings of the eleventh CIKM*, pages 356–364. ACM Press, 2002.

23. J. Devillers. *Neural networks in QSAR and Drug Design*. Acemedic Press, London, 1996.

24. dtp.nci.nih.gov. DTP AIDS antiviral screen dataset.

25. B. Dunkel and N. Soparkar. Data organizatinon and access for efficient data mining. In *Proc. of the 15th IEEE International Conference on Data Engineering*, March 1999.

26. H. Gao, C. Williams, P. Labute, and J. Bajorath. Binary quantitative structure-activity relationship (QSAR) analysis of estrogen receptor ligands. *Journal of Chemical Information and Computer Science*, 39(1):164–168, 1999.

27. J. Gasteiger, C. Rudolph, and J. Sadowski. Automatic generation of 3d-atomic coordinates for organic molecules. *Tetrahedron Computer Methodology*, 3:537–547, 1990.

28. T. A. Geissman. *Principles of Organic Chemistry*. W. H. Freeman and Company, 1968.

29. J. Gonzalez, L. Holder, and D. Cook. Application of graph based concept learning to the predictive toxicology domain. In *PTC, Workshop at the 5th PKDD*, 2001.

30. Anrew C. Good, Jonathan S. Mason, and Stephen D. Pickett. Pharmacophore pattern application in virtual screening, library design and QSAR. In H.J. Bohm and G. Schneider, editors, *Virtual Screening for Bioactive Molecules*, volume 10. Wiley-VCH, 2000.

31. L. H. Hall and L. B. Kier. Electrotopological state indices for atom types: A novel combination of electronic, topological, and valence state information. *Journal of Chemical Information and Computer Science*, 35(6):1039–1045, 1995.

32. Jeffrey S. Handen. The industrialization of drug discovery. *Drug Discovery Today*, 7(2):83–85, January 2002.

33. C. Hansch, P. P. Maolney, T. Fujita, and R. M. Muir. Correlation of biological activity of phenoxyacetic acids with hammett substituent constants and partition coefficients. *Nature*, 194:178–180, 1962.

34. C. Hansch, R. M. Muir, T. Fujita, C. F. Maloney, and Streich M. The correlation of biological activity of plant growth-regulators and chloromycetin derivatives with hammett constants and partition coefficients. *Journal of American Chemical Society*, 85:2817–1824, 1963.

35. L. Holder, D. Cook, and S. Djoko. Substructure discovery in the subdue system. In *Proceedings of the AAAI Workshop on Knowledge Discovery in Databases*, pages 169–180, 1994.

36. Akihiro Inokuchi, Takashi Washio, and Hiroshi Motoda. An apriori-based algorithm for mining frequent substructures from graph data. In *Proceedings of The 4th European Conference on Principles and Practice of Knowledge Discovery in Databases (PKDD'00)*, pages 13–23, Lyon, France, September 2000.

37. T. Joachims. *Advances in Kernel Methods: Support Vector Learning*, chapter Making large-Scale SVM Learning Practical. MIT-Press, 1999.

38. George Karypis. CLUTO a clustering toolkit. Technical Report 02-017, Dept. of Computer Science, University of Minnesota, 2002. Available at http://www.cs.umn.edu/~cluto.

39. Ross D. King, Stephen H. Muggleton, Ashwin Srinivasan, and Michael J. E. Sternberg. Strucutre-activity relationships derived by machine learning: The use of atoms and their bond connectivities to predict mutagenecity byd inductive logic programming. *Proceedings of National Acadamey of Sciences*, 93:438–442, January 1996.

40. Ross D. King, Stepher Muggleton, Richard A. Lewis, and J. E. Sternberg. Drug design by machine learning: The use of inductive logic programming to model the sturcture-activity relationships of trimethoprim analogues binding to dihydrofolate reductase. *Proceedings of National Acadamey of Sciences*, 89:11322–11326, December 1992.

41. S. Kramer, L. De Raedt, and C. Helma. Molecular feature mining in hiv data. In *7th International Conference on Knowledge Discovery and Data Mining*, 2001.

42. Michihiro Kuramochi and George Karypis. Frequent subgraph discovery. In *IEEE International Conference on Data Mining*, 2001. Also available as a UMN-CS technical report, TR# 01-028.

43. Michihiro Kuramochi and George Karypis. Discovering geometric frequent subgraph. In *IEEE International Conference on Data Mining*, 2002. Also available as a UMN-CS technical report, TR# 02-024.

44. Michihiro Kuramochi and George Karypis. An efficient algorithm for discovering frequent subgraphs. Technical Report TR# 02-26, Dept. of Computer Science and Engineering, University of Minnesota, 2002.

45. Paul Labute. Binary QSAR: A new method for the determination of quantitative structure activity relationships. *Pacific Symposium on Biocomputing*, 1999.

46. S. M. Le Grand and J. K. M. Merz. Rapid approximation to molecular surface area via the use of booleean logic look-up tables. *Journal of Computational Chemistry*, 14:349–352, 1993.

47. Andrew R. Leach. *Molecular Modeling: Principles and Applications*. Prentice Hall, Englewood Cliffs, NJ, 2001.

48. X. Q. Lewell, D. B. Judd, S. P. Watson, and M. M. Hann. RECAP retrosynthetic combinatorial analysis procedure: A powerful new technique for identifying privileged molecular fragments with useful applications in combinatorial chemistry. *Journal of Chemical Information and Computer Science*, 38(3):511–522, 1998.

49. Wenmin Li, Jiawei Han, and Jian Pei. Cmar: Accurate and efficient classification based on multiple class-association rules. In *IEEE International Conference on Data Mining*, 2001.

50. Bing Liu, Wynne Hsu, and Yiming Ma. Integrating classification and association rule mining. In *4th Internation Conference on Knowledge Discovery and Data Mining*, 1998.

51. D. J. Livingstone. Neural networks in QSAR and drug design. Academic Press, London, 1996.

52. D. J. Livingstone. The characterization of chemical structures using molecular properties. a survey. *Journal of Chemical Information and Computer Science*, 20(2):195–209, 2000.

53. Tom M. Mitchell. *Machine Learning*. Mc Graw Hill, 1997.

54. K. Morik, P. Brockhausen, and T. Joachims. Combining statistical learning with a knowledge-based approach - a case study in intensive care monitoring. In *International Conference on Machine Learning*, 1999.

55. S. Muggleton. Inverse entailment and Progol. *New Generation Computing*, 13:245–286, 1995.

56. Stephen Muggleton and L. De Raedt. Inductive logic programming: Theory and methods. *Journal of Logic Programming*, 19(20):629–679, 1994.

57. Stephen H. Muggleton and C. Feng. Efficient induction of logic programs. In Stephen Muggleton, editor, *Inductive Logic Programming*, pages 281–298. Academic Press, London, 1992.

58. C. A. Nicalaou, S. Y. Tamura, B. P. Kelley, S. I. Bassett, and R. F. Nutt. Analysis of large screening data sets via adaptively grown phylogenetic-like trees. *Journal of Chemical Information and Computer Science*, 42(5):1069–1079, 2002.

59. R. Nilakantan, N. Bauman, S. Dixon, and R. Venkataraghavan. Topological torsion: a new molecular descriptor for sar applications. comparison with other descriptors. *Journal of Chemical Information and Computer Science*, 27(2):82–85, 1987.

60. M. Otto. *Chemometrics*. Wiley-VCH, 1999.

61. S. D. Pickett, J. S. Mason, and I. M. McLay. Diversity profiling and design using 3d pharmacophores: Pharmacophore-derived queries (PDQ). *Journal of Chemical Information and Computer Science*, 1996.

62. F. Provost and T. Fawcett. Robust classification for imprecise environments. *Machine Learning*, 42(3), 2001.

63. J. Ross Quinlan. *C4.5: Programs for machine learning*. Morgan Kaufmann, San Mateo, CA, 1993.

64. Graham W. Richards. Virtual screening using grid computing: the screensaver project. *Nature Reviews: Drug Discovery*, 1:551–554, July 2002.

65. Andrew Rusinko, Mark W. Farmen, Christophe G. Lambert, Paul L. Brown, and Stanley S. Young. Analysis of a large structure/biological activity data set using recursive partitioning. *Journal of Chemical Information and Computer Scicncc*, 39(6):1017–1026, 1999.

66. Pradeep Shenoy, Jayant R. Haritsa, S. Sundarshan, Gaurav Bhalotia, Mayank Bawa, and Devavrat Shah. Turbo-charging vertical mining of large databases. In *Proc. of ACM SIGMOD Int. Conf. on Management of Data*, pages 22–33, May 2000.

67. R. P. Sheridan, M. D. Miller, D. J. Underwood, and S. J. Kearsley. Chemical similarity using geometric atom pair descriptors. *Journal of Chemical Information and Computer Science*, 36(1):128–136, 1996.

68. A. Srinivasan, R. D. King, S. H. Muggleton, and M. Sternberg. The predictive toxicology evaluation challenge. In *Proc. of the Fifteenth International Joint Conference on Artificial Intelligence (IJCAI-97)*, pages 1–6. Morgan-Kaufmann, 1997.

69. Ashwin Sriniviasan and Ross King. Feature construction with inductive logic programming: a study of quantitative predictions of biological activity aided by structural attributes. *Knowledge Discovery and Data Mining Journal*, 3:37–57, 1999.

70. Susan Y. Tamura, Patricia A. Bacha, Heather S. Gruver, and Ruth F. Nutt. Data analysis of high-throughput screening results: Application of multidomain clustering to the nci anti-hiv data set. *Journal of Medicinal Chemistry*, 45(14):3082–3093, 2002.

71. V. Vapnik. *Statistical Learning Theory*. John Wiley, New York, 1998.

72. D. Weininger. SMILES 1. introduction and encoding rules. *Journal of Chemical Information and Computer Sciences*, 28, 1988.

73. O.S Weislow, R. Kiser, D. L Fine, J. P. Bader, R. H. Shoemaker, and M. R. Boyd. New soluble fomrazan assay for hiv-1 cyopathic effects: appliication to high flux screening of synthetic and natural products for aids antiviral activity. *Journal of National Cancer Institute*, 1989.

74. Peter Willett. Chemical similarity searching. *Journal of Chemical Information and Computer Science*, 38(6):983–996, 1998.

75. S. Wold, E. Johansson, and M. Cocchi. 3d QSAR in drug design: Theory, methods and application. ESCOM Science Publishers B.V, 1993.

76. Xifeng Yan and Jiawei Han. gSpan: Graph-based substructure pattern mining. In *ICDM*, 2002.

77. Mohammed J. Zaki and Karam Gouda. Fast vertical mining using diffsets. Technical Report 01-1, Department of Computer Science, Rensselaer Polytechnic Institute, 2001.

78. Mohammed Javeed Zaki. Scalable algorithms for association mining. *Knowledge and Data Engineering*, 12(2):372–390, 2000.

79. J. Zupan and J. Gasteiger. *Neural Networks for Chemists*. VCH Publisher, 1993.

Sparse Component Analysis: a New Tool for Data Mining

Pando Georgiev[1], Fabian Theis[2], Andrzej Cichocki[3], and Hovagim Bakardjian[3]

[1] ECECS Department, University of Cincinnati
   Cincinnati, OH 45221 USA
   `pgeorgie@ececs.uc.edu`
[2] Institute of Biophysics, University of Regensburg
   D-93040 Regensburg, Germany
   `fabian@theis.name`
[3] Brain Science Institute, RIKEN, Wako-shi, Japan
   `{cia,hova}@bsp.brain.riken.go.jp`

Summary. In many practical problems for data mining the data $\mathbf{X}$ under consideration (given as $(m \times N)$-matrix) is of the form $\mathbf{X} = \mathbf{AS}$, where the matrices $\mathbf{A}$ and $\mathbf{S}$ with dimensions $m \times n$ and $n \times N$ respectively (often called mixing matrix or *dictionary* and source matrix) are unknown ($m \leq n < N$). We formulate conditions (SCA-conditions) under which we can recover $\mathbf{A}$ and $\mathbf{S}$ uniquely (up to scaling and permutation), such that $\mathbf{S}$ is *sparse* in the sense that each column of $\mathbf{S}$ has at least one zero element. We call this the *Sparse Component Analysis* problem (SCA). We present new algorithms for identification of the mixing matrix (under SCA-conditions), and for source recovery (under identifiability conditions). The methods are illustrated with examples showing good performance of the algorithms. Typical examples are EEG and fMRI data sets, in which the SCA algorithm allows us to detect some features of the brain signals. Special attention is given to the application of our method to the transposed system $\mathbf{X}^T = \mathbf{S}^T \mathbf{A}^T$ utilizing the sparseness of the mixing matrix $\mathbf{A}$ in appropriate situations. We note that the sparseness conditions could be obtained with some preprocessing methods and no independence conditions for the source signals are imposed (in contrast to Independent Component Analysis). We applied our method to fMRI data sets with dimension ($128 \times 128 \times 98$) and to EEG data sets from a 256-channels EEG machine.

Key words: Sparse Component Analysis, Blind Signal Separation, clustering.

1 Introduction

Data mining techniques can be divided into the following classes [3]:

1. Predictive Modelling: where the goal is to predict a specific attribute (column or field) based on the other attributes in the data.

2. Clustering: also called segmentation, targets grouping the data records into subsets where items in each subset are more "similar" to each other than to items in other subsets.

3. Dependency Modelling: discovering the existence of arbitrary, possibly weak, multidimensional relations in data. Estimate some statistical properties of the found relations.

4. Data Summarization: targets finding interesting summaries of parts of the data. For example, similarity between a few attributes in a subset of the data.

5. Change and Deviation Detection: accounts for sequence information in data records. Most methods above do not explicitly model the sequence order of items in the data.

In this chapter we consider the problem of linear representation or matrix factorization of a data set $\mathbf{X}$, given in the form of a $(m \times N)$-matrix:

$$\mathbf{X} = \mathbf{AS}, \quad \mathbf{A} \in \mathbb{R}^{m \times n}, \mathbf{S} \in \mathbb{R}^{n \times N}, \tag{1}$$

where n is the number of source signals, m is the number of observations and N is the number of samples. Such representations can be considered as a new class of data mining techniques (or a concrete subclass of the above described data mining technique 3). In (1) the unknown matrices $\mathbf{A}$ (dictionary) and $\mathbf{S}$ (signals) may have some specific properties, for instance:

1) the rows of $\mathbf{S}$ are as statistically independent as possible — this is the *Independent Component Analysis* (ICA) problem;

2) $\mathbf{S}$ contains as many zeros as possible — this is the sparse representation problem or *Sparse Component Analysis* (SCA) problem;

3) the elements of $\mathbf{X}, \mathbf{A}$ and $\mathbf{S}$ are nonnegative - this is *nonnegative matrix factorization* (NMF).

Such linear representations have several potential applications including decomposition of objects into "natural" components, learning the parts of the objects (e.g. learn from set of faces the parts a face consists of, i.e. eyes, nose, mouth, etc.), redundancy and dimensionality reduction, micro-array data mining, enhancement of images in nuclear medicine etc. (see [17, 10]).

There are many of papers devoted to ICA problems (see for instance [5, 15] and references therein) but mostly for the complete case $(m = n)$. We refer to [26, 4, 29, 1, 25] and reference therein for some recent papers on SCA and overcomplete ICA $(m < n)$.

A more general related problem is called *Blind Source Separation* (BSS) problem, in which we know *a priori* that a representation such as in equation (1) exists and the task is to recover the sources (and the mixing matrix) as accurately as possible. A fundamental property of the complete BSS problem (for $m = n$) is that such a recovery (under assumptions in 1 and non-Gaussianity of the sources) is possible up to permutation and scaling of the sources, which makes the BSS problem so attractive.

In this chapter we consider SCA as a special model of BSS problem in the overcomplete case ($m < n$ i.e. more sources than sensors), where the additional information compensating the lack of sensors is the *sparseness* of the sources. The task of the SCA problem is to represent the given (observed) data $\mathbf{X}$ as in equation (1) such that the matrix $\mathbf{S}$ (sources) is sparse in sense that each column of $\mathbf{S}$ has at least one zero element. We present conditions on the data matrix $\mathbf{X}$ (*SCA-conditions on the data*), under which the representation in equation (1) is unique up to permutation and scaling of the sources.

The task of BSS problem is to estimate the unknown sources $\mathbf{S}$ (and the mixing matrix $\mathbf{A}$) using the available data matrix $\mathbf{X}$ only. We describe conditions (*identifiability conditions on the sources*) under which this is possible uniquely up to permutation and scaling of the sources, which is the usual condition in the complete BSS problems using ICA.

In the sequel, we present new algorithms for solving the BSS problem using sparseness: matrix identification algorithms and source recovery algorithm, which recovers sparse sources (in sense that each column of the source matrix $\mathbf{S}$ has at least one zero). When the sources are sufficiently sparse (see the conditions of Theorem 2) the matrix identification algorithm is even simpler. We used this simpler form for separation of mixtures of images. We present several computer simulation examples which illustrate our algorithms, as well as application of our method to real data: EEG data set obtained by a 256 channels EEG machine, and fMRI data set with dimension $128 \times 128 \times 98$. In all considered examples the results obtained by our SCA method are better (for the computer simulated examples) and comparable and advantages with respect to the ICA method.

2 Blind Source Separation using sparseness

In this section we present a method for solving the BSS problem if the following assumptions are satisfied:

A1) the mixing matrix $\mathbf{A} \in \mathbb{R}^{m \times n}$ has the property that any square $m \times m$ submatrix of it is nonsingular;

A2) each column of the source matrix $\mathbf{S}$ has at least one zero element.

A3) the sources are sufficiently rich represented in the following sense: for any index set of $n - m + 1$ elements $I = \{i_1, ..., i_{n-m+1}\} \subset \{1, ..., n\}$ there exist at least m column vectors of the matrix $\mathbf{S}$ such that each of them has zero elements in places with indexes in I and each $m - 1$ of them are linearly independent.

Columns of $\mathbf{X}$ for which A2) is not satisfied are called outliers. We can detect them in some cases and eliminate from the matrix $\mathbf{X}$, if the condition A3) is satisfied for a big number of columns of $\mathbf{S}$.

2.1 Matrix identification

We describe conditions in the sparse BSS problem under which we can identify the mixing matrix uniquely up to permutation and scaling of the columns. We give two types of such conditions. The first one corresponds to the least sparsest case in which such identification is possible. Further, we consider the most sparsest (nontrivial) case (for small number of samples) as in this case the algorithm is much simpler.

General case – full identifiability

Theorem 1 [12] (Identifiability conditions - general case) *Assume that the representation* $\mathbf{X} = \mathbf{AS}$ *is valid, the matrix* $\mathbf{A}$ *satisfies condition A1), the matrix* $\mathbf{S}$ *satisfies conditions A2) and A3) and only the matrix* $\mathbf{X}$ *is known. Then the mixing matrix* $\mathbf{A}$ *is identifiable uniquely up to permutation and scaling of the columns.*

The proof of this theorem is contained in [12] and gives the idea for the matrix identification algorithm.

Algorithm 1: identification of the mixing matrix

1) Cluster the columns of $\mathbf{X}$ in $\binom{n}{m-1}$ groups $\mathcal{H}_p, p = 1, ..., \binom{n}{m-1}$ such that the span of the elements of each group $\mathcal{H}_p$ produces one hyperplane and these hyperplanes are different.

2) Cluster the normal vectors to these hyperplanes in the smallest number of groups $G_j, j = 1, ..., n$ (which estimates the number of sources n) such that the normal vectors to the hyperplanes in each group G_j lie in a new hyperplane $\hat{H}_j$.

3) Calculate the normal vectors $\hat{\mathbf{a}}_j$ to each hyperplane $\hat{H}_j, j = 1, ..., n$ (the one-dimensional subspace spanned by $\hat{\mathbf{a}}_j$ is the intersection of all hyperplanes in G_j). The matrix $\hat{\mathbf{A}}$ with columns $\hat{\mathbf{a}}_j$ is an estimation of the mixing matrix (up to permutation and scaling of the columns).

Remark. The above algorithm works for data for which we know a priori that they lie on hyperplanes (or near to hyperplanes).

A very suitable algorithm for clustering data near hyperplanes is the k-plane clustering algorithm of Bradley - Mangasarian [2]. In our case the data points are supposed to lie on hyperplanes passing trough zero, so their algorithm is simplified and has the following form:

We applied this algorithm for real data sets of fMRI images and EEG recordings in the last two sections. We noticed that the algorithms stops very often in local minima and need several re-initializations until a reasonably

Algorithm 2: simplified algorithm of Bradley – Mangasarian

Start with random $w_1^0, ..., w_k^0 \in \mathbb{R}^n$ with $\|w_i^0\|_2 = 1, i = 1, ..., k$. Having $w_1^j, ..., w_k^j$ at iteration j with $\|w_i^j\|_2 = 1, i = 1, ..., k$, compute $w_1^{j+1}, ..., w_k^{j+1}$ by the following two steps:

(a) Cluster Assignment: Assign each point to closest plane P_l. For each $A_i, i = 1, ..., m$, determine $l(i)$ such that

$$|A_i w_{l(i)}^j| = \min_{1 \leq j \leq k} |A_i w_l^j|.$$

(b) Cluster Update: Find a plane P_l that minimizes the sum of the squares of distances to each point in cluster l. For $l = 1, ..., k$, let A_l be the $m(l) \times n$ matrix with rows corresponding to all A_i assigned to cluster l. Define $B(l) = [A(l)]^T A(l)$. Set w_l^{j+1} to be an eigenvector of $B(l)$ corresponding to the smallest eigenvalue of $B(l)$. Stop whenever there is a repeated overall assignment of points to cluster planes or a nondecrease in the overall objective function.

good local (or global) minimum is found, measured by the nearness of the objective function to zero: the sum of the squared distances from the data points to the corresponding clustering hyperplanes should be near to zero.

Degenerate case – sparse instances

The following theorem is useful for identification of very sparse sources. Its proof can be found in [11].

Theorem 2 [11] (Identifiability conditions – locally very sparse representation) *Assume that (i) for each source $s_i := \mathbf{S}(i, .), i = 1, ..., n$ there are $k_i \geq 2$ time instances when all of the source signals are zero except s_i (so each source is uniquely present k_i times), and*

(ii) the set $\left\{ j \in \{1, ..., N\} : \mathbf{X}(., p) = c\mathbf{X}(., j) \text{ for some } c \in \mathbb{R} \right\}$, contains less than $\min_{1 \leq i \leq m} k_i$ elements for any $p \in \{1, ..., N\}$ for which $\mathbf{S}(., p)$ has more that one nonzero element.

Then the matrix $\mathbf{A}$ is identifiable up to permutation and scaling.

Below we include an algorithm for identification of the mixing matrix in the case of Theorem 2.

2.2 Identification of sources

Theorem 3 [12] (Uniqueness of sparse representation) *Let $\mathcal{H}$ be the set of all $\mathbf{x} \in \mathbb{R}^m$ such that the linear system $\mathbf{As} = \mathbf{x}, \mathbf{A} \in \mathbb{R}^{m \times n}$, has a solution with at least $n - m + 1$ zero components. If $\mathbf{A}$ fulfills A1), then there*

Algorithm 3: identification of the mixing matrix in the very sparse case

1) Remove all zero columns of $\mathbf{X}$ (if any) and obtain a matrix $\mathbf{X}_1 \in \mathbb{R}^{m \times N_1}$.

2) Normalize the columns $\mathbf{x}_i, i = 1, \ldots, N_1$ of $\mathbf{X}_1 : \mathbf{y}_i = \mathbf{x}_i/\|\mathbf{x}_i\|$ and set $\varepsilon > 0$.

Multiply each column $\mathbf{y}_i$ by -1 if the first element of $\mathbf{y}_i$ is negative.

3) Cluster $\mathbf{y}_i, i = 1, ..., N_1$ in $n = 1$ groups $G_1, ..., G_{n+1}$ such that for any $i = 1, ..., n$, $\|\mathbf{x} - \mathbf{y}\| < \varepsilon, \forall \mathbf{x}, \mathbf{y} \in G_i$ and $\|\mathbf{x} - \mathbf{y}\| \geq \varepsilon$ for any $\mathbf{x}, \mathbf{y}$ belonging to different groups.

4) Chose any $\mathbf{y}_i \in G_i$ and put $\mathbf{a}_i = \mathbf{y}_i$. The matrix $\mathbf{A}$ with columns $\{\mathbf{a}_i\}_{i=1}^n$ is an estimation of the mixing matrix, up to permutation and scaling.

exists a subset $\mathcal{H}_0 \subset \mathcal{H}$ with measure zero with respect to $\mathcal{H}$, such that for every $\mathbf{x} \in \mathcal{H} \setminus \mathcal{H}_0$ this system has no other solution with this property.

From Theorem 3 it follows that the sources are identifiable generically, i.e. up to a set with a measure zero, if they have level of sparseness grater than or equal to $n - m + 1$ (each column of $\mathbf{S}$ has at least $n - m + 1$ zeros) and the mixing matrix is known. Below we present an algorithm, based on the observation in Theorem 3.

Algorithm 4: source recovery algorithm

1. Identify the set of hyperplanes $\mathcal{H}$ produced by taking the linear hull of every subsets of the columns of $\mathbf{A}$ with $m - 1$ elements;

2. Repeat for $k = 1$ to N:

2.1. Identify the space $H \in \mathcal{H}$ containing $\mathbf{x}_k := \mathbf{X}(:, k)$, or, in practical situation with presence of noise, identify the one to which the distance from $\mathbf{x}_i$ is minimal and project $\mathbf{x}_k$ onto H to $\tilde{\mathbf{x}}_k$;

2.2. if H is produced by the linear hull of column vectors $\mathbf{a}_{k_1}, ..., \mathbf{a}_{k_{m-1}}$, then find coefficients $l_{k,j}$ such that

$$\tilde{\mathbf{x}}_k = \sum_{j=1}^{m-1} l_{k,j} \mathbf{a}_{k_j}.$$

These coefficients are uniquely determined if $\tilde{\mathbf{x}}_k$ doesn't belong to the set $\mathcal{H}_0$ with measure zero with respect to $\mathcal{H}$ (see Theorem 3);

2.3. Construct the solution $\mathbf{s}_k = \mathbf{S}(:, k)$: it contains $l_{k,j}$ in the place k_j for $j = 1, ..., m - 1$, the rest of the components are zero.

3 Sparse Component Analysis

In this section we describe sufficient conditions for the existence of solutions to the SCA problem. Note that the conditions are formulated only in terms of the data matrix $\mathbf{X}$. The proof of the following theorem can be found in [12].

Theorem 4 [12] (SCA conditions) *Assume that $m \leq n \leq N$ and the matrix $\mathbf{X} \in \mathbb{R}^{m \times N}$ satisfies the following conditions:*

(i) the columns of $\mathbf{X}$ lie in the union $\mathcal{H}$ of $\binom{n}{m-1}$ different hyperplanes, each column lies in only one such hyperplane, each hyperplane contains at least m columns of $\mathbf{X}$ such that each $m-1$ of them are linearly independent.

(ii) for each $i \in \{1, ..., n\}$ there exist $p = \binom{n-1}{m-2}$ different hyperplanes $\{H_{i,j}\}_{j=1}^{p}$ in $\mathcal{H}$ such that their intersection $L_i = \cap_{k=1}^{p} H_{i,j}$ is one dimensional subspace.

(iii) any m different L_i span the whole $\mathbb{R}^m$.

Then the matrix $\mathbf{X}$ is representable uniquely (up to permutation and scaling of the columns of $\mathbf{A}$ and $\mathbf{S}$) in the form $\mathbf{X} = \mathbf{AS}$, where the matrices $\mathbf{A} \in \mathbb{R}^{m \times n}$ and $\mathbf{S} \in \mathbb{R}^{n \times N}$ satisfy the conditions A1), A2), and A3) respectively.

4 Overdetermined Blind Source Separation

In this section we assume that $m > n$ and the identifiability conditions for the transposed matrix $\mathbf{A}^T$ are satisfied. So we have the model:

$$\mathbf{X}^T = \mathbf{S}^T \mathbf{A}^T, \tag{2}$$

but in order to apply Theorem 1 we select n rows of the matrices $\mathbf{X}^T$ and $\mathbf{S}^T$ (usually the first n, assuming that they (for $\mathbf{S}^T$) are linearly independent: this is true with "probability one", i.e. the matrices without this property form a set with measure zero). Denoting $\mathbf{X}_n = \mathbf{X}(:, 1 : n)$ and $\mathbf{S}_n = \mathbf{S}(:, 1 : n)$, we have

$$\mathbf{X}_n^T = \mathbf{S}_n^T \mathbf{A}^T. \tag{3}$$

By some of the matrix identification algorithms we identify firstly the matrix $\mathbf{S}_n^T$ and then we identify the matrix $\mathbf{A}$: $\mathbf{A} = \mathbf{X}_n \mathbf{S}_n^{-1}$. Now we recover the full matrix $\mathbf{S}$ from (10.1) by $\mathbf{S} = \mathbf{A}^+ \mathbf{X}$, where $\mathbf{A}^+$ means the Moore-Penrose pseudo-inverse of $\mathbf{A}$.

5 Computer simulation examples

5.1 Overdetermined Blind Source Separation – very sparse case

We consider the overdetermined mixture of two artificially created *non-independent* and *non-sparse* sources with 10 samples – see Figure 1. The

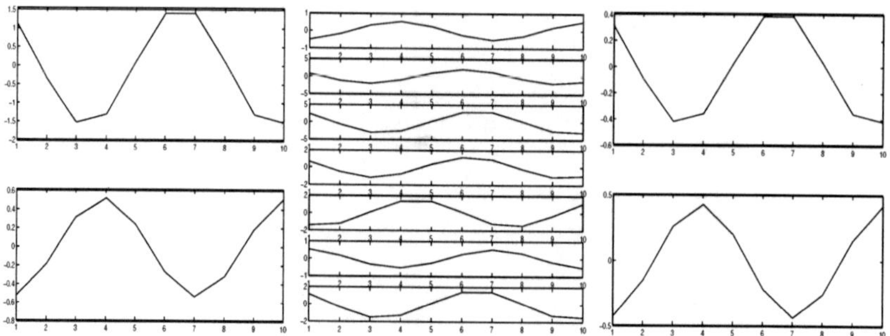

Fig. 1. Example 1. Left:Artificially created *non-independent* and *non-sparse* source signals. Middle: Their mixtures with matrix **A**. Right: Recovered source signals. The signal-to-noise ratio between the original sources and the recoveries is very high with 319 and 319 dB after permutation and normalization.

mixing matrix and the estimated matrix with the overcomplete blind source separation scheme (see section 4) are respectively

$$
\mathbf{A} = \begin{pmatrix} 0 & 1 \\ 2 & 3 \\ 2 & 0 \\ 1 & 1 \\ 1 & 5 \\ 0 & -1 \\ 1 & 0 \end{pmatrix} \quad \text{and} \quad \hat{\mathbf{A}} = \begin{pmatrix} 0 & 1.2 \\ 7.3 & 3.6 \\ 7.3 & 0 \\ 3.6 & 1.2 \\ 3.6 & 6.1 \\ 0 & -1.2 \\ 3.6 & 0 \end{pmatrix}.
$$

The mixtures and estimated sources are shown in Figure 1. In this case we applied Algorithm 3 for identification of the matrix $\mathbf{S}_2^T$ (the transposed of the first two rows of the source matrix $\mathbf{S}$, see (3)). After normalization of each row of $\hat{\mathbf{A}}$ we obtain the original matrix $\mathbf{A}$, which confirms the perfect reconstruction of the sources. The transposed matrix $\mathbf{A}^T$ (considered here as a new source matrix) satisfies the conditions of Theorem 2 and this is the reason for the perfect reconstruction of the sources.

5.2 Overdetermined Blind Source Separation - Sparse Case

Now let us consider the overdetermined mixture of 3 artificially created *non-independent* and *non-sparse* sources with only 10 samples (in fact only *three* are needed, as in the previous example only two were needed) — see Figure 2 (left).

The mixing matrix and the estimated matrix with the overcomplete blind source separation scheme (see section 4) are respectively

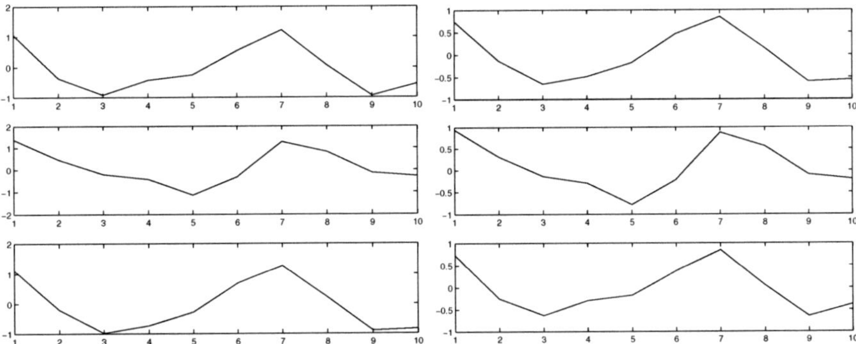

Fig. 2. Example 2. Left: Artificially created *non-independent* and *non-sparse* source signals. Right: Recovered source signals. The signal-to-noise ratio between the original sources and the recoveries is very high with 308, 293 and 307 dB after permutation and normalization.

$$
\mathbf{A} = \begin{pmatrix}
0.5287 & 0.5913 & 0 \\
0.2193 & -0.6436 & 0 \\
-0.9219 & 0.3803 & 0 \\
-2.1707 & 0 & 0.7310 \\
-0.0592 & 0 & 0.5779 \\
-1.0106 & 0 & 0.0403 \\
0 & 0.0000 & 0.6771 \\
0 & -0.3179 & 0.5689 \\
0 & 1.0950 & -0.2556
\end{pmatrix} , \quad
\hat{\mathbf{A}} = \begin{pmatrix}
0.0000 & 0.8631 & 0.7667 \\
-0.0000 & -0.9395 & 0.3180 \\
-0.0000 & 0.5552 & -1.3368 \\
1.0972 & -0.0000 & -3.1476 \\
0.8674 & -0.0000 & -0.0858 \\
0.0605 & 0.0000 & -1.4655 \\
1.0164 & 0.0001 & -0.0000 \\
0.8540 & -0.4640 & -0.0000 \\
-0.3837 & 1.5984 & 0.0000
\end{pmatrix} .
$$

Now we apply Algorithm 1 – note that only 9 samples are required by the identifiability theorem – Theorem 1 (due to condition A3)), and $\mathbf{A}^T$ has precisely 9 rows. The mixtures are shown in Figure 3, along with a scatter plot for a visualization of the matrix detection in this transposed case with the *very* low sample number of only 9, which is sufficient for a perfect recovery of (transposed) mixing matrix and the original sources (estimated sources are shown in Fig. 2 right).

5.3 Complete case

In this example for the complete case ($m = n$) of instantaneous mixtures, we demonstrate the effectiveness of our algorithm for identification of the mixing matrix in the case considered in Theorem 2. We mixed 3 images of landscapes (shown in Fig. 4) with a 3-dimensional randomly generated matrix $\mathbf{A}$ ($\det\mathbf{A} = 0.0016$). We transformed these three mixtures (shown in Fig. 5) by two dimensional discrete Haar wavelet transform and took only the 10-th row (160 points) of the obtained diagonal coefficients $cD\mathbf{X}$. As a result, since this transform is linear, the corresponding diagonal wavelet coefficients $cD\mathbf{S}$ of the source matrix $\mathbf{S}$ represented by the source images (as well as the

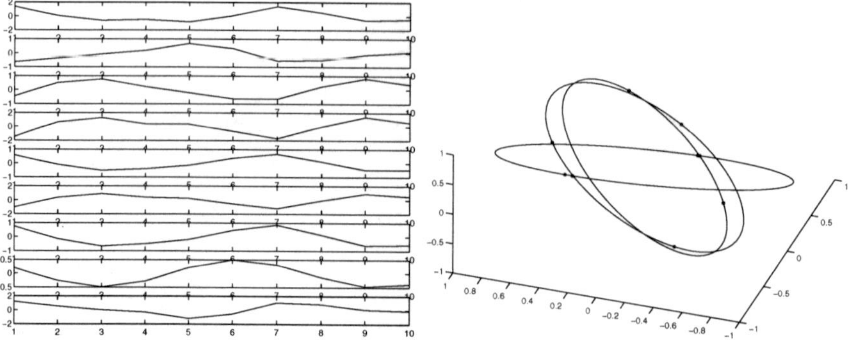

Fig. 3. Example 2. Left: mixed signals $\mathbf{X}$ (observed sources). Right: Scatterplot of the new 'observed sources' $\mathbf{X}_3^T$ (after transposition of $\mathbf{X}_3$ – the first 3 data samples) together with the hyperplanes on which they lie, indicated by their intersections with the unit sphere (circles).

horizontal and vertical ones) become very sparse (see Fig. 7) and they satisfy the conditions of Theorem 2. Using only one row (the 10-th or any other, with 160 points) of $cD\mathbf{X}$ appears to be enough to estimate very precisely the mixing matrix, and therefore, the original images. The estimated images are shown in Fig. 6.

Fig. 4. Original images

Fig. 5. Mixed (observed) images

Fig. 6. Estimated normalized images using the estimated matrix. The signal-to-noise ratios with the sources from Figure 1 are 232, 239 and 228 dB respectively.

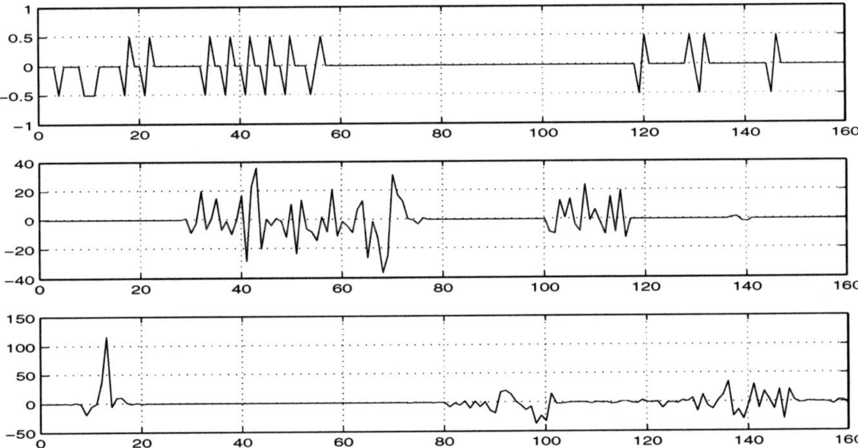

Fig. 7. Diagonal wavelet coefficients of the original images (displaying only the 10-th row of each of the three (120 × 160) matrixes). They satisfy the conditions of Theorem 1 and this is the reason for the perfect reconstruction of the original images, since our algorithm uses only the tenth row of each of the mixed images.

5.4 Underdetermined case

We consider a mixture of 7 artificially created sources (see Fig. 9 left) – sparsified randomly generated signals with at least 5 zeros in each column – with a randomly generated mixing matrix with dimension 3 × 7.

Figure 8 gives the mixed signals together with a normalized scatterplot of the mixtures – the data lies in $21 = \binom{7}{2}$ hyperplanes.

Applying the underdetermined matrix recovery algorithm (Algorithm 1) to the mixtures gives the recovered mixing matrix exactly, up to permutation and scaling. Applying the source recovery algorithm (Algorithm 4) we recover the source signals up to permutation and scaling (see Fig. 9, middle). This figure (right) shows also that the recovery by l_1-norm minimization (known as Basis Pursuit method of S. Chen, D. Donoho and M. Saunders [7]) does not perform well, even if the mixing matrix is perfectly known.

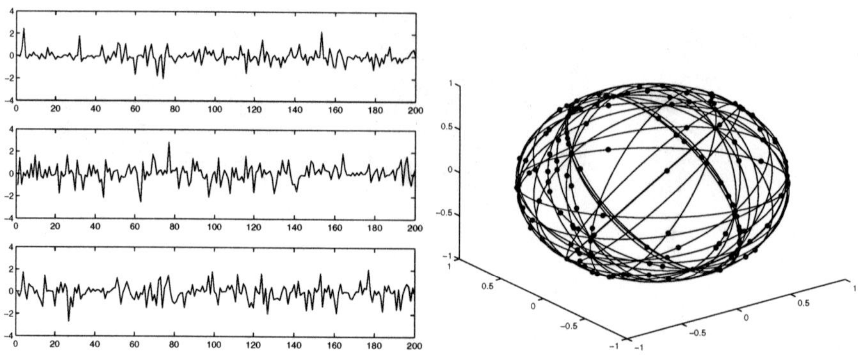

Fig. 8. Mixed signals (left) and normalized scatter plot (density) of the mixtures (right) together with the 21 data set hyperplanes, visualized by their intersection with the unit sphere in $\mathbb{R}^3$.

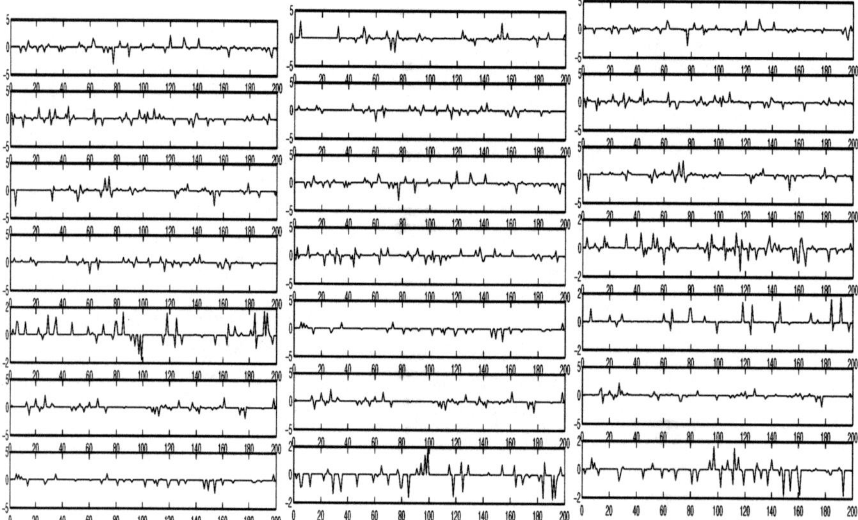

Fig. 9. The original source signals are shown in the left column. The middle column gives the recovered source signals — the signal-to-noise ratio between the original sources and the recoveries is very high (above 278 dB after permutation and normalization). Note that only 200 samples are enough for excellent separation. The right column shows the recovered source signals using l_1-norm minimization and known mixing matrix. Simple comparison confirms that the recovered signals are far from the original ones – the signal-to-noise ratio is only around 4 dB.

6 Extraction of auditory evoked potentials from EEG contaminated by eye movements by Sparse Component Analysis

6.1 Introduction

Ocular artifact contamination is very common in electroencephalographic (EEG) recordings. The electro-oculographic (EOG) signals are generated by the horizontal movement of the eyes, which act as charged electric dipoles with the positive poles at the cornea and the negative poles at the retina. These electric charges of the movement are picked up by frontal EEG electrodes. The EOG contamination is normally dealt with by instructing the subjects not to blink and not to move the eyes during an EEG experiment, as well as by trying to reject the affected data using voltage threshold criteria. Both of these measures leave a lot to be desired, because cognitive commands to subjects may introduce additional complexity, while at the same time very slow eye movements are difficult to identify only by voltage thresholding because their amplitudes may be comparable to those of the underlying electroencephalogram. Recent studies have proposed artifact removal procedures based on estimation of correction coefficients [8] and independent component analysis [13, 19, 14, 18, 20], etc. The goal of the present section is to demonstrate that the new Sparse Component Analysis (SCA) method extracts efficiently for further usage the underlying evoked auditory potentials masked by strong eye movements.

6.2 Methods

The electric potentials on the surface of the scalp of human subjects were measured with a geodesic sensor net using a 256-channel electroencephalographic (EEG) system (Electrical Geodesics Inc., Eugene, Oregon, USA). An on-screen pattern image was presented for scanning 20 times. During each presentation the subject had to scan 4 lines - two horizontal and two vertical. A button was pressed by the subject immediately before a line scan and another button - signaling that the line scan was completed. A 1000 Hz, 100ms, 100 dB sound accompanied the pattern image each time after the start button was activated. An eye tracking device (EyeGaze , LC Technologies, Inc.) was used for precision recording and control of all eye movements during the EEG experiments, scanning the subjects' screen gaze coordinates 60 times per second.

After aligning 20 single epochs of each eye movement type (horizontal left-to-right, diagonal down-left, horizontal right-to-left, diagonal up-left) with their corresponding sound stimulus onset times, EEG data was segmented into trials of 500ms lengths and averaged separately for each line type. We then preprocessed the segmented contaminated data by reducing its dimensionality from 256 to 4 using principal component analysis (PCA). A justification for such a reduction is shown in Fig. 11 which shows that the subspace spanned by the principal components corresponding to the biggest 4 singular values contain most of the information of the data. The new Sparse Component Analysis method was applied on this new data set and its performance was compared to basic ICA algorithms: Fast ICA algorithm and JADE algorithm (see [15] for reference about ICA methods and algorithms).

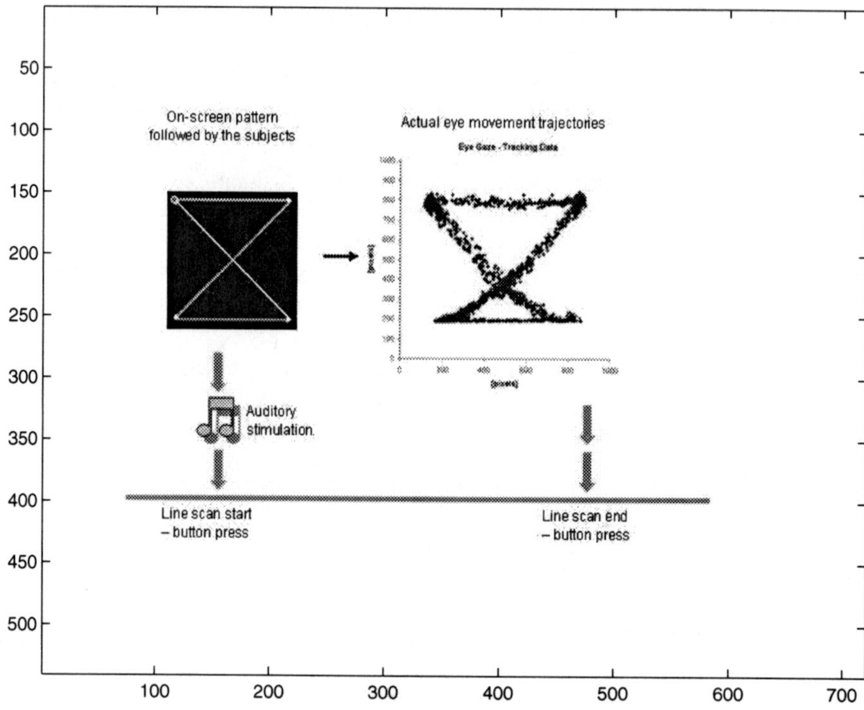

Fig. 10. Experimental design to extract auditory evoked potentials from high-density EEG data contaminated by eye movements. Ocular artifacts were controlled by an eye tracking device.

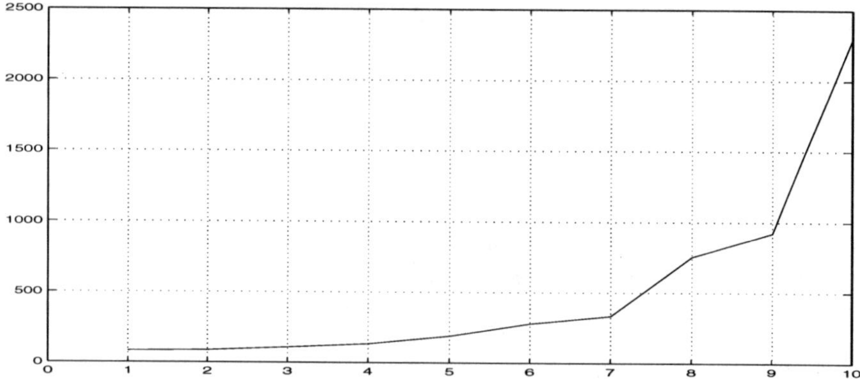

Fig. 11. The biggest 10 singular values of the data matrix from 256 channels EEG machine

6.3 Results

In this experiment we applied Algorithm 2 for matrix identification (using several re-initializations, until obtaining satisfactory local (or global) minimum of the cost function: the sum of the squared distances from the data points to the corresponding clustering hyperplanes should be small. For source recovery we apply either Algorithm 4, or inversion of the estimated matrix: the results are similar, and as in the the inversion matrix method the resulting signals are slightly more smooth.

The component related to the evoked auditory N1 potential [23] with a peak amplitude at 76-124ms [24] was identified (with some differences) by all applied algorithms. However, SCA algorithm gives the best result (Fig. 1 right, 4-th component) which correspond to the reality of the experiment, i.e. the auditory stimulus was silent after 150 ms. The Fast ICA and JADE algorithms (Fig. 2) show nonzero activity in the auditory component after 150 ms, which is false. The eye movements, however, were strongly mixed and masked, so that the various algorithms presented different performance capabilities. SCA's component 1 (Fig. 1 right) corresponded to the steady and continuous horizontal eye movement of the subject from the left side of the image to right side. The initial plateau (0–170 ms) was due to the subjective delay before the subject was able to start the actual movement. FastICA and JADE (Fig. 2, 3-rd left and 1-st right components respectively) were unable to reveal fully the underlying continuous potentials resulting from the eye movement. SCA component 3 was slightly different at 250-300 ms, but overall similar to component 4, which could have indicated that the real number of strong sources was 3 and this component was redundant. However, if that was not the case, then both this component, as well as SCA component 2 were either of eye movement origin and had been caused by acceleration jolt in response to the sound startle effect, or were related to the button press motor response potentials in cortex.

In order to verify or reject the hypothesis that the real number of strong sources was 3 (and SCA component 3 in Fig. 1, right, was redundant), we performed similar processing with just 3 input signals extracted by PCA signal reduction. The results are shown in Fig. 4 (right) and Fig. 5. Again, the auditory response was mixed with the eye movement potentials in the input data (Fig. 4 left) and all three algorithms were able to obtain the N1 evoked potential - SCA (Fig. 4 right, 3-rd component), FastICA and JADE (Fig. 5, 2-nd and 1-st components respectively), as those found by SCA is minimally deviated from zero in the period 150-500 ms. However, the steady eye movement ramp was most difficult to extract by the ICA methods FastICA (Fig. 5), while SCA (Fig. 4 right, 3-rd component) revealed again a clear basic trend potential without overlapping peaks. SCA component 2 (Fig. 4 right) was represented in a varying degree also by the ICA algorithms.

Our SCA method exhibited a best fit for the hypothesis with 3 sources of electrical potentials in the mixed auditory and eye movement data. Nev-

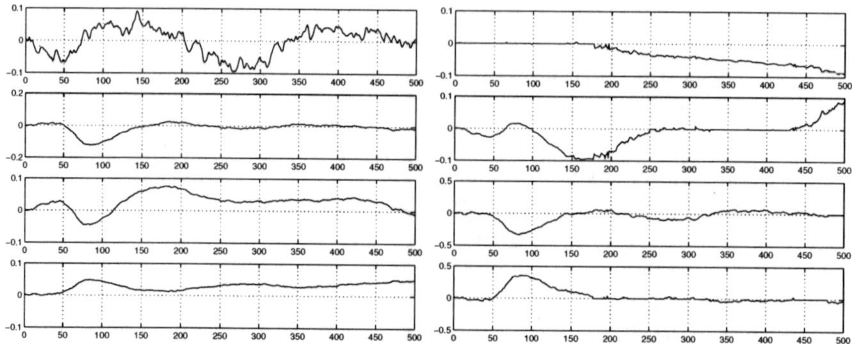

Fig. 12. Left: Input EEG data with dimensionality reduced from 256 channels to 4 principal components. This data was not sufficiently separated and still contained mixed information about the original cortical and eye dipole sources. Right: Sparse Component Analysis (SCA) results.

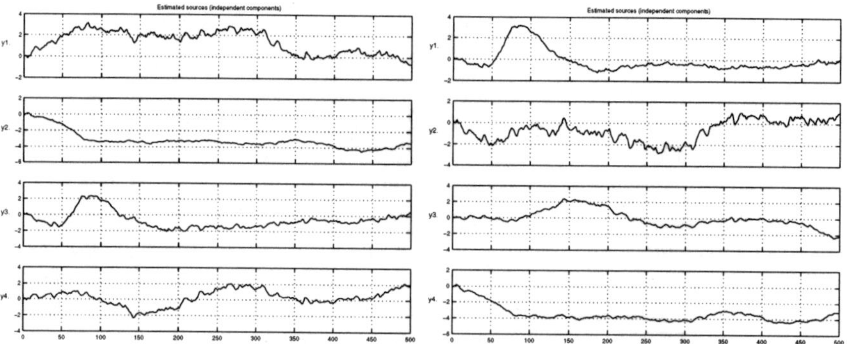

Fig. 13. Left: FastICA results. Right: JADE results.

ertheless, additional experiments may be needed to better reveal the rather complex structure of the eye movement signal.

7 Applications of Sparse Component Analysis to fMRI data

7.1 SCA applied to fMRI toy data

We simulated a low-dimensional example of fMRI data analysis. The typical setup of fMRI experiments is the following: NMR brain imaging techniques are used to record brain activity data over a certain span of time, during which the subject is asked to perform some kind of task (e.g. 5 seconds of activity in the motor cortex followed by 5 seconds of activity in the visual cortex; this

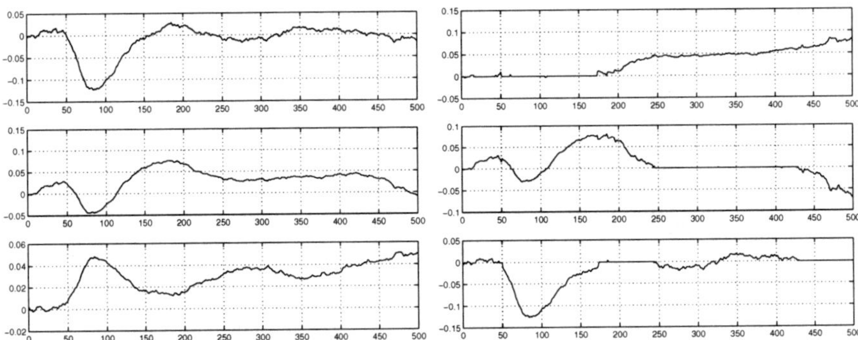

Fig. 14. Left: Input EEG data with dimensionality reduced from 256 channels to 4 principal components. This data was not sufficiently separated and still contained mixed information about the original cortical and eye dipole sources. Right: Sparse Component Analysis (SCA) results.

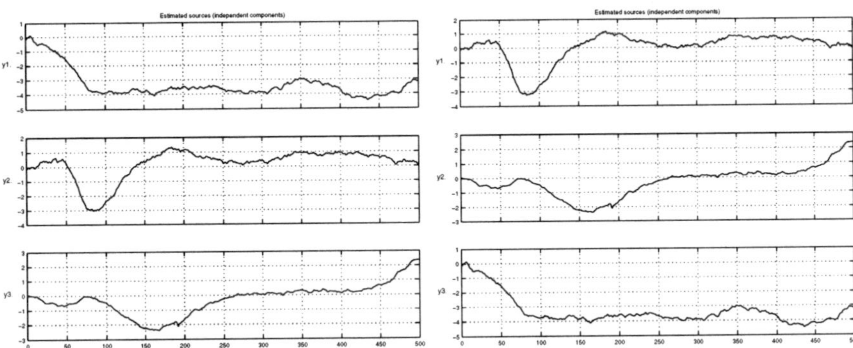

Fig. 15. Left: FastICA results. Right: JADE results.

iterative procedure is often called *block diagram*). The brain recordings show areas of high and of low brain activity (using the *BOLD effect*). Analysis is performed on the 2d-image slices recorded at the discrete time steps. General linear model (GLM) approaches or ICA-based fMRI analysis then decompose this data set into a certain set of *component maps* i.e. sets of (hopefully independent) images that are active at certain time steps corresponding to the block diagram.

In the following we simulate a low-dimensional example of such brain activity recordings. For this we mix three 'source component maps' (Fig. 16) linearly to three mixture images and add some noise.

These mixtures represent our recordings at three different time steps. From the recordings we want to recover the original components or component maps. We want to use an unsupervised approach (not GLM, which requires additional knowledge of the mixing system) but with a different contrast than

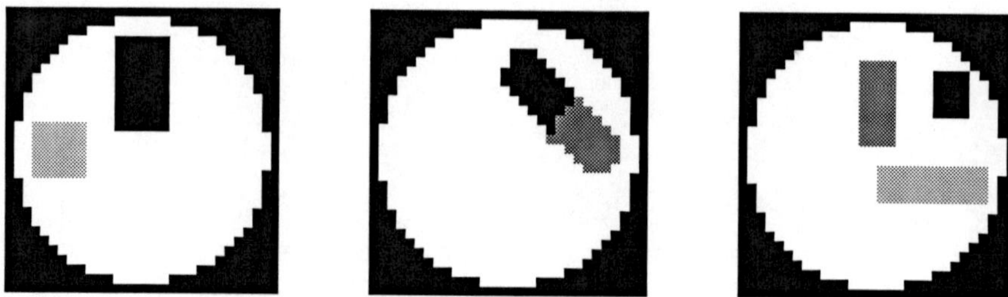

Fig. 16. Example: artificial *non-independent* and *non-sparse* source signals.

ICA. We believe that the assumption of *independence* of the component maps does not hold in a lot of situations, so we replace this assumption by *sparseness* of the maps, meaning that at a certain voxel, not all maps are allowed to be active (in the case of as many mixtures as sources).

We consider a mixture of 3 artificially created *non-independent* source images of size 30×30 — see Figure 16 — with the (normalized) mixing matrix

$$\mathbf{A} = \begin{pmatrix} -0.9069 & 0.1577 & 0.4726 \\ -0.2737 & -0.9564 & 0.0225 \\ -0.3204 & -0.2458 & -0.8810 \end{pmatrix}$$

and 4% of additive white noise. The mixtures are shown in Figure 17 together with their scatterplot after normalization to unit length.

Note that due to the circular 'brain region', we have to preprocess the data ('sparsification') by removing the non-brain voxels from the boundary. Then, we apply the matrix identification algorithm (Algorithm 1). This gives the recovered matrix (after normalization)

$$\hat{\mathbf{A}} = \begin{pmatrix} 0.9110 & 0.1660 & 0.4693 \\ 0.2823 & -0.9541 & 0.0135 \\ 0.3007 & -0.2494 & -0.8829 \end{pmatrix}$$

with low crosstalking error 0.12 and the recovered sources $\hat{\mathbf{S}}$ shown in Figure 18, with high signal-to-noise ratio of 28, 27 and 27 dB with respect to the original sources (after permutation and normalization).

This can be enhanced by applying a denoising algorithm to each image. Figure 19 shows the application of local PCA denoising with an MDL-parameter estimation criterion, which gives SNRs of 32, 31 and 29 dB, so a mean enhancement of around 4 dB has been achieved.

Note that if we apply ICA to the previous example (after sparsification as above — without sparsification ICA performs even worse), the algorithm cannot recover the mixing matrix

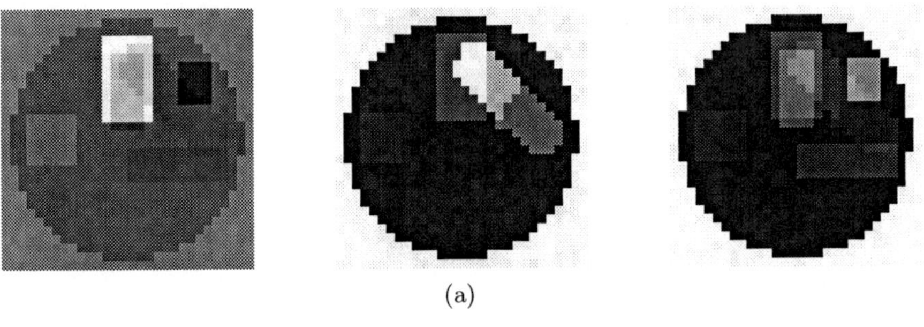

(a)

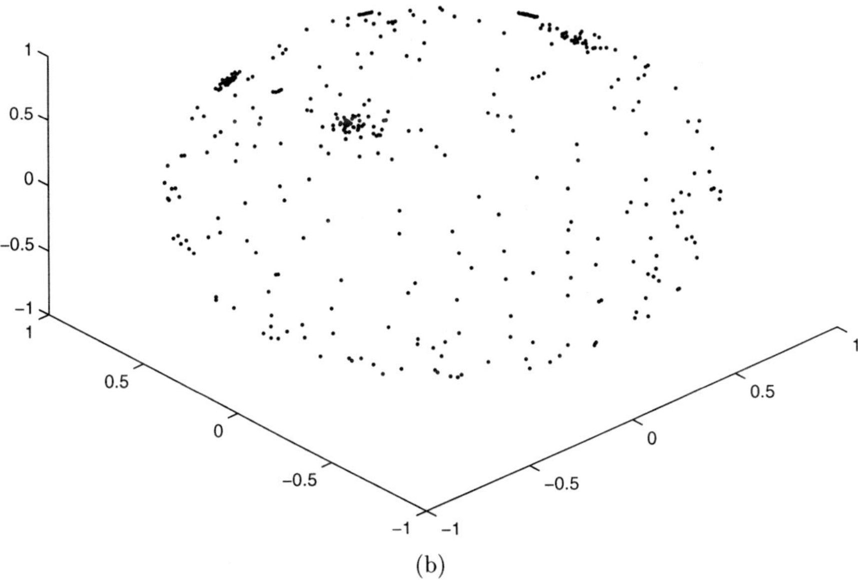

(b)

Fig. 17. Example: mixed signals with 4% additive noise (a), and scatterplot after normalization to unit length (b).

$$\bar{\mathbf{A}} = \begin{pmatrix} 0.6319 & -0.3212 & 0.8094 \\ -0.0080 & -0.8108 & -0.3138 \\ -0.7750 & -0.4893 & 0.4964 \end{pmatrix}$$

and has a very high crosstalking error of 4.7 with respect to $\mathbf{A}$. Figure 20 shows the poorly recovered sources; the SNRs with respect to the sources are only 3.3, 13 and 12 dB respectively. The reason for ICA not being able to

 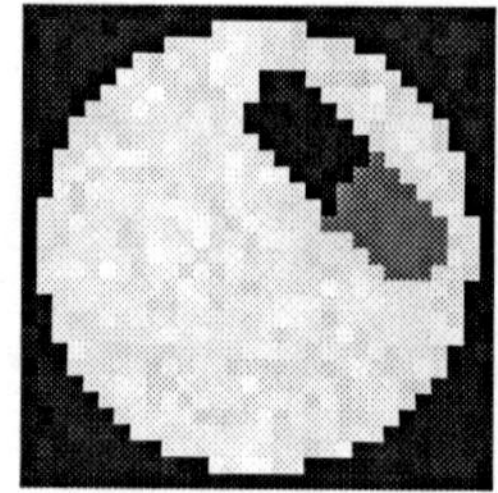 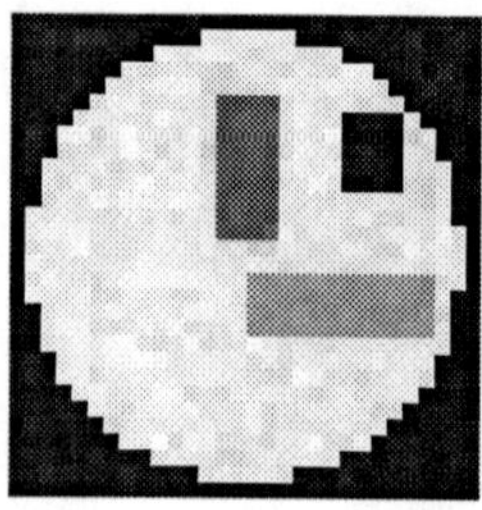

Fig. 18. Example: recovered source signals. The signal-to-noise ratio between the original sources (figure 16) and the recoveries is high with 28, 27 and 27 dB after permutation and normalization.

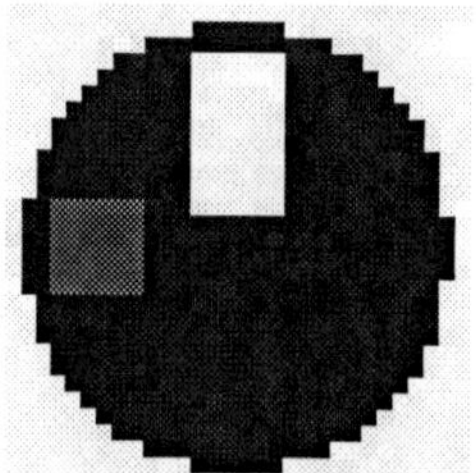 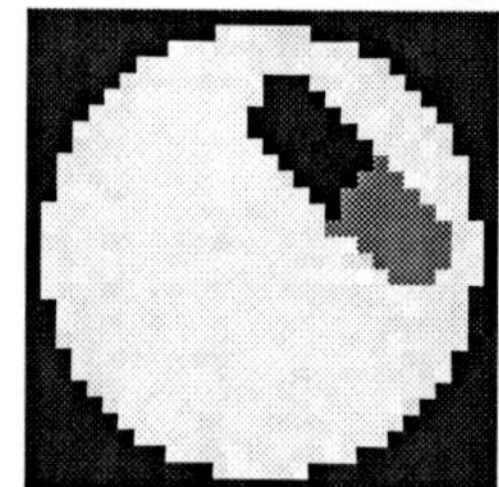 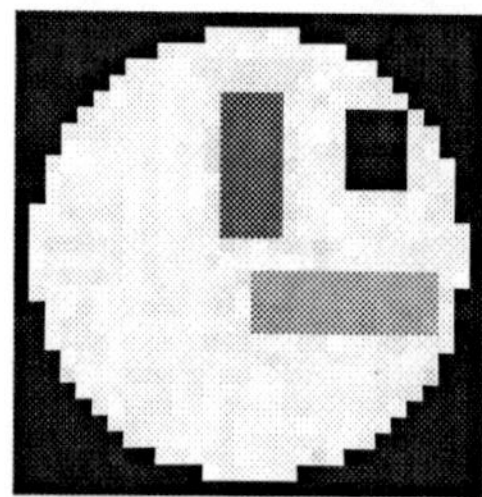

Fig. 19. Example: recovered denoised source signals. Now the SNR is even higher than in figure 18 (32, 31 and 29 dB after permutation and normalization).

recover the sources simply lies in the fact that they were not chosen to be independent.

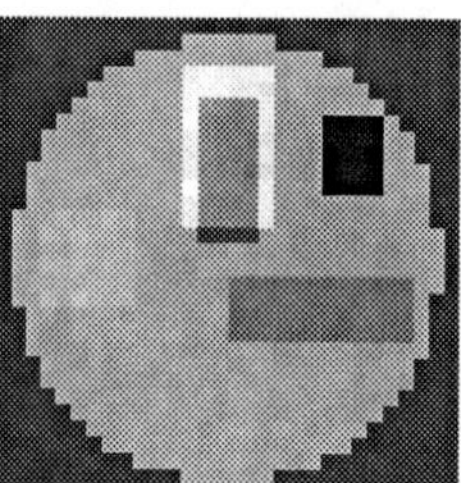 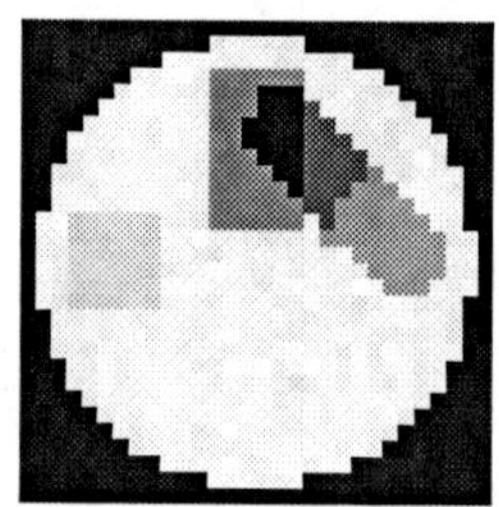 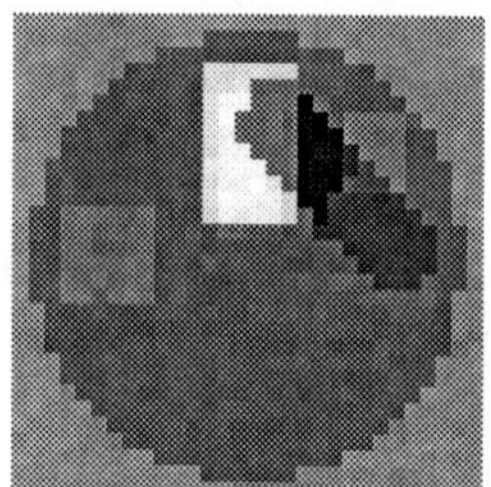

Fig. 20. Example: poorly recovered source signals using ICA. The signal-to-noise ratio between the original sources (figure 16) and the recoveries is very low with 3.3, 13 and 12 dB after permutation and normalization.

7.2 SCA applied to real fMRI data

We now analyze the performance of SCA when applied to real fMRI measurements. fMRI data were recorded from six subjects (3 female, 3 male, age 20–37) performing a visual task. In five subjects, five slices with 100 images (TR/TE = 3000/60 msec) were acquired with five periods of rest and five photic simulation periods with rest. Simulation and rest periods comprised 10 repetitions each, i.e. 30s. Resolution was $3 \times 3 \times 4$ mm. The slices were oriented parallel to the calcarine fissure. Photic stimulation was performed using an 8 Hz alternating checkerboard stimulus with a central fixation point and a dark background with a central fixation point during the control periods [27]. The first scans were discarded for remaining saturation effects. Motion artifacts were compensated by automatic image alignment (AIR, [28]).

Blind Signal Separation, mainly based on ICA, nowadays is a quite common tool in fMRI analysis (see for example [21, 22]). Here, we analyze the fMRI data set using as a separation criterion a spatial decomposition of fMRI data images to sparse component maps. Such an approach we consider as very reasonable and advantageous when the stimuli are sparse and dependent, and therefore the ICA methods couldn't give good results. Due to the availability of fMRI data, it appears that the results of our SCA method and ICA method give similar results, which itself we consider as a surprising fact. Here we use again Algorithm 2 for matrix identification and Algorithm 4 or matrix inversion of the estimated matrix, for estimation of the sources.

Figure 21 shows the performance of SCA method; see figure caption for interpretation. Using only the first 5 principal components, SCA could recover the stimulus component as well as detect additional components. It performs equally well as fastICA, Figure 22, which is interesting in itself: apparently the two different criteria, sparseness and independence, lead to similar results in this setting. This can be partially explained by noting that all components, mainly the stimulus component, have high kurtoses i.e. strongly peaked densities.

8 Conclusion

We rigorously defined the SCA and BSS problems of sparse signals and presented sufficient conditions for their solution. We presented four algorithms applicable to SCA: one for source recovery and three ones for identification of the mixing matrix – for the sparse and the very sparse cases and one based on a simplified Bradley-Mangasarian's k-plane clustering algorithm. We presented several experiments for confirmation of our methods, including applications in fMRI and EEG data sets.

Although it is a standard practice to cut those evoked-potentials in EEG data which are contaminated by eye movement artifacts, we have demonstrated that stimulus-related responses could be recovered successfully and

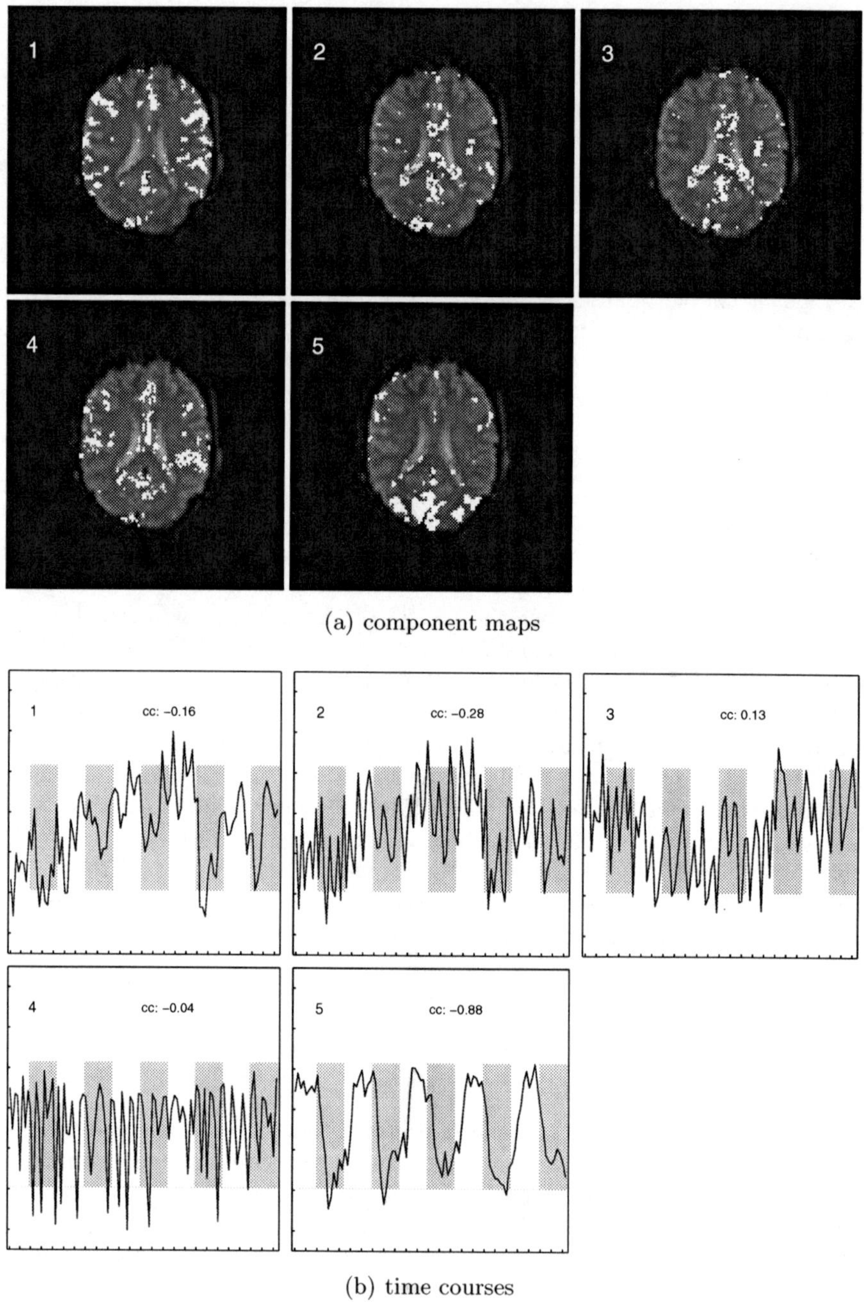

(a) component maps

(b) time courses

Fig. 21. SCA fMRI analysis. The data was reduced to the first 5 principal components. (a) shows the recovered component maps (white points indicate values stronger than 3 standard deviations), and (b) their time courses. The stimulus component is given in component 5 (indicated by the high crosscorrelation $cc = -0.86$ with the stimulus time course, delayed by roughly 2 seconds due to the BOLD effect), which is strongly active in the visual cortex as expected.

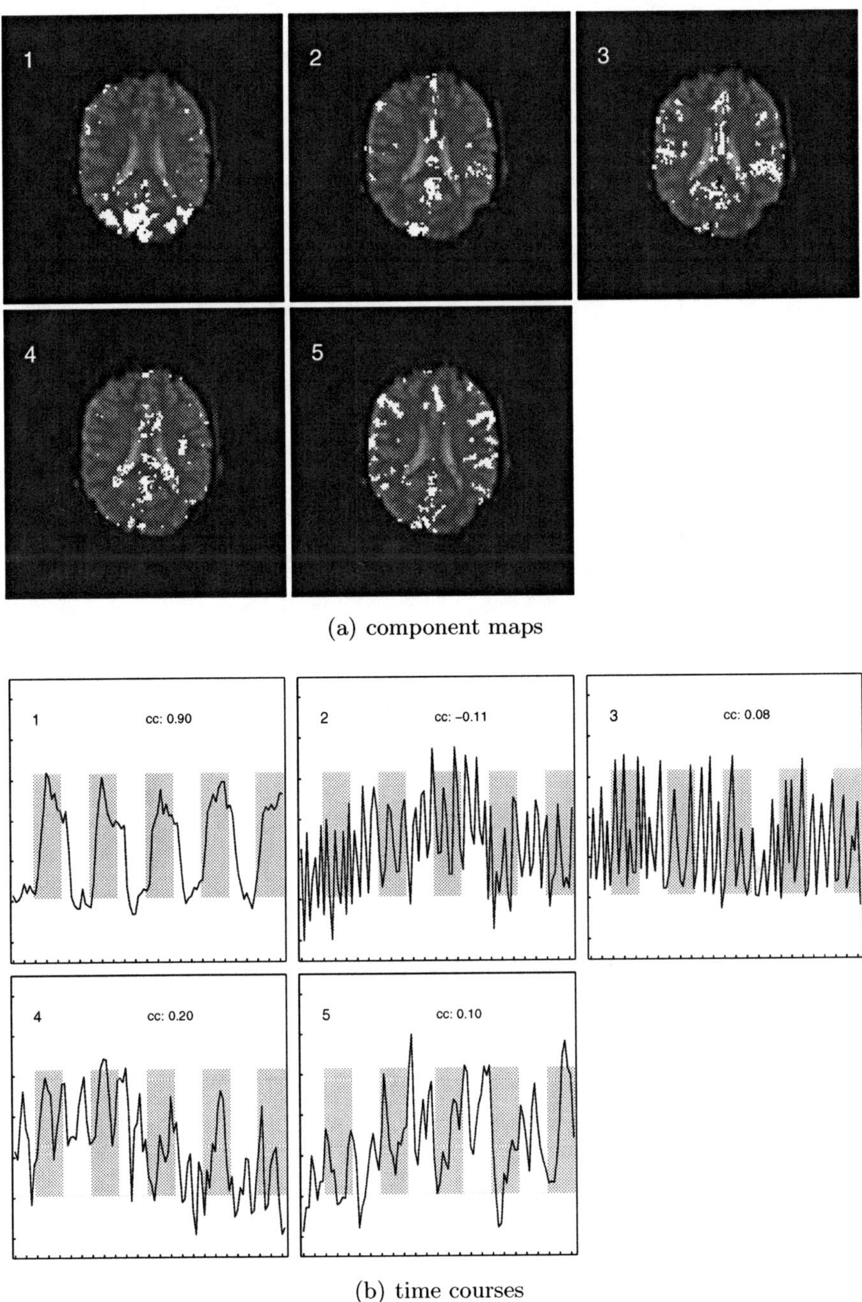

(a) component maps

(b) time courses

Fig. 22. FastICA result during fMRI analysis of the same data set as in figure 21. The stimulus component is given in component 1 with high stimulus cross-correlation $cc = 0.90$.

even better by the Sparse Component Analysis method. In addition, SCA has revealed a complex hidden structure of the dynamically accelerating eye movement signal, which could become a future basis for a new instrument to measure objectively individual psychological characteristics of a human subject in startle reflex-type experiments, exploiting sparseness of the signals rather than independence. We have also shown that our new method is a useful tool in separating the functional EEG components more efficiently in signal hyperspace than independent component analysis. Very promising are the results with real fMRI data images, which show that revealing the brain responses of sparse (and may be dependent) stimuli could be more successful by SCA than by ICA.

Acknowledgements

The authors would like to thank Dr. Dorothee Auer from the Max Planck Institute of Psychiatry in Munich, Germany, for providing the fMRI data, and Oliver Lange from the Department of Clinical Radiology, Ludwig-Maximilian University, Munich, Germany, for data preprocessing of fMRI data and visualization.

References

1. P. Bofill and M. Zibulevsky. Underdetermined Blind Source Separation using Sparse Representation. *Signal Processing*, 81(11): 2353-2362, 2001.

2. P.S. Bradley and O. L. Mangasarian. k-Plane Clustering. *Journal of Global Optimization*, 16(1): 23-32, 2000.

3. P.S. Bradley, U. M. Fayyad and O. L. Mangasarian. Mathematical progrmming for data mining: formulations and challenges. *INFORMS Journal on Computing*, 11(3): 217–238, 1999.

4. A.M. Bronstein, M.M. Bronstein, M. Zibulevsky, and Y. Y. Zeevi. Blind Separation of Reflections using Sparse ICA. In *Proceeding of the International Conference ICA2003*, Nara, Japan, pp.227-232, 2003.

5. A. Cichocki and S. Amari. *Adaptive Blind Signal and Image Processing*. John Wiley, Chichester, 2002.

6. A. Cichocki, S. Amari, and K. Siwek. ICALAB for signal processing package. http://www.bsp.brain.riken.jp.

7. S. Chen, D. Donoho, and M. Saunders. Atomic decomposition by basis pursuit. *SIAM Journal on Scientific Computing*, 20(1): 33–61, 1998.

8. R.J. Croft and R.J. Barry. Removal of ocular artifact from the EEG: a review. *Clinical Neurophysiology*, 30(1): 5-19, 2000.

9. D. Donoho and M. Elad. Optimally sparse representation in general (nonorthogonal) dictionaries via l^1 minimization. *Proceedings of the National Academy of Sciences*, 100(5): 2197–2202, 2003.

10. D. Donoho and V. Stodden, When Does Non-Negative Matrix Factorization Give a Correct Decomposition into Parts? Neural Information Processing Systems (NIPS) 2003 Conference, http://books.nips.cc, 2003.

11. P. G. Georgiev and A. Cichocki, Sparse component analysis of overcomplete mixtures by improved basis pursuit method. Accepted in 2004 IEEE International Symposium on Circuits and Systems (ISCAS 2004).

12. P.G. Georgiev, F. Theis and A. Cichocki, Blind Source Separation and Sparse Component Analysis of overcomplete mixtures. Accepted in ICASSP 2004 (International Conference on Acoustics and Statistical Signal Processing).

13. P. Georgiev, A. Cichocki, and H. Bakardjian. Optimization Techniques for Independent Component Analysis with Applications to EEG Data. In P.M. Pardalos et al., editors, *Quantitative Neuroscience: Models, Algorithms, Diagnostics, and Therapeutic Applications*, pages 53-68. Kluwer Academic Publishers, 2004.

14. I. Gorodnitsky and A. Belouchrani. Joint cumulant and correlation based signal separation with application of EEG data analysis. In Proceedings of the 3rd International Conference on Independent Component Analysis and Signal Separation, San Diego, California, Dec. 9-13, pp. 475-480, 2001.

15. A. Hyvärinen, J. Karhunen, and E. Oja. *Independent Component Analysis*, John Wiley & Sons, 2001.

16. T.-W. Lee, M.S. Lewicki, M. Girolami, and T.J. Sejnowski. Blind sourse separation of more sources than mixtures using overcomplete representaitons. *IEEE Signal Processing Letters*, 6(4): 87–90, 1999.

17. D. D. Lee and H. S. Seung. Learning the parts of objects by non-negative Matrix Factorization. *Nature*, 40: 788–791, 1999.

18. J. Iriarte, E. Urrestarazu, M. Valencia, M. Alegre, A. Malanda, C. Viteri, and J. Artieda. Independent component analysis as a tool to eliminate artifacts in EEG: a quantitative study. *Journal of Clinical Neurophysiology*, 20(4): 249-257, 2003.

19. T.P. Jung, S. Makeig, M. Westerfield, J. Townsend, E. Courchesne, and T.J. Sejnowski. Analysis and visualization of single-trial event-related potentials. *Human Brain Mapping*, 14(3): 166-85, 2001.

20. S. Krieger, J. Timmer, S. Lis, and H.M. Olbrich. Some considerations on estimating event-related brain signals. *Journal of Neural Transmission, General Section*, 99: 103–129, 1995.

21. M. McKewon, T. Jung, S. Makeig, G. Brown, S. Kindermann, A. Bell, and T. Sejnowski. Analysis of fMRI data by blind separation into independent spatial components. *Human Brain Mapping*, 6: 160–188, 1998.

22. M. McKeown, L. Hansen, and T. Sejnowski. Independent component analysis of functional MRI: what is signal and what is noise? *Current Opinion in Neurobiology*, 13: 620-629, 2003.

23. R. Naatanen and T. Picton. The N1 wave of the human electric and magnetic response to sound: a review and an analysis of the component structure. *Psychophysiology*, 24(4): 375-425, 1987.

24. G.F. Potts, J. Dien, A.L. Hartry-Speiser, L.M. McDougal, and D.M. Tucker. Dense sensor array topography of the event-related potential to task-relevant auditory stimuli. *Electroencephalography and Clinical Neurophysiology*, 106(5): 444-456, 1998.

25. F.J. Theis, E.W. Lang, and C.G. Puntonet. A geometric algorithm for overcomplete linear ICA. *Neurocomputing*, 56: 381-398, 2004.

26. K. Waheed and F. Salem. Algebraic Overcomplete Independent Component Analysis. In *Proceeding of the International Conference ICA2003*, Nara, Japan, pp. 1077–1082, 2003.

27. A. Wismüller, O. Lange, D. Dersch, G. Leinsinger, K. Hahn, B. Pütz, and D. Auer. Cluster Analysis of Biomedical Image Time–Series. *International Journal on Computer Vision*, 46(2): 102–128, 2002.

28. R. Woods, S. Cherry, and J. Mazziotta. Rapid automated algorithm for aligning and reslicing PET images. *Journal of Computer Assisted Tomography*, 16(8): 620–633, 1992.

29. M. Zibulevsky and B. A. Pearlmutter. Blind source separation by sparse decomposition in a signal dictionary. *Neural Computation*, 13(4): 863–882, 2001.

Data Mining Via Entropy and Graph Clustering

Anthony Okafor, Panos Pardalos, and Michelle Ragle

Department of Industrial and Systems Engineering
University of Florida,
Gainesville, FL, 32611

Summary. Data analysis often requires the unsupervised partitioning of the data set into clusters. Clustering data is an important but a difficult problem. In the absence of prior knowledge about the shape of the clusters, similarity measures for a clustering technique are hard to specify. In this work, we propose a framework that learns from the structure of the data. Learning is accomplished by applying the K-means algorithm multiple times with varying initial centers on the data via entropy minimization. The result is an expected number of clusters and a new similarity measure matrix that gives the proportion of occurrence between each pair of patterns. Using the expected number of clusters, final clustering of data is obtained by clustering a sparse graph of this matrix.

Key words: K-means clustering, Entropy, Bayesian inference, Maximum spanning tree, Graph Clustering.

1 Introduction

Data clustering and classification analysis is an important tool in statistical analysis. Clustering techniques find applications in many areas including pattern recognition and pattern classification, data mining and knowledge discovery, data compression and vector quantization. Data clustering is a difficult problem that often requires the unsupervised partitioning of the data set into clusters. In the absence of prior knowledge about the shape of the clusters, similarity measures for a clustering technique are hard to specify. The quality of a good cluster is application dependent since there are many methods for finding clusters subject to various criteria which are both ad hoc and systematic [9].

Another difficulty in using unsupervised methods is the need for input parameters. Many algorithms, especially the K-means and other hierarchical methods [7] require that the initial number of clusters be specified. Several authors have proposed methods that automatically determine the number of clusters in the data [5, 10, 6]. These methods use some form of cluster validity measures like variance, *a priori* probabilities and the difference of cluster centers. The obtained results are not always as expected and are data dependent [18]. Some criteria from information theory have also been proposed. The Minimum Descriptive Length (MDL) criteria evaluates the compromise between the likelihood of the classification and the complexity of the model [16].

In this work, we propose a framework for clustering by learning from the structure of the data. Learning is accomplished by randomly applying the K-means algorithm via entropy minimization (KMEM) multiple times on the data. The (KMEM) enables us to overcome the problem of knowing the number of clusters *a priori*. Multiple applications of the KMEM allow us to maintain a similarity measure matrix between pairs of input patterns. An entry a_{ij} in the similarity matrix gives the proportion of times input patterns i and j are co-located in a cluster among N clusterings using KMEM. Using this similarity matrix, the final data clustering is obtained by clustering a sparse graph of this matrix.

The contribution of this work is the incorporation of entropy minimization to estimate an approximate number of clusters in a data set based on some threshold and the use of graph clustering to recover the expected number of clusters.

The chapter is organized as follows: In the next section, we provide some background on the K-means algorithm. A brief introduction of entropy is presented in Section 3. K-Means via entropy minimization is outlined in Section 4. The graph clustering approach is presented in Section 5. The results of our algorithms are discussed in Section 6. We conclude briefly in Section 7.

2 K-Means Clustering

The K-means clustering [12] is a method commonly used to partition a data set into k groups. In the K-means clustering, we are given a set of n data points (patterns) $(x_1, ..., x_k)$ in d dimensional space R^d and an integer k and the problem is to determine a set of points (centers) in R^d so as to minimize the square of the distance from each data point to its nearest center. That is find k centers $(c_1, ..., c_k)$ which minimize:

$$J = \sum_k \sum_{x \in C_k} |d(x, c_k)|^2, \tag{1}$$

where the $C's$ are disjoint and their union covers the data set. The K-means consists of primarily two steps:

(1) The assignment step where based on initial k cluster centers of classes, instances are assigned to the closest class.

(2) The re-estimation step where the class centers are recalculated from the instances assigned to that class.

These steps are repeated until convergence occurs; that is when the re-estimation step leads to minimal change in the class centers. The algorithm is outlined in Figure 1.

The K-means Algorithm

Input

$\quad P = \{p_1,..., p_n\}$ (points to be clustered)

$\quad k$ (number of clusters)

Output

$\quad C = \{c_1,...c_k\}$ (cluster centers)

$\quad m : P \rightarrow \{1,...k\}$ (cluster membership)

Procedure K-means

1. Initialize C (random selection of P).

2. For each $p_i \in P$, $m(p_i) = \arg\min_{j \in k} \text{distance}(p_i, c_j)$.

3. If m has not changed, stop, else proceed.

4. For each $i \in \{1,...k\}$, recompute c_i as a center of $\{p \mid m(p) = i\}$.

5. Go to step 2.

Fig. 1. The K-Means Algorithm

Several distance metrics like the Manhattan or the Euclidean are commonly used. In this chapter, we consider the Euclidean distance metric. Issues that arise in using the K-means include: shape of the clusters, choosing, the number of clusters, the selection of initial clusters which could affect the final results and degeneracy. Degeneracy arises when the algorithm is trapped in a local minimum thereby resulting in some empty clusters. In this chapter we intend to handle the last threes problem via entropy optimization.

3 An Overview of Entropy Optimization

The concept of entropy was originally developed by the physicist Rudolf Clausius around 1865 as a measure of the amount of energy in a thermodynamic system [2]. This concept was later extended through the development of statistical mechanics. It was first introduced into information theory in 1948 by

Claude Shannon [15]. Entropy can be understood as the degree of disorder of a system. It is also a measure of uncertainty about a partition [15, 10].

The philosophy of entropy minimization in the pattern recognition field can be applied to classification, data analysis, and data mining where one of the tasks is to discover patterns or regularities in a large data set. The regularities of the data structure are characterized by small entropy values, while randomness is characterized by large entropy values [10]. In the data mining field, the most well known application of entropy is information gain of decision trees. Entropy based discretization recursively partitions the values of a numeric attribute to a hierarchy discretization. Using entropy as an information measure, one can then evaluate an attribute's importance by examining the information theoretic measures [10].

Using entropy as an information measure of the distribution data in the clusters, we can determine the number of clusters. This is because we can represent data belonging to a cluster as one bin. Thus a histogram of these bins represents cluster distribution of data. From entropy theory, a histogram of cluster labels with low entropy shows a classification with high confidence, while a histogram with high entropy shows a classification with low confidence.

3.1 Minimum Entropy and Its Properties

Shannon Entropy is defined as:

$$H(X) = -\sum_{i=1}^{n} (p_i \ln p_i) \tag{2}$$

where X is a random variable with outcomes $1, 2, ..., n$ and associated probabilities $p_1, p_2, ..., p_n$.

Since $-p_i \ln p_i \geq 0$ for $0 \leq p_i \leq 1$ it follows from (2) that $H(X) \geq 0$, where $H(X) = 0$ iff one of the p_i equals 1; all others are then equal to zero. Hence the notation $0 \ln 0 = 0$. For continuous random variable with probability density function $p(x)$, entropy is defined as

$$H(X) = -\int p(x) \ln p(x) dx \tag{3}$$

This entropy measure tells us whether one probability distribution is more informative than the other. The minimum entropy provides us with minimum uncertainty, which is the limit of the knowledge we have about a system and its structure [15]. In data classification, for example the quest is to find minimum entropy [15]. The problem of evaluating a minimal entropy probability distribution is the global minimization of the Shannon entropy measure subject to the given constraints. This problem is known to be NP-hard [15].

Two properties of minimal entropy which will be fundamental in the development of KMEM model are *concentration* and *grouping* [15]. Grouping implies moving all the probability mass from one state to another, that is, reduce the number of states. This reduction can decrease entropy.

Proposition 1. *Given a partition* $\text{\ss} = [B_a, B_b, A_2, A_3, ...A_N]$, *we form the partition* $\mathring{A} = [A_1, A_2, A_3, ...A_N]$ *obtained by merging* B_a *and* B_b *into* A_1, *where* $p_a = P(B_a)$, $p_b = P(B_b)$ *and* $p_i = P(A_i)$, *we maintain that:*

$$H(\mathring{A}) \leq H(\text{\ss}) \tag{4}$$

Proof. The function $\varphi(p) = -p \ln p$ is convex. Therefore for $\lambda > 0$ and
$p_1 - \lambda < p_1 < p_2 < p_2 + \lambda$
we have:

$$\varphi(p_1 + p_2) < \varphi(p_1 - \lambda) + \varphi(p_2 + \lambda) < \varphi(p_1) + \varphi(p_2) \tag{5}$$

Clearly,

$$H(\text{\ss}) - \varphi(p_a) - \varphi(p_b) = H(\mathring{A}) - \varphi(p_a + p_b)$$

because each side equals the contribution to $H(\text{\ss})$ and $H(\mathring{A})$ respectively due the to common elements of ß and $\mathring{A}$. Hence, (4) follows from (5).

Concentration implies moving probability mass from a state with low probability to a state with high probability. Whenever this move occurs, the system becomes less uniform and thus entropy decreases.

Proposition 2. *Given two partitions* $\text{\ss} = [b_1, b_2, A_3, A_4, ...A_N]$ *and* $\mathring{A} = [A_1, A_2, A_3, ...A_N]$ *that have the same elements except the first two. We maintain that if* $p_1 = P(A_1)$, $p_2 = P(A_2)$ *with* $p_1 < p_2$ *and* $(p_1 - \lambda) = P(b_1) \leq (p_2 + \lambda) = P(b_2)$, *then*

$$H(\text{\ss}) \leq H(\mathring{A}) \tag{6}$$

Proof. Clearly,

$$H(\mathring{A}) - \varphi(p_1) - \varphi(p_2) = H(\text{\ss}) - \varphi(p_1 - \lambda) - \varphi(p_2 + \lambda)$$

because each side equals the contribution to $H(\text{\ss})$ and $H(\mathring{A})$ respectively due to the common elements of $\mathring{A}$ and ß. Hence, (6) follows from (5).

3.2 The Entropy Decomposition Theorem

Another attractive property of entropy is the way in which aggregation and disaggregation are handled [4]. This is because of the property of additivity of entropy. Suppose we have n outcomes denoted by $X = \{x_1, ..., x_n\}$, with probability $p_1, ..., p_n$. Assume that these outcomes can be aggregated into a smaller number of sets $C_1, ..., C_K$ in such a way that each outcome is in only one set C_k, where $k = 1, ...K$. The probability that outcomes are in set C_k is

$$p_k = \sum_{i \in C_k} p_i \tag{7}$$

The entropy decomposition theorem gives the relationship between the entropy $H(X)$ at level of the outcomes as given in (2) and the entropy $H_0(X)$ at the level of sets. $H_0(X)$ is the between group entropy and is given by:

$$H_0(X) = -\sum_{k=1}^{K} (p_k \ln p_k) \tag{8}$$

Shannon entropy (2) can then be written as:

$$
\begin{aligned}
H(X) &= -\sum_{i=1}^{n} p_i \ln p_i \\
&= -\sum_{k=1}^{K}\sum_{i \in C_k} p_i \ln p_i \\
&= -\sum_{k=1}^{K} p_k \sum_{i \in C_k} \frac{p_i}{p_k} \left(\ln p_i + \ln \frac{p_k}{p_i} \right) \\
&= -\sum_{k=1}^{K} (p_k \ln p_k) - \sum_{k=1}^{K} p_k \sum_{i \in C_k} \frac{p_i}{p_k} \ln \frac{p_i}{p_k} \\
&= H_0(X) + \sum_{k=1}^{K} p_k H_k(X)
\end{aligned}
\tag{9}
$$

where

$$H_k(X) = -\sum_{i \in C_k} \frac{p_i}{p_k} \ln \frac{p_i}{p_k} \tag{10}$$

A property of this relationship is that $H(X) \geq H_0(X)$ because p_k and $H_k(X)$ are nonnegative. This means that after data grouping, there cannot be more uncertainty (entropy) than there was before grouping.

4 The K-Means via Entropy Model

In this section we outline the K-means via entropy minimization. The method of this section enables us to perform learning on the data set, in order to obtain the similarity matrix and to estimate a value for the expected number of clusters based on the clustering requirements or some threshold.

4.1 Entropy as a Prior Via Bayesian Inference

Given a data set represented as $X = \{x_1,...,x_n\}$, a clustering is the partitioning of the data set to get the clusters $\{C_j, j = 1,...K\}$, where K is usually less than n. Since entropy measures the amount of disorder of the system, each cluster should have a low entropy because instances in a particular cluster should be similar. Therefore our clustering objective function must include some form of entropy. A good minimum entropy clustering criterion has to reflect some relationship between data points and clusters. Such relationship information will help us to identify the meaning of data, i.e. the category of data. Also, it will help to reveal the components, i.e. clusters and components of mixed clusters. Since the concept of entropy measure is identical to that of probabilistic dependence, an entropy criterion measured on *a posteriori* probability would suffice. The Bayesian inference is therefore very suitable in the development of the entropy criterion.

Suppose that after clustering the data set X, we obtain the clusters $\{C_j, j = 1,...K\}$ by Bayes rule, the posterior probability $P(C_j|X)$ is given as;

$$P(C_j|X) = \frac{P(X|C_j)P(C_j)}{P(X)} \propto P(X|C_j)P(C_j) \tag{11}$$

where $P(X|C_j)$ given in (12) is the likelihood and measures the accuracy in clustering the data and the prior $P(C_j)$ measures consistency with our background knowledge.

$$P(X|C_j) = \prod_{x_i \in C_j} P(x_i|C_j) = e^{\sum_{x_i \in C_j} \ln p(x_i|C_j)} \tag{12}$$

By the Bayesian approach, a classified data set is obtained by maximizing the posterior probability (11). In addition to three of the problems presented by the K-means which we would like to address: determining number of clusters, selecting initial cluster centers and degeneracy, a fourth problem is, the choice of the prior distribution to use in (11). We address these issues below.

Defining the Prior Probability

Generally speaking, the choice of the prior probability is quite arbitrary [21]. This is a problem facing everyone and no universal solution has been found. For our our application, we will define the prior as an exponential distribution, of the form;

$$P(C_j) \propto e^{\beta \sum_{i=1}^{k} p_i \ln p_i} \tag{13}$$

where $p_j = |C_j|/n$ is the prior probability of cluster j, and $\beta \geq 0$ refers to a weighting of the *a priori* knowledge. Hence forth, we call β the entropy constant.

Determining Number of Clusters

Let k^* be the final unknown number of clusters in our K-means algorithm (KMEM). After clustering, the entropy

$$H(X) = -\sum_{i=1}^{k^*} p_i \ln p_i$$

will be minimum based on the clustering requirement. From previous discussions, we know that entropy decreases as clusters are merged. Therefore if we start with some large number of clusters $K > k^*$, our clustering algorithm will reduce K to k^* because clusters with probability zero will vanish. Note that convergence to k^* is guaranteed because the entropy of the partitions is bounded below by 0. A rule of thumb on the value of initial number of clusters is $K = \sqrt{n}$ [3].

The KMEM Model

The K-Means algorithm works well on a data set that has spherical clusters. Since our model (KMEM) is based on the K-means, we make the assumption that the each cluster has Gaussian distribution with mean values $c_j, i = (1, ..., k)$ and constant cluster variance. Thus for any given cluster C_j,

$$P(x_i|C_j) = \frac{1}{\sqrt{2\pi\sigma^2}} e^{\left(-\frac{(x_i-c_j)^2}{2\sigma^2}\right)} \tag{14}$$

Taking natural log and omitting constants, we have

$$\ln P(x_i|C_j) = -\frac{(x_i - c_j)^2}{2\sigma^2} \tag{15}$$

Using equations (12) and (13), the posterior probability (11) now becomes:

$$P(C_j|X) \propto \exp \sum_{x_i \in C_j} (\ln p(x_i|C_j)) \exp \left[\beta \sum_{i=1}^{k^*} p_i \ln p_i\right] \propto \exp(-E) \tag{16}$$

where E is written as follows:

$$E = -\sum_{x_i \in C_j} \ln p(x_i|C_j) - \beta \sum_{i=1}^{k^*} p_i \ln p_i \tag{17}$$

If we now use equation (14), equation (17) becomes

$$E = \sum_{i=1}^{k^*} \sum_{x_i \in C_j} \frac{(x_i - c_j)^2}{2\sigma^2} - \beta \sum_{i=1}^{k^*} p_i \ln p_i \tag{18}$$

or

$$E = \sum_{i=1}^{k^*} \sum_{x_i \in C_j} \frac{(x_i - c_j)^2}{2\sigma^2} + \beta H(X) \tag{19}$$

Maximizing the posterior probability is equivalent to minimizing (19). Also, notice that since the entropy term in (19) is nonnegative, equation (19) is minimized if entropy is minimized. Therefore (19) is the required clustering criterion.

We note that when $\beta = 0$, E is identical to the cost function of the K-Means clustering algorithm.

The Entropy K-means algorithm (KMEM) is given in figure 2. Multiple runs of KMEM are used to generate the similarity matrix. Once this matrix is generated, the learning phase is complete.

Entropy K-means Algorithm

1. Select the initial number of clusters k and a value for the stopping criteria ε.

2. Randomly initialize the cluster centers $\theta_i(t)$, and the *a priori* probabilities p_i, $i = 1, 2, ..., k$, β, and the counter $t = 0$.

3. Classify each input vector x_j, $j = 1, 2, ..., n$ to get the partition C_i such that for each $x_j \in C_r$, $r = 1, 2, ..., k$
$$[x_j - \theta_r(t)]^2 - \frac{\beta}{n} \ln(p_r) \leq [x_j - \theta_i(t)]^2 - \frac{\beta}{n} \ln(p_i)$$

4. Update the cluster centers
$$\theta_i(t+1) = \frac{1}{|C_i|} \sum_{j \in C_i} x_j$$
and the *a priori* probabilities of clusters
$$p_i(t+1) = \frac{|C_i|}{n}$$

5. Check for convergence; that is see if
$$\max_i |\theta_i(t+1) - \theta_i(t)| < \varepsilon$$
if it is not, update $t = t+1$ and go to step 3.

Fig. 2. The Entropy K-means Algorithm

This algorithm iteratively reduces the numbers of clusters as some empty clusters will vanish.

5 Graph Matching

The rationale behind our approach for structure learning is that any pair of patterns that should be co-located in a cluster after clustering must appear together in the same cluster a majority of the time after N applications of KMEM.

Let $G(V, E)$ be the graph of the similarity matrix where each input pattern is a vertex of G and V is the set of vertices of G. An edge between a pair of patterns (i, j) exists if the entry (i, j) in the similarity matrix is non-zero. E is a collection of all the edges of G. Graph matching is next applied on the maximum spanning tree of the sparse graph $G'(V, E) \subset G(V, E)$. The sparse graph is obtained by eliminating inconsistent edges. An inconsistent edge is an edge whose weight is less than some threshold τ. Thus a pattern pair whose edge is considered inconsistent is unlikely to be co-located in a cluster. To understand the idea behind the maximum spanning tree, we can consider the minimum spanning tree which can be found in many texts, for example [20] pages 278 and 520. The minimum spanning tree (MST) is a graph theoretic method, which determines the dominant skeletal pattern of points by mapping the shortest path of nearest neighbor connections [19]. Thus given a set of input patterns $X = x_1, ..., x_n$ each with edge weight d_{ij}, the minimum spanning tree is an acyclic connected graph that passes through all input patterns of X with a minimum total edge weight. The maximum spanning tree on the other hand is a spanning with a maximum total weight. Since all of the edge weight in the similarity matrix are nonnegative, we can negate these values and the apply the minimum spanning tree algorithm.

6 Results

The KMEM and the graph matching algorithms were tested on some synthetic image and data from the UCI data repository [22]. The data include the Iris data, wine data and heart disease data. The results for the synthetic images and iris data are given in 6.1 and 6.2. The KMEM algorithm was run 200 times in order to obtain the similarity matrix and the average number of clusters k_{ave}.

6.1 Image Clustering

For the synthetic images, the objective is to reduce the complexity of the grey levels. Our algorithm was implemented with synthetic images for which the ideal clustering is known. Matlab and Paint Shop Pro were used for the image processing in order to obtain an image data matrix. A total of three test images were used with varying numbers of clusters. The first two images, test1 and test2, have four clusters. Three of the clusters had uniformly distributed values with a range of 255, and the other had a constant value. Test1 had clusters of varying size while test2 had equal sized clusters. The third synthetic image, test3, has nine clusters each of the same size and each having values uniformly distributed with a range of 255. We initialized the algorithm with the number of clusters equal to the number of grey levels, and the value of cluster centers equal to the grey values. The initial probabilities (p_i) were computed from the image histogram. The algorithm was able to correctly detect the number of clusters. Different clustering results were obtained as the value of the entropy constant was changed, as is shown in Table 1. For the image test3, the correct number of clusters was obtained using a β of 1.5. For the images test1 and test2, a β value of 5.5 yielded the correct number of clusters. In Table 1, the optimum number of clusters for each synthetic image are bolded.

Table 1. The number of clusters for different values of β

β	Images		
	test1	test2	test3
1.0	10	10	13
1.5	6	8	**9**
3.5	5	5	6
5.5	**4**	**4**	5

6.2 Iris Data

Next we tested the algorithm on the different data obtained from the UCI repository and got satisfactory results. The results presented in this section are on the Iris data. The Iris data is well known [1, 8] and serves as a benchmark for supervised learning techniques. It consists of three types of Iris plants: *Iris Versicolor, Iris Virginica, and Iris Setosa* with 50 instances per class. Each datum is four dimensional and consists of a plants' morphology namely *sepal width, sepal length, petal width, and petal length*. One class *Iris Setosa* is well separated from the other two. Our algorithm was able to obtain the three-cluster solution when using the entropy constant β's of 10.5 and 11.0. Two cluster solutions were also obtained using entropy constants of 14.5, 15.0, 15.5 and 16.0 Table 2 shows the results of the clustering.

To evaluate the performance of our algorithm, we determined the percentage of data that were correctly classified for three cluster solution. We compared it to the results of direct K-means. Our algorithm had a 91% correct classification while the direct K-means achieved only 68% percent correct classification, see Table 3. Another measure of correct classification is entropy. The entropy of each cluster is calculated as follows:

$$H(C_j) = -\sum_{j=1}^{k} \frac{n_j^i}{n_j} \ln \frac{n_j^i}{n_j} \tag{20}$$

where n_j is the size of cluster j and n_j^i is the number of patterns from cluster i that were assigned to cluster j. The overall entropy of the clustering is the sum of the weighted entropy of each cluster and is given by

$$H(C) = \sum_{j=1}^{k} \frac{n_j}{n} H(C_j) \tag{21}$$

where n is the number of input patterns. The entropy is given in table 3. The lower the entropy the higher the cluster quality.

We also determined the effect of β and the different cluster sizes on the average value of k obtained. The results are given in tables 4, 5 and 6. The tables show that for a given β and different k value the average number of clusters converge.

Table 2. The number of clusters as a function of β for the Iris Data

β	10.5	11.0	14.5	15.0	15.5	16
k	3	3	2	2	2	2

Table 3. Percentage of correct classification of Iris Data

k	3.0	3.0	2.0	2.0	2.0	2.0
%	90	91	69	68	68	68
$Entropy$	0.31	0.27	1.33	1.30	1.28	1.31

Table 4. The average number of clusters for various k using a fixed $\beta = 2.5$ for the Iris Data

k	10	15	20	30	50
k_{ave}	9.7	14.24	18.73	27.14	42.28

Table 5. The average number of clusters for various k using a fixed $\beta = 5.0$ for the Iris Data

k	10	15	20	30	50
k_{ave}	7.08	7.10	7.92	9.16	10.81

Table 6. The average number of clusters for various k using a fixed $\beta = 10.5$ for the Iris Data

k	10	15	20	30	50
k_{ave}	3.25	3.34	3.36	3.34	3.29

7 Conclusion

The KMEM provided good estimates for the unknown number of clusters. We should point out that whenever the clusters are well separated, the KMEM algorithm is sufficient. Whenever that was not the case, further processing by the graph clustering produced the required results. Varying the entropy constant β allows us to vary the final number of clusters in KMEM. However, we had to empirically obtain values for β. Further work will be how to estimate the value of β based on the some properties of the data set. Our approach worked well on the data that we tested, producing the required number of clusters. While our results are satisfactory, we observed that our graph clustering approach sometimes matched weakly linked nodes, thus combining clusters. Therefore, further work will be required to reduce this problem. Such a result would be very useful in image processing and other applications.

References

1. R.O. Duba and P.E. Hart. *Pattern Classification and Scene Analysis.* Wiley-Interscience, New York, NY, 1974.
2. S. Fang, J.R. Rajasekera, and H.-S. J. Tsao. *Entropy Optimization and Mathematical Programming.* Kluwer Academic Publishers, 1997.
3. M. Figueiredo and A.K. Jain. Unsupervised learning of finite mixture models. *IEEE Transactions on Pattern Analysis and Machine Intelligence*, 24(3): 381-396, 2002.
4. K. Frenken. Entropy Statistics and Information Theory. In H. Hanusch and A. Pyka, editors, *The Elgar Companion to Neo-Schumpeterian Economics.* Edward Elgar Publishing (in press).
5. D. Hall and G Ball. ISODATA: A Novel Method of Data Analysis and Pattern Classification. Technical Report, Stanford Research Institute, Menlo Park, CA, 1965.
6. G. Iyengar, and A. Lippman. Clustering Images using Relative Entropy for Efficient retrieval. *IEEE Computer Magazine*, 28(9): 23-32, 1995.
7. A. Jain and M. Kamber. *Algorithms for Clustering.* Prentice Hall, 1998.
8. M. James. *Classification Algorithms.* Wiley-Interscience, New York, NY, 1985.
9. T. Kanungo, D.M. Mount, N.S. Netayahu, C.D. Piako, R. Silverman, and A.Y. Wu. An Efficient K-Means Clustering Algorithm: Analysis and Implementation. *IEEE Transactions on Pattern Analysis and Machine Intelligence*, 24(7): 881-892, 2002.
10. J.N. Kapur and H.K. Kesaven. *Entropy Optimization Principle with Applications*, Ch.1. London Academic, 1997.
11. Y.W. Lim and S.U. Lee. On the Color Image Segmentation Algorithm based on Thresholding and Fuzzy C-means Techniques. *Pattern Recognition*, 23: 935-952, 1990.
12. J.B. McQueen. Some Methods for Classification and Analysis of Multivariate Observations. In *Proceedings of the Fitfth Symposium on Math, Statistics, and Probability*, pages 281-297. University of California Press, Berkeley, CA, 1967.
13. D. Miller, A. Rao, K. Rose, and A. Gersho. An Information Theoretic Framework for Optimization with Application to Supervised Learning. IEEE International Symposium on Information Theory, Whistler, B.C., Canada, September 1995.
14. B. Mirkin. *Mathematical Classification and Clustering – Nonconvex Optimization and its Applications*, v11. Kluwer Academic Publishers, 1996.
15. D. Ren. An Adaptive Nearest Neighbor Classification Algorithm. Available at www.cs.ndsu.nodak.edu/ dren/papers/CS785finalPaper.doc
16. J. Rissanen. A Universal prior for integers and Estimation by Minimum Description Length. *Annals of Statistics*, 11(2): 416-431, 1983.
17. J.T. Tou and R.C. Gonzalez. *Pattern Recognition Principles.* Addison-Wesley, 1994.
18. M.M. Trivedi and J.C. Bezdeck. Low-level segmentation of aerial with fuzzy clustering. *IEEE Transactions on Systems, Man, and Cybernetics*, SMC-16: 589-598, 1986.

19. H. Neemuchawala, A. Hero, and P. Carson. Image Registration using entropic graph-matching criteria. *Proceedings of Asilomar Conference on Signals, Systems and Computers*, 2002.
20. R.K. Ahuja, T.L. Magnanti, and J.B. Orlin. *Network Flows: Theory, Algorithms, and Applications*. Prentice Hall, 1993.
21. N. Wu. *The Method of Maximum Entropy*. Springer, 1997.
22. C.L. Blake and C.J. Merz. UCI Repository of machine learning databases http://www.ics.uci.edu/ mlearn/MLRepository.html. Irvine, CA: University of California, Department of Information and Computer Science, 1998.

Molecular Biology and Pooling Design

Weili Wu[1], Yingshu Li[2], Chih-hao Huang[2], and Ding-Zhu Du[1]

[1] Department of Computer Science,
University of Texas at Dallas,
Richardson, TX 75083, USA
{weiliwu,dzdu}@utdallas.edu
[2] Department of Computer Science and Engineering,
University of Minnesota,
Minneapolis, MN 55455, USA
{yili,huang}@cs.umn.edu

Summary. The study of gene functions requires a high-quality DNA library. A large amount of testing and screening needs to be performed to obtain a high-quality DNA library. Therefore, the efficiency of testing and screening becomes very important. Pooling design is a very helpful tool, which has developed a lot of applications in molecular biology. In this chapter, we introduce recent developments in this research direction.

1 Molecular Biology and Group Testing

One of the recent important developments in biology is the success of Human Genome Project. This project was done with a great deal of help from computer technology, which made molecular biology a hot research area conjugated with computer science. Bio-informatics is a new born research area that grows very rapidly from this conjugation.

The technology for obtaining sequenced genome data is getting more developed as and more and more sequenced genome data is available to the scientific research community. Based on those data, the study of gene functions has become a very important research direction. This requires high-quality gene libraries. The high-quality gene libraries are obtained from extensive testing and screening of DNA clones, that is, identifying clones used in the libraries. Therefore, the efficiency of DNA screening is very important. For example, in 1998, the Life Science Division of Los Alamos National Laboratories [14] was dealing with a dataset of 220,000 clones. Individual testing those clones requires 220,000 tests. However, they used only 376 tests with a technology called *group testing*.

The group testing takes advantage of small percentage of clones containing target probes. It tests subsets of clones called *pools*, instead of testing each of them individually. For example, in the above mentioned testing at Los Alamos National Laboratories, each pool contained about 5,000 clones. The technology of group testing was started from Wassernan-type blood test in World War II. A very simple design that was used in the earlier stage is as follows: Divide each blood sample into two parts. First, mix all first parts into a pool and test the pool. If the outcome is positive, i.e., there is a presence of syphilitic antigen, then test the second part individually. Otherwise, all men in the pool passed the test. During the past 60 years, more efficient designs have been developed. These designs have gained more and more attention due to significant applications in the study of genome.

A typical application of pooling designs is DNA library screening. A DNA library is a collection of cloned DNA segments usually taken from a specific organism. Those cloned DNA segments are called *clones*. Given a DNA library, the problem is to identify whether each clone contains a probe from a given set of probes. A *probe* is a piece of DNA labeled with radioisotope or fluorescence. The probe is often used to detect specific DNA sequences by hybridization. A clone is said to be *positive* if it contains a given probe and *negative* otherwise. A pool is *positive* if it contains a positive clone and *negative* otherwise. In a group testing algorithm a clone may appear in two or more pools. Therefore, making copies is a necessary preprocessing procedure.

Hybridization is one of the techniques to reproduce clones or perform *DNA cloning*. To better understand the concept of hybridization, let us describe the composition of DNA. DNA is a large molecule with double helix structure that consists of two nucleic acids which in turn are strings of nucleotides. There are four types of nucleotides A (adenine), T (thymine), G (guanine) and C (cytosine). Thus, each nucleic acid can be seen as a string of four symbols A, T, G, C. When two nucleic acids are joined into a double helix, A must bond with T and G must bond with C. Heating can break the DNA into two separated nucleic acids. Through the action of an enzyme each nucleic acid may be jointed with a probe and consequently the probe would grow into a dual nucleic acid. This process is referred to as *hybridization*.

By repeating hybridization we can clone unlimited number of copies of any piece of DNA. This approach is called Polymerase Chain Reaction (PCR). It is a cell-free, fast, and inexpensive technique. Another technique for DNA cloning is cell-based. It contains four steps:

(1) Insert the DNA fragment (to be cloned) into an agent called *vector*. This step results in a recombinant.

(2) Put the recombinant DNA into a host cell to proliferate. This step is called *transformation*.

(3) Reproduce the transformed cell.

(4) Isolate the desired DNA clones from the cells obtained from (3).

In general, there are two conditions that need to be satisfied for group testing: (a) copies of items are available and (b) testing on a subset of items is available. In DNA library screening both conditions are available due to DNA cloning, especially hybridization.

2 Pooling Design

There are two types of group testing, sequential and non-adaptive. To explain them, let us look at two examples of group testing algorithms.

Consider a set of nine clones with one positive clone. In the first example, the method is sequential. At each iteration, bisect the positive pool into two equal or almost equal pools and test each of the obtained two pools until only one positive clone is found in the pool. In the worst case, this method takes at most six tests to identify the positive clone. In general, for a set of n clones with one positive clone, the bisection would take at most $2\lceil \log_2 n \rceil$ tests to identify the positive one.

In the second example, the method is to put the nine clones into a 3×3 matrix. Each row and each column represent a test. Since there is only one positive clone, there is exactly one positive row and one positive column. Their intersection is the positive clone. In general, for n clones that include a positive one, this method takes $O(\sqrt{n})$ tests. For large n, this method needs more tests than the first one. However, all tests in this method are independent. They can be performed simultaneously. This type of group testing is called *non-adaptive group testing*.

Group testing in molecular biology is usually called *pooling design*. The pooling design is often non-adaptive [3, 8]. This is due to the time consuming nature of tests in molecular biology. Therefore, we may simply refer to the pooling design as the non-adaptive group testing. Hence, every pooling design can be represented as a binary matrix by indexing rows with pools and columns with clones and assigning 1 to cell (i,j) if and only if the ith pool contains the jth clone.

A positive clone would imply the positivity of all pools containing it. Therefore, d positive clones would result in the positivity of all pools containing any of them. If we consider each column (clone) as a set of pools with 1-entry in the column, then the union of the d columns represents the testing outcome when those d clones form the set of all positive clones. Therefore, if a binary matrix representing a pooling design can identify up to d positive clones, all unions of up to d columns should be distinct. A binary matrix with this property is called $\bar{d}$-*separable*.

For a $\bar{d}$-separable matrix, a naive way for decoding a given testing outcome vector to find all positive clones is to compare it with all unions of up to d columns. This takes $O(n^d)$ time. Is it possible to do better? The following result of Li mentioned in [18] gives a negative answer.

Theorem 1 *Decoding for $\bar{d}$-separable matrix can be done in polynomial time with respect to n and d if and only if the hitting set problem is polynomial-time solvable, i.e., if and only if P=NP.*

Indeed, decoding is equivalent to finding a subset of at most d clones hitting every positive pool. By a set hitting another set, we mean that the intersection of two sets is nonempty. Note that every clone in a negative pool is negative. Therefore, the input size of this hitting problem is controlled by the union of negative pools. The following result gives an interesting condition on the size of this union.

Theorem 2 *For a $\bar{d}$-separable matrix, the union of negative pools always contains at least $n - d - k + 1$ clones if and only if no d-union contains a k-union, where a d-union means a union of d columns.*

When $k = 1$, the union of negative pools contains at least $n - d$ clones. Thus, the number of clones that are not in any negative pool is at most d, and hence they form a hitting set of at most d clones, which should be the solution. The binary matrix with the property that no column is contained in any d-union is said to be *d-disjunct*. For any d-disjunct matrix, decoding can be done in $O(n)$ time.

3 Simplicial Complex and Graph Properties

Finding the best d-disjunct matrix is an intractable problem for computer science. So far, its computational complexity is unknown. Therefore, we can only make approximate designs with various tools, including classical combinatorial designs, finite geometry, finite fields, etc. Recently, the construction of pooling designs using simplicial complexes was developed. A simplicial complex is an important concept in geometric topology [15, 18].

A *simplicial complex* Δ is a family of subsets of a finite set E such that $A \in \Delta$ and $B \subset A$ imply $B \in \Delta$. Every element in E is called a *vertex*. Every member in the family Δ is called a *face* and furthermore called a *k-face* if it contains exactly k vertices. Motivated by the work of Macula [12, 13], Park *et al.* [15] construct a binary matrix $M(\Delta, d, k)$ for a simplicial complex Δ by indexing rows with all d-faces and columns with all k-faces ($k > d$) and assigning 1 to cell (i, j) if and only if the ith d-face is contained in the jth k-face. They proved the following theorem.

Theorem 3 $M(\Delta, d, k)$ *is d-disjunct.*

An important family of simplicial complexes is induced by monotone graph properties. A graph property is *monotone increasing* if every graph containing a subgraph having this property also has this property. Similarly, a graph property is *monotone decreasing* if every subgraph of a graph with this property has this property. If one fixes a vertex set and considers edge sets of all graphs satisfying a monotone decreasing property, they will form a simplicial complex. Since graphs not satisfying a monotone increasing property form a monotone decreasing property, every monotone increasing property is also associated with a simplicial complex.

Matching is an example of a monotone decreasing property. Let Δ_m be the the simplicial complex consisting of all matchings in a complete graph of order m. Then k-matching (a matching of k edges) is a k-face of Δ_m. There is an error tolerance result for matching [7].

Theorem 4 *If k-matching is perfect, then $M(\Delta_m, d, k)$ is a d-error detecting d-disjunct matrix.*

Here, by a *d-error detecting* matrix, we mean that if there exist at most d erroneous tests, the matrix is still able to identify all positive clones.

Park *et al.* [15] also generalized this result to the case of a simplicial complex.

Theorem 5 *If for any two k-faces A and B $|A \setminus B| \geq 2$, then $M(\Delta, d, k)$ is a d-error detecting d-disjunct matrix.*

Huang and Weng [10] generalized Theorem 3 to a class of partial ordering sets, including lattices.

4 Error-Tolerant Decoding

Error-tolerant decoding is a very interesting issue in various pooling design models. To see it, let us study a so-called inhibitor model.

In fact, in some situations, a clone can be negative, positive or anti-positive. An *anti-positive* clone can cancel the positivity of a pool, that is, a test outcome on a pool containing an anti-positive clone must be negative, even if the pool contains a positive clone. An anti-positive clone is also called an *inhibitor*. If we know a positive clone, then all inhibitors can be identified by testing all pairs of clones consisting of the known positive clone and all clones in negative pools. However, if no positive clone is known, it is not so easy to identify inhibitors. Therefore, it is an interesting problem to decode all positive clones without knowing inhibitors.

Du and Hwang [4] developed the following method.

For each clone j and a possible subset I of inhibitors, compute $t(j, I)$, the number of negative pools containing j, but disjoint from I. Set $T(j) = \min t(j, I)$ over all possible subsets I.

They proved the following theorem.

Theorem 6 *For a $(d+r+e)$-disjunct matrix, if the input sample contains at most r inhibitors and at most d positive clones, and testing contains at most e erroneous tests, then $T(j) < T(j')$ for any positive clone j and any negative clone j'.*

Consequently, the following results can be formulated.

Theorem 7 (Du and Hwang [4]) *A $(d + r + e)$-disjunct matrix can identify all positive clones for every sample with d positive clones and at most r inhibitors subject to at most e erroneous tests.*

Theorem 8 (Hwang and Liu [9]) *A $(d + r + 2e)$-disjunct matrix can identify all positive clones for every sample with at most d positive clones and at most r inhibitors subject to at most e erroneous tests.*

The inhibitor model was proposed by Farach *et al.* [6]. De Bonis and Vaccaro [1] developed a sequential algorithm for this model and raised an open problem of finding non-adaptive algorithm in this model. While D'yachkov *et al.* [5] solved the error-free case, Hwang and Liu [9] gave a general solution.

5 Future Research

The development of error-tolerant pooling designs is very important in practice. Theorems 3 and 4 established connections between error-tolerant designs and simplicial complexes. Since all monotone graph properties induce simplicial complexes, these connections may open a new research direction joint with graph theory to develop efficient designs.

There are many issues that we need to consider when constructing a pooling design. For example, after receiving test outcomes on all pools, the question to be addressed is how to decode this data to obtain information on each clone. The different designs have different computational complexity for decoding. One can find some interesting contributions and open problems in this area in [17].

In practice, DNA screening is closely related to information retrieval and data mining. In fact, database systems have already employed the technique of group testing. This opens an opportunity to attack some problems in data processing by applying our new designs. Therefore, our research work can be widely extended into different areas of computer science.

References

1. A. De Bonis and U. Vaccari. Improved algorithms for group testing with inhibitors. *Information Processing Letters*, 65: 57-64, 1998.
2. D.-Z. Du and F. K. Hwang. *Combinatorial Group Testing and Its Applications (2nd ed.)*, World Scientific, Singapore, 1999.
3. D.-Z. Du and F. K. Hwang. Pooling Designs: Group Testing in Biology, manuscript.
4. D.-Z. Du and F. K. Hwang. Identifying d positive clones in the presence of inhibitors, manuscript.
5. A. G. D'ychkov, A. J. Macula, D. C. Torney, and P. A. Vilenkin. Two models of nonadaptive group testing for designing screening experiments. In *Proceedings of the 6th International Workshop on Model-Oriented Designs and Analysis*, pages 63-75, 2001.
6. M. Farach, S. Kannan, E. Knill, and S. Muthukrishnan. Group testing problem with sequences in experimental molecular biology. In *Proceedings of the Compression and Complexity of Sequences*, pages 357-367, 1997.
7. H. Q. Ngo and D.-Z. Du. New constructions of non-adaptive and error-tolerance pooling designs. *Discrete Mathematics*, 243: 161-170, 2002.
8. H. Q. Ngo and D.-Z. Du. A survey on combinatorial group testing algorithms with applications to DNA library screening. In D.-Z. Du, P.M. Pardalos, and J. Wang, editors, *Discrete Mathematical Problems with Medical Applications*, pages 171–182. American Mathematical Society, Providence, RI, 2000.
9. F. K. Hwang and Y. C. Liu. Error tolerant pooling designs with inhibitors. *Journal of Computational Biology*, 10: 231-236, 2003.
10. T. Huang and C.-W. Weng. A note on decoding of superimposed codes. *Journal of Combinatorial Optimization*, 7: 381-384, 2003.
11. F.K. Hwang. On Macula's error-correcting pooling design, to appear in *Discrete Mathematics*, 268: 311–314, 2003.
12. A.J. Macula. A simple construction of d-disjunct matrices with certain constant weights. *Discrete Mathematics* 162: 311–312, 1996.
13. A. J. Macula. Error correcting nonadaptive group testing with d^e-disjunct matrices. *Discrete Applied Mathematics*, 80: 217–222, 1997.
14. M. V. Marathe, A. G. Percus, and D. C. Torney. Combinatorial optimization in biology, manuscript, 2000.
15. H. Park, W. Wu, Z. Liu, X. Wu, and H. Zhao, DNA screening, pooling designs, and simplicial complex. *Journal of Combinatorial Optimization*, 7(4): 389–394, 2003.
16. W. W. Paterson. *Error Correcting Codes*, MIT Press, Cambridge, MA, 1961.
17. W. Wu, C. Li, X. Wu, and X. Huang. Decoding in pooling designs. *Journal of Combinatorial Optimization*, 7(4): 385–388, 2003.
18. W. Wu, C. Li, and X. Huang. On error-tolerant DNA screening, submitted to *Discrete Applied Mathematics*.

An Optimization Approach to Identify the Relationship between Features and Output of a Multi-label Classifier

Musa Mammadov, Alex Rubinov, and John Yearwood

Centre for Informatics and Applied Optimization
University of Ballarat, Victoria, 3353, Australia
{m.mammadov,a.rubinov,j.yearwood}@ballarat.edu.au

Summary. Multi-label classification is an important and difficult problem that frequently arises in text categorization. The accurate identification of drugs which are responsible for reactions that have occurred is one of the important problems of adverse drug reactions (ADR). In this chapter we consider the similarities of these two problems and analyze the usefulness of drug reaction relationships for the prediction of possible reactions that may occur. We also introduce a new method for the determination of responsibility for subsets of drug(s), out of all drugs taken by a particular patient, in reactions that have been observed. This method is applied for the evaluation of the level of correctness of suspected drugs reported in Cardiovascular type reactions in the ADRAC database. The problem of interaction of drugs is also considered.

Key words: Knowledge representation, adverse drug reaction, text categorization, multi-label classification, suspected drugs.

1 Introduction

In general the problem of classification is to determine the classes from a set of predefined categories that an object belongs to, based on a set of descriptors of the object. For example the text categorization task is to label an incoming message (document) with the label of one or more of the predefined classes. There have been a number of approaches to solving categorization problems by finding linear discriminant functions. In these approaches there are assumptions that each class has a Gaussian distribution. Least squares fit has also been used. Without any distributional assumptions a linear separator can be found by using a perceptron with minimization of the training error. Another approach that has been used in text categorization and information retrieval is logistic regression and this is closely related to support vector machines which have recently had much success in text categorization.

In text categorization problems, although techniques have been developed for *feature selection*, interest has been primarily in classification and there has not been much interest in determining the features (words) that are responsible for assigning a particular document to a particular class.

In studies of adverse drug reactions (ADRs) on patients, a patient record consists of a list of the drugs that have been taken and the reactions that have been experienced. The question that is of interest is "Which drugs are responsible for each reaction?" Certainly this is a classification problem, but the interest is focused on determining the features (drugs) that are most important in determining the class (reaction), rather than simply determining the class based on the set of drugs that the patient took.

In this chapter we consider a situation in which we have n records. In the case of text categorization these would be documents and in the case of ADRs each record would represent a patient with a number of drugs taken and various reactions that were observed or reported. Let the records or objects be $x_1, ..., x_{n,}$, where n is the number of records (documents/patients);

Each record (document/patient) is a vector of terms (words/drugs): $x_i = (x_{i1}, ..., x_{i,m})$, $i = 1, ..., n$. So m is the number of terms (words/drugs). $x_{ij} = 1$ if the word (drug) j is used in record i, $x_{ij} = 0$ if not.

In a classification task there may be two disjoint classes (binary classification), many disjoint classes (multi-class classification) or multiple overlapping classes (multi-label classification). In many cases multi-class classification problems are reduced to many binary classification problems. Below we look further at binary classification and multi-label classification.

A. Binary classification: Two classes $y \in \{-1, 1\}$.

Each document (patient) belongs to one of these classes. We denote by y_i the class for x_i, $i = 1, ..., n$; that is, $y_i = -1$ or $y_i = 1$. The problem is to find a weight vector $w = (w_1, ..., w_m)$; $w_i \in \mathbb{R}$ (that is, weight for each word) such that, the values wx_i and y_i $(i = 1, ..., n)$ are close overall. We will denote the scalar product of two vectors a and b by ab. Closeness can be assessed in many ways but considering the least squares fit or the logistic regression approach leads to the optimization problems below.

1. Least squares fit (LLSF) aims to solve:

$$w = \arg \inf_{w} \frac{1}{n} \sum_{i=1}^{n} (wx_i - y_i)^2. \tag{1}$$

2. Logistic regression (LR) aims to solve:

$$w = \arg \inf_{w} \frac{1}{n} \sum_{i=1}^{n} \ln(1 + e^{-wx_i y_i}). \tag{2}$$

B. Multi-label classification:

Let c be the number of classes. Each record (document/patient) can belong to a number of these c classes (reactions). We denote by $y_i = (y_{i1}, ..., y_{ic})$ the vector of classes for x_i, $i = 1, ..., n$. We will consider two different versions for the representation of the vector y_i.

1) $y_{ij} = 1$, if x_i belongs to the class j, and $y_{ij} = -1$ if not.

2) $y_{ij} = 1$, if x_i belongs to the class j, and $y_{ij} = 0$ if not.

In this chapter we consider some alternative approaches to solving the classification problem. In particular we focus on the Australian Drug Reactions Database and the problem of determining the drugs responsible for the various reaction classes.

2 Optimization Approaches to Feature Weighting

The problem is to find a weight matrix $W = (w_1, ..., w_m)$; that is, a weight vector $w_j = (w_{1j}, ..., w_{cj})$ for each feature (word/drug) j;   such that, the vectors Wx_i and y_i ($i = 1, ..., n$) are close overall. Problems (1) and (2) can be generalized as

1. Least squares fit (LLSF) aims to solve (see [20]):

$$W_{llsf} = \arg\inf_{W} \frac{1}{n} \sum_{i=1}^{n} (\|Wx_i - y_i\|)^2. \tag{3}$$

2. Logistic regression (LR) aims to solve:

$$W_{lr} = \arg\inf_{W} \frac{1}{n} \sum_{i=1}^{n} \ln(1 + e^{-y_i W x_i}). \tag{4}$$

In fact, the existence of different numbers of classes for different patients/ documents (one may belong to one class, another may belong to many classes) means that the errors described by (3) and (4) may not be comparable (the error obtained for multi-class patient/document can be equal to the sum of errors from many single class patients/documents). This is very important, at least, for ADR problems. As a result, the weights may not be best defined as a solution to (3) or (4).

Therefore, we suggest that it makes sense to consider the following formula for determining a classifier so that the errors are more effectively measured (the versions of algorithm $A(p)$, $p = 0, 1, 2$, see section 6):

$$W_p = \arg\inf_W \frac{1}{n} \sum_{i=1}^{n} (\|y_i\|)^{-p} \cdot \sum_{j=1}^{c} \left(\frac{\|y_i\|}{\sum_{l=1}^{c} H_{i,l}} H_{ij} - y_{ij} \right)^2 ; \qquad (5)$$

where $H_{ij} = \sum_{q=1}^{m} w_{jq} x_{iq}$, and $\|y_i\|$ is the number of positive coordinates in the vector $(y_{i1}, ..., y_{ic})$,.

In the application of this approach to the Australian Adverse Drug Reaction (ADRAC) database, $\|y_i\|$ is the number of reactions that occurred for patient i). To explain the advantage of this formula we consider one example.

Example 2.1   Assume that there is just one word/drug ($m = 1$), there are 5 classes ($c = 5$) and there are 4 patients/documents ($n = 4$), where, the first document belongs to the all 5 classes, and the other 3 documents belong to the first class. We need to find an optimal weight matrix (vector in this example) $W = (w_1, w_2, ..., w_5)$. We note that, this is a typical situation in ADR problems, where one drug can cause different reactions (and in different combinations).

Consider both cases **1)** and **2)**. In the calculations, the weight for the first class is set to be one.

Version 1). In this case the vectors for each class are: $y_1 = (1, 1, 1, 1, 1)$, $y_i = (1, -1, -1, -1, -1)$ for $i = 2, 3, 4, 5$. We have:

LLSF: $W = (1, 0, 0, 0, 0)$;
LR: $W = (1, 0, 0, 0, 0)$;
$A(0):$ $W = (1, 0.42, 0.42, 0.42, 0.42)$;
$A(1):$ $W = (1, 0.06, 0.06, 0.06, 0.06)$;
$A(2):$ $W = (1, 0, 0, 0, 0)$.

We observe that, in this version, the methods LLSF and LR have failed to produce sensible results. For this version function (5), for $p = 0, 1$, works well.

Version 2). In this case the class vectors (or vectors of reactions) are: $y_1 = (1, 1, 1, 1, 1)$, $y_i = (1, 0, 0, 0, 0)$ for $i = 2, 3, 4, 5$. We have:

LLSF: $W = (1, 0.25, 0.25, 0.25, 0.25)$;
LR: $W = (1, 1, 1, 1, 1)$;
$A(0):$ $W = (1, 0.62, 0.62, 0.62, 0.62)$;
$A(1):$ $W = (1, 0.25, 0.25, 0.25, 0.25)$;
$A(2):$ $W = (1, 0.06, 0.06, 0.06, 0.06)$.

We see that, LR fails in the second version too, whilst all others work well. In this simple example, LLSF = A(1), but in general, when there are combinations of more than one word/drugs are involved, the outcome will be different.

Comparison of the weights for the algorithms $A(p)$, $p = 0, 1, 2$, shows that, the difference between weights for the first class and the others is minimal for $A(0)$, and is maximal for $A(2)$. Which one is better? Of course, we cannot answer this question; for different situations different versions could be better. That is why, it is very useful to consider different versions $p = 0, 1, 2$.

This example shows that, it is more convenient to represent the vector of classes as in the second version; moreover, formula (5) is more preferable than formulae (3) and (4). This is the way we will encode class membership in this chapter.

Using functions like (5) requires a more complex solution for the corresponding optimization problem, but the choice of more reasonable distance functions is important.

In the next sections of this chapter we shall concentrate on the application of this approach to the study of the drugs that are the most suspect or responsible for the adverse drug reactions observed in patients.

3 Adverse Drug Reactions

An Adverse Drug Reaction (ADR) is defined by the WHO as: "a response to a drug that is noxious and unintended and occurs at doses normally used in man for the prophylaxis, diagnosis or therapy of disease, or for modification of physiological function" [18]. ADRs are estimated to be the fourth leading cause of death in the USA [11], and the amount of published literature on the subject is vast [1]. Some of the problems concerning ADRs are discussed in our research report [8]. Many approaches have been tried for the analysis of adverse reaction data, such as: Fisher's Exact Test and matched pair designs (McNemar's test) [15], Reporting Odds Ratio (ROR), and Yule's Q [16]. One approach that has had some success is the Proportional Reporting Ratios (PRR) for generating signals from data in the United Kingdom. The Norwood-Sampson Model has been applied to data in the United States of America and approved by the Food and Drug Administration. A common approach to the assessment of ADRs uses the Bayesian method [4]. For example, the Bayesian confidence propagation neural network (BCPNN) [2], and an empirical Bayesian statistical data mining program called a Gamma Poisson Shrinker (GPS) [5], and the Multi-item Gamma Poisson Shrinker (MGPS) [13], which have been applied to the United Sates Food and Drug Administration Spontaneous Reporting System database. The Bayesian method has met with success, but is very exacting regarding the quantification of expectations [7].

Each method has its own advantages and disadvantages with respect to applicability in different situations and possibilities for implementation. In [8, 9] a new approach was developed where the main goal was to study, for each drug, the possible reactions that can occur; that is, to establish drug-reaction relationships. In this work the ADR problem was formulated as a text categorization problem having some peculiarities. This approach was applied to the Australian Adverse Drug Reaction Advisory Committee (ADRAC) database.

One of the main problems of ADR is the following: given a patient (that is, the sets of drugs and reactions) identify the drug(s) which are responsible for the adverse reactions experienced. In the ADRAC database drugs thought to be responsible for the reactions are labelled as "suspected" drugs. The accurate definition of suspected drugs for each report has a very significant impact on the quality of the database for the future study of drug-reaction relationships. In this chapter, we develop the approach introduced in [8, 9] for the study of suspected drugs.

The ADRAC database has been developed and maintained by the Therapeutic Goods Administration (TGA) with the aim to detect signals from adverse drug reactions as early as possible. It contains 137,297 records collected from 1972 to 2001. A more detailed account of the ADRAC database is given in [8].

In ADRAC there are 18 System Organ Class (SOC) reaction term classes, one of which is the Cardiovascular SOC. The Cardiovascular class consists of four sub-classes. In this chapter we will consider the part of the ADRAC data related to the cardiovascular type of reactions. We collect all records having at least one reaction from these four sub-groups. We call this dataset **Card20**. In this dataset some records may have a reaction from outside the Cardiovascular group. We define four classes according to these four subgroups and additionally a fifth class that contains reactions belonging to the other 17 SOCs. For the number of records see Table 1 below.

The information about each patient consist of mainly two sets of information: individual patient information and information about drug(s) and reaction(s). In this chapter we will use only the second set of information. By understanding the drug-reaction relationship in the absence of information about other factors influencing this relationship, we expect to be able to establish a clearer relationship between drugs and reactions. Another reason for focussing primarily on drugs and reactions relates to the inconsistent quality and quantity of relevant data on factors which also play a role in the drug-reaction association. This is largely due to the voluntary nature of the ADRAC reporting system. Some of the problems of such a reporting system are discussed in [3, 6, 10, 14, 17].

Therefore, we consider drug-reaction relationships not involving any other patient information. In other words we define for each drug a vector of weights which indicate the probability of occurrence of each reaction. This problem can be considered as a text categorization problem, where each patient is considered as one document, and the set of drug(s) taken by this patient is considered as a text related to this document; that is, each drug is considered as a word. For a review of some of the issues in text categorization see [12, 19, 20]. In this chapter, together with the algorithm $A(p)$, described below, we applied the algorithm Boostexter (version AdaBoost.MH with real-valued predictions [12]) which seems to be suitable for drug-reaction representations purposes.

4 Drug-Reaction Representations

We denote by $\mathcal{X}$ the set of all patients and by $\mathcal{D}$ the set of all drugs used by these patients. Let c be a finite number of possible reactions (classes). Given patient $x \in \mathcal{X}$, we denote by $\mathcal{Y}(x) = (\mathcal{Y}_1(x), \mathcal{Y}_2(x), \cdots, \mathcal{Y}_c(x))$ a c-dimensional vector of reactions observed for this patient; where $\mathcal{Y}_i(x) = 1$ if the reaction i has occurred, and $\mathcal{Y}_i(x) = 0$ if it has not. Let $D(x)$ be the set of all drugs taken by the patient x. In the ADRAC data, the number of drugs reported for a patient is restricted to 10. Some of these drugs are reported as suspected drugs responsible in the reactions $\mathcal{Y}(x)$. Therefore, we divide the set $D(x)$ into two parts: $DS(x)$ - the set of suspected drugs and $DN(x)$ - the set of non-suspected drugs. Clearly $D(x) = DS(x) \cup DN(x)$, and it may be $DN(x) = \emptyset$. We also note that, in the ADRAC data, for some patients, suspected drugs are reported in the form of interaction.

The goal of the study of drug-reaction relationships is to find a function $h : \mathcal{D} \to R_+^c$, where given drug $d \in \mathcal{D}$ the components h_i of the vector $h(d) = (h_1, h_2, \ldots, h_c)$ are the weights ("probabilities") of the occurrence of the reactions $i = 1, 2, \ldots, c$. Here R_+^c is the set of all c-dimensional vectors with non-negative coordinates.

In the next step, given a set of drugs $\Delta \subset \mathcal{D}$, we need to define a vector

$$H(\Delta) = (H_1(\Delta), H_2(\Delta), \ldots, H_c(\Delta)), \tag{6}$$

where the component $H_i(\Delta)$ indicates the probability of occurrence of the reaction i after taking the drugs Δ. In other words, we need to define a function $H : S(\mathcal{D}) \to R_+^c$, where $S(\mathcal{D})$ is the set of all subsets of $\mathcal{D}$.

Let $\Delta \subset \mathcal{D}$. The vectors $h(d)$ show what kind of reactions are caused by the drugs $d \in \Delta$. Therefore the vector $H(\Delta)$ can be considered as potential reactions which could occur after taking the drugs Δ. But what kind of reactions will occur? This will depend upon the individual characteristics of the patient as well as external factors. Different patients can have different predispositions for different reactions. Some reactions which have potentially high degrees of occurrence may not be observed because of the strong resistance of the patient to developing these reactions. But the existence of these potential reactions could have an influence on the patient somehow. The results obtained in [8] have shown that the information about the existence of potential reactions (but which were not reported to ADRAC) helps to make prediction of reaction outcomes (*bad* and *good*) more precise.

The function H can be defined in different ways and it is an interesting problem in terms of ADR(s). We will use the linear (sum) function $H(\Delta)$ (see [8]), where the components $H_i(\Delta)$ are defined as follows:

$$H_i(\Delta) \;=\; \sum_{d \in \Delta} h_i(d), \quad i = 1, \ldots, c. \tag{7}$$

The use of this function means that, we accumulate the effects from different drugs. For example, if $h_i(d_n) = 0.2$ $(n=1,2)$ for some reaction i, then there exists a potential of 0.4 for this reaction; that is, the two small effects (i.e. 0.2) become a greater effect (i.e. 0.4). This method seems more natural, because physically both drugs are taken by the patient, and the outcome could even be worse if there were drug-drug interaction(s).

Given patient $x \in \mathcal{X}$, we can define potential reactions $\mathcal{H}(x) = H(\Delta)$ corresponding to the set of drugs $\Delta \subset D(x)$. If $\Delta = D(x)$, then we have $\mathcal{H}(x) = \mathcal{H}^A(x) \doteq H(D(x))$, which means that all the drugs taken by the patient x are used in the definition of potential reactions; whereas, if $\Delta = DS(x)$, then $\mathcal{H}(x) = \mathcal{H}^S(x) \doteq H(DS(x))$, which means that we use only $suspected$ drugs neglecting all the others. We can also consider the potential reactions $\mathcal{H}(x) = \mathcal{H}^N(x) \doteq H(DN(x))$.

Therefore, drug-reaction relationships will be represented by vectors $h(d)$, $d \in \mathcal{D}$. The definition of these vectors depends on the drugs that are used in the calculations: we can use either all drugs or only suspected drugs. The evaluation of different drug-reaction representations can be defined by the closeness of two vectors: $\mathcal{H}(x)$, the vectors of potential (predicted) reactions, and $\mathcal{Y}(x)$, the vectors of observed reactions. We will use the evaluation measure $Average\ Precision$, presented in Section 5, to describe the closeness of these reaction vectors.

Our main goal in this chapter is to study the usefulness and correctness of the suspected drugs reported in ADRAC data.

The usefulness of suspected drugs is examined in the prediction of reactions. For this, first we define vectors $h(d)$ by using all drugs, and then by using only suspected drugs. In this way, we obtain potential reactions $\mathcal{H}^A(x)$ and $\mathcal{H}^S(x)$, respectively. We evaluate the closeness of these reactions to the observed reactions $\mathcal{Y}(x)$ over the all training and test sets. The calculations are made by the algorithms BoosTexter ([12]) and $A(p)$ described below. Note that, in all cases, the suspected drugs (that is, potential reactions $\mathcal{H}^S(x)$) provided better results. This means that the suspected drugs in the ADRAC data are identified "sufficiently correctly".

Then, we aim to evaluate correctness. We consider the case when the drug-reaction relations $h(d)$ are defined by using only suspected drugs. We calculate potential reactions related to the suspected and non-suspected drugs. To evaluate the correctness of suspected drugs, we find the convex combination of these two vectors of potential reactions which provides the minimal distance to the observed reactions. The weighting of the suspected drugs, in this optimal combination, is taken as an evaluation value of the correctness of suspected drugs. The calculations are made only by the algorithm $A(p)$. Note that the algorithm BoosTexter could not be used to evaluate correctness.

5 Evaluation Measure: Average Precision

To evaluate the accuracy of established drug-reaction relations by a given classifier (h, H); that is, to evaluate the closeness of the two vectors $\mathcal{H}(x)$ (predicted reactions) and $\mathcal{Y}(x)$ (observed reactions) we will use the *Average Precision* measure considered in [12]. Note that, this measure allow us to achieve more completely evaluation in multi-label classification problems.

Let $Y(x) = \{l \in \{1, \ldots, c\} : \mathcal{Y}_l(x) = 1\}$ be the set of reactions that have been observed for the patient x and $\mathcal{H}(x) = \{\mathcal{H}_1(x), \cdots, \mathcal{H}_c(x)\}$ be potential reactions calculated for this patient. We denote by $\mathcal{T}(x)$ the set of all ordered reactions $\tau = \{i_1, \ldots, i_c\}$ satisfying the condition

$$\mathcal{H}_{i_1}(x) \geq \ldots \geq \mathcal{H}_{i_c}(x);$$

where $i_k \in \{1, \ldots, c\}$ and $i_k \neq i_m$ if $k \neq m$.

In the case, when the numbers $\mathcal{H}_i(x)$, $i = 1, \cdots, c$, are different, there is just one order τ satisfying this condition. But if there are reactions having the same potential reactions then we can order potential reactions in different ways; that is, in this case the set $\mathcal{T}(x)$ contains more than one order.

Given order $\tau = \{\tau_1, \ldots, \tau_c\} \in \mathcal{T}(x)$, we define the rank for each reaction $l \in Y(x)$ as $rank_\tau(x; l) = k$, where the number k satisfies $\tau_k = l$. Then *Precision* is defined as:

$$P_\tau(x) = \frac{1}{|Y(x)|} \sum_{l \in Y(x)} \frac{|\{k \in Y(x) : \ rank_\tau(x; k) \leq rank_\tau(x; l)\}|}{rank_\tau(x; l)}.$$

Here, we use the notation $|S|$ for the cardinality of the set S. This measure has the following meaning. For instance, if all observed reactions $Y(x)$ have occurred on the top of ordering τ then $P_\tau(x) = 1$. Clearly the number $P_\tau(x)$ depends on order τ. We define

$$P_{best}(x) = \max_{\tau \in \mathcal{T}(x)} P_\tau(x) \quad \text{and} \quad P_{worst}(x) = \min_{\tau \in \mathcal{T}(x)} P_\tau(x),$$

which are related to the "best" and "worst" ordering. Therefore, it is sensible to define the *Precision* as the midpoint of these two versions: $P(x) = (P_{best}(x) + P_{worst}(x))/2$.

Average Precision over all records $\mathcal{X}$ will be defined as:

$$P_{av} = \frac{1}{|\mathcal{X}|} \sum_{x \in \mathcal{X}} P(x).$$

6 The Algorithm $A(p)$

Given a vector $V = (V_1, \cdots, V_c)$, with nonnegative coordinates, we will use the notation

$$\|V\| = \sum_{i=1}^{c} V_i. \tag{8}$$

Let $x \in \mathcal{X}$. We define the distance between predicted potential reactions $\mathcal{H}(x) = (\mathcal{H}_1(x), \ldots, \mathcal{H}_c(x))$ and observed reactions $\mathcal{Y}(x) = (\mathcal{Y}_1(x), \ldots, \mathcal{Y}_c(x))$ as:

$$dist\,(\mathcal{H}(x), \mathcal{Y}(x)) = \sum_{i=1}^{c} (\bar{\mathcal{H}}_i(x) - \mathcal{Y}_i(x))^2; \tag{9}$$

where the sign "bar" stands for a normalization with respect to the number of observed reactions $\|\mathcal{Y}(x)\|$:

$$\bar{\mathcal{H}}_i(x) = \begin{cases} \frac{\|\mathcal{Y}(x)\|}{\|\mathcal{H}(x)\|}\,\mathcal{H}_i(x) & \text{if } \|\mathcal{H}(x)\| > 0; \\ 0 & \text{if } \|\mathcal{H}(x)\| = 0. \end{cases} \tag{10}$$

The Algorithm $A(p)$ uses the following distance measure (we assume that $\|\mathcal{Y}(x)\| > 0$):

$$dist_p\,(\mathcal{H}(x), \mathcal{Y}(x)) = \|\mathcal{Y}(x)\|^{-p} \cdot dist\,(\mathcal{H}(x), \mathcal{Y}(x)), \quad p = 0, 1, 2. \tag{11}$$

Note that, these distance functions are slightly different from the Linear Least Squares Fit (LLSF) mapping function considered in [19], [20].

We explain the difference between distances $dist_p$, $p = 0, 1, 2$. Consider the case $\|\mathcal{H}(x)\| > 0$. The following representation is true:

$$dist_p\,(\mathcal{H}(x), \mathcal{Y}(x)) = \sum_{i=1}^{c} (a_i - b_i)^2;$$

where $a_i = \frac{\|\mathcal{Y}(x)\|^{1-\frac{p}{2}}}{\|\mathcal{H}(x)\|}\,\mathcal{H}_i(x)$, $b_i = \|\mathcal{Y}(x)\|^{-\frac{p}{2}}\,\mathcal{Y}_i(x)$, and clearly

$$\sum_{i=1}^{c} a_i = \sum_{i=1}^{c} b_i = \|\mathcal{Y}(x)\|^{1-\frac{p}{2}}.$$

In the distance $dist_0$ (that is, $p = 0$) the sums $\sum_i a_i$ and $\sum_i b_i$ are equal to the number of reactions $\|\mathcal{Y}(x)\|$. For $dist_1$ and $dist_2$ the corresponding sums are equal to $\sqrt{\|\mathcal{Y}(x)\|}$ and 1, respectively. $dist_1$ can be considered as a middle version, because the number of reactions $\|\mathcal{Y}(x)\| \geq 1$ and therefore

$$1 \leq \sqrt{\|\mathcal{Y}(x)\|} \leq \|\mathcal{Y}(x)\|.$$

It is not difficult to observe that the following property holds:

$$dist_p\,(\lambda\,\mathcal{H}(x), \mathcal{Y}(x)) = dist_p\,(\mathcal{H}(x), \mathcal{Y}(x)), \quad \text{for all } \lambda > 0. \tag{12}$$

The algorithm $A(p)$ aims to define drug-reaction relations $h(d)$ minimizing the average distance $dist_p\left(\mathcal{H}(x),\mathcal{Y}(x)\right)$ over all training examples. In other words, we consider the following optimization problem:

$$E_{av}^{p} = \frac{1}{|\mathcal{X}|}\sum_{x\in\mathcal{X}} dist_p\left(\mathcal{H}(x),\mathcal{Y}(x)\right) \;\rightarrow\; \min; \tag{13}$$

$$\text{subject to}: \quad h_i(d) \geq 0, \quad i = 1,...,c, \;\; d \in \mathcal{D}. \tag{14}$$

Here $|\mathcal{X}|$ stands for the cardinality of the set $\mathcal{X}$. Note that, taking different numbers $p = 0, 1, 2$, we get different versions $A(p)$, $p = 0, 1, 2$, which generate different drug-reaction representations $h(d)$.

6.1 Calculation of Weights for Each Drug. A Solution to the Optimization Problem (13),(14).

The function in (13) is non-convex and non-linear, and therefore may have many local minimum points. We need to find the global optimum point. The number of variables is $|\mathcal{D}| \cdot c$. For the data Card20, that we will consider, $|\mathcal{D}| = 3001$ and $c = 5$. Thus we have a global optimization problem with 15005 variables, which is very hard to handle using existing global optimization methods. Note that, we also tried to use local minimization methods which were unsuccessful. This means that there is a clear need to develop new optimization algorithms for solving problem (13),(14), taking into account some peculiarities of the problem.

In this chapter we suggest one heuristic method for finding a "good" solution to the problem (13),(14). This method is based on the proposition given below.

We denote by S the unit simplex in R^c; that is,

$$S = \{h = (h_1,\ldots,h_c): \;\; h_i \geq 0, \; h_1 + \ldots h_c = 1\}.$$

In this case for each $h(d) \in S$ the component $h_i(d)$ indicates simply the probability of the occurrence of the reaction i.

Given drug d we denote by $X(d)$ the set of all records in $\mathcal{X}$, which used just one drug - d. Simply, the set $X(d)$ combines all records where the drug d was used alone.

Consider the problem:

$$\sum_{x\in X(d)} \|\mathcal{Y}(x)\|^{-p} \cdot \sum_{j=1}^{c}(\|\mathcal{Y}(x)\| h_j(d) - \mathcal{Y}_j(x))^2 \;\rightarrow\; \min, \tag{15}$$

$$h(d) = (h_1(d),\ldots,h_c(d)) \in S. \tag{16}$$

Proposition 6.1 *A point* $h^*(d) = (h_1^*(d), \ldots, h_c^*(d))$, *where*

$$h_j^*(d) = \left(\sum_{x \in X(d)} \|\mathcal{Y}(x)\|^{2-p} \right)^{-1} \cdot \sum_{x \in X(d)} \|\mathcal{Y}(x)\|^{1-p} \, \mathcal{Y}_j(x), \quad j = 1, \ldots, c, \quad (17)$$

is the global minimum point for the problem (15),(16).

Now, given drug d, we consider the set $X_{all}(d)$ which combines all records that used the drug d. Clearly $X(d) \subset X_{all}(d)$. The involvement of other drugs makes it impossible to solve the corresponding optimization problem similar to (15), (16). In this case, we will use the following heuristic approach to find a "good" solution.

(S) (Single). The set $X(d)$ carries very important information, because here the drug d and reactions are observed in a pure relationship. Therefore, if the set $X(d)$ contains a "sufficiently large" number of records, then it will be reasonable to define the weights $h_j(d)$, $(j = 1, \ldots, c)$ only by this set neglecting all the mixed cases.

We consider two numbers: $|X(d)|$ - the number of cases where the drug is used alone, and $P(d) = 100|X(d)|/|X_{all}(d)|$ - the percentage of these cases. To determine whether the set $X(d)$ contains enough records we need to use both numbers. We will consider a function $\phi(d) = a|X(d)| + bP(d)$ to describe how large the set $X(d)$ is.

Therefore, if the number $\phi(d) \geq p^*$, where p^* is a priori given number, then we use only the set $X(d)$ to calculate weights $h(d)$; in other words, we use formula (17) which provides a global minimum $h(d) = h^*(d)$ for the part of data $X(d) \subset \mathcal{X}$.

We denote by $\mathcal{D}'$ the set of all drugs from $\mathcal{D}$ for which the weights are calculated in this way.

(M) (Mixed). If the set $X(d)$ is not "sufficiently large"; that is, $\phi(d) < p^*$, then we have to use the set $X_{all}(d)$ which contains patients $x \in \mathcal{X}$ having more than one drug taken. In this case we use $h(d) = h^{**}(d)$; where

$$h_j^{**}(d) = \left(\sum_{x \in X(d)} \|\mathcal{Y}(x)\|^{2-p} \right)^{-1} \cdot \sum_{x \in X(d)} \|\mathcal{Y}(x)\|^{1-p} \frac{rem\,(\mathcal{Y}_j(x))}{|\Delta''(x)|}, \quad j = 1, \ldots, c.$$

$$(18)$$

Here, given a patient x, the set $\Delta''(x) = \Delta(x) \setminus \mathcal{D}'$ combines all drugs the weights for which are not calculated in the first step. Note that, $\Delta(x)$ is the set of drugs corresponding to the patient x, and we will consider either $\Delta(x) = D(x)$ (all drugs) or $\Delta(x) = DS(x)$ (suspected drugs).

$rem\,(\mathcal{Y}_j(x))$, $j \in \{1, \cdots, c\}$, stands for the "remaining" part of the reaction $\mathcal{Y}_j(x)$, associated with the drugs $\Delta''(x)$. For the calculation of $rem\,(\mathcal{Y}_j(x))$ see Section 6.2.

This formula has the following meaning. If $|\Delta''(x)| = 1$ for all $x \in X_{all}(d)$, then, given $rem\,(\mathcal{Y}_j(x)),\ j = 1, \cdots, c,$ formula (18) provides a global minimum solution (similar to (17)).

If $|\Delta''(x)| > 1$, for some patient $x \in X_{all}(d)$, then we use the assumption that all suspected drugs are responsible to the same degree; that is, for this patient, we associate only the part $1/|\Delta''(x)|$ of the reactions $rem\,(\mathcal{Y}_j(x))$ to this drug.

Therefore, we define $h(d) = (h_1(d), \ldots h_c(d))$ as follows:

$$h(d) = \begin{cases} h^*(d) & \text{if } \phi(d) \geq p^*; \\ h^{**}(d) & \text{otherwise;} \end{cases} \tag{19}$$

where $h^*(d)$ and $h^{**}(d)$ are defined by (17) and (18), respectively.

Remark 6.1 *We note that the weight $h_i(d)$ is not exactly a probability of the occurrence of the reaction i; that is, the sum $\sum_{i=1}^{c} h_i(d)$ does not need to be equal to 1.*

6.2 Calculation of $rem\,(\mathcal{Y}(x))$

Consider a particular patient $x \in \mathcal{X}$. We divide the set of drugs $\Delta(x)$ (which could be all drugs or suspected drugs used by this patient) into two parts:

$$\Delta(x) = \Delta'(x) \cup \Delta''(x);$$

where for each drug d, in the set $\Delta'(x)$, the vector $h(d)$ has already been defined by (17), and the set $\Delta''(x)$ combines all the other drugs. Note that it may be $\Delta'(x) = \emptyset$.

We set $\mathcal{H}(x) = G(x) + Z(x)$ where the sum $G(x) = \sum_{d \in \Delta'(x)} h(d)$ defines a part of potential reactions associated with the drugs $\Delta'(x)$, and $Z(x)$ the other (unknown) part which will be defined by the drugs $\Delta''(x)$: $Z(x) = \sum_{d \in \Delta''(x)} h(d)$.

We will use a reasonable assumption that all drugs in $\Delta''(x)$ are responsible in the observed reactions $\mathcal{Y}(x)$ at the same degree; that is, we associate equal parts $1/|\Delta''(x)|$ of $Z(x)$ to each drug in $\Delta''(x)$. Therefore, after accepting such an assumption, we need only to find $Z(x)$ which is an optimal solution to the problem

$$dist_p\,(\mathcal{H}(x), \mathcal{Y}(x)) \to \min.$$

This problem is equivalent (see (11)) to

$$dist(\mathcal{H}(x), \mathcal{Y}(x)) = \sum_{i=1}^{c} (\bar{\mathcal{H}}_i(x) - \mathcal{Y}_i(x))^2 \to \min, \tag{20}$$

where the "bar" stands for a normalization with respect to the reactions $\mathcal{Y}(x)$ (see (10)).

As we consider a particular patient x, for sake of simplicity, we will drop the sign x. Therefore, to find the vector $Z = (Z_1, \cdots, Z_c)$, we need to solve the following problem:

$$\sum_{i=1}^{c} \left(\frac{\|\mathcal{Y}\| (G_i + Z_i)}{\sum_{j=1,\ldots,c}(G_j + Z_j)} - \mathcal{Y}_i \right)^2 \to \min, \qquad (21)$$

$$\text{subject to:} \quad 0 \le Z_i, \quad G_i + Z_i \le Z^{max}, \quad i = 1,\ldots,c. \qquad (22)$$

We denote by $\phi(Z^{max})$ the optimal value of the objective function in the problem (21),(22).

Proposition 6.2 *The vector* $Z^* = (Z_1^*, \cdots, Z_c^*)$, *where*

$$Z_i^* = (Z^{max} - G_i)\mathcal{Y}_i, \quad i = 1, \cdots, c,$$

is the optimal solution to the problem (21),(22). Moreover, $\phi(Z^{max}) \to 0$ *as* $Z^{max} \to \infty$.

This proposition shows that for the optimal solution Z^* the the sums $Z_i^* + G_i$, $i = 1, \cdots, c$, $\mathcal{Y}_i = 1$ are constant being equal to Z^{max}.

We also note that we can decrease the distance $\phi(Z^{max})$ by increasing the umber Z^{max}. Note that, Z^{max} serves to restrict the values Z_i in order to get

$$\max_{i=1,\cdots,c} (G_i + Z_i) = 1, \qquad (23)$$

which means that the patient x would be taken into account with the weight 1 (like the patients in $X(d)$). Therefore, we need to chose a number Z^{max} close to 1.

We will define the number Z^{max} as follows. Denote $G^0 = \max\{G_i : i = 1, \cdots, c, \ \mathcal{Y}_i = 0\}$, $G^1 = \max\{G_i : i = 1, \cdots, c, \ \mathcal{Y}_i = 1\}$. Then, we set

$$Z^{max} = \max\{1, \ G^0 + \varepsilon, \ G^1\}, \quad \text{where } \varepsilon > 0.$$

The choice of such a number Z^{max} has the following meaning. First, we note that if $G^0 < 1$ and $G^1 \le 1$ then there is a number $\varepsilon > 0$ such that $Z^{max} = 1$; that is, (23) holds. On the other hand, if $G^0 \ge 1$ and $G^0 \ge G^1$, then the weights $Z_i^* + G_i$, corresponding to the occurred reactions i (that is, $\mathcal{Y}_i = 1$) will be grater than the weights $Z_i^* + G_i$, corresponding to the non-occurred reactions. In this case, choosing the number $\varepsilon > 0$ smaller, we get more closer approximation to (23).

Therefore, $rem(\mathcal{Y}_j(x))$ will be defined as

$$rem\,(\mathcal{Y}_j(x)) = Z_j^*, \quad j = 1, \cdots, c. \qquad (24)$$

7 Evaluation of Correctness of Suspected Drugs Reported

Drug-reaction representations in the form of a vector of weights allow us to evaluate the correctness of suspected drugs reported.

Consider a particular patient x and let $D(x)$ be the set of drugs used by this patient and $\mathcal{Y}(x)$ be the set of observed reactions. The set $D(x)$ consists of suspected drugs $DS(x)$ and non-suspected drugs $DN(x)$. Our aim in this section is to evaluate how correctly suspected drugs are identified.

The method of evaluation is based on distance measure (9). Assume that for each drug $d \in \mathcal{D}$ the vector of weights $h(d)$ are calculated. Then we can define potential reactions $\mathcal{H}^S(x)$ and $\mathcal{H}^N(x)$, corresponding to the sets of suspected drugs and non-suspected drugs, respectively. We have

$$\mathcal{H}_i^S(x) = H_i(DS(x)) = \sum_{d \in DS(x)} h_i(d), \quad i = 1, \cdots, c;$$

$$\mathcal{H}_i^N(x) = H_i(DN(x)) = \sum_{d \in DN(x)} h_i(d), \quad i = 1, \cdots, c.$$

The method, used in this chapter for the evaluation of suspected drugs, can be identified as "all suspected drugs versus all non-suspected drugs". For this aim we consider convex combinations of these two group of drugs and try to find the optimal combination which provides the maximal closeness to the observed vector of reactions. In other words we are looking for a combination of suspected and non-suspected drugs which is optimal in the sense of distance (9). Before considering convex combinations we need to be careful about the "comparability" of the vectors $\mathcal{H}^S(x)$ and $\mathcal{H}^N(x)$ in the sense of scaling. For this reason, it is meaningful to consider convex combinations of normalized (see (10)) vectors $\bar{\mathcal{H}}^S(x)$ and $\bar{\mathcal{H}}^N(x)$. Therefore we define

$$\bar{\mathcal{H}}(x,\mu) = \mu\,\bar{\mathcal{H}}^S(x) + (1-\mu)\,\bar{\mathcal{H}}^N(x), \quad 0 \le \mu \le 1. \tag{25}$$

Note that, $\|\bar{\mathcal{H}}^S(x)\| = \|\bar{\mathcal{H}}^N(x)\| = \|\mathcal{Y}(x)\|$ and, therefore, $\|\bar{\mathcal{H}}(x,\mu)\| = \|\mathcal{Y}(x)\|$ for all $\mu \in [0,1]$.

The number μ indicates the proportion of suspected and non-suspected drugs in the definition of potential reactions. Clearly, $\bar{\mathcal{H}}(x,1) = \bar{\mathcal{H}}^S(x)$ and $\bar{\mathcal{H}}(x,0) = \bar{\mathcal{H}}^N(x)$, which implies

$$dist(\bar{\mathcal{H}}(x,1), \mathcal{Y}(x)) = dist(\mathcal{H}^S(x), \mathcal{Y}(x)) \quad \text{and}$$

$$dist(\bar{\mathcal{H}}(x,0), \mathcal{Y}(x)) = dist(\mathcal{H}^N(x), \mathcal{Y}(x)).$$

The combination of all drugs with the same weight; that is, the vector $\mathcal{H}^A(x) = H(D(x)) = \mathcal{H}^S(x) + \mathcal{H}^N(x)$ is also included in (25). To confirm this, it is sufficient to consider the case $\|H^S(x)\| > 0$ and $\|H^N(x)\| > 0$.

In this case $\|H^A(x)\| = \|H^S(x)\| + \|H^N(x)\| > 0$. Then we take $\mu' = \|\mathcal{H}^S(x)\|/\|\mathcal{H}^A(x)\| \in (0,1)$ and get (see (10))

$$\bar{\mathcal{H}}(x,\mu') = \mu' \frac{\|\mathcal{Y}(x)\|}{\|\mathcal{H}^S(x)\|} \mathcal{H}^S(x) + (1-\mu') \frac{\|\mathcal{Y}(x)\|}{\|\mathcal{H}^N(x)\|} \mathcal{H}^N(x)$$

$$= \frac{\|\mathcal{Y}(x)\|}{\|H^A(x)\|} \mathcal{H}^S(x) + \frac{\|\mathcal{Y}(x)\|}{\|\mathcal{H}^A(x)\|} \mathcal{H}^N(x) = \bar{\mathcal{H}}^A(x);$$

which implies

$$dist(\bar{\mathcal{H}}(x,\mu'), \mathcal{Y}(x)) = dist(\bar{\mathcal{H}}^A(x), \mathcal{Y}(x)) = dist(\mathcal{H}^A(x), \mathcal{Y}(x)).$$

Consider the following minimization problem with respect to μ;

$$f(\mu) \doteq dist\left(\bar{\mathcal{H}}(x,\mu), \mathcal{Y}(x)\right) = \sum_{i=1}^{c}(\bar{\mathcal{H}}_i(x,\mu) - \mathcal{Y}_i(x))^2 \;\rightarrow\; \min; \quad 0 \le \mu \le 1.$$

$$(26)$$

The optimal solution μ^* to problem (26) gives an information about the correctness of definition of suspected drugs. For instance, if $\mu^* = 1$ then we see that the suspected drugs provide the better approximation to the observed reactions than if we involve the other drugs. We refer this situation as 100 percent correctness. Whereas, if $\mu^* = 0$ then non-suspected drugs provide better approximation to the observed reactions and we can conclude that in this case suspected drugs are defined completely wrong. Therefore, the optimal value μ^* can be considered as an evaluation measure for the correctness of suspected drugs.

From (11) we obtain:

Proposition 7.1 *The optimal solution μ^* to the problem (26) is optimal with respect to the all distance measures $dist_p$, $p = 0, 1, 2$; that is, given vectors of weights $h(d)$, $d \in D(x)$, for all $p = 0, 1, 2$ the following inequality holds:*

$$dist_p(\bar{\mathcal{H}}(x,\mu^*), \mathcal{Y}(x)) \le dist_p(\bar{\mathcal{H}}(x,\mu), \mathcal{Y}(x)), \quad \text{for all} \quad \mu \in [0,1].$$

This proposition shows that given patient $x \in \mathcal{X}$ and given vectors of weights $h(d)$, the definition of correctness of suspected drugs, as an optimal value μ^*, does not depend on choice of distance functions $dist$ and $dist_p$, $p = 0, 1, 2$.

It is clear that problem (26) can have many optimal solutions μ^*; that is, different proportions of suspected and non-suspected drugs can provide the same closeness to the observed reactions. In this case we will define the correctness of suspected drugs, as the maximal value among the all optimal solutions μ^*:

$$\mu^*(x) = \max\{\mu^* : \; \mu^* \text{ is an optimal solution to (26)}\}. \tag{27}$$

The reason for such a definition can be explained; for instance, if $\mu^* = 1$ (only suspected drugs) and $\mu^* = 0$ (only non-suspected drugs) are the two different optimal solutions, giving the closest approximation to the observed reactions, then there would be no reason to doubt about the correctness of suspected drugs.

Problem (26) can be easily solved. Let

$$A = \sum_{j=1}^{c} \left(z_i \, \|\mathcal{Y}(x)\| - z \, \mathcal{Y}_i(x) \right) \left(z_i \, \|\bar{\mathcal{H}}^N(x)\| - z \, \bar{\mathcal{H}}_i^N(x) \right);$$

$$B = \sum_{j=1}^{c} \left(\bar{\mathcal{H}}_i^N(x) \, \|\mathcal{Y}(x)\| - \|\bar{\mathcal{H}}^N(x)\| \mathcal{Y}_i(x) \right) \left(z_i \, \|\bar{\mathcal{H}}^N(x)\| - z \, \bar{\mathcal{H}}_i^N(x) \right);$$

where $z_i = \bar{\mathcal{H}}_i^S(x) - \bar{\mathcal{H}}_i^N(x)$, $z = \|\bar{\mathcal{H}}^S(x)\| - \|\bar{\mathcal{H}}^N(x)\|$. Then, we find the derivative of the function $f(\mu)$, defined by (26), in the following form:

$$f'(\mu) = \frac{2}{(z\mu + \|\bar{\mathcal{H}}^N(x)\|)^4} (A\mu + B). \tag{28}$$

From (28) we have

Proposition 7.2 *The optimal solution $\mu^*(x)$ to the problem (26) can be found as follows.*
1) Let $A = 0$. Then

$$\mu^*(x) = \begin{cases} 0 \text{ if } B > 0; \\ 1 \text{ otherwise.} \end{cases}$$

2) Let $A > 0$. Then

$$\mu^*(x) = \begin{cases} 0 & \text{if } B > 0; \\ \min\{1, \ -B/A\} & \text{otherwise.} \end{cases}$$

3) Let $A < 0$. Then

$$\mu^*(x) = \begin{cases} 0 \text{ if } f(0) < f(1); \\ 1 \quad \text{otherwise.} \end{cases}$$

Therefore, we have defined the correctness of suspected drugs for a particular patient x. Given set of patients $\mathcal{X}$, *Average Correctness* of suspected drugs will be calculated as

$$P_{sus} = \frac{1}{|\mathcal{X}|} \sum_{x \in \mathcal{X}} \mu^*(x). \tag{29}$$

7.1 Remark

The above definition of correctness of suspected drugs can be considered as a method where the group of suspected drugs (already defined) are taken versus the group of non-suspected drugs. In fact, drug-reaction representations in the form of vectors of weights allow us to consider more general statements of this problem. Here, we formulate some of them.

Consider a patient $x \in \mathcal{X}$. Let $D(x) = \{d_1, d_2, \cdots, d_{|D(x)|}\}$ be the set of drugs that have been taken, and $h(d)$, $d \in D(x)$, be the vectors of weights for these drugs. We introduce a $|D(x)|$−dimensional vector $v = (v_1, v_2, \cdots, v_{|D(x)|})$, where the i-th component v_i indicates the degree of responsibility of the drug d_i in the observed reactions. In this case, potential reactions can be defined as follows:

$$\mathcal{H}(x, v) = \sum_{i=1}^{|D(x)|} v_i \, h(d_i).$$

Then we can consider the following optimization problem:

$$dist\,(\mathcal{H}(x, v), \mathcal{Y}(x)) \quad \rightarrow \quad \min; \quad 0 \le v_i \le 1, \quad i = 1, \cdots, |D(x)|. \qquad (30)$$

If, for instance, the optimal value for a particular drug i is 0.6 (that is, $v_i = 0.6$) then we say that the degree of responsibility of this drug in the observed reactions is 0.6. Such information about drugs is more complete than just saying "suspected" or "non-suspected". But the application of this method encountered the absence of any such kind of classification of suspected drugs in the ADRAC data.

We can also consider a special case when each variable v_i takes only two values: 1 (which means that i is a suspected drug) and 0 (which means that i is a non-suspected drug). In this case, we obtain a combinatorial optimization problem which is to find an optimal subset of drugs that provides the closest approximation to the observed reactions.

The application of problem (30) allows us to study suspected drugs in each report more precisely. It is our opinion that, the determining of responsibility of each drug from the set of drugs have been taken, is a very important problem in terms of ADRs. But there are some issues that should be mentioned.

Such a precise statement of the problem should be accomplished with more precise definitions of function $H(x)$ (in this chapter we use (7)) and then weights $h(d)$ which have even more impact on the results. First of all the times of starting and withdrawing drugs should be taking into account. Such information is presented in the ADRAC data but more research needs to be done in this area. The other factor that could be helpful for more precise definitions of weights $h(d)$, relates to the amount of general use of each drug, and the difficulty of getting such information is the major factor in ADR problems.

8 Interaction of Drugs

Interaction of drugs is one of the main problems of ADR. In [8] this problem was considered from a statistical point of view. Interaction of drugs was defined as a case when these drugs together cause a reaction which is different from the reactions that could have occurred if they were used alone. In this chapter we aim to study the possibility of using vectors of weights $h(d)$ calculated for each drug d, for drug-drug interactions. In other words, we aim to check the closeness of potential reactions to the observed reactions for patients having interactions of drugs. In this way we establish the accuracy with which the potential reactions could be used for the prediction of reactions in drug-drug interaction cases.

We will use the following two methods for the evaluation of correctness.

1). First we will use the methodology developed in the previous section. We divide the set of drugs $D(x)$, taken by a particular patient x, into two subsets: $I(x)$ is the set of drugs which are reported as interaction, $O(x)$ is the set of all other drugs.

As we are interested in drug-drug interactions, we will consider only records x where the set $I(x)$ contains at least two drugs and the set $O(x)$ is not empty.

As in the previous section, we define potential reactions $\mathcal{H}^I(x)$ and $\mathcal{H}^O(x)$, corresponding to the drugs $I(x)$ and $O(x)$, respectively. Then we consider convex combinations of these vectors:

$$\mathcal{H}(x,\mu) = \mu\,\mathcal{H}^I(x) + (1-\mu)\,\mathcal{H}^O(x).$$

Similar to (26) and (27), the maximal optimal value $\mu = \mu^{**}(x)$ which minimizes the distance $dist\,(\mathcal{H}(x,\mu), \mathcal{Y}(x))$, will be taken as the degree of responsibility of the drugs $I(x)$ in the observed reactions $\mathcal{Y}(x)$.

Given a set of patients the $\mathcal{X}$, *Average Responsibility* of drugs in interaction will be calculated as

$$P_{int} = \frac{1}{|\mathcal{X}|}\sum_{x\in\mathcal{X}}\mu^{**}(x). \tag{31}$$

Then we apply the evaluation measure presented in Section 5. This will provide a precision $P(x)$ calculated for each patient x having interaction effects of drugs.

The numbers $\mu^{**}(x)$ and $P(x)$ give some information about each interaction case. For instance, if $\mu^{**}(x) = 1$ and $P(x) = 1$, (that is, 100 percent) then we can conclude that the potential reactions defined by the drugs $I(x)$ provide 100 percent correct prediction of reactions. Therefore, in this case, we can say that the potential reactions could be used for reaction predictions in the case of interactions.

9 Results of Numerical Experiments

For our analysis we use two algorithms. The first algorithm $A(p)$ is described above. The second algorithm that we use is BoosTexter which has shown good performance in text categorization problems. These two algorithms produce the weighted vector $H(x)$ for each patient x. The methods of calculating the vectors $H(x)$ are quite different: $A(p)$ uses only drugs have been taken by the patient x, whilst BoosTexter uses all drugs in the list of ("weak hypotheses" generated (that is, the drugs that have not been taken by the patient x, are used for the calculation of the vector $H(x)$). Our hope is that, the application of these two quite different methods can make the results obtained more accurate.

We will consider three versions of the algorithm $A(p)$, corresponding to the distance functions $dist_p$, $p = 0, 1, 2$, respectively. Each of these versions tends to minimize the average distance calculated by its own distance measure.

The weights for each drug are calculated by formula (19). We used a function $\phi(d) = |X(d)| + P(d)$ to describe the informativeness of the set $X(d)$. We also need to set a number p^*. The calculations show that the results are not essentially changed for different values of p^* in the region $p^* \geq 30$. We set $p^* = 80$ in the calculations.

The second algorithm that we will use is the well known text categorization algorithm BoosTexter, version AdaBoost.MH with real-valued predictions ([12]). The main reason for using this algorithm is that it produces predictions in the form $H(x) = (H_1(x), \ldots, H_c(x))$, where the numbers $H_i(x)$ are real values which can be positive or negative. In other words, this algorithm defines potential reactions that we are interested in.

To apply the distance measure described above, we need to make all weights calculated by BoosTexter non-negative. Let $H_{min}(x) = \min_{i=1,\ldots,c} H_i(x)$. Then we set $\mathcal{H}(x) = H(x)$, if $H_{min}(x) \geq 0$; and

$$\mathcal{H}(x) = (H_1(x) - H_{min}(x), \ldots, H_c(x) - H_{min}(x), \quad \text{if} \quad H_{min}(x) < 0.$$

In the calculations below, we ran this algorithm with the number of rounds set at 3000. Note that BoosTexter defines a weak hypothesis using one drug at each round. In the Data there are 2896 suspected drugs. Therefore, choosing the number of rounds 3000, alows the possibility of using all suspected drugs.

9.1 New Drugs and Events

We define *a new drug* (in the test set) as a case when this drug either is a new drug which has not occurred in the training set or has never been considered as a suspected drug in the training set. For all such new drugs d, we set $h_i(d) = 0, i = 1, \ldots, c$. It is possible that for some new (test) example all suspected drugs are new. We call this case s *new event*. This situation mainly relates to the fact that, new drugs are constantly appearing on the market. Obviously, to make predictions for such examples does not make sense. Therefore, in the calculations below, we will remove all new events from test sets.

9.2 Training and Test Sets

In the calculations below we take as a test set records sequentially from each year, starting from 1996 until 2001. For example, if records from 1999 are taken as a test set, then all records from years 1971-1998 form a training set. In Table 1 we summarized the number of records in test and training sets, and, also, the number of new events removed. In the second part of this table we presented the number of records in training and test sets having at least two drugs have been used.

Table 1. Card20. *The training and test sets. 'Removed' means the number of records removed from the test set. For example, in 1996 there are 1147 records and 98 of them are new events. Then, the number of records in the test set for this year is 1049 (=1147-98)*

Test	Number of Records			Records with $\geq$ 2 drugs	
Year	Training	Test	Removed	Training	Test
1996	12600	1049	98	6270	597
1997	13747	1091	163	6905	552
1998	15001	1418	265	7513	673
1999	16684	1746	169	8290	494
2000	18599	1988	158	8801	519
2001	20749	1054	65	9329	433

9.3 The Effectiveness of Using Suspected Drugs for Reaction Predictions. Prediction of Reactions

We consider calculations for two cases. First we consider all drugs as suspected; that is, we do not involve the suspected drugs reported. In the second case we consider only suspected drugs. The results obtained by the algorithms $A(p)$ and BoosTexter are presented in Table 2. The results for Average Precision are presented in percentages. In these tables "*All*" means that all drugs were used for definition of the reaction weights for drugs, and "*Sus*" means that only suspected drugs were used.

From Table 2 we observe that, in all cases, the results obtained for training sets are better if only the suspected drugs are considered. This means that,

Table 2. The results obtained for Average Precision (P_{av}) by using all (*All*) drugs and suspected (*Sus*) drugs. *The algorithm BoosTexter2_1 [12] was set to run 3000 training rounds. Average Precision is presented in percent*

Test Year	drugs	BoosTexter		$A(0)$		$A(1)$		$A(2)$	
		Training	Test	Training	Test	Training	Test	Training	Test
1996	*All*	83.90	79.07	81.45	79.97	82.68	80.07	83.74	79.78
	Sus	84.15	80.42	82.91	80.25	83.96	80.07	84.51	79.61
1997	*All*	83.83	79.15	81.53	79.05	82.69	79.48	83.83	79.23
	Sus	84.17	80.66	82.80	79.79	83.97	80.20	84.51	79.98
1998	*All*	83.77	77.32	81.58	77.86	82.69	78.64	83.70	78.47
	Sus	84.10	78.23	82.82	78.02	83.94	78.59	84.49	78.52
1999	*All*	83.55	81.17	81.52	80.41	82.58	80.61	83.55	80.96
	Sus	83.90	81.30	82.72	80.91	83.79	80.94	84.37	80.89
2000	*All*	83.41	77.80	81.69	77.57	82.72	77.85	83.65	77.45
	Sus	83.86	78.46	82.72	77.88	83.79	78.26	84.35	77.67
2001	*All*	83.13	77.70	81.35	76.87	82.51	77.43	83.44	77.58
	Sus	83.64	77.91	82.32	76.85	83.52	77.29	84.09	77.50

definition of weights $h(d)$ by using only suspected drugs provides more accurate approximation to the observed reactions. The weights obtained in this way also work, in general, better in test sets. This emphasizes the effectiveness of determining suspected drugs in each adverse drug reaction case.

The next problem is to define suspected drugs more accurately. The fact that using only suspected drugs provided better results allows us to conclude that, in the ADRAC data (at least records related to the cardiovascular type of reactions) suspected drugs are defined "sufficiently correctly". In the next section we aim to evaluate this correctness.

The algorithms BoosTexter and $A(p)$ define the potential reactions $\mathcal{H}(x)$ in quite different ways. There are some important points that make using the algorithm $A(p)$ preferable for the study of drug-reaction associations.

First we note that, the algorithm $A(p)$ calculates weights for each drug, which is very important because in this case we establish drug-reaction relations for all drugs. BoosTexter does not calculate weights for each drug. Moreover, BoosTexter classifies examples so that drugs that are not used are still assigned weights in the function $H(x)$. In the other words, reactions are predicted not only by drugs actually used, but also, drugs which were not taken. This leads to the situation where we could say that, for example, patient x has the first reaction, because he/she did not take some drugs (which are in the list of "week hypothesis" generated by BoosTexter). But anyhow, applying the algorithm BoosTexter is very useful for having some idea about the possible "maximal" accuracy that could be achieved in reaction predictions.

One of the advantages of the algorithm $A(p)$ includes the determination of weights for each drug, and, then the classification of reactions, observed for each patient, on the basis of drugs actually used by this patient. This advantage allows us to use the algorithm $A(p)$ to study the identification of suspected drugs and of drug-drug interactions.

9.4 Evaluation of Correctness of Suspected Drugs Reported

In this section we will evaluate the correctness of suspected drugs reported. The methodology is described in Section 7. As mentioned above, BoosTexter can not be used for this. So only the algorithm $A(p)$ is used.

The results obtained in the previous section have shown that potential reactions calculated by using suspected drugs provide more accurate predictions of reactions. Therefore, in the calculations below weights for each drugs will be calculated only by suspected drugs.

The case when a patient uses only one drug, is not interesting to consider, because in this case there is no doubt that the drug used should be a suspected drug. That is why, we consider records having two or more drugs that have been taken. The number of patients in training and test sets are presented in Table 1.

Table 3. Evaluation of correctness of suspected drugs (P_{sus}) obtained by Algorithm $A(p)$

Test	$A(0)$		$A(1)$		$A(2)$	
Year	Training	Test	Training	Test	Training	Test
1996	78.1	72.0	78.7	71.9	78.4	72.6
1997	78.0	72.2	78.6	71.3	78.2	71.3
1998	77.7	71.3	78.2	71.8	77.8	71.0
1999	77.3	85.0	77.9	85.4	77.5	85.7
2000	77.8	90.4	78.2	90.3	77.6	89.3
2001	78.5	70.2	78.8	69.4	78.4	67.8

The results are presented in Table 3. We see that, the suspected drugs reported in the ADRAC data are determined with sufficiently high accuracy. For instance, the accuracy 78.0 means that, in the optimal combination of suspected and non-suspected drugs which provides the closest approximation to the observed reactions, the suspected drugs are used with weight 0.78 (non-suspected - 0.22). This could be considered as a high degree of "responsibility".

Note that, the correct identification of suspected drugs in each new report is a very important problem. The method described here provides us an alternative method which can be used for this aim.

9.5 Interaction of Drugs

As mentioned before the study of interactions of drugs is one of the interesting problems. We consider here the possibility of using vectors of weights in drug-drug interactions. In other words, we aim to evaluate the closeness of potential reactions (calculated by a vector of weights) to the observed reactions in drug-drug interaction cases. For the evaluation of closeness we will use two measures: *Average Responsibility - P_{int}* and *Average Precision P_{av}*.

For our analysis we consider the records having more than 3 drugs, where some of drugs were reported as an interaction (in ADRAC data the value 2 was associated to this drugs) and the others were reported as non-suspected (the value 0 was used in this case). Of course, to make the problem of evaluation of drug-drug interactions meaningful, we need to consider the records for which both parts are not empty sets.

Table 4. Evaluation of drug-drug interactions by using a vector of weights obtained by Algorithm $A(p)$

Test	Number of	P_{int}				P_{av}			
Year	records	$A(0)$	$A(1)$	$A(2)$	max	$A(0)$	$A(1)$	$A(2)$	max
1996	4	71.8	83.3	78.2	90.6	89.6	89.6	89.6	89.6
1997	17	70.8	73.9	75.5	80.9	70.3	79.6	83.5	83.5
1998	16	69.3	71.3	65.2	72.0	61.0	64.1	71.4	71.9
1999	6	66.7	66.7	66.7	66.7	62.5	70.8	64.4	70.8
2000	5	63.3	70.4	85.6	85.6	73.3	73.3	90.0	90.0
2001	14	77.1	73.1	63.0	77.1	85.6	83.8	89.1	92.7
	62	70.9	72.7	70.2	77.4	72.1	75.8	80.7	82.3

The results obtained are presented in Table 4. Training sets are used for the calculating of weights for each drug. AS in the previous section, the weights are calculated by using suspected drugs. Then the evaluation of interaction of drugs is made only using the test sets, because, in the training sets, the interaction of drugs (as suspected) are used for the calculation of weights. The number of cases in the test sets are also presented in Table 4.

In the last row of Table 4, we present the average results obtained by all test sets which combines 62 cases. The results obtained by the algorithm $A(2)$ is: $P_{sus} = 70.2$, $P_{av} = 80.7$. The first number means that, in the observed reactions, the "degree of responsibility" of the drugs, in interaction cases, is 70.2 percent. The second number indicates high accuracy in the prediction of these reactions. This emphasizes that, drug-drug interaction cases could be successfully explained by the weights calculated for each drug.

In fact the accuracy of this method could be much higher if we could calculate weights more "correctly". To show this, we did the following.

First we note that, the numbers P_{int} and P_{av} are the average values of $\mu^{**}(x)$ and $P(x)$ calculated for each patient x. Different versions $A(p)$ provide different values $\mu^{**}(x)$ and $P(x)$. We take the corresponding maximal values obtained by different versions, and then calculate the average responsibility and precision. The results obtained are presented in the columns "max" in Table 4. We see that these results are much better than the results obtained by a particular version.

10 Conclusion

In this chapter, we have used a new optimization approach to study multi-label classification problems. In particular we have focussed on drug-reaction relations in the domain of the Cardiovascular group of reactions from the ADRAC data. The suggested method of representation for drug-reaction relations in the form of a vector of weights is examined in the prediction of reactions. In particular it was shown that, the suspected drugs reported in the ADRAC data provide more accurate drug-reaction information. The suggested method was applied for the evaluation of *correctness* of suspected drugs. The results obtained have shown that the reactions that occurred in the cases of interaction of drugs reported in the ADRAC data, could be predicted by this method with sufficiently high accuracy.

References

1. J.K. Aronson, S. Derry, and Y.K. Loke. Adverse drug reactions: keeping up to date. *Fundamental & Clinical Pharmacology*, 16:49–56, 2002.
2. A. Bate, M. Lindquist,I.R. Edwards, S. Olsson, O. Orre, A. Lansner, and R. De Freitas. A Bayesian neural network method for adverse drug reaction signal generation. *European Journal of Clinical Pharmacology*, 54(4):315–321, 1998.
3. S.D. Brown Jr. and F.J. Landry. Recognizing, Reporting, and Reducing Adverse Drug Reactions. *Southern Medical Journal*, 94: 370-374, 2001.
4. D.M. Coulter, A. Bate, R.H.B. Meyboom, M. Lindquist, and I.R. Edwards. Antipsychotic drugs and heart muscle disorder in international pharmacovigilance: data mining. *British Medical Journal*, 322(7296):1207–1209, 2001.
5. W. DuMouchel. Bayesian data mining in large frequency tables, with an application to the FDA spontaneous reporting system. *American Statistician*, 53(3):177–190, 1999.
6. E. Heely, J. Riley, D. Layton, L.V. Wilton, and S.A.W. Shakir. Prescription-event monitoring and reporting of adverse drug reactions. *The Lancet*, 358: 182-184, 2001.
7. T.A. Hutchinson. Bayesian assessment of adverse drug reactions. *Canadian Medical Association Journal*, 163(11):1463-1466, 2000.
8. M.A. Mamedov and G.W. Saunders. An Analysis of Adverse Drug Reactions from the ADRAC Database. Part 1: Cardiovascular group. University of Ballarat School of Information Technology and Mathematical Sciences, Research Report 02/01, Ballarat, Australia, 1-48, February 2002. [http://www.ballarat.edu.au/itms/research-papers/paper2002.shtml]
9. M. Mamedov, G. Saunders, and J. Yearwood. A Fuzzy Derivative Approach to Classification of outcomes from the ADRAC database. *International Transactions in Operational Research*, 11(2):169-180, 2004.
10. M. Pirmohamed, A.M. Breckenridge, N.R. Kitteringham, and B.K. Park. Adverse drug reactions. *British Medical Journal*, 316: 1294-1299, 1998.
11. W.S. Redfern, I.D. Wakefield, H. Prior, C.E. Pollard, T.G. Hammond, and J.-P. Valentin. Safety pharmacology – a progressive approach. *Fundamental & Clinical Pharmacology*, 16:161–173, 2002.
12. R.E. Schapire and Y. Singer. Boostexter: A boosting-based system for text categorization. *Machine Learning*, 39: 135-168, 2000.
13. A. Szarfmann, S.G. Machado, and R.T. O'Neill. Use of Screening Algorithms and Computer Systems to Efficiently Signal Higher-Than-Expected Combinations of Drugs and Events in the US FDA's Spontaneous Reports Database. *Drug Safety*, 25(6): 381-392, 2002.
14. W.G. Troutman and K.M. Doherty. Comparison of voluntary adverse drug reaction reports and corresponding medical records. *American Journal of Health-System Pharmacy*, 60: 572-575, 2003.
15. P. Tubert-Bitter and B. Begaud. Comparing Safety of Drugs. *Post Marketing Surveillance*, 7: 119-137, 1993.
16. E.P. van Puijenbroek, A. Bate, H.G.M. Leufkens, M. Lindquist, R. Orre, and A.C.G. Egberts. A comparison of measures of disproportionality for signal detection in spontaneous reporting systems for adverse drug reactions. *Pharmacoepidemiology and Drug Safety*, 11(1): 3-10, 2002.

17. E.P. van Puijenbrock, W.L. Diemount, and K. van Grootheest. Application of Quanitative Signal Detection in the Dutch Spontaneous Reporting System for Adverse Drug Reactions. *Drug Safety*, 26: 293-301, 2003.
18. WHO Technical Report No 498, 1972 and Note for Guidance on Clinical Safety Data Management: Definitions and Standards for Expedited Reporting (CPMP/ICH/377/95).
19. Y. Yang and X. Liu. A Re-examination of Text Categorization Methods. *Proceedings of SIGIR-99, 22nd ACM International Conference on Research and Development in Information Retrieval*, Vol. 39, pages 42-49, 1999.
20. Y. Yang, J. Zhang, and B. Kisiel. A Scalability Analysis of Classifiers in Text Categorization. *SIGIR'03, Toronto, Canada, July 28-August 1, 2003.*

Classifying Noisy and Incomplete Medical Data by a Differential Latent Semantic Indexing Approach

Liang Chen[1], Jia Zeng[1], and Jian Pei[2]

[1] Computer Science Department
University of Northern British Columbia
Prince George, BC, Canada V2N 4Z9
{chenl,zeng}@unbc.ca
[2] School of Computing Science
Simon Fraser University
Burnaby, BC Canada V5A 1S6
jpei@cs.sfu.ca

Summary. It is well-recognized that medical datasets are often noisy and incomplete due to the difficulties in data collection and integration. Noise and incompleteness in medical data post substantial challenges for accurate classification. A *differential latent semantic indexing (DLSI)* approach which is an improvement of the standard LSI method has been proposed for information retrieval and demonstrated improved performance over standard LSI approach. The key idea is that DLSI adapts to the unique characteristics of individual record/document. By experimental results on real datasets, we show that DLSI outperforms the standard LSI method on noisy and incomplete medical datasets. The results strongly indicate that the DLSI approach is also capable of medical numerical data analysis.

Key words: Medical data classification, Latent semantic indexing, Differential latent semantic indexing

1 Introduction

It is well-recognized that medical datasets are often noisy and incomplete. For example, different breast cancer patients may take different sets of examinations. On the other hand, the laboratory tests may also introduce certain noise in probability, though persistent efforts have been committed to the control. Noise and incompleteness in medical datasets post substantial challenges to data analysis. In general, error-tolerant methods are strongly desirable. By observing the ways how medical data is obtained and fact that a physician's diagnosis is always based on a patient's integrated symptoms, we believe that the attributes of medical data are always correlated in a certain way, which makes error-tolerant methods possible.

In this chapter, we are particularly interested in the problem of classification on noisy and incomplete medical datasets. Classification is a well researched area, with many well developed methods, such as SVM, C4.5, neural network methods, and so on. Many previously developed methods can handle a limited amount of noise. However, many of the previous methods are not robust to incomplete data. For example, the Neural Network and Support Vector Machine (SVM) approaches, which have been widely used in the field of automated learning, take each attribute as an independent element so that their results might be invalid if a data record is incomplete. A vector space model-based method, Latent Semantic Indexing (LSI), which relies on the constituent terms to suggest the document's semantic content [1], is popularly used for content-based text information retrieval and document classification. By using Singular Value Decomposition (SVD), relationships between terms can be discovered in the vector space model [2]. The LSI method has been successfully extended to the area of content-based image data analysis. However, like all global projection schemes, LSI also encounters a difficulty in adapting to the unique characteristics of each document [3]. Differential Latent Semantic Indexing (DLSI) approach, exploiting both the distances to and the projections on a reduced document space improves the performance and the robustness of the classifier [3]. By simply using a posteriori calculation of the intra- and extra-attribute vector statistics, this new method demonstrates advantage over standard LSI approach.

In this chapter, we investigate the application of DLSI approach on medical numerical data analysis. The rest of the chapter is organized as follows. We present the preliminaries and the algorithm of DLSI approach in Section 2. A case study on the Wisconsin Breast Cancer Data (WBCD) is reported in Section 3. We conclude the chapter and project our future work in Section 4.

2 The DLSI Approach

The DLSI approach takes a data object as a vector of attributes, and projects attribute vectors into dimension-reduced spaces as in most of the vector space models. However, DLSI uses not only the projections of vectors on the reduced space but also the distances from the attribute vectors to the reduced space. In this section, we will first provide the preliminaries of the method, and then describe the algorithm in detail.

2.1 Preliminaries

For an object in a dataset, the attribute vector can be simply formed by exploiting the attribute values of the object.

Technically, consider a dataset has n objects and each object has m attributes. For each object j, we assign an *attribute vector* $(a_{1j}, a_{2j}, \cdots, a_{mj})^T$, where a_{ij} is the value of attribute i in object j. For a class of objects whose attribute vectors are $I_1, I_2, \cdots, I_k$, we calculate the *mean vector* $(s_1, s_2, \cdots, s_m)^T$ of the member objects in the class as

$$(s_1, s_2, \cdots, s_m)^T = \frac{1}{k} \sum_{j=1}^{k} I_j.$$

A *differential attribute vector* is defined as $I_i - I_j$, where I_i and I_j are two attribute vectors. A *differential intra-attribute vector* D_I is the differential attribute vector defined as $I_i^{int} - I_j^{int}$, where I_i^{int} and I_j^{int} are two attribute vectors belonging to the same class. A *differential extra-attribute vector* D_E is the differential attribute vector defined as $I_i^{ext} - I_j^{ext}$, where I_i^{ext} and I_j^{ext} are two attribute vectors belonging to two different classes. Note that the mean vector of a class is also regarded as an attribute vector of the class. Therefore we could use it to construct the differential attribute vector as well.

The corresponding differential intra-attribute matrices D_I and D_E are defined as matrices, each column of which comprises a differential intra- and extra-attribute vectors, respectively.

2.2 The Algorithm

The algorithm consists of two phases, namely the construction of the DLSI-based classifier and the classification using the DLSI-based classifier. They are described as follows.

Setting Up the DLSI-Based Classifier

Based on every object in the database, the classifier sets up the parameters of the posteriori function for each class. Details of this procedure are described as below.

(1) Construct attribute vector for every item in the database;
(2) Construct the differential intra-attribute matrix $D_I^{m \times n_I}$, such that each of its columns is a differential intra-attribute vector. For a class with s elements, we may include at most $(s-1)$ differential intra-attribute vectors in D_I to avoid linear dependency among columns.
(3) Decompose D_I, by an SVD algorithm, into $D_I = U_I S_I V_I^T$, $S_I = diag(\delta_{I,1}, \delta_{I,2}, \dots)$. Find an appropriate k_I. The way to choose value k is not fixed. Although reduction in k can help removing noise, keeping too few dimensions may lose important information as well. Therefore, only by applying experiments on certain dataset and observing its performance on different k, can we get the most appropriate value. We apply

D_I to get an approximate matrix D_{I,k_I}, where $D_{I,k_I} = U_{k_I} S_{k_I} V_{k_I}^T$. Then evaluate the likelihood function:

$$P(x|D_I) = \frac{n_I^{1/2} exp(-\frac{n_I}{2} \sum_{i=1}^{k_I} \frac{y_i^2}{\delta_{I,i}^2}) exp(-\frac{n_I \varepsilon^2(x)}{2\rho_I})}{(2\pi)^{n_I/2} \prod_{i=1}^{k_I} \delta_{I,i} \cdot \rho_I^{(r_I-k_I)/2}}, \tag{1}$$

where $y = U_{k_I}^T x$, $\varepsilon^2(x) = (\|x\|)^2 - \sum_{i=1}^{k_I} y_i^2$, $\rho_I = \frac{1}{r_I-k_I} \sum_{i=k_I+1}^{r_I} \delta_{I,i}^2$, and r_I is the rank of matrix D_I;

(4) Construct the differential extra-attribute matrix $D_E^{m \times n_E}$, such that each of its columns is a differential extra-attribute vector;

(5) Decompose D_E by an SVD algorithm into $D_E = U_E S_E V_E^T$, $S_E = diag(\delta_{E,1}, \delta_{E,2}, \dots)$. Find an appropriate k_E, and apply it to get an approximate matrix D_{E,k_E}, where $D_{E,k_E} = U_{k_E} S_{k_E} V_{k_E}^T$. Then calculate the likelihood function:

$$P(x|D_E) = \frac{n_E^{1/2} exp(-\frac{n_E}{2} \sum_{i=1}^{k_E} \frac{y_i^2}{\delta_{E,i}^2}) exp(-\frac{n_E \varepsilon^2(x)}{2\rho_E})}{(2\pi)^{n_E/2} \prod_{i=1}^{k_E} \delta_{E,i} \cdot \rho_E^{(r_E-k_E)/2}}, \tag{2}$$

where $y = U_{k_E}^T x$, $\varepsilon^2(x) = (\|x\|)^2 - \sum_{i=1}^{k_E} y_i^2$, $\rho_E = \frac{1}{r_E-k_E} \sum_{i=k_E+1}^{r_E} \delta_{E,i}^2$, and r_E is the rank of matrix D_E;

(6) Define the posteriori function as

$$P(D_I|x) = \frac{P(x|D_I)P(D_I)}{P(x|D_I)P(D_I) + P(x|D_E)P(D_E)}, \tag{3}$$

where $P(D_I)$ is set to $1/n_c$, and n_c is the number of classes in the database and $P(D_E)$ is set to $1 - P(D_I)$.

SVD transform is the most time-consuming step in the above algorithm. In general, for a database of n objects with m attributes, the computational complexity of the algorithm is $O(nm^2)$.

Automatic Classification

In order to classify a new object into the most proper class, the classifier evaluates the Bayesian posteriori function based on the analysis of the new object. The procedure can be described as follows.

(1) For a new object to be classified, set up its attribute vector Q by assigning its attributes' values to the vector. For each class in the database, repeat the procedure of objects (2)-(4) below;

(2) Construct a differential attribute vector $x = Q - S$, where S is the mean vector of the class;

(3) Calculate the intra-attribute likelihood function $P(x|D_I)$ and the extra-attribute likelihood function $P(x|D_E)$ for the object;

(4) Evaluate the Bayesian posteriori probability function $P(D_I|x)$;

(5) Select the class having the largest $P(D_I|x)$ as the recall candidate.

3 A Case Study

In this section, we will describe a case study. We apply the DLSI classifier on a real dataset.

3.1 The WBCD Dataset

To examine our algorithm, we did experiments on a well-known benchmark database, WBCD, to test the effectiveness of the classifier. The database is available from public domain ftp://ftp.cs.wisc.edu/math-prog/cpo-dataset/machine-learn/. WBCD is a breast cancer sample collection periodically collected by Dr. Wolberg in his clinical cases. Each sample is assigned a 9-dimensional vector, whose components are integers between 1 and 10, with value 1 corresponding to a normal state and 10 to a most abnormal state [4]. The aim of classification is to distinguish between benign and malignant cancer objects. The original database has 699 objects. Its class distribution is of 65% benign objects and 35% malignant objects.

3.2 DLSI Space-Based Classifier

There are two classes in this dataset, denoted as C_1 and C_2, respectively. Firstly, we set up the attribute vector for every sample object in the dataset and calculate the mean vectors for C_1 and C_2, respectively. Then, we construct the differential attribute matrix by assigning each of its columns to be $I_{i,j1} - S_{j2}$, where $I_{i,j1}$ represents the i-th attribute vector in class j_1 and S_{j2} represents the mean vector for class j_2. For D_I, $j_1 = j_2$ holds, while for D_E, $j_1 \neq j_2$ holds. Using the SVD algorithm, we decompose them into $D_I = U_I S_I V_I^T$ and $D_E = U_E S_E V_E^T$.

The dataset includes 16 objects with missing attributes and 1 outlying object. These sample data objects are discarded before training, as it was carried out by other methods. We chose $k_I = 5$ and $k_E = 2$ in this example, which reserve 88.2% and 92.0% of the original matrices, respectively. Using Equations (1), (2) and (3), we calculate $P(x|D_I)$, $P(x|D_E)$ and finally $P(D_I|x)$ for each differential attribute vector $x = Q - S_j$ (j=1,2). The class C_i having a larger $P(D_I|Q - S_i)$ value is chosen as the class to which the new object Q belongs. Table 1 is the classification accuracy we get based on the classifier trained by subsets of the original database. The training subsets range from 20% to 80% of the whole database, the testing datasets are the corresponding complement sets.

Table 1. DLSI Classifier Over Partial Training Set

Percentage Over Database For Training Set	20%	30%	40%	50%	60%	70%	80%
Accuracy On Testing Set	95.8%	96.6%	95.1%	97.6%	97.1%	96.6%	97.1%

In order to demonstrate the ability of our classifier to deal with incomplete data objects in test dataset (i.e., some attribute values are missing), we artificially add evenly distributed noises into the original database to get the noise-contaminated databases. Each time we replace the value of attribute i of every sample object in the original database by a randomly picked integer whose range is from 1 to 10, to get DB_i.[3] Table 2 shows its classification performance over the database with noises.

Table 2. DLSI Classifier Over Noise-Contaminated Testing Set

Testing Set	DB_1	DB_2	DB_3	DB_4	DB_5	DB_6	DB_7	DB_8
Accuracy	96.0%	96.8%	96.9%	97.1%	96.8%	92.4%	97.4%	96.8%

Testing Set	DB_9
Accuracy	96.5%

3.3 Classifier Evaluation

To date, much research work has been done on the WBCD dataset. Methods such as the Rule Generation approach, the Fuzzy-Genetic approach, and the Neural Network approach have been presented in the literature. These methods achieve good classification performance on the WBCD dataset as well. However, we would like to point out two common features of these methods. One is that there is always some limitation on the application of the methods and their good results highly depend on the success of carefully examining the network structure and extracting rules. Taha and Gosh proposed a method using a Neural Network to extract rules [6]. But it can be applied to data with binary attributes only. Setiono proposed a method based on finding a set of concise rules by using pruned Neural Network [7], his classifier achieves good performance but the extraction of rules is manually processed, which involves much human intervention. Another feature these methods share is that the incomplete sample objects in the original data base have been simply discarded [8].

---

[3] DB_i denotes database whose attribute i is affected by noise.

In our method, the classifier obtains an average of 96.6% accuracy for the dataset where all sample objects are complete, as shown in Table 1. An accuracy of 97.6% is obtained when we use 50% of the original database to train the system and test it on the remaining 50% data. The classification performance still achieves 95.8% accuracy when using only 20% data (that is, 136 samples) for training. As far as we know, no known method has ever reached such a high performance with less than 200 training samples. The only comparable result using small training set can be found in Wolberg & Mangasarian's paper [9], where an accuracy of 93.5% is achieved by using 185 samples for training.

As discussed earlier, in the field of medicine, it is common to encounter dataset with incomplete entries. Most of the researchers ignore the cases with missing attributes. Experiments displayed in Table 1.2 clearly show that DLSI approach is extremely stable for incomplete data.

4 Concluding Remarks

We presented a classification method using the DLSI approach for breast cancer data. Overall, it achieves very good performance. It also gets good classification accuracy for the noise-contaminated databases. Compared to some of the best known methods to date, our classifier not only attains high performance with a much smaller training sample set, but also demonstrates high robustness by achieving promising results on incomplete test sets.

In our current method, the attribute vector is constructed by the numerical values of the attributes of the objects, without considering the medical meaning of the attributes. Based on the understanding that one of the advantages of DLSI approach is its ability in computing with concepts, we believe that we should take into account the literal description of the data in the future work.

References

1. T. A. Letsche, and M. W. Berry. Large-Scale Information Retrieval with Latent Semantic Indexing. *Information Sciences - Applications*, 100: 105-137, 1997.
2. M. W. Berry, Z. Drmac, and E. R. Jessup. Matrices, Vector Spaces, and Information Retrieval. *SIAM Review*, 41(2): 335-362, 1999.
3. L. Chen, N. Tokuda, and A. Nagai. A New Differential LSI Space-Based Probabilistic Document Classifier. *Information Processing Letters*, 88: 203-212, 2003.
4. O. L. Mangasarian and W. H. Wolberg. Cancer Diagnosis via Linear Programming. *SIAM News*, 23(5): 1-18, 1990.
5. K. P. Bennett and O. L. Mangasarian. *Neural Network Training via Linear Programming*. Elsevier Science, 1992.
6. I. Taha and J. Gosh. Characterization of the Wisconsin Breast Cancer Database Using a Hybrid Symbolic-Connectionist System. *Tech. Report UT-CVISS-TR-97-007*, the Computer and Vision Research Center, University of Texas, Austin, 1996.
7. R. Setiono. Extracting Rules from Pruned Neural Network for Breast Cancer Diagnosis. *Artificial Intelligence in Medicine*, 8: 37-51, 1996.
8. R. Setiono. Generating Concise and Accurate Classification Rules for Breast Cancer Diagnosis. *Artificial Intelligence in Medicine*, 18: 205-219, 2000.
9. W. H. Wolberg and O. L. Mangasarian. Multisurface method of pattern separation for medical diagnosis applied to breast cytology. *Proceedings of the National Academy of Sciences*, 87: 9193–9196, 1990.

Ontology Search and Text Mining of MEDLINE Database

Hyunki Kim and Su-Shing Chen

Computer and Information Science and Engineering Department,
University of Florida,
Gainesville, Florida 32611, USA
{hykim,suchen}@cise.ufl.edu

Summary. With the explosion of biomedical data, information overload and users' inability of expressing their information needs may become more serious. To solve those problems, this chapter presents a text data mining method that uses both text categorization and text clustering for building concept hierarchies for MEDLINE citations. The approach we propose is a three-step data mining process for organizing MEDLINE database: (1) categorizations according to MeSH terms, MeSH major topics, and the co-occurrence of MeSH descriptors; (2) clustering using the results of MeSH term categorization; and (3) visualization of categories and hierarchical clusters. The hierarchies automatically generated may be used to construct multiple viewpoints of a collection. Providing multiple viewpoints of a document collection and allowing users to move among these viewpoints will enable both inexperienced and experienced searchers to more fully exploit the information contained in a document collection. User interfaces with multiple viewpoints for this underlying system are also presented.

Key words: Data mining, MEDLINE, Document Clustering, Self-Organizing Map, Multiple Viewpoints

1 Introduction

MEDLINE, developed by the U.S. National Library of Medicine (NLM), is a database of indexed bibliographic citations and abstracts. It contains over 4,600 biomedical journals [16]. MEDLINE citations and abstracts are searchable via PubMed or the NLM Gateway.

The NLM produces the MeSH (Medical Subject Headings) for the purposes of subject indexing, cataloging and searching journal articles in MEDLINE with an annual update cycle. MeSH consists of descriptors (or main headings), qualifiers (or subheadings), and supplementary concept records. It contains more than 19,000 descriptors which are used to describe the subject topic of an article. It also provides less than 100 qualifiers which are used to express a certain aspect of the concept represented by the descriptor. MeSH terms are arranged both alphabetically and in a hierarchical tree, in which specific subject categories are arranged beneath broader terms. MeSH terms provide a consistent way of retrieving information regardless of different terminology used by the authors in the original articles. By using MeSH terms, the user is able to narrow the search space in MEDLINE. As a result, by adding more MeSH terms to the query, retrieval performance may be improved [9].

However, there are inherent challenges, as well. There may be information overload [18], and users may be unable to express their information needs, in order to take full advantage of the MEDLINE database. MEDLINE contains over 12 million article citations. Beginning in 2002, it began to add over 2,000 new references on a daily basis [16]. Although the user may be able to limit the search space of MEDLINE with MeSH terms, keyword searches often result in a long list of results. For instance, when the user queries the term "Parkinson's Disease" by limiting it to the MeSH descriptors, PubMed returns over 21,000 results. Here, there is a problem of information overload, with the user having difficulty finding relevant information.

The inability of users to express information needs may become more serious, unless users have a precise knowledge in their area of interest, or an understanding of MeSH and its structure. The use of common abbreviations, technical terms, and synonyms in biomedical articles prevent users from articulating their information needs accurately. To avoid the vocabulary problem, MeSH may be used. However, it is difficult for an unfamiliar user to locate appropriate descriptors and/or qualifiers, since MeSH is a very complex thesaurus. Furthermore, new terms are added, some are modified, and others are removed each year as biomedical fields change. An imprecise query usually results in a long list of irrelevant hits [5]. Under such circumstances, a better mechanism is needed to organize information in order to help users explore within an organized information space [7].

In order to arrange the contents in a useful way, text categorization and text clustering have been researched extensively. Text categorization is a boiling down of the specific content of a document into a set of one or more pre-defined labels [10]. Text clustering can group similar documents into a set of clusters based on shared features among subsets of the documents [5,12].

In this chapter, we present a text data mining method that uses both text categorization and text clustering for building a concept hierarchy for MEDLINE citations. The approach we propose is a three-step data mining process for organizing MEDLINE database: (1) categorizations according to MeSH terms, MeSH major topics, and the co-occurrence of MeSH descriptors, (2) clustering using the results of MeSH term categorization, and (3) visualization of categories and hierarchical clusters. The hierarchies automatically generated may be used to support users in browsing behavior as well as help them identify good starting points for searching. An interface for this underlying system is also presented.

The remainder of the chapter is organized as follows. We describe the data mining procedure and the implementation of the SOM (Self-Organizing Map) algorithm in Section 2. Section 3 describes the user interfaces with multiple viewpoints, the MeSH major topic view, the MeSH term view, the MeSH co-occurrence view, and the subject-specific concept view. In Section 4, we briefly introduce related work on document clustering. Section 5 explains the problems encountered when we implemented the system. Conclusions are given in Section 6.

2 Data Mining Method for Organizing MEDLINE Database

In this Section, we will explain the data mining method proposed in detail. We used MySQL to store MEDLINE citations and additional data that was generated by the data mining process.

2.1 The Data

For the following experiment, we extracted a total of 1,736 citations encoded in XML (eXtensible Markup Language) from the query "Secondary Parkinson Disease", limiting the results to the MeSH major topic field and to citations with abstracts in MEDLINE.

2.2 Text Categorization

Categorization refers to an algorithm or procedure which results in the assignment of categories to documents [10]. We chose the MeSH major topic, the MeSH descriptor and qualifier, and a co-occurrence of MeSH descriptors as a feature to be used in classification. To categorize the collection according to the selected features, we first parsed the data collection encoded in XML using SAX (Simple API for XML). After extracting the MeSH major topics, the MeSH descriptors, and the co-occurrence of MeSH descriptors for each citation, we inserted the data into the corresponding MySQL tables.

2.3 Text Clustering using the Results of MeSH descriptor Categorization

Since many MeSH terms may be assigned to a citation and vice versa, categorization with the MeSH terms or the co-occurrence of MeSH terms often results in a large list or hierarchy. Some categories may contain a large number of documents. Simply listing categories associated with documents is inadequate for organizing data [10].

To alleviate this problem, the approach we propose here is to cluster the results of MeSH descriptor categorization using the hierarchical Self-Organizing Map (SOM). We chose only those MeSH descriptor categories whose document frequencies are over a predetermined threshold for clustering. Document frequency is the number of documents in which a term occurs. Terms are extracted and selected using category dependent document frequency thresholding from the categories chosen. There are two ways that document frequency is calculated: category independent term selection and category dependent term selection [6]. In category independent term selection, document frequency of each term is computed from all the documents in the collection and the selected set of terms are used on each category. In category dependent term selection, document frequency of each term is calculated from only those documents belonging to that category. Thus, different sets of terms are used for different categories.

After the feature selection and extraction, and the SOM clustering, a concept hierarchy is obtained, by relying on the MeSH descriptors for the top layer, and by using feature vectors extracted from the titles and abstracts for the sub-layer.

Feature Extraction and Selection

To produce a concept hierarchy using the SOM, documents must be represented by a set of features. For this purpose, we use full-text indexing to extract a list of terms (words or phrases). The input vector is constructed by indexing the title and abstract elements of the collection. We then weight these terms using the vector space model in Information Retrieval [19]. In the vector space model, documents are represented as term vectors using the product of the term frequency (TF) and the inverse document frequency (IDF). Each entry in the document vector corresponds to the weight of a term in the document. We used normalized TF x IDF term weighting scheme, best fully weighted scheme [19], so that longer documents are not given more weight and all values of a document vector are distributed in the range of 0 to 1. Thus, weighted word histogram can be viewed as the feature vector describing the document [13].

The preprocessing procedure is mainly divided into two stages: noun phrase extraction and term weighting. In the noun phrase extraction phase, we first fetched the MEDLINE identifier, the title and abstract elements from the collection and then tokenized the title and abstract elements based on Penn Treebank tokenization scheme to detect sentence boundaries, and to separate extraneous punctuations from the input text. The MEDLINE identifier was used as a document identifier. We then automatically assigned part of speech tags to words reflecting their syntactic category by using the rule-based part of speech tagger [2,3]. After recognizing the chunks that consist of noun phrases from the tagged text, we extracted a set of noun phrases for each citation. At this stage, we removed common terms by consulting a list of 906 stop words.

We computed document frequency of all terms using category dependent term selection for those MeSH descriptor categories whose document frequencies were over a predetermined threshold (in this experiment, greater than 100 times). We then eliminated terms from the feature space whose document frequency was less than a predetermined threshold (in this experiment, less than 10 times). Finally, we weighted the terms indexed using the best fully weighted scheme [19], and assigned corresponding term weights to each document for each category selected. Thus, the weighted term vector set can be used as the input vector set for the SOM.

Construction of a Concept Hierarchy

Document clustering is defined as grouping similar documents into a cluster. To improve retrieval efficiency and effectiveness, related documents should be collected together in the same cluster based on some notion of similarity

The Self-Organizing Map is an unsupervised learning neural network algorithm for the visualization of high-dimensional data. The SOM defines a mapping from the input data space onto a two-dimensional array of nodes. Every node i is represented by a model vector, also called *reference vector*, $m_i = [m_{i1}, m_{i2}, \ldots, m_{in}]$, where n is input vector dimension. Our algorithm is different from other SOM-variant algorithms, in that each sub-layer SOM dynamically reconstructs a new input vector from an upper-level input vector. The following algorithm describes how to construct a subject-specific concept hierarchy.

1. *Initialize network by using the subject feature vector as the input vector*: Create a two-dimensional map and randomly initialize model vectors m_i in the range of 0 to 1 to start from an arbitrary initial state.

2. *Present input vector in sequential order*: Cyclically present the input vector $x(t)$, the weighted input vector of an n-dimensional space, to all nodes in the network. Each entry in the input vector corresponds to the weight of a noun phrase in the document; zero means the term has no significance in the document or it simply doesn't exist in the document.

3. *Find the winning node by computing the Euclidean distance for each node*: In order to compare the input and weight vectors each node computes the Euclidean distance between its weight vector and the input vector. The smallest of the Euclidean distance identifies the best-matching node that is chosen as the winning node for that particular input vector. The best-matching node, denoted by the subscript c, is

$$\|x - m_c\| = \min_i \{\|x - m_i\|\}.$$

4. *Update weights of the winning node and its topological neighborhoods*: The
   update rule for the model vector of node i is

$$m_i(t+1) = m_i(t) + \alpha(t)h_{ci}(t)\left[x(t) - m_i(t)\right],$$

where t is the discrete-time coordinate, $\alpha(t)$ is the adaptation coefficient,
and $h_{ci}(t)$ is the neighborhood function, a smoothing kernel centered on
the wining node.

5. *Repeat steps 2-4 until all iterations have been completed.*

6. *Label nodes of the trained network with the noun phrases of the subject
   feature vectors*: For each node, we determine the dimension with the great-
   est value, and label the node with a corresponding noun phrase for that
   node, and aggregate nodes with the same noun phrase into groups. Thus,
   a subject-specific top-tier concept map is generated.

7. *Repeat steps 1-6 by using the description feature vector as the input vector
   for each grouped concept region*: For each grouped concept region containing more than k documents (e.g. 100), recursively create a sub-layer SOM
   and repeat steps 1-6 by using the description feature vector as the input
   vector. At this point new input feature vector of the sub-layer SOM is dynamically created by selecting only those items that belong to the concept
   region represented by its parent SOM from the description feature vector.
   Thus, different sets of feature vectors are used for different clusters and
   this reduces the training time significantly.

For each MeSH descriptor category containing more than 100 documents,
we generated a concept hierarchy using the SOM, limiting the maximum level
of hierarchy to 3. We built a 10 x 10 SOM, and presented each input vector 100
times to the SOM. We then recursively built the sub-layer concept hierarchy
by training a new 10 x 10 SOM with a new input vector, which is dynamically
constructed by selecting only a document feature vector contained in the
concept region from the upper-level feature vector. The concept hierarchy
generated contains two kinds of information: category labels extracted from
the MeSH descriptors for the top-level, and the concept hierarchy using the
SOM for the sub-layer. We inserted this information into the MySQL database
to build an interactive user interface.

2.4 Results

For the results of categorization, we extracted 2,210 distinct MeSH descriptors,
70 distinct MeSH qualifiers, 269 distinct MeSH major topics, and 60,192 co-
occurring MeSH descriptors from the collection. On average, each citation in
the collection contains 14 MeSH descriptors, 10 MeSH qualifiers, and 4 MeSH
major topics.

For text clustering, we identified a total of 20,367 distinct terms from the
collection after the stop word removal. A total of 22 categories containing

more than 100 citations were identified from the results of MeSH descriptor categorization. After the category dependent document frequency thresholding, an average of 66 terms were selected per category, ranging from 14 terms for one category to 260 terms for another category. After the hierarchical SOM clustering, 193 distinct concepts were generated from 22 categories.

3 User Interfaces

We provided four different views, three category hierarchies and one clustering hierarchy for the users. We represented this hierarchy information as hierarchical trees to help users understand MeSH qualifiers and descriptors, so that they could find a set of documents of interest, and locate good starting points for searching.

3.1 MeSH Major Topic Tree and MeSH Term Tree

The MeSH term tree displays the categorized information space, arranged by first descriptors and then qualifiers. Figure 1 shows the interface of the MeSH term tree.

In each level of hierarchy, MeSH terms are listed in alphabetical order, along with their document frequencies. When the user clicks on a category label that is either a descriptor or a qualifier on the left pane, the associated document set is displayed on the right pane. At this point, if the category is a descriptor, the associated qualifiers in the collection are also expanded as its children in the tree. Users can see more detailed information of a document by clicking on the title of a document that is shown on the right pane. To help users better understand the meaning of an ambiguous MeSH term, the corresponding descriptor data and context in the MeSH tree may be displayed by clicking on the link "MeSH Descriptor Data & Tree Structures" within each level of the tree.

In some cases, the user may want to see the category arranged by only MeSH major topics. The MeSH major topic tree provides the same information as the MeSH term tree except that it shows the category hierarchy arranged by only MeSH major topics.

3.2 MeSH Co-occurrence Tree

The MeSH co-occurrence tree provides the co-occurrence of MeSH descriptors, along with their co-occurrence frequency in the collection. Since an average of 14 MeSH descriptors are assigned to each citation in the collection, there are a large number of nodes in the co-occurrence tree. To better organize the co-occurrence tree, the interface allows the user to select the co-occurrence frequency range. Thus, the user can easily identify co-occurring semantic types in the collection.

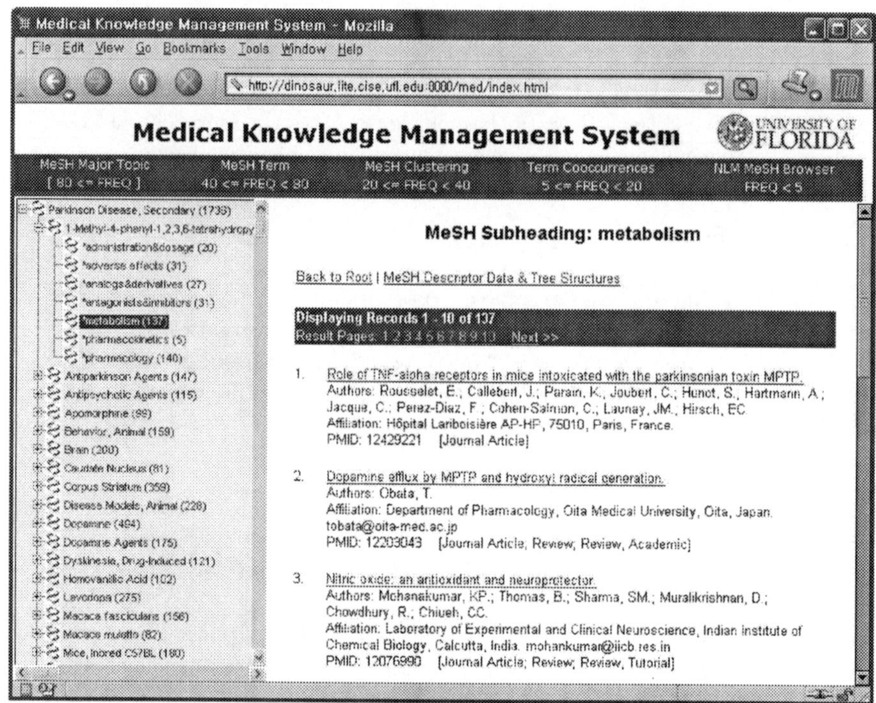

Fig. 1. Interface of MeSH Major Topic View

3.3 SOM Tree

The SOM tree was constructed for each MeSH descriptor whose document frequency was less than some predetermined threshold. Typically, 10 to 12 MeSH descriptors are assigned to each MEDLINE citation. Thus, some categories associated with a large number of citations do not characterize the information in a way that is of interest to the user [10]. To solve this problem, we further arrange those categories hierarchically using the SOM. In some cases, clustering seems useful in helping users filter out sets of documents that are clearly not relevant and should be ignored [10]. Figure 2 shows the interface for browsing the SOM tree.

4 Related Work

Data mining is defined as the nontrivial extraction of implicit, previously unknown, and potentially useful information from data [2,3]. To improve retrieval

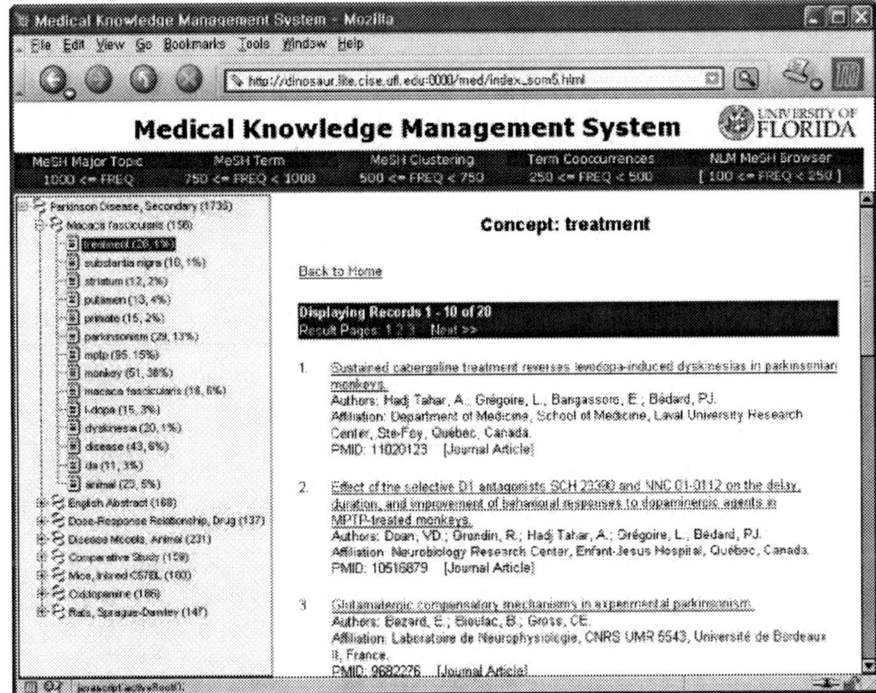

Fig. 2. Interface of SOM Tree View

efficiency and effectiveness, data mining uses document classification, document clustering, machine learning, and visualization technologies. We discuss related work on document clustering approaches in this section.

Document clustering is used to group similar documents into a set of clusters [16]. To improve retrieval efficiency and effectiveness, related documents should be collected together in the same cluster based on shared features among subsets of the documents.

In general, document clustering methods are divided into two ways: hierarchical and partitioning approaches [20]. The hierarchical clustering methods build a hierarchical clustering tree called *a dendrogram*, which shows how the clusters are related.

There are two types of hierarchical clustering: agglomerative (bottom-up) and divisive (top-down) approaches [20]. In agglomerative clustering, each object is initially placed in its own cluster. The two or more most similar clusters are merged into a single cluster recursively. A divisive clustering initially places all objects into a single cluster. The two objects that are in the same cluster but are most dissimilar are used as seed points for two clusters. All objects in this cluster are placed into the new cluster that has the closest seed. This procedure continues until a threshold distance, which is used to determine when the procedure stops, is reached.

Partitioning methods divide a data set into a set of disjoint clusters. Depending on how representatives are constructed, partitioning algorithms are subdivided into k-means and k-medoids methods. In k-means, each cluster is represented by its centroid, which is a mean of the points within a cluster. In k-medoids, each cluster is represented by one data point of the cluster, which is located near its center. The k-means method is minimizing the error sum of squared Euclidean distances whereas the k-medoids method is instead using dissimilarity. These methods are either minimizing the sum of dissimilarities of different clusters or minimizing the maximum dissimilarity between the data point and the seed point.

Partitioning methods are better than hierarchical ones in the sense that they do not depend on previously found clusters [20]. On the other hand, partitioning methods make implicit assumptions on the form of clusters and cannot deal with the tens of thousands of dimensions [20]. For example, the k-means method needs to define the number of final clusters in advance and tends to favor spherical clusters. Hence, statistical clustering methods are not suitable for handling high dimensional data, reducing the dimensionality of a data set, or visualization of the data. A new approach to addressing clustering and classification problems is based on the connectionist, or neural network computing [5,8,12,13,14]. The Self-Organizing Map (SOM) is an artificial neural network algorithm is especially suitable for data survey because it has prominent visualization and abstraction properties [5,8,12,13,14].

5 Lessons Learned and Discussion

We have also proposed the multi-layered Self-Organizing Map algorithm for building a subject-specific concept hierarchy using two input vector sets constructed by indexing the MEDLINE citations. The proposed SOM algorithm is different from other SOM-variant algorithms. First, it uses two different input vectors to cluster MEDLINE database more meaningfully. Second, after constructing the top-level concept map and aggregating nodes with the same concept on the map into a group, it dynamically reconstructs input vector by selecting only those items that are contained for each concept region from input vector of the higher level to generate the sub-layer map. Thus, new input vector would reflect only the contents of the region and not the all collection for each SOM. The concept hierarchy generated by the SOM can be used for building an interactive concept browsing service with multiple viewpoints.

To be more efficient, out system may be improved in several directions. However, when new documents are added, the SOM processing for new data is not feasible. This is because we have to recalculate input vectors of SOM and retrain the SOM with new input vectors. A Further limitation when using the SOM is that the size and lattice type of the map should be determined in advance [9]. It is difficult to choose optimal parameters for the SOM without knowledge of the type and organization of the documents. Therefore, to obtain the best SOM with the minimum quantization error, we have to repeat training procedures several times with different parameter settings. This process is often very time-consuming for large collections of documents. Finally, documents can be assigned with more than one concept in SOM clustering. Although documents may contain several topics, users may be confused when finding overlapped documents over several concepts in some cases.

6 Conclusions

We have proposed a three-step data mining process for organizing MEDLINE database: (1) categorizations according to MeSH terms, MeSH major topics, and the co-occurrence of MeSH descriptors; (2) clustering using the results of MeSH term categorization; and (3) visualization of categories and hierarchical clusters.

The proposed SOM algorithm is different from other SOM-variant algorithms. First, it uses the results of categorization. Second, after constructing the top-level concept map and aggregating nodes with the same concept on the map into a group, it dynamically reconstructs input vector by selecting only terms that are contained for each concept region from the input vector of the higher level and re-computing their weights to generate the sub-layer map. Thus, the new input vector would reflect only the contents of the region and not the all collection for each SOM.

References

1. R. Baeza-Yates and B. Ribeiro-Neto. *Modern Information Retrieval.* Addison Wesley, 1999.
2. E. Brill. A Simple Rule-based Part of Speech Tagger. *Proceedings of the 3rd Conference on Applied Natural Language Processing,* Trento, Italy, 1992.
3. E. Brill. Some advances in transformation-based part of speech tagging. *Proceedings of the 12th National Conference on Artificial Intelligence,* Seattle, WA, 1994.
4. S. Chakrabarti. Data mining for hypertext: A tutorial survey. *ACM SIGKDD Explorations,* 1(2): 1-11, 2000.
5. H. Chen, C. Schuffels, and R. Orwig. Internet Categorization and Search: A Self-Organizing Approach. *Journal of Visual Communication and Image Representation:* 7(1): 88-102, 1996.
6. H. Chen and T.K. Ho. Evaluation of Decision Forests on Text Categorization. *Proceedings of the 7th Conference on Document Recognition and Retrieval* pp. 191-199, 2000.
7. S. Chen. *Digital Libraries: The Life Cycle of Information.* Better Earth Publisher, 1998.
8. M. Dittenbach, D. Merkl, and A. Rauber. The Growing Hierarchical Self-Organizing Map. *Proceedings of the International Joint Conference on Neural Networks (IJCNN 2000),* Vol. 6, pp. 15-19, 2000.
9. J.C. French, A.L. Powell, F. Gey, and N. Perelman. Exploiting a Controlled Vocabulary to Improve Collection Selection and Retrieval Effectiveness. *Proceedings Tenth International Conference on Information and Knowledge Management (CIKM),* pp. 199-206, November 2001.
10. M.A. Hearst. The Use of Categories and Clusters for Organizing Retrieval Results. In T. Strzalkowski, editor, *Natural Language Information Retrieval,* pages 333-374. Kluwer Academic Publishers, Dordrecht, The Netherlands, 1999.
11. H. Kim, C. Choo, and S. Chen. An Integrated Digital Library Server with OAI and Self-Organizing Capabilities. *Proceedings of the 7th European Conference on Research and Advanced Technology for Digital Libraries (ECDL 2003),* Trondheim, Norway, August 2003.
12. T. Kohonen. Self-Organization of Very Large Document Collection: State of the Art. *Proceedings of ICANN98, the 8th International Conference on Artificial Neural Networks,* Skovde, Sweden, 1998.
13. T. Kohonen, S. Kaski, K. Lagus, J. Salojärvi, J. Honkela, V. Paatero, and A. Saarela. Self Organizing of a Massive Document Collection. *IEEE Transactions on Neural Networks,* 11(3): 574-585, 2000.
14. T. Kohonen. *Self-Organizing Maps,* 3rd Edition. Springer-Verlag, Berlin, Germany, 2001.
15. C. Lin, H. Chen, and J.F. Nunamaker. Verifying the Proximity Hypothesis for Self-organizing Maps. *Journal of Management Information Systems,* 16(3): 61-73, 1999-2000.
16. National Library of Medicine. MEDLINE Fact Sheet. http://www.nlm.nih.gov/pubs/factsheets/medline.html.

17. A. Powell, and J.C. French. The Potential to Improve Retrieval Effectiveness with Multiple Viewpoints. Technical Report CS-98-15, Department of Computer Science, University of Virginia, 1998.

18. W. Pratt, L. Fagan. The Usefulness of Dynamically Categorizing Search Results. *Journal of the American Medical Informatics Association (JAMIA)*, 7(6): 605-617, 2000.

19. G. Salton and C. Buckley. Term-Weighting Approaches in Automatic Text Retrieval. *Information Processing and Management*, 24(5): 513-523, 1988.

20. J. Vesanto and E. Alhoniemi. Clustering of the Self-Organizing Map. *IEEE Transactions on Neural Networks*, 11(3): 586-600, 2000.

## Part II

Data Mining Techniques in Disease Diagnosis

Logical Analysis of Computed Tomography Data to Differentiate Entities of Idiopathic Interstitial Pneumonias

M.W. Brauner[1], N. Brauner[2], P.L. Hammer[3], I. Lozina[3], and D. Valeyre[4]

[1] Department of Radiology, Fédération MARTHA, UFR Bobigny, Université Paris 13 et Hôpital Avicenne AP-HP, 125, route de Stalingrad, 93009 Bobigny Cedex, France
`michel.brauner@wanadoo.fr`
[2] Laboratoire Leibniz-IMAG, 46 av. Felix Viallet, 38031 GRENOBLE Cedex, France
`Nadia.Brauner@imag.fr`
[3] RUTCOR, Rutgers University, 640 Bartholomew Rd., Piscataway NJ, 08854-8003 USA
`{hammer,ilozina}@rutcor.rutgers.edu`
[4] Department of Pneumology, Fédération MARTHA, UFR Bobigny, Université Paris 13 et Hôpital Avicenne AP-HP, 125, route de Stalingrad, 93009 Bobigny Cedex, France

Summary. The aim of this chapter is to analyze computed tomography (CT) data by using the Logical Analysis of Data (LAD) methodology in order to distinguish between three types of idiopathic interstitial pneumonias (IIPs). The chapter demonstrates that LAD can distinguish different forms of IIPs with high accuracy. It shows also that the patterns developed by LAD techniques provide additional information about outliers, redundant features, the relative significance of attributes, and makes possible the identification of promoters and blockers of various forms of IIPs.

1 Introduction

Idiopathic interstitial pneumonias (IIPs) are a heterogeneous group of non-neoplastic disorders resulting from damage to the lung parenchyma by varying patterns of inflammation and fibrosis. A new classification of IIPs was established in 2001 by an International Consensus Statement defining the clinical manifestations, pathology and radiological features of patients with IIPs [4]. Various forms of IIP differ both in their prognoses and their therapies, but are not easily distinguishable using clinical, biological and radiological data, and therefore frequently require pulmonary biopsies to establish the diagnosis. The aim of this chapter is to analyze computed tomography (CT) data by techniques of biomedical informatics to distinguish between three types of IIPs:

- Idiopathic Pulmonary Fibrosis (IPF)
- Non Specific Interstitial Pneumonia (NSIP)
- Desquamative Interstitial Pneumonia (DIP)

2 Patients and Methods

This study deals with the CT scans in patients with IIPs referred to the Department of Respiratory Medicine, Avicenne Hospital, Bobigny, France, for medical advice on diagnosis and therapy. The diagnosis was established on clinical, radiographic and pathologic (i.e. biopsy–based) data. The 56 patients included 34 IPFs, 15 NSIPs, and 7 DIPs.

We reviewed the CT examination of the chest from these patients. CT scans were evaluated for the presence of signs and a score was established for the 2 main lesions, ground-glass attenuation and reticulation. Pulmonary disease severity on thin section CT scans was scored semi-quantitatively in upper, middle and lower lung zones. The six areas of the lung were defined as follows: the upper zones above the level of the carina; the middle zones between the level of the carina and the level of the inferior pulmonary veins, and the lower zones under the level of the inferior pulmonary veins. The profusion of opacities was recorded separately in the six areas of the lung to yield a total score of parenchymal opacities. The severity was scored in each area according to four basic categories: 0 = normal, 1 = slight, 2 = moderate, and 3 = advanced (total: 0-18)

The data consisted of the binary attributes 1, 2, . . . ,10, and the numerical attributes 11, 12, and 13 listed bellow:

1. IIT intra-lobular interstitial thickening
2. HC honeycombing
3. TB traction bronchiectasis
4. GG1 ground-glass attenuation
5. BRVX peri-bronchovascular thickening
6. PL polygonal lines
7. HPL hilo-peripheral lines
8. SL septal lines
9. AC airspace consolidation
10. N nodules
11. GG2 ground-glass attenuation score
12. RET [5] reticulation score
13. GG2/RET ground-glass attenuation/reticulation score

---

[5] RET is a generic term which includes the three main fibrotic lesions : ITT, HC and TB

The analysis of this dataset was carried out using the combinatorics, optimization and logic based methodology called the *Logical Analysis of Data* (LAD) proposed in [7,8]. Detailed description of this methodology appears in [5]. Also, a brief outline of LAD appears in this volume (in [3]). Among previous studies dealing with applications of LAD to medical problems we mention [1,2,10].

The choice of LAD for analyzing the IIP data is due on the one hand to its proven capability to provide highly accurate classifications, and on the other hand to the usefulness of LAD patterns in analyzing the significance and nature of attributes.

The conclusions of LAD have been confirmed by other methods used in bioinformatics (neural networks, decision trees, support vector machine, etc.). An additional result of the study was the identification by LAD of two outliers, which turned out to have complete medical explanation.

3 Outliers

We have constructed 3 different LAD models to distinguish between IPF, NSIP or DIP patients:

- model I to distinguish IPF patients (considered to be the positive observations in this model) from non-IPF patients (negative observations);
- model II to distinguish NSIP patients (positive in this model) from non-NSIP patients (negative observations);
- model III to distinguish DIP patients (positive in this model) from non-DIP patients (negative observations).

These models use only pure patterns. Their degrees are at most 4, and their prevalences range between 40% and 85.7%.

a) Two suspicious observations

The classification given by 3 LAD models for 56 observations in the dataset is shown in Table 1. It can be seen that all the 56 classifications are correct, but only 54 of them are precise. In fact the classifications of the observations s003 and s046 are vague. Since observation s003 is classified as either a DIP or an NSIP patient, we have built an additional model to distinguish between these two classes. It turns out that the model contains only one pattern covering observation s003. This pattern shows (correctly) that s003 is a DIP patient, however it does not cover any other observation, i.e. its prevalence is so low that it cannot be considered reliable. A very similar argument concerning the observation s046 shows that in a model distinguishing IPF/NSIP cases, it is classified as being an NSIP case; however, this classification is based only on extremely weak patterns, whose reliability is low. The above facts raise suspicions about the specific nature of these two observations and the question of whether they should be included at all in the dataset.

b) Medical confirmation

In view of the suspicions related to these two observations, the medical records of these two patients have been re-examined. It was found that patient s003 was exposed to asbestos, and therefore its classification as DIP is uncertain. Asbestosis may be responsible for a pathologic aspect similar to that of IPF, but very different from DIP. It is also possible that the pathologic result on the biopsy of a very small area of the lung was wrong. Also, it was found that the data of patient s046 are highly atypical in all the features (age, clinical data and lung pathology). Based on the clinical, radiographic and pathologic data, this patient does not seem to belong to any of the three classes in the initial classification before CT analysis, and it was suggested that in view of these reasons, (s)he should be considered non-classable and removed from the dataset.

c) Improving classification accuracy by removing outliers

The medical confirmation of the suspicions raised by the inability of the LAD models to classify the two unusual observations, have led us to check the ways in which the accuracy of various classification methods changes when these two observations are removed from the dataset. In order to evaluate these changes, we have applied five classification methods taken from the WEKA package (*http: //www.cs.waikato.ac.nz/~ml/weka/index.html*) separately to the original dataset of 56 observations and to the dataset of 54 observations obtained by removing the two suspicious ones. The 5 methods used for this purpose were: artificial neural networks ("Multilayer Perceptron" in WEKA), linear logistic regression classifier ("Simple Logistic" in WEKA), support vector machine classifier ("SMO" in WEKA), nearest-neighbor classifier ("IB1" in WEKA), and decision trees ("J48" in WEKA).

Twenty 3-folding experiments were carried out for each of the 3 classification problems (IPF/non-IPF, NSIP/non-NSIP, DIP/non-DIP). In each of the experiments the dataset was randomly partitioned into three approximately equal parts, two of which were used as the training set, and the third one as the testing set. By rotating the subset taken as the test set, each experiment in fact consisted of three tests, i.e. a total of 60 experiments were carried out for each of the three classification problems. The average accuracy of these 1800 experiments (i.e. five methods applied 60 times to original and reduced datasets of three problems) measured on the test sets is shown in Table 2.

It can be seen from Table 2 that by removing the two outliers, the accuracy of every single classification method was improved for each of the 3 models.

In conclusion, the suspicions generated by the weakness of the coverage with patterns of two of the observations, lead to the identification of these two patients as outliers, and eventually to medical explanations of the inappropriateness of maintaining them in the dataset. The "cleaned" dataset obtained by eliminating these two outliers was shown to allow a substantial improvement in the accuracy of all the tested classification methods.

Table 1. Classification given by 3 LAD models for 56 observations in the dataset

| Observations | Given Classification | Classification by LAD Models | | | Conclusion |
		IPF/ non-IPF	NSIP/ non-NSIP	DIP/ non-DIP	
s001	DIP	0	?	1	DIP
s002	DIP	0	0	1	DIP
s003	DIP	0	?	?	NSIP or DIP
s004	DIP	0	0	1	DIP
s005	DIP	0	?	1	DIP
s006	DIP	0	?	1	DIP
s007	DIP	0	0	1	DIP
s008	IPF	1	0	0	IPF
s009	IPF	1	?	0	IPF
s010	IPF	1	?	0	IPF
s011	IPF	1	0	0	IPF
s012	IPF	1	0	0	IPF
s013	IPF	1	0	0	IPF
s014	IPF	1	0	0	IPF
s015	IPF	1	0	0	IPF
s016	IPF	1	?	0	IPF
s017	IPF	1	?	0	IPF
s018	IPF	1	0	0	IPF
s019	IPF	1	0	0	IPF
s020	IPF	1	0	0	IPF
s021	IPF	1	?	0	IPF
s022	IPF	1	0	0	IPF
s023	IPF	1	0	0	IPF
s024	IPF	1	0	0	IPF
s025	IPF	1	0	0	IPF
s026	IPF	1	0	0	IPF
s027	IPF	1	0	0	IPF
s028	IPF	1	0	0	IPF
s029	IPF	1	0	0	IPF
s030	IPF	1	0	0	IPF
s031	IPF	1	0	0	IPF
s032	IPF	1	0	0	IPF
s033	IPF	1	0	0	IPF
s034	IPF	1	0	0	IPF
s035	IPF	1	0	0	IPF
s036	IPF	1	0	0	IPF
s037	IPF	1	0	0	IPF
s038	IPF	1	0	0	IPF
s039	IPF	1	0	0	IPF
s040	IPF	1	0	0	IPF
s041	IPF	1	0	0	IPF
s042	NSIP	0	1	0	NSIP
s043	NSIP	0	1	0	NSIP
s044	NSIP	0	1	0	NSIP
s045	NSIP	0	1	0	NSIP
s046	NSIP	?	?	0	IPF or NSIP
s047	NSIP	0	1	0	NSIP
s048	NSIP	0	1	?	NSIP
s049	NSIP	0	1	?	NSIP
s050	NSIP	0	1	0	NSIP
s051	NSIP	0	1	0	NSIP
s052	NSIP	0	1	0	NSIP
s053	NSIP	0	1	0	NSIP
s054	NSIP	0	1	0	NSIP
s055	NSIP	0	1	0	NSIP
s056	NSIP	0	1	0	NSIP

Table 2. Classification Accuracies Before/After Elimination of Outliers

	Dataset	Multilayer Perceptron	Simple Logistic	SMO	IB1	J48	Average change in accuracy
NSIP/non-NSIP	Original	72.00%	72.19%	75.35%	66.96%	71.32%	+6.25%
	Reduced	79.26%	78.33%	79.63%	75.37%	76.48%	
IPF/non-IPF	Original	80.27%	81.78%	81.25%	70.66%	82.74%	+2.08%
	Reduced	82.87%	84.07%	82.04%	72.87%	85.28%	
DIP/non-DIP	Original	84.07%	87.81%	88.23%	84.58%	85.97%	+3.29%
	Reduced	89.07%	88.80%	90.37%	88.15%	90.74%	

4 Support Sets

a) Set covering formulation

Although the dataset involves 13 variables, some of them may be redundant. Following the terminology of LAD [6,7,8], we shall call an irredundant set of *variables* or *attributes* or *features* a *support set* of the dataset if there is be no overlap between the 3 different types of IIPs after projecting the 13-dimensional vector representing the patients on this subset.

The determination of a minimum size support set was formulated as a set covering problem. The basic idea of the set covering formulation of this problem consists in the simple observation that a subset S is a support set if and only if the projections on S of the positive and the negative observations in the dataset are disjoint.

In order to illustrate this reduction we shall identify a minimum size subset of the variables in the dataset which are capable of distinguishing IPF observations from non-IPF observations. We shall assume that the three numerical variables x_{11}, x_{12}, x_{13} have been *"binarized"*, i.e. each of them had been replaced by one or several 0-1 variables, as proposed in [5, 6]. The binarized variables are associated to so-called *cut-points*. For instance, there are 2 cut-points (5.5 and 6.5) associated to the numerical variable x_{11}, and the corresponding binary variables $x_{11}^{5.5}$ and $x_{11}^{6.5}$ are then defined in the following way:

$$x_{11}^{5.5} = 1 \text{ if } x_{11} \geq 5.5, \text{ and } x_{11}^{5.5} = 0 \text{ if } x_{11} < 5.5,$$

$$x_{11}^{6.5} = 1 \text{ if } x_{11} \geq 6.5, \text{ and } x_{11}^{6.5} = 0 \text{ if } x_{11} < 6.5.$$

Similarly, two cut-points (7.5, 8.5) are introduced for x_{12}, along with two associated binary variables. The variable x_{13} is binarized using four 0-1 variables associated to the cut-points 0.5, 1, 1.05 and 1.2.

Using the original 10 binary variables along with the 8 binarized variables (which replace the numerical variables x_{11}, x_{12}, x_{13}), we shall now represent the observations as 18 dimensional binary vectors $(x_1,\ldots,x_{10}, x_{11}^{5.5}, x_{11}^{6.5}, x_{12}^{7.5}, x_{12}^{8.5}, x_{13}^{0.5}, \ldots, x_{13}^{1.2})$. For example, the positive (i.e. IPF) observation s008 = (0, 1, 1, 1, 0, 1, 1, 0, 0, 0, 2, 9, 0.22) will become in this way the binary vector b008 = (0, 1, 1, 1, 0, 1, 1, 0, 0, 0, 0, 0, 1, 1, 0, 0, 0, 0). Similarly the negative (i.e. non-IPF) observation s006 = (0, 0, 0, 1, 0, 1, 1, 0, 0, 0, 5, 2, 2.5) becomes the binary vector b006 = (0, 0, 0, 1, 0, 1, 1, 0, 0, 0, 0, 0, 0, 0, 1, 1, 1, 1).

Clearly, the positive binarized observation b008 and the negative binarized observation b006 differ only in the following 8 components: x_2, x_3, $x_{12}^{7.5}, x_{12}^{8.5}, x_{13}^{0.5}, \ldots, x_{13}^{1.2}$. It follows that any support set S must include at least one of these variables, since otherwise the projections of the positive observation b008 and the negative observation b006 on S could not be distinguished. Therefore, if we denote by $(s_1, \ldots, s_{10}, s_{11}^{5.5}, s_{11}^{6.5}, s_{12}^{7.5}, s_{12}^{8.5}, s_{13}^{0.5}, \ldots, s_{13}^{1.2})$ the characteristic vector of S, one of the necessary conditions for S to be a support set will be

$$s_2 + s_3 + s_{12}^{7.5} + s_{12}^{8.5} + s_{13}^{0.5} + \ldots + s_{13}^{1.2} \geq 1.$$

A similar inequality can be written for every pair consisting of a positive (IPF) and a negative (non-IPF) observation in the binarized dataset. The $34 \times 22 = 748$ pairs of positive-negative observations define the constraints of a set covering problem for finding a minimum size support set. Since our dataset consists of a rather limited number of observations, in order to increase the accuracy of the models to be built on the support sets obtained in this way, we have further strengthened the above set covering-type constraints, by replacing the 1 on their right-hand side, by 3 (the choice of 3 is based on empirical considerations, the basic idea being simply to sharpen the requirements of separating positive and negative observations).

Clearly the objective function of this set covering type problem is simply the sum

$$s_1 + \ldots + s_{10} + s_{11}^{5.5} + s_{11}^{6.5} + s_{12}^{7.5} + s_{12}^{8.5} + s_{13}^{0.5} + \ldots + s_{13}^{1.2}.$$

b) Three minimum support sets

By solving this problem we found that the binary variables x_3, x_4, x_9, x_{10} are redundant, and that a minimum size support set (using the original binary and numerical variables) consists of the attributes 1, 2, 5, 6, 7, 8, 11, 12 and 13.

In a similar way we can see that a minimum support set distinguishing DIP observations from non-DIP ones consists of the 6 original attributes: 1, 2, 3, 5, 12 and 13, while a minimum support set distinguishing NSIP patients from non-NSIP ones consists of the 8 original attributes: 1, 2, 5, 6, 7, 8, 11 and 12.

c) Accuracy of classification on minimum support sets

It is important to point out that the elimination of redundant variables does not reduce the accuracy of classification. In order to demonstrate the qualities of the minimum support sets obtained for the IPF/non-IPF, DIP/non-DIP and NSIP/non-NSIP problems we have carried out 20 three-folding classification experiments on these 3 problems using 5 different classification methods from the WEKA package.[6] These experiments first used the original 13 variables, and after that the support sets of 9, 6, and 8 variables respectively, obtained above for these 3 problems. The results of these experiments are presented in Table 3.

Table 3. Classification Accuracies on all Original Variables and on Support Sets

	Support Set	Multilayer Perceptron	Simple Logistic	SMO	IB1	J48	Average change in accuracy
NSIP/non-NSIP	Original	79.26%	78.33%	79.63%	75.37%	76.48%	+2.48%
	Reduced	85.19%	78.80%	79.35%	81.57%	76.57%	
IPF/non-IPF	Original	82.87%	84.07%	82.04%	72.87%	85.28%	+0.98%
	Reduced	85.28%	83.33%	81.20%	76.39%	85.83%	
DIP/non-DIP	Original	89.07%	88.80%	90.37%	88.15%	90.74%	+1.89%
	Reduced	91.76%	90.28%	89.44%	92.87%	92.22%	

In conclusion we can see from Table 3 that the elimination of those features which were identified as redundant not only maintains the accuracy of classification, but also increases it in each of the three models.

5 Patterns and Models

Using the support sets developed in the previous section, now we shall apply the LAD methodology to this dataset for generating patterns and classification models. It turns out that in spite of the very small size of this dataset, some surprisingly strong patterns can be identified in it. For example in the IPF/non-IPF model, 14 (i.e. 70%) of the 20 non-IPF patients satisfy the simple pattern "GG2/RET $\geq$ 1.2"; moreover none of the 34 IPF patients satisfy this condition. While this simple pattern involves a single variable, other more complex patterns exist and are capable of explaining the IPF or non-IPF character of large groups of patients. For instance, the negative pattern

$$\text{"RET} \leq 8 \text{ and GG2/RET} > 1\text{"}$$

is satisfied by 70% of the non-IPF patients, and by none of the IPF patients. As an example of a positive pattern, we mention

$$\text{"HC} = 1,\ \text{HPL} = 0 \text{ and GG2/RET} \leq 1.2\text{"};$$

24 (i.e. 70.6%) of the 34 IPF patients satisfy all the 3 constraints of this pattern, and none of the non-IPF patients satisfy these 3 conditions simultaneously.

---

[6] *http: //www.cs.waikato.ac.nz/~ml/weka/index.html*

While the above patterns can distinguish large groups of patients having a certain type of IIP from those of other types of IIP, larger collections of patterns constructed by LAD can collectively classify the entire set of 54 observations in the dataset. We shall first illustrate the way the classification works by considering the problem of distinguishing IPF and non-IPF patients.

In Table 4, we present a model consisting of 20 positive and 20 negative patterns allowing the accurate classification of IPF/non-IPF patients. Note that the equality of the numbers of positive and negative patterns in this model is a simple coincidence.

The way in which the model allows the classification of a new observation is the following. First, if an observation satisfies all the conditions describing some positive (negative) patterns, but does not satisfy all the conditions describing any one of the negative (positive) pattern, then the observation is classified as positive (negative); this classification is shown in the tables as "1" (respectively, "0"). Second, if an observation does not satisfy all the defining conditions of any positive or negative pattern, then it remains "unclassified"; this is shown in the tables as "?". Third, if an observation satisfies all the defining conditions of some positive and also of some negative patterns in the model, then a weighting process is applied to decide on the appropriate classification; the process of finding weights for such classification is described in [3].

Besides the IPF/non-IPF model discussed above, we have also constructed a model to distinguish the 14 NSIP patients from the 40 non-NSIP patients, and another model to distinguish the 6 DIP patients from the 48 non-DIP patients. The NSIP/non-NSIP model is built on the support set of 8 attributes described in the previous section, and includes 16 positive and 4 negative patterns. The DIP/non-DIP model is built on the support set of 6 attributes described in the previous section, and includes 7 positive and 15 negative patterns.

The combination of the three models allows one to draw additional conclusions. For example, if the results of the three classifications are "0", "0" and "?" respectively, and one knows that each patient is exactly one of the types of IIP, one can conclude that the "?" in the classification of the third condition can be replaced by "1".

The results of the classification of 54 patients given by the three models, along with the conclusions derived from the knowledge of all the three classifications are presented in Table 5. The accuracy of this classification is 100%.

It is usually said that the CT diagnosis of NSIP is difficult. In a recent study [9] experienced observers considered the CT pattern indistinguishable from IPF in 32% of cases. In another investigation the author assessed the value of CT in the diagnosis of 129 patients with histologically proven idiopathic interstitial pneumonias [11]. Two independent observers were able to make a correct first choice diagnosis in more than 70% of IPF cases, in more than 60% of DIP, but only in 9% of NSIP cases. In that study, NSIP was

Table 4. IPF/non-IPF model

Pattern	attr.1 IIT	attr.2 HC	attr.5 BRV	attr.6 XPL	attr.7 HPL	attr8 SL	attr.11 GG2	attr.12 RET	attr.13 GG2/RET	Pos. Prevalence	Neg. Prevalence
P1		1			0				≤ 1.2	70.6%	0
P2		1			0			≥ 8		47.1%	0
P3	1	1					≥ 4		≤ 1.2	47.1%	0
P4		1			0	0		≥ 6		47.1%	0
P5	1	1			0			≥ 6		47.1%	0
P6	1	1					≥ 4	≥ 8		41.2%	0
P7		1	0						≤ 0.5	41.2%	0
P8		1	0					≤ 8	≤ 1.2	41.2%	0
P9	1	1							>0.5, ≤ 1.2	38.2%	0
P10				1					≤ 1.2	32.4%	0
P11	1	1						≥ 8	>0.5	32.4%	0
P12	1	1						≥ 9		29.4%	0
P13		1	0				≤ 3			29.4%	0
P14		1					≥ 4	≤ 8	≤ 1.2	26.5%	0
P15				0				≥ 8	≤ 0.5	26.5%	0
P16			0	1				≥ 6		26.5%	0
P17	0							≤ 8	≤ 1.2	20.6%	0
P18						1		≥ 8		20.6%	0
P19				1			≤ 3			20.6%	0
P20				1	0					17.6%	0
N1								≤ 8	>1	0	70.0%
N2									>1.2	0	70.0%
N3		0				0	≥ 4			0	50.0%
N4								≤ 5		0	50.0%
N5		0				0			>0.5	0	50.0%
N6		0		0	0					0	45.0%
N7		0		0				≤ 7		0	45.0%
N8		0			0				>0.5	0	40.0%
N9		0			0		≥ 4			0	40.0%
N10					0				>1	0	40.0%
N11		0		0	0					0	40.0%
N12	0								>1	0	35.0%
N13	1	0					≥ 4	≤ 7		0	30.0%
N14					1	0		≤ 8	>0.5	0	30.0%
N15					1	0	≥ 4	≤ 8		0	30.0%
N16				0	1	0		≤ 8		0	20.0%
N17						1			>1	0	15.0%
N18	0				1	0	≥ 4			0	15.0%
N19			1		1	0		≤ 8		0	15.0%
N20	0				1	0			>0.5	0	15.0%

confused most often with DIP, and less often with IPF. It seems that LAD makes possible to distinguish NSIP from the other entities in the majority of cases.

6 Validation

It has been shown in the previous section (Table 5) that the accuracy of classifying by LAD the 54 patients is 100%. It should be added however that

Table 5. Results of classification of 54 patients by 3 models

| Observations | Given Classification | Classification by LAD Models | | | Conclusion |
		IPF/ non-IPF	NSIP/ non-NSIP	DIP/ non-DIP	
s001	DIP	0	?	1	DIP
s002	DIP	0	0	1	DIP
s004	DIP	0	0	1	DIP
s005	DIP	0	?	1	DIP
s006	DIP	0	?	1	DIP
s007	DIP	0	0	1	DIP
s008	IPF	1	0	0	IPF
s009	IPF	1	?	0	IPF
s010	IPF	1	?	0	IPF
s011	IPF	1	0	0	IPF
s012	IPF	1	0	0	IPF
s013	IPF	1	0	0	IPF
s014	IPF	1	0	0	IPF
s015	IPF	1	0	0	IPF
s016	IPF	1	?	0	IPF
s017	IPF	1	0	0	IPF
s018	IPF	1	0	0	IPF
s019	IPF	1	0	0	IPF
s020	IPF	1	0	0	IPF
s021	IPF	1	?	0	IPF
s022	IPF	1	0	0	IPF
s023	IPF	1	0	0	IPF
s024	IPF	1	0	0	IPF
s025	IPF	1	0	0	IPF
s026	IPF	1	0	0	IPF
s027	IPF	1	0	0	IPF
s028	IPF	1	0	0	IPF
s029	IPF	1	0	0	IPF
s030	IPF	1	0	0	IPF
s031	IPF	1	0	0	IPF
s032	IPF	1	0	0	IPF
s033	IPF	1	0	0	IPF
s034	IPF	1	0	0	IPF
s035	IPF	1	0	0	IPF
s036	IPF	1	0	0	IPF
s037	IPF	1	0	0	IPF
s038	IPF	1	0	0	IPF
s039	IPF	1	0	0	IPF
s040	IPF	1	0	0	IPF
s041	IPF	1	0	0	IPF
s042	NSIP	0	1	0	NSIP
s043	NSIP	0	1	?	NSIP
s044	NSIP	0	1	0	NSIP
s045	NSIP	0	1	0	NSIP
s047	NSIP	0	1	0	NSIP
s048	NSIP	0	1	?	NSIP
s049	NSIP	0	1	0	NSIP
s050	NSIP	0	1	0	NSIP
s051	NSIP	0	1	?	NSIP
s052	NSIP	0	1	0	NSIP
s053	NSIP	0	1	0	NSIP
s054	NSIP	0	1	0	NSIP
s055	NSIP	0	1	0	NSIP
s056	NSIP	0	1	0	NSIP

this result represents only the correctness of the proposed classification model when the entire dataset is used both as a training set, and as a test set. In order to establish the reliability of these classifications they have to be validated. Because of the very limited size of the dataset (in particular because of the availability of data for only 6 DIP patients and only 14 NSIP patients) the traditional partitioning of the dataset into a training and a test set would produce extremely small subsets, and therefore highly unreliable conclusions. In view of this fact, we shall test the accuracy of the LAD classification by cross-validation, using the so-called "jackknife" or "leave-one-out" method. As an example, the cross-validation of the classification results for the IPF/non-IPF model will be presented in the next section.

The basic idea of the "leave-one-out" method is very simple. One of the observations is temporarily removed from the dataset, a classification method is "learned" from the set of all the remaining observations, and it is applied then to classify the extracted observation. This procedure is then repeated separately for every one of the observations in the dataset. For example in the case of the IPF/non-IPF model we have to apply this procedure 54 times.

Table 6 shows the results of the "leave-one-out" procedure applied to this model. The table includes the results of directly applying leave-one-out experiments to the 3 models (IPF/non-IPF, NSIP/non-NSIP, DIP/non-DIP) and the resulting combined classifications. The combined classifications are then used to derive the final conclusion about the IPF/non-IPF character of each observation; the correctness of the conclusion (compared with the given classification) is presented in the last column of Table 6 ("evaluation").

It can be seen that out of 54 observations, 44 are classified correctly, there are 6 errors (the IPF patients s009, s010 and s021 are classified as non-IPF, and the non-IPF patients s042, s047 and s053 are classified as IPF), two patients (s007 and s052) are unclassified, and for two other patients (s016 and s055) the classifications ("IPF or NSIP") are imprecise.

If one considers every unclassified and every imprecisely classified patient as an error, the accuracy of the classification in the leave-one-out experiment is 81.48%. However, if we use the formula established in [3] for accuracy, this turns out to be 85.80%.

Given that the size of the dataset is very small, the results of the leave-one-out tests can be viewed as extremely encouraging.

7 Attribute Analysis

a) Importance of attributes

A simple measure of the importance of an attribute is the frequency of its inclusion in the patterns appearing in the model. For example, attribute 1 (IIT) appears in 11 (i.e. in 27.5%) of the 40 patterns of the IPF/non-IPF model in Table 4.

Table 6. Validation by Leave-One-Out of IPF/non-IPF Classification

Obs.	Given Classification	Classification by Leave-One-Out			Derived Classification	Conclusion	
		IPF/ non-IPF	NSIP/ non-NSIP	DIP/ non-DIP		IPF/ non-IPF	Evaluation
s001	DIP	0	?	1	DIP	0	correct
s002	DIP	0	0	1	DIP	0	correct
s004	DIP	0	0	1	DIP	0	correct
s005	DIP	0	1	1	DIP or NSIP	0	correct
s006	DIP	0	1	1	DIP or NSIP	0	correct
s007	DIP	0	0	0	?	?	unclassified
s008	IPF	1	0	0	IPF	1	correct
s009	IPF	0	?	0	NSIP	0	error
s010	IPF	0	1	0	NSIP	0	error
s011	IPF	1	0	0	IPF	1	correct
s012	IPF	1	0	0	IPF	1	correct
s013	IPF	1	0	0	IPF	1	correct
s014	IPF	1	0	0	IPF	1	correct
s015	IPF	1	0	0	IPF	1	correct
s016	IPF	1	1	0	IPF or NSIP	?	imprecise
s017	IPF	1	0	0	IPF	1	correct
s018	IPF	1	0	0	IPF	1	correct
s019	IPF	1	0	0	IPF	1	correct
s020	IPF	1	0	0	IPF	1	correct
s021	IPF	0	1	0	NSIP	0	error
s022	IPF	1	0	0	IPF	1	correct
s023	IPF	1	0	0	IPF	1	correct
s024	IPF	1	0	0	IPF	1	correct
s025	IPF	1	0	0	IPF	1	correct
s026	IPF	1	0	0	IPF	1	correct
s027	IPF	1	0	0	IPF	1	correct
s028	IPF	1	0	0	IPF	1	correct
s029	IPF	1	0	0	IPF	1	correct
s030	IPF	1	0	0	IPF	1	correct
s031	IPF	1	0	0	IPF	1	correct
s032	IPF	1	0	0	IPF	1	correct
s033	IPF	1	0	0	IPF	1	correct
s034	IPF	?	0	0	IPF	1	correct
s035	IPF	1	0	0	IPF	1	correct
s036	IPF	1	0	0	IPF	1	correct
s037	IPF	1	0	0	IPF	1	correct
s038	IPF	1	0	0	IPF	1	correct
s039	IPF	1	0	0	IPF	1	correct
s040	IPF	1	0	0	IPF	1	correct
s041	IPF	1	0	0	IPF	1	correct
s042	NSIP	1	0	0	IPF	1	error
s043	NSIP	0	1	?	NSIP	0	correct
s044	NSIP	0	1	0	NSIP	0	correct
s045	NSIP	0	1	0	NSIP	0	correct
s047	NSIP	1	?	0	IPF	1	error
s048	NSIP	0	?	1	DIP	0	correct
s049	NSIP	0	1	0	NSIP	0	correct
s050	NSIP	0	1	0	NSIP	0	correct
s051	NSIP	0	1	?	NSIP	0	correct
s052	NSIP	0	0	0	?	?	unclassified
s053	NSIP	1	0	0	IPF	1	error
s054	NSIP	0	1	0	NSIP	0	correct
s055	NSIP	1	1	0	NSIP or IPF	?	imprecise
s056	NSIP	0	1	0	NSIP	0	correct

The frequencies of all the 13 attributes in the models are shown in Table 7 for the 3 LAD models considered, along with the averages of these 3 indicators.

Table 7. Frequencies of Attributes in Models

Attributes	IPF/non-IPF	NSIP/non-NSIP	DIP/non-DIP	Average
IIT	0.275	0.25	0.343	0.289
HC	0.525	0.813	0.357	0.565
TB	0	0	0.238	0.079
GG1	0	0	0	0.000
BRVX	0.125	0.219	0.381	0.242
PL	0.2	0.094	0	0.098
HPL	0.4	0.375	0	0.258
SL	0.3	0.156	0	0.152
AC	0	0	0	0.000
N	0	0	0	0.000
GG2	0.25	0.688	0	0.313
RET	0.5	0.5	0.376	0.459
GG2/RET	0.475	0	0.662	0.379

Two of the most important conclusions which can be seen from this table are:

- the most influential attributes are honeycombing (HC), reticulation score (RET),   ground-glass attenuation/reticulation score (GG2/RET), and ground-glass attenuation score (GG2);
- the attributes ground-glass attenuation (GG1), airspace consolidation (AC) and nodules (N) have no influence on the classification.

b) Promoting and Blocking Attributes

We shall illustrate the promoting or blocking nature of some attributes on the IPF/non-IPF model shown in Table 4. It can be seen from the table that every positive pattern which includes a condition on HC (honeycombing) requires that HC=1. Conversely, every negative pattern which includes a condition on HC requires that HC=0. This means that if a patient is known to be a non-IPF case with HC=1, and all the attributes of another patient have identical values except for HC which is 0, then this second patient is certainly not an IPF case. This type of monotonicity simply means that HC is a "promoter" of IPF. It is easy to see that the attribute PL(polygonal lines) has a similar property.

On the other hand, the attribute BRVX (peri-bronchovascular thickening) appears to have a converse property. Indeed, every positive pattern which includes this attribute requires that BRVX=0, while the only negative pattern (N19) which includes it requires that BRVX=1. Therefore if a patient's BRVX would change from 1 to 0, the patient's condition would not change from IPF to non-IPF (assuming again that none of the other attributes change their values). Similarly to the previous case, this type of monotonicity simply means that BRVX is a "blocker" of IPF.

In this way the IPF/non-IPF model allows the identification of two promoters and of one blocker. None of the other attributes in the support set appear to be promoters or blockers.

A similar analysis of the DIP/non-DIP model shows that intralobular interstitial thickening (IIT) and traction bronchiectasis (TB) are blockers of DIP. Also, the analysis of the NSIP/non-NSIP model shows that peribronchovascular thickening (BRVX) is a promoter of NSIP, while honeycombing (HC), polygonal lines (PL) and septal lines (SL) are blockers of NSIP.

To conclude, in Table 8 we show the promoters and blockers which have been identified for the three forms of idiopathic interstitial pneumonias.

Table 8. Promoters and blockers for three forms of idiopathic interstitial pneumonias

	Idiopathic Pulmonary Fibrosis	Desquamative Interstitial Pneumonia	Non Specific Interstitial Pneumonia
honeycombing	promoter		blocker
polygonal lines	promoter		blocker
peri-bronchovascular thickening	blocker		promoter
intralobular interstitial thickening		blocker	
traction bronchiectasis		blocker	
septal lines			blocker

8 Conclusions

We have shown that it is possible to use a computational technique (LAD) for analyzing CT data for distinguishing with high accuracy different entities (IPF, NSIP and DIP) of idiopathic interstitial pneumonias (IIPs). This is particularly important for NSIP which is yet poorly defined. It was also shown that the patterns developed by LAD techniques provide additional information about outliers, redundant features, the relative significance of the attributes, and allow one to identify promoters and blockers of various forms of IIPs. These encouraging results will form the basis of a forthcoming study of a broader population of IIPs, which will include not only CT data, but also clinical and biological ones.

Acknowledgements

Peter L. Hammer and Irina Lozina gratefully acknowledge the partial support of NSF through Grant # NSF-IIS-0312953.

References

1. G. Alexe, S. Alexe, P.L. Hammer, L. Liotta, E. Petricoin, and M. Reiss. Logical Analysis of Proteomic Ovarian Cancer Dataset. *Proteomics*, 4: 766-783, 2004.

2. S. Alexe, E. Blackstone, P.L. Hammer, H. Ishwaran, M.S. Lauer, and C.E.P. Snader. Coronary Risk Prediction by Logical Analysis of Data. *Annals of Operations Research*, 119: 15-42, 2003.

3. S. Alexe and P.L. Hammer. Pattern-Based Discriminants in the Logical Analysis of Data. *In this Volume.*

4. American Thoracic Society / European Respiratory Society International Multidisciplinary Consensus. Classification of the Idiopathic Interstitial Pneumonias. *American Journal of Respiratory and Critical Care Medicine*, 165: 277-304, 2002.

5. E. Boros, P.L. Hammer, T. Ibaraki, and A. Kogan. Logical Analysis of Numerical Data. *Mathematical Programming*, 79: 163-190, 1997.

6. E. Boros, P.L. Hammer, T. Ibaraki, A. Kogan, E. Mayoraz, and I. Muchnik. An Implementation of the Logical Analysis of Data. *IEEE Transactions on Knowledge and Data Engineering*, 12(2): 292-306, 2000.

7. Y. Crama, P.L. Hammer, and T. Ibaraki. Cause-Effect Relationships and Partially Defined Boolean Functions. *Annals of Operations Research*, 16: 299-326, 1988.

8. P.L. Hammer. Partially Defined Boolean Functions and Cause-Effect Relationships. *International Conference on Multi-Attribute Decision Making Via OR-Based Expert Systems*, University of Passau, Passau, Germany, 1986.

9. T.E. Hartman, S.J. Swensen, D.M. Hansell, T.V. Colby, J.L. Myers, H.D. Tazelaar, A.G. Nicholson, A.U. Wells, J.H. Ryu, D.E. Midthun, R.M. du Bois, and N.L. Muller. Nonspecific Interstitial Pneumonia: Variable Appearance at High-Resolution Chest CT. *Radiology*, 217(3): 701-705, 2000.

10. M.S. Lauer, S. Alexe, C.E.P. Snader, E. Blackstone, H. Ishwaran, and P.L. Hammer. Use of the "Logical Analysis of Data" Method for Assessing Long-Term Mortality Risk After Exercise Electrocardiography. *Circulation*, 106: 685-690, 2002.

11. T. Johkoh, N.L. Muller, Y. Cartier, P.V. Kavanagh, T.E. Hartman, M. Akira, K. Ichikado, M. Ando, and H. Nakamura. Idiopathic Interstitial Pneumonias: Diagnostic Accuracy of Thin-Section CT in 129 Patients. *Radiology*, 211(2): 555-560, 1999.

Diagnosis of Alport Syndrome by Pattern Recognition Techniques

Giacomo Patrizi[1]*, Gabriella Addonisio[1], Costas Giannakakis[2], Andrea Onetti Muda[2], Gregorio Patrizi[3], and Tullio Faraggiana[2]

[1] Dipartimento di Statistica, Probabilità e Statistiche Applicate
Università di Roma "La Sapienza", Italy
[2] Dipartimento di Medicina Sperimentale e Patologia
Università di Roma "La Sapienza", Italy
[3] Dipartimento di Scienze Chirurgiche
Università di Roma "La Sapienza", Italy

Summary. Alport syndrome is a genetic multi-organ disorder, primarily linked with the X-Chromosome, although autosomal forms have been reported. Ultrastructural observations in some cases indicate highly characteristic lesions and mutations in certain genes. Thus the symptomatology of the syndrome is complex and diverse. The aim of this chapter is to present a pattern recognition algorithm to diagnose with high precision patients who may be subject to Alport syndrome. Images of the epidermal basement membrane are studied and a rule to classify them precisely is presented. Theoretical and experimental results are given regarding the possibility of solving this problem.

1 Introduction

The diagnosis of pathologies is often carried out through clinical expertise. A clinician will use a number of instruments to determine symptoms and pathological states to assess the evidence and impart a diagnosis.

The scientific method requires an evidence-based management consisting of an interpersonal and inter-temporal invariance of replicated instances. If other experts should repeat the examination, the same diagnosis and treatment should be proposed. For similar cases (replicated) the experts should assess identical diagnoses and treatment every time. What constitutes "similar cases" is controversial and extremely difficult to determine before the outcome is revealed. Thus it is usually mandatory to perform suitable clinical trials to determine that this latter aspect is fulfilled and that the treatment is beneficial.

* Corresponding author. Email: `g.patrizi@caspur.it`, Fax:+39 06 49 59 241

If, through the subjective evaluations by the physician, a higher precision rate in diagnoses is reached than with the best formal methods, then clearly the aspects grasped by the clinician are not reproducible by formal methods, although these may be useful for other purposes. Instead, if diagnoses by formal methods are at least as accurate, then they should be favored since they are free from subjective judgements, replicable and they can be formulated from standard invariant definitions and measurements (at least to a predetermined level of approximation) [6].

There are many pathologies that are difficult to diagnose or their diagnosis requires an invasive method. Therefore formal diagnostic methods should be available for these and be highly precise, so that the patient can enjoy an early diagnosis, while the clinician receives better information.

The Alport syndrome is a genetic multi-organ disorder, primarily X-Chromosome linked, although autosomal forms and sporadic cases have been reported [9]. Ultra-structural observations indicate highly characteristic lesions and mutations in certain genes are likely to occur. Thus the symptomatology of the syndrome is complex and multi-varied.

The aim of this chapter is to present a pattern recognition algorithm to diagnose with high precision patients who may be subject to the Alport syndrome . Images of the epidermal basement membrane are studied and a rule to classify the image precisely is presented. Theoretical and experimental results are given regarding the possibility of solving this problem.

The outline of the chapter is as follows: in the next section the pattern recognition algorithm to be applied called **T.R.A.C.E.** (**T**otal **R**ecognition by **A**daptive **C**lassification **E**xperiments) is described and its properties are made evident. In Section 3, the Alport syndrome is presented and the standard diagnostic procedures, as well as the current ones, are described. In Section 4, the experimental results obtained with this pattern recognition algorithm are described. Finally, the conclusions follow in Section 5.

2 The Pattern Recognition Algorithm for Diagnosis

In Subsection 1, the pattern recognition algorithm will be described. Its main properties regarding the convergence of the training phase in which the criteria of diagnosis are derived will be presented in Subsection 2. In Subsection 3, results related to the classification phase in which the generalization capabilities of the algorithm and its potential level of precision for random samples of patterns to be classified are stated. A number of experimental aspects to obtain precise results will be indicated in Subsection 3.

2.1 Description of the algorithm

Consider a training set of objects which have been characterized by a set of common attributes whose classification is known. This is called a training set [12].

The data set which includes the training set and the classification set, must be coherent, as defined in [12]. This means that the training set must have objects whose class membership is mutually exclusive, collectively exhaustive and such that no two identical objects are assigned to different membership classes. There are various ways to check that this is so, both for the training set and for the classification set. Further, the membership class of the objects in the training set may be assigned precisely or imprecisely (this occurs when there are errors in their classification). This aspect is important for the coherence of the data set.

The iterative training procedure may be described in this manner. Given a training set which is coherent and for which the class of each pattern is known, mean vectors, called barycenters, can be formed for each class. The distance (Euclidean or general) of each pattern from the mean vector of each class is determined. Out of all the patterns which are closer to a barycenter of another class, the one that is the furthest from the barycenter of its own class is chosen. This is selected as the seed of a new barycenter. For some class, there will be now two barycenters. All the patterns of that class can be reassigned to one or to the other of the barycenters of the class, depending on the distance of each pattern of that class from the barycenters of that class. The patterns in this class will now form two subgroups based on a least distance criterion.

Each class will have a barycenter vector, except for one class which has two barycenters. Calculate all distances of the patterns from all the barycenters anew; and, using the same criterion determine a new pattern to be used to form a new barycenter and repeat the whole procedure as described above. The procedure is repeated again and again, until all patterns result nearer to a barycenter of their own class than to one of another class. The algorithm is then said to have converged. If the training set is piecewise separable, this will always be so, and the set of barycenters formed can be used in classification [12].

If a classifier is defined according to this method, given that the training set is piecewise separable and the algorithm is allowed to run until convergence, a completely correct partition of the training set will be returned. It can also be shown that for a large enough training set the error rate in the classification of the data set can be bounded by a constant, if it is allowed to run until convergence [12]. The description of the algorithm **T.R.A.C.E.** is indicated in the flow diagram in Figure 1.

2.2 Properties of the Algorithm

In this subsection, we define in a more precise way the entities involved and show the convergence properties of the algorithm and its generalization accuracy.

Definition 1. *A subset of a data set is termed a training set if every entity in the training set has been assigned a class label .*

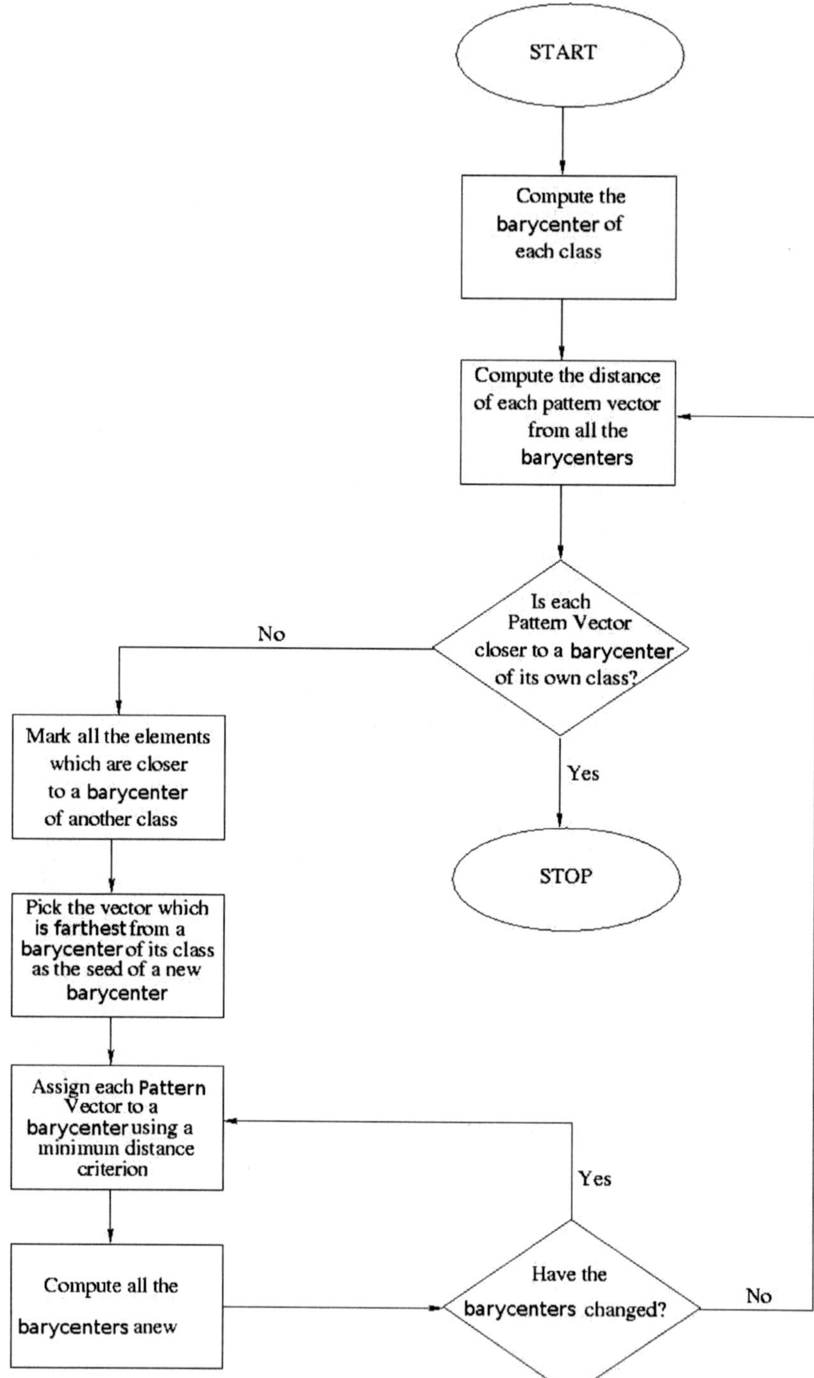

Fig. 1. Flow-chart of **T.R.A.C.E.** algorithm

Definition 2. *Suppose there is a set of entities E and a set $P = \{P_1, P_2, ..., P_n\}$ of subsets of the set of entities, i.e. $P_j \subseteq E, j \in J = \{1, 2, ..., n\}$. A subset $\hat{J} \subseteq J$ forms a cover of E if $\bigcup_{j \in \hat{J}} P_j = E$. If, in addition for every $k, j \in \hat{J}, j \neq k, P_j \cap P_k = \emptyset$ it is a partition .*

Definition 3. *The data set is coherent if there exists a partition that satisfies the following properties:*

1. *The relationships defined on the training set and in particular, the membership classes defined over the dataset are disjoint unions of the subsets of the partition.*
2. *Stability: the partition is invariant to additions to the data set. This invariance should apply both to the addition of duplicate entities and to the addition of new entities obtained in the same way as the objects under consideration.*
3. *Extendability: if further attributes are added to the representation of the objects of the data set so that the dimension of the set of attributes is augmented to p+1 attributes, then the partition obtained by considering the smaller set, will remain valid even for the augmentation, as long as it does not alter the existing relationships defined on the data set.*

Definition 4. *A data set is linearly separable if there exist linear functions such that the entities belonging to one class can be separated from the entities belonging to the other classes. It is pairwise linearly separable if every pair of classes is linearly separable. Finally, a set is piecewise separable if every element of each class is separable from all the other elements of all the other classes [5].*

Clearly, if a set is linearly separable, it is pairwise linearly separable and piecewise separable, but the converse may not be true [17].

Theorem 1. *If a data set is coherent then it is piecewise separable.*

Proof: By the definition 3, a partition exists for a coherent data set and therefore there exists subsets $P_j \subseteq E, j \in J = \{1, 2, ..., n\}$ such that for every $j \neq k \in J$, $P_j \cap P_k = \emptyset$, as indicated by definition 2.

A given class is formed from distinct subsets of the partition, so no pattern can belong to two classes. Therefore each pattern of a given class will be separable from every pattern in the other subsets of the partition. Consequently the data set is piecewise separable.

Theorem 2. *Given a set which does not contain two identical patterns assigned to different classes then a classifier can be constructed which defines a correct partition on the training set.*

Proof: The proof is trivial. If the data set does not contain two identical patterns that belong to different classes, each pattern or group of identical patterns can be assigned to different subsets of the partition. This classifier is necessarily correct and on this basis subsets can be aggregated, as long as the aggregated subsets of different classes remain disjoint.

Corollary 1. *Given that the training set does not contain two or more identical patterns assigned to different classes, the given partition yields a completely correct classification of the patterns in training.*

The avoidance of the juxtaposition property, i.e. two identical patterns belonging to different classes, entails that the Bayes error is zero [4]. In general, this does not mean that in any given neighborhood of a pattern there cannot be other patterns of other classes, but only that they cannot lie in the same point. Thus, the probability distribution of the patterns may overlap with respect to the classes, although they will exhibit discontinuities in the overlap region if juxtaposition is to be avoided.

The algorithm may also be formulated as a combinatorial problem with binary variables, see [13] regarding the relationship between the two implementations. This will avoid the possible formation of more subsets of the partition than required, due to the order of processing; although the rule adopted to choose the next misclassified pattern should ensure that this will not happen.

Suppose that a training set is available with n patterns, represented by appropriate feature vectors indicated by x_i $i = 1, 2, ..., n$ and grouped in c classes. An upper bound is selected for the number of barycenters that may result from the classification. This can be taken "ad abundantiam" as m, or on the basis of a preliminary run of the previous algorithm.

The initial barycenter matrix will be an $n \times mc$ matrix which is set to zero. The barycenter, when calculated, will be written in the matrix. Thus, a barycenter of class k will occupy a column of the matrix between $(m(k-1)+1)$ and mk.

Since we consider a training set, the feature vectors can be ordered by increasing class label. Thus, the first n_1 columns of the training set matrix consist of patterns of class 1; from $n_1 + 1$ to n_2 of class 2; and in general from $n_{k-1} + 1$ to n_k of class k.

Consider the following optimization problem:

$$Min \ Z = \sum_{j=1}^{mc} z_j \tag{1}$$

$$s.t. \ \sum_{j=km+1}^{m(k+1)} y_{ij} = 1 \qquad k = 0, 1, ..., (c-1); \forall i = n_{k-1} + 1, ..., n_k \tag{2}$$

$$\sum_{1=1}^{n} y_{ij} - M z_j \leq 0 \qquad \forall j = 1, 2, ..., mc \tag{3}$$

$$\left(\sum_{i=1}^{n} y_{ij}\right) \times m_j - \sum_{i=1}^{n} x_i y_{ij} = 0 \qquad \forall j = 0, 1, ..., mc \tag{4}$$

$$(x_i - m_j)^T (x_i - m_j) \times y_{ij} - (x_i - m_k)^T (x_i - m_k) \leq 0$$

$$\forall j = 1, 2, ..., mc; \ k = 0, 1, ..., c - 1; \ \forall i = 1, 2, ..., n \tag{5}$$

$$z_j, y_{ij} \in \{0, 1\} \ integer$$

This optimization problem determines the least number of barycenters (1) which will satisfy the stated constraints. The n constraints (2) state that each feature vector from a pattern in a given class must be assigned to some barycenter vector of that class. As patterns and barycenters have been ordered by class, the summation should be run over the appropriate index sets. The mc constraints (3) impose that no pattern be assigned to a non-existing barycenter. Instead the constraints (4) determine the value of the barycenter vector by summing over all the elements of the feature vector. Notice that x_i is a vector, so the number of equations will be mc times the number of elements in the feature vector. Finally, the last set of equations (5) indicate that each feature vector must be nearer to the assigned barycenter of its own class than to any other barycenter. Should the barycenter be null, this is immediately verified; while, if it is non-zero, this must be imposed.

Theorem 3. *Given a set which does not contain two identical patterns assigned to different classes, a correct classifier will be determined by solving the problem (1) - (5).*

PROOF: If there is no juxtaposition of the patterns belonging to different classes, a feasible solution will always exist to the problem (1) - (5). Such a solution assigns a unique barycenter to every pattern.

Given that a feasible solution exists and that the objective function has a lower bound formed from the mean vectors to each class, an optimal solution to the problem (1) - (5) must exist.

The binary programming problem (1) - (5) with the control variables z_j, y_{ij}, may be solved by suitable branch and bound methods, which will determine an optimal solution to the problem (1) - (5) [11], which is assured by the above theorem.

2.3 Classification and Precision Properties

Suppose a training set is available, defined over a suitable representation space, which is piecewise separable and coherent, as well as a data set in which all the relevant attributes are known, but not the class to which each entity belongs. The algorithm **T.R.A.C.E.** considered in either form, will determine the classification rules to impose on the data set based on the partition which has been found for the training set, so that to each entity in the data set a class is assigned. If the training set forms a random sample and the data set which includes the training set is coherent, then this classification can be performed to any desired degree of accuracy by extending the size of the training sample.

Thus, in classification, new patterns of unknown class belonging to the data set are given. For every pattern the distance is determined from each of the available barycenter vectors, and then the pattern is assigned to the class of the barycenter to which it is closest.

As it can be seen, this algorithm may be slow in training but it is extremely fast in recognition. This is because only a matrix vector product is involved, followed by the determination of the smallest element.

Theorem 4. *Suppose that the data set is coherent. Then the data set can be classified correctly.*

PROOF: Follows by the definition of a coherent data set, definition 3 and by theorems 1 2 and corollary 1.

To obtain correct classification results, the training set must be a representative sample of the data set, and the data set must be coherent.

So consider a data set $\{(x_1, y_1), (x_2, y_2), ..., (x_n, y_n)\}$, where x_i is the feature vector of pattern i and its membership class is given by y_i.

Without loss of generality assume that two-class classification problems are considered, so that eventually a series of such problems must be considered for a polytomous classification problem. Also, assume that the patterns are independently identically distributed with function $F(z)$, where $z_i = (x_i, y_i)$.

Let $f(x, \alpha) : R^n \rightarrow \{0, 1\}$ $\alpha \in A$ be the classifier, where A is the set of parameters identifying the classification procedure from which the optimal parameters must be selected. The loss function of the classifier is given by:

$$L\left(y, f(x, \alpha)\right) = \begin{cases} 0 & if & y = f(x, \alpha) \\ 1 & if & y \neq f(x, \alpha) \end{cases} \tag{6}$$

The misclassification error over the population in this case, is given by the risk functional:

$$R(\alpha) = \int L\left(y, f(x, \alpha)\right) dF(x, y) \tag{7}$$

Thus, the value of $\alpha \in A$, say α^*, which minimizes the expression (7), must be chosen. Hence, for any sample, the misclassification error will be:

$$R_n(\alpha^*) = \frac{1}{n} \sum_{i=1}^{n} L\left(y_i, f(x_i, \alpha^*)\right) \tag{8}$$

It will depend on the actual sample, its size n and the classifier used.

To avoid having to introduce distributional properties on the data set considered, the empirical risk minimization inductive principle may be applied [15]:

1. The risk functional $R(\alpha)$, given in equation (7), is replaced by the empirical risk functional $R_n(\alpha)$, given by equation (8), constructed purely on the basis of the training set.
2. The function that minimizes the risk is approximated by the function that minimizes the empirical risk.

Definition 5. *A data set is stable, according to Definition 3, with respect to a partition and a population of entities, if the relative frequency of misclassification is $R_{emp}(\alpha^*) \geq 0$ and*

$$\lim_{n \to \infty} pr\{R_{emp}(\alpha^*) > \epsilon\} = 0 \tag{9}$$

where α^* is the classification procedure applied and $\epsilon > 0$ for given arbitrary small value and $pr\{.\}$ is the probability of the event included in the braces.

In some diagnostic studies the set of attributes considered has no significant relationship with the outcome or classification attribute of the entity. Typically, the classes could be the eye colour and the attributes be the weight, height and sex of a person. Such a classification would be spurious, since there is no relation between the eye colour and the body indices.

In general, consider smaller and smaller subsets of the attribute space X. If there exists a relationship between the attributes and the classes of the entities, the frequency of the entities of a given class for certain subsets will increase to the upper limit of one; while in other subsets it will decrease to the lower limit of zero. Thus, for a very fine subdivision of the attribute space, each subset will tend to include entities only of a given class.

Definition 6. *A proper subset S_k of the attribute space X of the data set will give rise to a spurious classification if the conditional probability of a pattern to belong to a given class c is equal to unconditional probability over the attribute space. The data set is spurious if this holds for all subsets of the attribute space X and the data set is non-spurious otherwise*

$$pr\{y_i = c \mid (y_i, x_i) \cap S_k\} = pr\{y_i = c \mid (y_i, x_i) \cap X\} \tag{10}$$

Theorem 5. *Consider a training set of patterns randomly selected and assigned to two classes, where the unconditional probability of belonging to class 1 is p, with the set being of size n greater than a, a suitable large number, such that $(n > a)$. Let the training set form b_n barycenters, then under the algorithm* **T.R.A.C.E.**, *this training set will provide a spurious classification, if*

$$\frac{b_n}{n} \geq (1 - p) \qquad \forall n > a \tag{11}$$

PROOF: From the definition 6 a classification is spurious if the class assigned to the entity is independent of the values of the set of attributes considered.

Any pattern will be assigned to the barycenter which is nearest to it, which without loss of generality, may be considered to be a barycenter of class 1, being composed of entities in class 1. The probability that the pattern considered will result not of class 1 is $(1 - p)$ which is the probability that a new barycentre will be formed. As the number of patterns are n, the result follows.

Theorem 6. *Let the probability that a pattern belongs to class 1 be p. The number of barycenters formed from the application of the* **T.R.A.C.E.** *algorithm, which are required to partition correctly a subset S, containing $n_s > a$ non-spurious patterns is $b_s < n_s$, $\forall n_s > a$.*

PROOF: If the classification is not spurious, by definition 6, without loss of generality, the following relationship between the conditional and unconditional probabilities holds for one or more subsets $S_k, S_h \in X, S_h \cap S_k = \emptyset$:

$$pr\{y_i = 1 \mid (x_i, y_i) \cap S_k\} > pr\{y_i = 1 \mid (x_i, y_i) \cap X\} = p \tag{12}$$

$$pr\{y_i = 0 \mid (x_i, y_i) \cap S_h\} < pr\{y_i = 0 \mid (x_i, y_i) \cap X\} = (1 - p) \tag{13}$$

Thus on the basis of the algorithm, for the subsets $S_k \cap X$ the probability that a new barycenter of class 1 will be formed, because one or more patterns result closer to a pattern of class zero, is less than (1 - p). In the set $S_h \cap X$, the probability that patterns of class one will appear, is less than p, so that the probability that a pattern will be formed is less than p.

Thus if the number of patterns present in the dominant subsets $S_k \cap X$ is n_k while the number of patterns present in the subsets $S_h \cap X$ is n_h, the total number of barycentres for the patterns of class 1 will be:

$$b_s < (1 - p)n_k + pn_h \tag{14}$$

As $n_s = n_k + n_h$, there results $b_s < n_s$, $\forall n_s > a$.

Corollary 2. *The Vapnik-Cervonenkis dimension (VC dimension), $s(C, n)$ for the class of sets defined by the* **T.R.A.C.E.** *algorithm, restricted to the classification of a non-spurious data set which is piecewise separable, with $n_s > a$ elements and two classes, is less than 2^{n_s} [15].*

PROOF: By theorem 6 the number of different subsets formed is $b_s < n_s < 2^{n_s}$ whenever $n_s > a$ and the data set is not spurious .

Theorem 7. *Let C be a class of decision functions and ψ_n^* be a classifier restricted to the classification of a data set, which is not spurious and returns a value of the empirical error equal to zero based on the training sample $(z_1, z_2, ..., z_n)$. Thus, $Inf_{\psi \in C} L(\psi) = 0$, i.e. the Bayes decision is contained in C. Then [4]*

$$pr\{L(\psi_n^*) > \epsilon\} \leq 2s(C, 2n)2^{\frac{-n\epsilon}{2}} \tag{15}$$

By calculating bounds on the VC dimension, the universal consistency property can be established for this diagnostic algorithm applied to the classification of a data set which is not spurious.

Corollary 3. *A classification problem with a piecewise classifier and a piecewise separable training set is strongly universally consistent [12]*

2.4 Experimental Considerations

A spurious collection of entities in which there are no similarity relations may occur and should be recognized. With the **T.R.A.C.E.** algorithm, this occurrence is easily determined, since many barycenters are formed, almost one per object. Such spuriousness may arise even in the presence of some meaningful relationships in the data which are swamped by noise. In this case data reduction techniques may be useful [12, 16].

The experimental setup used in this chapter must consider a training set and a verification set, whose class membership should be known, so that the accuracy in classification can be determined. Thus, to verify the algorithm, the training set, can be split by non-repetitive random sampling into two subsets, one used in training and the other for verification. Often a split 90% for training and 10% for verification is a nice balance which yields appropriate standard errors of the estimates. To obtain the probable error estimate, the process is replicated 150 times.

The data coherence problem [12] may arise when the training set is very small. In such a set, an object may form a distinct barycenter set if it is very different from other objects with the same classification. If this item falls in the training sample, then it will be set aside to form a distinct barycenter set which will be composed of a singleton element. If instead it appears in the verification set, then it will be classified wrongly because there will be no other similar barycenter in the training set since it is a singleton and will be nearer to a barycenter of another class. If this were not so, in training it would not have formed a new barycenter. Thus, because of the reduced size of the sample, classification errors will be made, which would not be made if the training set were larger.

This phenomenon, akin to under-sampling, is due to the structure of the training and data sets that have randomly occurred which may not be collectively exhaustive. Thus, the training set should be consistent with regard to the patterns in both sets. The presence of different similarity structures in the two sets certainly indicates that changes have occurred in the sets. Hence, if small verification samples are involved, the verification instances should be selected by a stratified sample routine instead of a simple random sample selection process. The sample design should sample less than proportionally those objects that appear in very small groups. In practice, instead of defining these probabilities for each barycenter group, a biased verification sample is formed by excluding from the selection those objects which are assigned to singleton barycenters and restricting to not more than half those objects which are assigned to these subgroups with low consistency [7]. This sampling method will provide an asymptotic precision rate as the sample increases if the data set is coherent [3, 12].

In classification, the whole training set is used to determine the classifier. When a new pattern of unknown class is given, its distance is determined from each of the available barycenter vectors and then it is assigned to the class of the barycenter which is the closest to it. As it can be seen, this algorithm may be slow in training, but it is extremely fast in recognition since only a matrix vector product operation followed by the determination of the smallest element is involved.

A further refinement can be applied. Since the number of objects, which define every barycenter, is known, relative certainty levels may be assigned to this classification. Thus, if an unknown pattern is assigned to a class on the basis of the closest barycenter, as required, and the latter is formed from a consistent number of objects, then this pattern is said to be classified in the certain category. If the number of objects defining this barycenter is limited, then it can be classified in the probable category and so on.

An iterative correction procedure can be enacted in the presence of imprecise classifications. The classes in the training set may be assigned because of sure attributions, such as when a person has died, or the autopsy has been performed. On the other hand, in many diagnostic procedures, there may be some imprecision in their classification. If the training set is large, the verification error rate will be small and purely due to sampling variation because the sample set is limited.

By performing the replication of the data set as described above, objects will appear in the verification set about fifteen times in every 150 replications. Thus, if the object should result misclassified with respect to its actual classification, say two-thirds of the times, then the actual classification can be considered imprecise and the class can be corrected to the class to which it has most frequently been assigned by the algorithm in the verification process. When all misclassified patterns have been corrected, the training and verification procedures are applied anew on the corrected data set [3]. The whole process is known as classification with an imprecise teacher.

Finally, the patterns with unknown classification may in certain circumstances be assigned also by a majority rule obtained by the replication of a number of different classification procedures or different sampling procedures [14].

By using **T.R.A.C.E.** this way for sufficiently large samples, precise classification results can be obtained and the changes in the characteristics of the data set can be followed.

3 Alport Syndrome

The Alport syndrome (AS) is a hereditary disease of the basement membranes with a prevalent genetic inheritance link with the X-chromosome. This disease occurs in families that are autosomic recessive, or rarely that have a dominant inheritance. Sporadic cases have also been reported [1, 9, 10].

The basement membrane damage is mostly evident in renal glomeruli. The usual diagnostic process consists of electron microscopy of kidney biopsies. The syndrome, in fact, arises from a defect of the encoding of genes for isoforms of type IV collagen α-chains [2]:

- COL4A5 gene on chromosome-X and encoding for COL IV α_5-chain (X-linked AS),
- COL4A3 and COL4A4 genes on chromosome 2 and encoding for COL IV α_3 and α_4-chains (autosomic AS).

The Alport syndrome has a variable clinical expression and consists of the following:

- childhood haematuria,
- progressive renal failure,
- high tone sensorial deafness,
- minor ocular lesions.

The severity of the disease varies among different families and the pathology is difficult to diagnose. A careful analysis of renal biopsies though, based on a number of morphological criteria identified through the electronic microscope, is helpful even though these are not pathognomonic to the disease.

To diagnose the pathology, the following morphological changes to the renal glomerular basement membrane should be considered [9]:

1. Thinning of the membrane. The glomerular basement membrane thickness is age related, so this criterion must be carefully applied and is not applicable to children under 3 years of age;
2. Non-specific thickening, which is defined by the presence of peculiar lamina densa changes;
3. Splitting or lamellation defined as the repentine forking of the lamina densa to form two or three unconnected parallel layers;
4. Basket weaving or reticulation, characterized by an irregular thickening of the glomerular basement membrane with a complex replication of the lamina densa, transformed into a heterogeneous network of membraneous strands.

None of these electron microscope findings are pathognomonic of AS, although splitting and basket weaving of the glomerular basement membrane are considered major changes and are assumed to be cardinal for a diagnosis.

The possibility of these alterations leads to the definition of three levels of diagnosis for this pathology, based on the features of renal glomerular basement membrane indicated above:

1. Features consistent with AS (presence of AS):
   - all 4 features regarding the glomerular basement membrane are present,
   - or 3 or more of these major features are extensively found.
2. Features partly consistent with AS (segmental AS):
   - 2 of the major features are lacking,

- or lesions present have a segmental distribution.
3. No morphological evidence of AS:
    - none of the major changes indicated above are present,
    - or thinning and/or non-specific thickening found sometimes.

As it can be seen, diagnosing AS is a difficult task due to the non-specificity of the ultra-structural investigation of the renal biopsies and of the molecular genetic analysis of the α-chains [2], although in combination this diagnostic process may be useful even though it is rather complex.

The specificity of the diagnoses varies with age, sex and family history but the combination of the ultra-structural investigations of kidney biopsies and the molecular genetic analysis of the α-chains leads to a sensitivity precision of about 92%.

Regarding this syndrome, there is a form of a paradox, referred to as the skin paradox. The α-chains are present both in the kidneys and in the skin, but at the epidermal level no morphological defects are encountered. Immunofluorescence investigations using antibodies against $\alpha_5(IV)$-chains is routinely used as an additional tool to diagnose the disease. Absence or segmental distribution of the signal is considered to be highly suggestive of the Alport Syndrome. It has recently been shown that the absence or a segmental distribution of $\alpha_5(IV)$-chain along the epidermal basement membrane is associated with an increased intensity of a fluorescent signal using a monoclonal antibody against collagen type VII, which is usually present along the dermal epidermal junction [8].

With immunofluorescence, the abundance of collagen VII can be quantified. This may be related to the diagnostic levels of the syndrome, as it appears to substitute the missing type IV collagen in patients who are afflicted with this pathology. Thus there must be apparent differences in the texture of the skin if this relationship holds.

If it is true that collagen type VII substitutes in the skin for collagen IV for patients afflicted with the Alport syndrome, then the pathology should be diagnosable at the electron microscopic level using skin samples [1]. Since there must be apparent and recognizable differences in the texture of the skin, these differences should be recognizable by a suitable classification algorithm.

4 Experimental Results

This section describes the methodology and results of the experimentation of **T.R.A.C.E.** regarding the classification of patients into three diagnostic classes of the Alport Syndrome.

Specimens of skin were considered, which included the epidermal basement membrane. Samples were prepared in order to take images under the electronic microscope.

Briefly, punch biopsies of the skin were fixed in 4 % buffered formaldehyde, post-fixed in a 1 % buffered Osmium Tetraoxide and embedded in Epon. Thin

$(120 - 150$ nm ($nanometers$)) sections were cut with Reichert-Jung ultramicrotomer and collected onto copper grids, stained with uranyl acetate and lead citrate and observed under a Philips CM10 Transmission electron microscope. All samples were observed and photographed at the same magnification $(11500\times)$.

All the photographs were then analyzed by the pathologists but no significant differences were detected. Therefore, **T.R.A.C.E.** was applied.

4.1 Training and Verification

Nine patients were considered and grouped as follows:

- three patients were diagnosed with the presence of AS,
- three patients were diagnosed with the presence of segmental AS,
- three individuals without AS were used as controls.

From each individual a skin specimen was examined. Approximately seven images were taken of different areas of the biopsy, so that there were obtained 27 distinct images representing a diagnosis of the presence of AS, 28 images representing a diagnosis of the presence of segmented AS, and 21 images representing a diagnosis of normality. From each image 5 sub-images were randomly selected. Thus, 354 sub-images were formed, that is, there were 129, 125 and 100 sub-images respectively in each category.

From the image pixel maps, central moments were obtained in the horizontal and vertical direction. The first moment was identically zero, so patterns were formed with 5, 10, 15, moments in each direction. A better feature extraction procedure turned out to be the definition of histograms of the pixel distribution, which are bivariate frequency functions. Central moments were determined on these structures [1].

A series of experiments was performed with simple non-repetitive sampling on various patterns with different moments and on various feature extraction implementations. The results of these experiments for some selected feature extraction methods are given in Table 1.

In column 1 of Table 1, various feature extraction procedures are indicated. For each procedure in column 2 the mean precision over 150 trials is presented, while in column 3 the standard error of the mean precision is given. In columns 4 and 5 the best and the worst precision results over the run are given. The results indicate that in all cases the precision is significantly different, since a random assignment would exhibit a mean precision of 0.3333 with three classes. Thus, **T.R.A.C.E.** algorithm does in fact classify the patterns, although with a significant error. This is not surprising as the sample of individuals is very small. There were nine individuals and 71 images from which 354 sub-images were drawn. Hence, we can expect sampling variation to be high. Many other pattern recognition techniques were also tried, but to little avail providing results similar to those obtained with **T.R.A.C.E.** [1].

Table 1. Classification results for the verification of Alport syndrome sample derived from non-repetitive random sampling.

	Mean precision	Standard error	Best	Worst
Pixel distribution (5 moments)	0.4425	0.0066	0.6111	0.2308
Pixel distribution (10 moments)	0.4367	0.0066	0.6000	0.2500
Pixel distribution (15 moments)	0.4226	0.0065	0.5833	0.2500
histogram (5 moments)	0.4970	0.0066	0.6750	0.2500
histogram (10 moments)	0.5000	0.0067	0.7200	0.2600
histogram (15 moments)	0.4734	0.09066	0.7027	0.3243

As it can be seen from the table, the histogram feature extraction technique fares somewhat better than the central moment in the orthogonal directions of the moments of the distribution of the pixels. Although the results were statistically significant, it was deemed that the proper coherency transformation was worthless in this environment. Hence, the specimens were subjected to a stratified sampling procedure, and the obtained results are reported in Table 2 using the same format as in Table 1.

Table 2. Classification results for the verification of Alport syndrome sample derived from non-repetitive stratified sampling.

	Mean precision	Standard error	Best	Worst
3 moments	0.6465	0.0070	0.8667	0.4546
5 moments	0.7145	0.0067	0.9355	0.5161
15 moments	0.7060	0.0067	0.9091	0.4194
histogram (5 moments)	0.8215	0.0056	0.9643	0.6563

Although the results are satisfactory, they can be improved. The principal requirement is to enlarge the sample, as the whole structure of the classification problem shows that there is still abundant variability.

Thus, ten specimen were extracted from the sample randomly. It was ensured, however, that three were selected from each diagnostic category, while the tenth specimen was of the non-pathological class. The iterative classification procedure was attempted on 150 trials as indicated, choosing a ten percent verification sample every time. The ten preselected instances appeared in classification about 15 times on average, so they were assigned by the majority rule to a class. The results of this experiment are reported in Table 3.

This technique yields better results with an 8% improvement over the previous procedure. The obtained result is satisfactory, although efforts are

Table 3. Classification by the iterative procedure of 10 patients with regard to the Alport syndrome

	Class 1	Class 2	Class3	Assigned class	Correct class
Specimen 1 (pattern 51)	0.19	0.26	0.55	3	1
Specimen 2 (pattern 61)	0.86	0.07	0.07	1	1
Specimen 3 (pattern 94)	0.65	0.0	0.35	1	1
Specimen 4 (pattern 215)	0.39	0.61	0.0	2	2
Specimen 5 (pattern 224)	0.30	0.70	0.0	2	2
Specimen 6 (pattern 226)	0.0	0.64	0.36	2	2
Specimen 7 (pattern 307)	0.01	0.10	0.89	3	3
Specimen 8 (pattern 318)	0.17	0.0	0.83	3	3
Specimen 9 (pattern 333)	0.03	0.16	0.81	3	3
Specimen 10 (pattern 343)	0.10	0.09	0.81	3	3

being enacted to further improve the precision to higher levels; and in the limit for a sufficiently large sample, achieve recognition with probability one [12].

4.2 Classification in two classes of a sample with unknown diagnoses

To further study the classification performance of this algorithm, an additional series of tests were performed. First, it was felt that the reduction to two classes by merging one of the three classes might lead to a more meaningful classification.

Table 4. Classification results for 2 classes from the verification of Alport syndrome sample derived from non-repetitive random sampling

	Mean precision	Standard error	Best	Worst
3 moments(12),(3)	0.6384	0.0074	0.8333	0.4286
3 moments(1),(23)	0.6328	0.0074	0.8333	0.3793
3 moments(13),(2)	0.6299	0.0074	0.8571	0.4167
histogram (5 moments)(12),(3)	0.6130	0.0076	0.8056	0.4571
histogram (5 moments)(1),(23)	0.6073	0.0077	0.7949	0.4000
histogram (5 moments)(13),(2)	0.7542	0.0063	0.9024	0.5833

Table 4 presents the results of the classification of the sample when it has been aggregated into two classes. In column 1 the feature extraction method is indicated and then the grouping of the three classes into two classes is reported. Thus, the obvious notation (12), (3) indicates that class 2 has been merged with class 1 and class 3 is considered alone.

All the results are significantly different from equiprobability, which in this case is one-half since there are only two classes. But the most interesting feature is the result for the classification of the sample when the diagnosis of the presence of AS and presence of segmented AS are merged in a pathological category to be contrasted to a normal category. The results appear in the last row of Table 4.

Another set of thirteen patients was proposed for diagnosis. The methodology was identical to the one described, except that the whole training set of 354 patterns was used. On the basis of the barycenters formed from this classification, the thirteen new patients, represented by seven images for each biopsy from which 465 sub-images in total were formed and classified.

Table 5. Classification of 13 patients with regard to the Alport syndrome in 2 classes on the basis of the majority rule.

	class 1	class 2	total	assigned class
Patient f2	19	16	35	1
Patient f3	23	12	35	1
Patient f4	28	12	40	1
Patient f5	27	8	35	1
Patient f6	25	10	35	1
Patient f7	27	8	35	1
Patient f8	10	25	35	2
Patient f9	15	25	40	2
Patient f10	14	21	35	2
Patient f12	29	6	35	1
Patient f13	29	6	35	1
Patient f14	22	13	35	1
Patient f15	23	12	35	1

The classification of this blind set on the basis of the majority rule gave the results reported in Table 5. Patients were given a code number and 35 or 40 sub-images were obtained for each from the skin biopsy by an identical procedure as the one described above. The number of times that the images of a patient were classified in class 1 (pathological diagnosis) or class 2 (diagnosis of normality) are given in columns 2 and 3 respectively. The total number of images classified for each patient are reported in column 4, while in column 5 the class assigned by the majority rule is given.

Subsequently, the correct classification was communicated for part of the classification sample considered, which is presented in Table 6. As it can be seen on the basis of these results, the precision rate was 0.6250.

Table 6. Comparison of the assigned class and the correct class for Alport syndrome for part of the classification sample.

	f7	f8	f9	f10	f12	f13	f14	f15
assigned class	1	2	2	2	1	1	1	1
correct class	1	1	2	2	2	2	1	1

Table 7. Classification of 5 patients with regard to the Alport syndrome in 2 classes on the basis of the majority rule.

	class 1	class 2	total	assigned class
Patient f2	18	17	35	1
Patient f3	20	15	35	1
Patient f4	35	5	40	1
Patient f5	30	5	35	1
Patient f6	27	8	35	1

The training set was enlarged to 639 patterns with these new images, and the rest of the classification sample was classified again. The results are given in Table 7. The precision in this case was equal to 1.0000. The patterns f2 and f3, though, would appear to be less sure based on intuition. However, the process is nonlinear, so no significance can be given to the distance criterion.

Table 8. Classification results for 2 classes from the verification of Alport syndrome sample derived from non-repetitive random sampling

	Mean precision	St. error	Best	Worst	n.patterns
histogram (5 moments)(13),(2)	0.6651	0.0068	0.8154	0.4849	639
histogram (5 moments)(13),(2)	0.8510	0.0055	0.9294	0.7073	819

The last experiment was performed. The original training sample was merged with the new samples. This was accomplished by increasing first the items in the training set to 639 and then increasing the number of patterns to 819, by making use of all the images available.

The algorithm was run again 150 times by first removing ten percent of the sample to form the verification set and then performing the training of the classifier with the rest of the sample. The results are presented in Table 8. In this table, the results are reported in an identical way to those in Tables 1 and 4. Notice that the classification accuracy of the sample of 639 instances in Table 8 is lower than the result obtained with just a sample of 354 instances, as indicated in Table 4. The classification accuracy increases substantially when the sample is enlarged to 819 patterns. This is not exceptional and indicates that sampling variability is still present, due to the small number of considered cases (only 22 cases total).

5 Conclusions

The Alport Syndrome is a severe pathology which is difficult to diagnose and if not identified early leads inevitably to death. The traditional difficulties involved in diagnosing such a pathology have been indicated in this chapter. Essentially, the malfunctioning of chromosomes on certain genes is responsible for the pathology. This malfunctioning may not trigger a pathological state, but may lead to the gradual development of the pathology which will progress eventually towards end stage renal disease.

It is imperative that potential patients are periodically tested for this disease and if diagnosed the whole family group must be subjected to the same tests, as this pathology is family related. Moreover, to have a chance to provide an early cure, invasive tests must be performed periodically even on young children. It may be that this can be avoided, through the technique described, in which a fragment of skin is used to determine the condition of the person. This procedure can be repeated as often as desired. At the moment, if the estimates are fulfilled there may be about a two percent approximate difference in the sensitivity of the techniques. The invasive technique has a precision of about 92% and the image recognition technique has a precision of 90%. The work in progress is to increase the sample size, which should increase the precision [12], and to apply other transformations to stabilize the data set without losing beneficial information.

It may turn out that a series of genetic disorders are at work, but these give rise to more or less the same set of symptoms. In this case it might be advisable to proceed to a different classification of the disease, i.e. by considering the presence of each diagnostic criteria and their combinations as different pathological classes.

If, for a sufficiently large training set, this information is available, one could run **T.R.A.C.E.** on this data set and assess the diagnostic analysis. In this way a whole series of diagnostic experiments can be repeated by determining the classes in alternative ways. Some might be fruitful and others might not, since they would disregard the important aspect of clinical insight.

Pattern recognition methods are valuable if a high precision is reached, so as to provide a high precision in the classification results. Thus, only those techniques that give demonstrable precise results both in theory and in practice, supported by a complete theoretical analysis of its properties without extraneous assumptions, such as specific parametric distributions, should be applied [12], since a reliable analysis which should save lives is at stake.

The important contribution of this chapter to Alport Syndrome studies is that it has been shown that, indeed, the absence of α_5-chain in the skin is associated with the modification in the epidermal basement membrane of such patients and that these changes can be detected by a classification algorithm, although apparently not by expert clinicians directly.

References

1. G. Addonisio. Tecnica di classificazione per la diagnosi della sindrome di Alport. Technical report, Laurea Thesis Universita di Roma La Sapienza Facoltà di Statistica, 2002.

2. P. Barsotti, A. Onetti Muda, G. Mazzucco, L. Masella, B. Basolo, M. De Marchi, G. Rizzoni, G. Monga, and T. Faraggiana. Distribution of alpha-chains of type IV collagen in glomerular basement memebranes with ultrastructural alterations suggestive of Alport syndrome. *Nephrology Dialysis Transplantation*, 16:945–952, 2001.

3. G. Bonifazi, P. Massacci, L. Nieddu, and G. Patrizi. The classification of industrial sand-ores by image recognition methods. In *Proceedings of 13th International Conference on Pattern Recognition Systems, Vol.4: Parallel and Connectionist Systems*, pages 174–179, Los Alamitos, CA, 1996. IEEE Computer Society Press.

4. L. Devroye, L. Gyorfi, and G. Lugosi. *A Probabilistic Theory of Pattern Recognition*. Springer-Verlag, Berlin, 1996.

5. R. O. Duda and P. E. Hart. *Pattern Recognition and Scene Analysis*. Wiley, New York, 1973.

6. E. P. Goss and G. S. Vozikis. Improving health care organizational management through neural network learning. *Health Care Management Science*, 5:221–227, 2002.

7. H. S. Konjin. *Statistical Theory of Sample Design and Analysis*. North Holland, Amsterdam, 1973.

8. L. Massella, K. Giannakakis, A. Onetti Muda, A. Taranta, G. Rizzoni, and T. Faraggiana. Type VII colagen in Alport syndrome. *Journal of the American Society of Nephrology*, 13:309, 2002.

9. G. Mazzucco, P. Barsotti, A. Onetti Muda, M. Fortunato, M. Mihatsch, L. Torri-Tarelli, A. Renieri, T. Faraggiana, M. De Marchi, and G. Monca. Ultrastructural and immunohistochemical findings in Alport's syndrome: A study of 108 patients from 97 italian families with particular emphasis on col4a5 gene mutation correlations. *Journal of Nephrology*, 9:1023–1031, 1998.

10. S. Meleg-Smith, S. Magliato, M. Cheles, R.E. Garola, and C.E. Kashtan. X-linked Alport syndrome in females. *Human Pathology*, 29:404–408, 1998.

11. G.L. Nemhauser and G.L. Wolsey. *Integer and Combinatorial Optimization*. Wiley, New York, 1988.

12. L. Nieddu and G. Patrizi. Formal properties of pattern recognition algorithms: A review. *European Journal of Operational Research*, 120:459–495, 2000.

13. G. Patrizi. Optimal clustering properties. *Ricerca Operativa*, 10:41–64, 1979.

14. G. Patrizi, L. Nieddu, P. Mingazzini, F. Paparo, Gr. Patrizi, C. Provenza, F. Ricci, and L. Memeo. Algoritmi di supporto alla diagnosi istopatologica delle neoplasie del colon. *Associazione Italiana per l'Intelligenza Artificiale (AI*IA)*, 2:4–14, 2002.

15. V. N. Vapnik. *Learning Theory*. Wiley, New York, 1998.

16. S. Watanabe. *Pattern Recognition: Human and Mechanical*. Wiley, New York, 1985.

17. T. Y. Young and W. Calvert. *Classification, Estimation and Pattern Recognition*. Elsevier, New York, 1974.

Clinical Analysis of the Diagnostic Classification of Geriatric Disorders

Giacomo Patrizi[1]*, Gregorio Patrizi[2], Luigi Di Cioccio[3], and Claudia Bauco[4]

[1] Dipartimento di Statistica, Probabilità e Statistiche Applicate
   Università degli Studi "La Sapienza", Rome, Italy
[2] Dipartimento di Scienze Chirurgiche,
   Università degli Studi "La Sapienza", Rome, Italy
[3] Area Dipartimentale Geriatrica, ASL Frosinone, Frosinone, Italy
[4] Unita Operativa Geriatria, Ospedale "G. De Bosis", Cassino, Italy

Summary. The aim of this chapter is to present a classification algorithm and its application to an initial set of 156 patients afflicted with dementia syndromes and classified by clinicians in the categories of probable Alzheimer, possible Alzheimer or vascular dementia pathologies. It will be shown that the diagnoses of dementia patients by this method is very accurate, and that the classification criteria can be transformed into suitable clinical factors, which can then be interpreted by clinicians. This formal implementation suggests that recent research on the general diagnosis of dementia can be confirmed.

1 Introduction

Diagnostic procedures should apply as widely as possible, to recognize the pathological instances and correctly refute all those instances that have some of the symptoms, but not the pathology. In accordance with the scientific method, such procedures should be formal, devoid of value judgements and replicable across place and time.

The efficacy of a diagnostic procedure is usually studied through an appropriate statistical design of experiments, [3, 9] identifying summary measures of the pathology or treatment effects without chasing after the outliers or the vagrant outcome. The objective of statistical analysis is to summarize the principal characteristics of the objects being studied in a few meaningful measures or by functional relationships [24]. The emphasis is on summary descriptors and not on interpreting all possible outcomes. Thus it would be surprising if statistics were to satisfy medical needs completely.

In clinical analysis it is desired to diagnose correctly every patient which appears to have the pathology. In pursuing the pathologies wherever and

---

* Corresponding author. Email: `g.patrizi@caspur.it`

however they occur, knowledge is obtained and one is better able to help patients in their recovery. The objective in medicine therefore regards high diagnostic precision, but this ideal is rarely achieved and in particular for the diagnosing of Alzheimer's disease and vascular dementia, there are many conceptual and implementation difficulties.

The aim of this chapter is to present an algorithm for accurate classification of patients afflicted with dementia syndromes and apply it to a set of patients to show how the formal characteristics obtained from the application can be transformed into clinical terminology and indicate various aspects of recent research, which are confirmed by this methodology and these results.

Alzheimer's disease is the most common form of a complex set of symptoms leading to dementia which involve motor, cognitive and physical degeneration. Ultimately it causes the slow destruction of brain cells [10]. Considered a rare pathology in the past, today Alzheimer's disease represents an important epidemiological, clinical and social phenomenon. The disease reduces life expectancy as the average expected survival time is 6-8 years. The cause of death is most commonly recurrent pathologies (e.g., bronchopulmonary pathologies).

Although the symptomatological framework is typical of patients afflicted with dementia, the actual symptoms presented vary greatly from individual to individual, often due to other related pathologies which may affect the patient. Many of the principal characteristics of Alzheimer's disease are not specific, so that a diagnosis can only be made by evaluating the available symptoms in a complex way, as the symptoms are conditional on other pathological states which afflict the patient [46].

Previously, it was thought that the disease affected all the associative areas of the brain cortex in a diffused and undifferentiated way. Lately, neuropathological studies have shown that both the initial localization of the histopathological lesions, typical of Alzheimer's disease, and their progression in time present very specific aspects [45].

The characteristics that differentiate a brain of a healthy elderly person from that of one afflicted by Alzheimer's disease are purely quantitative. In both cases, there is a reduction in the mass and the volume of the organ, a dilation of the ventricular cavities at the level of the cortex, an enlargement of the grooves on the surface and a thinning of the cerebral circular components. In Alzheimer's disease, there is an exasperation of the loss of cerebral matter in the reduction of the hemispheric volume and mass of the organ with regard to controls of similar age. A number of specific degenerative aspects have been reported in the brain of Alzheimer patients, such as diffuse plaques [34], neocortical neurofibrillatory tangles [5] and neuronal loss [17], but these vary with the progression of the disease and do not correlate across patients. Moreover these are diffused within the cerebral matter, so a precise analysis of their state and their progression can only be determined after death [8].

There are a number of characteristics that are thought to be risk factors of this disease, such as age, genetic traits and familiarity [1]. The putative risk factors, which are thought to affect the process, are sex (more females than males are affected) [42], level of instruction: people with lower levels of education may be affected more than proportionally [45], cardiovascular diseases, smoke, alcohol, depression and head injuries [42].

Thus the diagnosis of Alzheimer's disease remains primarily dependent on clinical assessment [49]. Historically, clinical diagnostic accuracy has varied due to the use of multiple sets of broadly based diagnostic criteria. A notable advancement occurred with the introduction of standardized clinical criteria [29] which contributed to improved diagnostic accuracy and allowed meaningful comparisons of results from therapeutic trials and other clinical investigations [49]. Further, recent results indicate that the use of a computerized system with binary decisions (yes/no) increased the specificity of diagnoses when compared to standard clinical diagnosis [21]. It is held that the improvement observed is due to the fact that in normal practice, the clinician's overall impression of a case may lead to a less strict application of the criteria [21].

It will be shown in this chapter that computer programs can achieve sizeable improvements in diagnostic accuracy and that the major confounding factor is the abundance of noisy partial information. The presence of too many diagnostic elements that must be considered will induce a clinician to rely on his overall impression, since he cannot individually weigh many partially conflicting diagnostic elements which contain noise.

The consideration of all these aspects will give rise to a model of diagnosis, exceedingly complex, often with contradicting evidence [28] difficult to formalize and thus open to subjective evaluations. On the other hand, all these symptoms may be connected in some form, so that a few aspects extracted from the set of neuropsychological tests are sufficient to provide the necessary diagnostic evidence.

Accordingly, in the next section on Methods and Materials, the clinical characteristics of the sample will be presented and the procedure to carry out this selection of independent symptoms will be described, called **T.R.A.C.E.** (**T**otal **R**ecognition by **A**daptive **C**lassification **E**xperiments), originally proposed [14], whose convergence results were proven [38]. In the section on Results, the outcomes of these experiments will be examined. Then in the Discussion section, the transformations to carry out on the medical diagnostic techniques and some other aspects will be specified. Finally, the appropriate conclusion will be drawn.

It will be seen that the method presented is powerful enough to reveal new aspects and relationships and thus lead to knowledge discovery.

2 Methods and Materials

The study was conducted on 156 patients, 103 females and 53 males who endured an initial visit during the period 1999-2001 at the clinical facilities south of Rome (Aquino and Sora), under the auspices of a major Italian research project called "the Cronos Project" [15].

2.1 The Diagnostic Procedure

Dementia is a group of clinical syndromes characterized by multiple cognitive deficits in an alert patient sufficient to interfere with daily activities and quality of life. Dementia represents a decline from previously higher levels of cognitive function. Operationally, a demented patient conducts customary activities poorly relative to past performance because of cognitive loss [49].

Many diseases are known to cause dementia and new ones are still being recognized. The term "dementia" does not imply a prognosis. Some dementia are fully treatable and others do not have effective treatment at the present time. Therefore it is increasingly important to accurately diagnose dementia so that reversible and treatable dementia are appropriately managed [23].

The most common forms of dementia is Alzheimer's disease and vascular dementia, with the former being more common in Western countries, while the latter is more common in the Far East [27, 11].

For patients with dementia symptoms, but which had not been assigned to other categories of dementia, such as dementia due to Parkinson's disease, dementia with Lewy bodies, HIVS/AIDS dementia, Pick's disease etc. [18], three diagnostic categories were considered, based on specific criteria formulated by the National Institute of Neurological and Communicative Disorders and Stroke and The Alzheimer's Disease and Related Disorders Association (NINCDS-ADRDA) [29, 49].

Probable Alzheimer's Disease:

This represents the most confident level of antemortem diagnosis and is diagnosed when subjects present the typical course, including a gradual onset and progression of memory and other cognitive problems [49].

- Criteria for the clinical diagnosis of Probable Alzheimer's Disease are [29]:

  - Dementia established by clinical examination and documented by the Mini/Mental test; Blessed Dementia Scale, or some similar examination and confirmed by neuropsychological tests;
  - deficits in two or more areas of cognition;
  - progressive worsening of memory and other cognitive functions;
  - no disturbances of consciousness;
  - absence of systematic disorders or other brain diseases that in and of themselves could account for the progressive deficits in memory and cognition.
- The diagnosis of Probable Alzheimer's disease is supported by:
  - progressive deterioration of specific cognitive functions such as language (aphasia), motor skills (apraxia) and perception (agnosia);

- impaired activities of daily living and altered patterns of behavior;
- family history of similar disorders, particularly if confirmed neuropathologically;
- laboratory results should include:
  - normal lumbar puncture as evaluated by standard techniques;
  - normal pattern, or non-specific changes, in EEG such as increased slow-wave activity;
  - evidence of cerebral atrophy on computed tomography with progression documented by serial observation.
- Clinical features consistent with the diagnosis of Probable Alzheimer's disease, after exclusion of causes of dementia other than Alzheimer's disease, include:
  - plateaus in the course of progression of the illness;
  - associated symptoms of depression, insomnia, incontinence, delusions, illusions, hallucinations, catastrophic verbal, emotional or physical outbursts, sexual disorders, or weight loss;
  - other neurologic abnormalities in some patients, especially with more advanced disease and including motor signs such as increased muscle tone, myoclonus or gait disorder;
  - seizures in advanced disease;
  - computed tomography normal for age.
- Features that make the diagnosis of Probable Alzheimer's disease uncertain or unlikely, include:
  - sudden apoplectic onset;
  - focal neurological findings such as hemiparesis, sensory loss, visual fields deficits and incoordination early in the course of the illness;
  - seizures or gait disturbances at the onset or very early in the course of the illness.

Possible Alzheimer's Disease:

This is diagnosed when the patient presents an atypical course of dementia (e.g. language problems as an early feature) or when there is a coexistent potentially dementing illness, although Alzheimer's Disease is thought to be the primary cause of the progressive dementia [49].

- evident dementia syndrome;
- absence of neurological, psychiatric, or systematic disorder attributable to the dementia;
- onset of variations of the disorders during the appearance and development of the illness;
- presence of a secondary systematic or cerebral illness, susceptible to produce a demential syndrome, but which is not considered as the cause of this dementia;

- may be used as a category in clinical research when a severe and progressive cognitive deficit is identified in the absence of the identification of other causes.

The NINCDS-ADRDA criteria have been adopted by the multi-center Consortium to establish a Registry for Alzheimer's Disease (CERAD) [33] in which the clinical diagnosis of Alzheimer's disease has been validated in 90% of the cases, after death, when the autopsy can be performed [16]. The use of standardized diagnostic criteria along with informant-based history has routinely yielded diagnostic accuracy rates of 85% or greater [4, 30, 16]. Note that none of these criteria utilize a biological marker for Alzheimer's Disease, while recent evidence indicates that such markers could be helpful in diagnosis, although none yet appear sufficiently reliable to warrant diagnostic use in the traditional way [37].

Vascular Dementia:

This is considered as probable vascular dementia according to the National Institute of Neurological and Communicative Disorders and Stroke and Association Internationale pour la Recherche et l'Enseignement en Neurosciences, (NINCDS-AIREN) criteria [43].

- Criteria for the clinical diagnosis of probable vascular dementia include all the following:
  - Dementia defined by cognitive decline from a previously higher level of functioning and manifested by impairment of memory and of two or more cognitive domains (orientation, attention, language, visual-spatial functions executive functions motor control and praxis).
    - Inclusion criteria:
      - Dementia established by clinical examination and documented by neuropsychological testing;
      - Dementia of sufficient severity to interfere with activities of daily living not due to physical effects of stroke alone.
    - Exclusion criteria:
      - cases with disturbance of consciousness, delirium, psychosis, severe aphasia or major sensorimotor impairment precluding neuropsychological testing.
      - Systematic disorders or other brain disease (such as Alzheimer's Disease) that could account for deficits in memory and cognition by themselves.
  - Cerebrovascular disease defined by the presence of focal signs on neurologic examination, such as:
    - hemiparesis, lower facial weakness, Babinski sign, sensory deficit, hemianopia and dysarthria consistent with stroke (with or without history of stroke);

- · evidence of relevant cardiovascular disease (CVD) by brain imaging (computed tomography or magnetic resonance imaging) including large-vessel infarcts or a single strategically placed infarct (angular gyrus, thalamus, basal forebrain, or posterior cerebral artery (PCA) or the anterior cerebral artery (ACA) territories;
  - · multiple basal ganglia, and white matter lacunes or extensive periventricular white matter lesions, or a combination thereof;
  - A relationship between the above two disorders manifested or inferred by the presence of one or more of the following:
    - · onset of dementia within three months of a recognized stroke;
    - · abrupt deterioration in cognitive functions;
    - · fluctuating or stepwise progression of cognitive deficits.
- Clinical features consistent with the diagnosis of Probable vascular dementia include the following:
  - early presence of gait disturbance (small-step gait or marche à petits pas, or magnetic, apraxic-ataxic or parkinson gait);
  - history of unsteadiness and frequent unprovoked falls;
  - early increase in urinary frequency, urgency and other urinary symptoms not explained by urologic disease;
  - pseudobulbar palsy;
  - personality or mood changes, abulia, depression, emotional incontinence or other subcortical deficits including psychomotor retardation and abnormal executive function.
- Features that make the diagnosis of vascular dementia uncertain or unlikely, include:
  - early onset of memory deficit, progressive worsening of memory and other cognitive functions such as language (transcortical sensory aphasia), motor skills (apraxia) and perception (agnosia) in the absence of corresponding focal lesions on brain imaging;
  - absence of focal neurologic signs other than cognitive disturbances;
  - absence of cerebrovascular lesions on brain computed tomography or magnetic resonance imaging.

Of particular importance to differentiate Alzheimer's type dementia from multi-infarct dementia is one of the alternative or modified Hachinski scales. The Hachinski ischaemia score is given by assigning the full score of the clinical features considered to an individual or assigning zero in the absence of that feature. No intermediate scores are contemplated for the 2-point features: either two or zero. The features and their scores are given in Table 1.

It is generally believed that a Hachinski score value less than or equal to two is a strong indication of the patient suffering from Alzheimer's disease, while a score value greater than four is considered as an indication of vascular dementia if these scores are accompanied in both cases by other supportive evidence.

Table 1. Hachinski Ischaemia score

feature	score	feature	score
Abrupt onset	2	emotional incontence	1
Stepwise deterioration	1	history of hypertension	1
fluctuatung course	2	history of strokes	2
nocturnal confusion	1	evidence of associated atherosclerosis	1
relative preservation of personality	1	focal neurological symptoms	2
Depression	1	focal neurological signs	2
somatic complaints	1		

Neuropathological studies of patients with dementia have revealed that cerebrovascular disease and Alzheimer's disease frequently coexist, suggesting that pure vascular dementia may be relatively rare and that mixed dementia may be more common than previously recognized [11, 16, 7, 20].

It has also been shown that the concomitant cerebrovascular disease helps to determine the severity of dementia among patients with Alzheimer's disease and thus the synergetic effects of the two pathologies may influence the cognitive decline of the patient [20]. Furthermore, even among relatively pure cases of vascular dementia and Alzheimer's disease, the clinical manifestations of those two dementia subtypes may not be fully distinct [32].

These criteria were applied to the sample considered and it was found that 111 patients were diagnosed with a form of dementia attributable to Alzheimer's disease (75 women and 36 men) and the remaining 45 (28 women and 17 men) had a form of dementia originating from vascular dementia. Furthermore, of the 111 patients afflicted with Alzheimer's disease, 69 received a diagnosis of Possible Alzheimer and 42 one of Probable Alzheimer.

Each patient was administered the complete set of physical, neuropsychological laboratory and clinical tests to allow the clinicians to make a diagnosis among the three categories considered and assign a therapy.

From the data set of each patient, a number of responses were singled out for the purpose of applying the classification algorithm. These are:

- 7 variables of the Mini Mental State Examination (MMSE),
- 7 variables from the Activities of Daily Living index (ADL),
- 9 variables from the Instrumental Activities of Daily Living index(IADL),
- the Hachinski Scale value,
- the Global Deterioration Scale value (GBS),
- 3 social variables (sex, age and level of schooling),
- 18 variable comorbility (IDS) tests, which are not administered to patients thought to have vascular dementia.

Thus 28 variables are available, for all groups of pathologies, while up to 46 variables had been collected for the Alzheimer patients.

Except in some preliminary analysis, this additional data for a group of patients was of little use, since the main results of this chapter indicate the need for appropriate variable reduction procedures, which are better handled in the data set containing 28 variables per patient.

It is to be noticed that in line with the exclusion criteria for vascular dementia diagnosis, patients with pronounced symptoms of Alzheimer's disease were not considered for vascular dementia, and vice versa. This aspect is, however, rather controversial, as recent research indicates that perhaps this exclusion principle should not be maintained.

The ability to recognize distinctive patterns of cognitive deficits, attributable to cerebrovascular disease or to a cognitive syndrome is very important not only because it provides information regarding the clinical impact of varied brain lesions, but also because it facilitates the determination of the dementia subtype (e.g. vascular dementia versus Alzheimer's disease) which can be helpful in patient management [12].

A recurrent problem in clinical diagnosis regards the rater reliability, or if comparisons are made, the interrater reliability [22]. This problem arises in decision processes, such as clinical diagnosis, in which human judgement must be exercised, which may make diagnoses not free of value judgements and therefore not replicable. To remedy this subjectivism, a decision tree approach can also be tried, as indicated in [21], where it resulted that the interrater consistency was much higher because of the formal procedures.

The formal algorithm proposed below solves these problems in part and interrater reliability is high. By formal means another problem can be solved, also, the consistency of the criteria used and interpreted.

The criteria applied may be contradictory. One such criterion has been indicated above, but there may be others and recent research results indicate that this is indeed so [11, 48].

Moreover, the replies may be coherent but they may contain noise. The satisfaction of many criteria is tantamount to the multiplication of a number of measurements subject to noise together, which can be taken as a percentage of the measurement. However, such multiplication measurements with noise lead to the addition of the noise percentages, which will reduce the precision these evaluations may have.

It will be seen that these problems are very important, but with the proper use of formal methods, such as the ones considered here, they can be reduced to manageable proportions, if not eliminated completely.

2.2 The Classification Algorithm

The algorithm indicated as **T.R.A.C.E.** (**T**otal **R**ecognition by **A**daptive **C**lassification **E**xperiments) is a statistical pattern recognition procedure which converges under quite general conditions with precise results if certain properties of the data set are fulfilled [38, 36], as presented in this book in [39] regarding a companion application of the algorithm.

However, in that application the data sets considered are pixel image maps, converted to frequency distributions, so that the mapping from the input space to the feature space is one to one. In this application, instead, the data sets consist of noisy measurements, permitting different patients with different diagnoses to have similar patterns. Also, in certain cases, the noise element may swamp the data set, so all patterns receive approximately the same values.

Thus instead of a syntactical description of the algorithm, which has been presented in [39], here we shall deal with the semantic aspects of the algorithm. This is concerned with the transformation of the data so as to retain its semantic meaning. Accordingly, the pattern vector (the vector of measured characteristics), is in homomorphic relation to the feature vector (the reduced or transformed vector of characteristics), to maintain a homomorphic relation with the classification classes and thus make recognition possible.

General Properties of the Algorithm

The problem to be solved is to assign the patients represented by a series of appropriate measurements, based on the classification criteria indicated above, which will be called an instance, to the appropriate class of pathology, when it is not known. This classification is based on a similar set of instances whose class membership is however known (the training set). Thus we wish to learn the classification rules to apply to these entities, purely on the basis of their attributes, by using those that are known and available.

Often, to verify the precision of the implementation, part of the training set is sacrificed to be used as a verification set, usually 10% of the total sample, by selecting the instances to be used in training and in the verification set with a random procedure and repeating this many times, usually 150.

Suppose that a training set is given with a correct classification (this will be generalized below). Then from the patterns of each class, a mean vector, called a barycenter, for each class can be formed. The distance (Euclidean or general) of each pattern from the mean vector of each class can be determined. Out of all the patterns which are closer to a barycenter of another class, choose the one that is furthest from the barycenter of its own class and select this as a new barycenter.

Thus for some class, there are now two barycenters and all the patterns of that class can be reassigned to one or other barycenter, depending on the distance of each pattern of the class from the barycenters of that class. The patterns in this class can then be repartitioned on the basis of their closeness to the two barycenters, so as to determine two new mean vectors for the two resulting subclasses of that class.

Each class has a barycenter vector, except for one class which now has two barycenters. Calculate all distances anew and use the same criterion to determine a new pattern to be used to form a new barycenter, and repeat the procedure again and again until all patterns result nearer to a barycenter of their own class than to one of another class. The algorithm is then said to have converged and this will happen every time if some very mild conditions on the data set are satisfied and if the algorithm is allowed to run until convergence [36].

If verification is being implemented, the assigned class is compared to the actual class of that instance and thus for every trial the proportion of correct classification can be determined.

The accuracy of the classification of each instance can now be checked by comparing the classification assigned by **T.R.A.C.E.** to the actual classification.

In classification, a new pattern of unknown class is given, its distance is determined from each of the available barycenter vectors and then it is assigned to the class of the barycenter which results closest to it.

It can be shown that, for a large enough training set, the error rate in the classification of the data set can be bounded by a constant if it is allowed to run until convergence. In practice, given a finite sample, the precision in classification will depend on its size and the actual representation used, since some representations may require larger training samples for a sufficiently precise classification [36].

In the presence of some meaningful relationships in the data set, which may be however swamped by noise, the precision may fall drastically, because of the large component of confusion attached to the meaningful underlying relationships for a given size of sample. Still the convergence of the classification to a sufficiently precise result can be proven. In this case, with such large interclass variances, due to noise, convergence may require extremely large training sets. Thus, in the presence of noise, suitable formal methods should be considered to isolate this noise component and use the meaningful part of the data to formulate the diagnosis [36].

Feature Selection

Data samples may contain many characteristics of the patient or more generally of the object to be classified. These characteristics may be subject to imprecise measurements and so be subject to noise.

Originally, questionnaires completed by the patient, his kin, and the clinician are available, since any psychophysical test can be cast in this form and from these a pattern vector for each instance is defined through a suitable procedure to transform the attributes recorded into a pattern vector. Thus:

Definition 1. *A pattern set is obtained from an attribute set by a transformation of its elements so as to render the structure of a vector space, with a similarity measure defined on it or a binary relation defined on its components [38, 36].*

The difference between the attribute space and the pattern space is that the latter satisfies the triangular inequality, while the former may not, so any two vectors in that space may be incommensurable [38].

The pattern space that emerges from this transformation may be too large, highly redundant and swamped by noise. Hence, it is considered worthwhile to carry out a further transformation to remove these inconveniences as much as possible.

It is to be noted that this is strictly speaking superfluous and is only valuable computationally, as defining a mapping from a pattern space to a feature space and then defining another mapping from a feature space to the outcome space, is formally equivalent to defining a direct mapping from the pattern space to the outcome space. Computationally, the results may be different as the ill-conditioning of matrices may be avoided, as well as redundancy and total random error due to error in the variables, both of which may be substantially reduced.

Definition 2. *A feature vector is obtained from a pattern vector by a (non)linear transformation, which is applied to all entities of the data set. When the transformation is just a selection of certain elements of the pattern set, it is called feature selection, otherwise it is known as feature extraction [36].*

There are many feature selection and extraction procedures [13, 50], but only three were considered in this chapter:

- feature selection procedures:
  - Stepwise discriminant analysis [44].
    Given a training set, the relevant pattern elements are chosen sequentially by determining the pattern elements either in forward or backward selection, using one of a number of statistical discrimination criteria. Thus a reduced pattern vector set is defined composed of just the elements that have been selected.
  - Classification and Regression Trees (CART) [31]. In most general terms, logical trees are built from the data to determine a set of if-then logical (split) conditions so as to build a logical tree with branching conditions at each node depending on the element that is split and a depth which is determined by the user depending on the desired precision.
    When this method yields good results, it has a number of advantages over other techniques, which are:
    - simplicity: the interpretation of the results are straight forward, consisting in determining the value of the elements to be considered and establishing to which branch each element must be assigned. Reading the tree by descending a branch path leads to the determination of the intervals which must include the values of the pattern elements for that diagnosis to result.
    - tree methods are non parametric and nonlinear: as it is evident from the above description. The diagnosis does not depend on any implicit assumption between the predictor variables and the dependent variables and continuous discrete or categorical variable can be considered just as easily.
- feature extraction procedures:

– Principal Component Analysis [13, 24].
It consists of a very well known technique, which acts on the variance and covariance matrix of the data set and effects a diagonalisation determining its eigenvalues and its eigenvectors. A certain number of eigenvalues and their corresponding eigenvectors are retained. Through these eigenvectors the data matrix is rebuilt, but this will have orthogonal columns and various other important properties.
The significance of the transformation is that most of the information content of the data set is retained, while all partial duplication of the information spread in various rows and columns is eliminated, as well as eventual residual noise. The amount of the latter discarded depends on the values and the number of eigenvalues discarded.
However, the actual pattern elements considered in the transformed data set are usually a complex linear combination of the original pattern elements and this is considered a complicating factor which should be avoided if possible.

All feature selection and extraction procedures are attempts to reduce duplication and noise in the pattern vectors and if there are non systematic errors in the pattern elements to remove both as much as possible, so that there will be as little accumulation as possible in the classification process.

Imprecise Classification

In many classification experiments, classes assigned to the instances may be imprecise due to human error or to accumulated noise. In the diagnosing of dementia, this may be particularly true, since the diagnostic criteria are so numerous and should be considered interpretations rather than precise measurements of characteristics.

If the training set is not too small, the training sample should be adequate and the classification should be precise, with errors made in the verification sample purely due to sampling variation.

If the verification set is composed of 10% of the training set, objects will appear in the verification set about 10% of the number of replications. If 150 trials are made, first by selecting randomly 10% of the training set and then using the rest as a training set, every object will appear on average 15 times in the verification set.

It is expected that the object will have been misclassified very few times and per contra, if it results misclassified with respect to its assigned classification, say 2/3 of the times, then it would appear to be a misclassified object which requires that the class to which it was assigned be altered to the class it has been most frequently assigned in these trials.

Once this has been performed on all apparently misclassified patterns, so as to correct their class designation, the verification algorithm is applied anew on the corrected data set and the results recorded. In this way the imprecise classification can be checked and corrected, see [6] for details. The whole process is known as classification with an imprecise teacher.

Finite Sample Approximation

A further problem in verification may occur because of the data coherence problem, often due to the small training sample size available. Consider a small training set which may contain only one object which forms a distinct barycenter, quite unlike the other objects with the same classification. If this item falls in the training sample, then it will be set aside to form a distinct barycenter set, which will be composed of this instance only as a singleton element. Nothing more should happen and, of course, because of the singleton sample the barycenter vector, understood as the mean vector of a set of like objects, will be not a good estimate of the population values.

On the other hand, if this object, which would constitute a singleton barycenter set in training, falls in the verification set, it will be assigned to the wrong class because it will find no opportune barycenter, since in the training set it would have figured as a singleton and thus it will result nearer to a barycenter of another class, for otherwise in training it would not have formed a new barycenter. Thus because of reduced size of the sample, classification errors will be made, which would not be made if the training set was larger.

It is often worthwhile to sample the training set with a stratified sample instead of a simple random sample. To do this it is advantageous to sample those objects that appear in very small subgroups less than proportionally. Thus a very small probability of being chosen in verification is given to those objects that in classifying the complete training set turn out to form a singleton barycenter and are the only member of that subgroup. Such a probability should be so small, to all intents and purposes, as to exclude that pattern from being chosen in verification and a slightly larger probability is chosen if a doubleton barycenter set is formed, so that on the average one object of the couple may be chosen but not both. In the same way, stratified sampling probabilities should be chosen for three, four and five-element barycenter sets, so that on the average not more than half of the constituent elements are likely to be chosen.

In practice, instead of defining these probabilities, a biased verification sample is formed by excluding those objects which are assigned to singleton barycenters from selection in the verification set and restricting to not more than half those objects which are assigned to these subgroups with low consistency [25]. These small subsamples are termed labeled.

When enough confidence has been obtained on the diagnostic experiment, through the use of verification and replication, to perform the actual classification all the instances are used in the training set and as new objects arise they are classified by assigning the class membership which belongs to that barycenter which is nearest to it. Periodically, perhaps, when the actual class of these objects has been ascertained the extended training set can be retrained and verification can be performed to check the precision.

3 Results

Although the clinical factors studied in the diagnosis of dementia are very numerous, each may be related to the pathology in a tenuous way and its measurement may be highly noisy, so that if many factors are used to categorize the sample, the result may be extremely varied. Hence noise reduction procedures of various types should be applied by identifying missing factors and proxy variables. This yields some relevant results regarding possible categories of Alzheimer's disease.

The aim of this section is to carry out this analysis and then determine clinical factors, which have a low noise to effect ratio and are sufficient to account for all of the diagnostic results.

An initial analysis was performed with the entire data set and all three classes. Without data reduction techniques, the results obtained are statistically non significant, at a confidence level of 5% [26], while as less and less principal components were used, the precision of the diagnosis becomes significant. This may be confirmed by using data reduction techniques, such as stepwise discriminant analysis to explore the pattern space and indicate the 3, 5, 7, ..., best variables in terms of the variance explained increment, as shown in Table 2 [44].

In column 2 of this table, the explained variance by that set of components is given, while in column 3, the average precision ratio for a verification set drawn from the training set by considering a 10% non-repetitive random sample replicated 150 times is presented for each set of principal components on applying the algorithm **T.R.A.C.E.**. In the next column, similar results for the verification set are formed from a stratified sample of the training set, as described above. In the last column, the percentage of instances which were labeled and so subject to less than proportional sampling are indicated.

Although statistically significant, in most cases the results given are not satisfactory. There are ambiguities in the data set, partly due to the relatively small size of the sample compared to the number of features considered for each pattern. This leads to an excessive number of labeled patterns.

Close analysis of the individual confusion matrices per replication shows that the Alzheimer categories tend to be confused with the vascular data, so it is suggested to try to perform the classification in two phases.

The reasons to separate the data set arise from the results of the application of the algorithm, but they are also confirmed by the considerations indicated in Section 2.1 and confirmed by recent research results.

In the first phase, the data is classified with respect to the Alzheimer syndrome as compared to the vascular one. Secondly all the patients with

Table 2. Classification results for T.R.A.C.E.: precision in the verification samples for various sampling procedures, all Alzheimer patients, 3 classes, 28 variables per instance

	Explained Variance	Random sample	Stratified Sample	Labels (%)
3 principal components	0.5256	0.4945	0.8726	75
5 principal components	0.6229	0.4650	0.6857	72
7 principal components	0.7060	0.4823	0.6319	61
9 principal components	0.7643	0.4776	0.7121	64
12 principal components	0.8414	0.4473	0.6181	61

Alzheimer syndrome are classified in two Alzheimer categories. The rationale for this subdivision lies in the following consideration. To distinguish probable from possible Alzheimer we require a given subset of the attributes, while to distinguish between Alzheimer and vascular afflicted patients another set is required. The inclusion of both sets of variables in the attributes to be applied leads to the incorporation of too much noise into the data set. This is why in Table 2, the random sample precision is so low and the accuracy falls in the stratified sample with an increase in the number of features considered for each pattern.

Thus consider the sample set of 111 patients afflicted with possible Alzheimer's disease (69 patients) or with probable Alzheimer's disease (42 patients). In Table 3, the data for 28 attributes, is used for each instance and was first subjected to a principal component analysis and a certain number of principal components were considered. The results for such a small sample can be considered very good.

Table 3. Classification results for T.R.A.C.E.: precision in the verification samples for various sampling procedures, Alzheimer patients, 2 classes, 28 variables per instance

	Explained Variance	Random sample	Stratified Sample	Labels (%)
3 principal components	0.4433	0.6213	0.8667	71
5 principal components	0.5591	0.5798	0.7700	60
7 principal components	0.6547	0.5633	0.6870	61
8 principal components	0.6925	0.5617	0.7532	64
12 principal components	0.8156	0.5468	0.8001	64

Other classification techniques were used for comparison, but these fare significantly less well, which indicates that the training to complete precision has an important effect. The analysis above can be repeated with two other feature extraction techniques: stepwise discrimination methods [44] and the CART method [31], which were applied to attribute selection and used in various routines, namely: Linear Discriminant Analysis (LDA) and Classification and Regression Trees (CART). The results are indicated in Table 4.

Table 4. Classification results for T.R.A.C.E.: precision in the verification samples for various selection techniques and sampling procedures, Alzheimer patients, two classes, 28 variables per instance

	LDF	CART	Trace random	Trace Stratified	Labels (%)
6 stepwise discrimant	0.6858	0.6211	0.6255	0.7635	71
4 CART variables	0.6667	0.5730	0.5904	0.7535	60

With Linear Discriminant Analysis [44] the formation of groups based just on their labels yields a set of barycenters which are comparable to those obtained in **T.R.A.C.E.**. With the former method, training yields an imprecise classification of objects, while with the latter there is complete precision, since this is a termination requirement in the latter but not in the former. If Linear Discriminant Analysis provides equivalent precision results, this means that the algorithm **T.R.A.C.E.** just partitions noise. This was in fact so [26].

The stepwise discriminant variable selection technique and the CART technique have a lower precision than the principal component variable selection technique. However, this procedure does have the advantage that the attributes selected have direct medical relevance, while with the principal component procedure all the measurements in the battery of tests are required to extract the desired features.

In the stepwise discriminant analysis, the number of variables selected were six which included:

- The value on the Hachinski scale, which evaluates the ischemal capacity loss.
- The MMSE_RM value, which indicates the capacity of the patient to register concepts in memory.
- The age of the patient.
- The IADL_SP value on a 2 point nominal scale, which indicates the ability of the patient to move outside of his home.
- The MMSE_R value on a 3 point ordered scale, which indicates the capacity of the patient to remember the names of objects listed in the MMSE_RM measurement phase.
- The IADL_C value on a 2 point nominal scale, which indicates the ability of the patient to order his home.

On the other hand, the variables chosen in the CART technique were:

- Age of the patient.
- The value on the Hachinski scale.

- The MMSE_TC value: A four digit variable, with one decimal digit that indicates the global cognitive value of the patient, corrected for age and education.
- The MMSE_RM values.

The results obtained seem to indicate that in determining patients afflicted with possible or probable Alzheimer the distinction is not completely unambiguous. Perhaps correction for an imprecise teacher should be used. Certainly the original 46 attributes measured contain a lot of duplication, so that nothing is lost if only a subset of 28 variables are chosen and these in their turn can be reduced to 4-5 variables, without significant losses in precision. As for the previous case, the best policy is to use first 3 principal components obtained from the full 28 variable data set.

The classification of Alzheimer's disease and vascular dementia patients is also important and it is composed of 111 patients suffering from Alzheimer and 45 patients suffering from vascular dementia. The results of this classification are given in Table 5.

Table 5. Classification results for T.R.A.C.E.: precision in the verification samples for various sampling procedures, Alzheimer and vascular dementia patients, 2 classes, 28 variables per instance

	Explained Variance	Random sample	Stratified Sample	Labels (%)
3 principal components	0.5256	0.6997	0.8502	45
5 principal components	0.6229	0.7075	0.8252	50
7 principal components	0.7060	0.7594	0.8235	34
8 principal components	0.7643	0.7598	0.8638	37
12 principal components	0.8414	0.7488	0.8401	38

Even for such a small sample, the classification of Alzheimer patients and vascular dementia patients based on the common set of neuropsychological tests can be effected with a high accuracy.

It is also important to study the selection of an actual subset of variables. Thus in Table 6 the results are given for various selections of variables based on the stepwise discriminant variable selection technique and the CART technique. Good results are obtained for all considered pattern recognition techniques and for the T.R.A.C.E. algorithm in both versions. Again, the tendency of the stepwise discriminant analysis is to incorporate many variables, while for the CART selection only 3 variables are required, which gives highly accurate values in this case.

From Table 6, there results that the given stepwise selection of variables does not require the finer analysis of the **T.R.A.C.E.** algorithm. Once selected these attributes are good classifiers of the type of pathology based on

Table 6. Classification results for T.R.A.C.E.: precision in the verification samples for various selection techniques and sampling procedures, Alzheimer patients, two classes, 28 variables per instance

	LDF	CART	Trace random	Trace Stratified	Labels (%)
16 stepwise discriminant	0.9066	0.9057	0.8046	0.8363	38
19 stepwise discriminant	0.9021	0.9057	0.7858	0.8314	38
3 CART variables	0.7898	0.9057	0.8522	0.9383	32

the determination of the means of the classes and the classification of instances on the basis of the least distance criterion. On the other hand, if the variables are selected with the CART variable selection method, then only three variables are required, namely:

- value on the Hachinski scale,
- MMSE_O value on an interval scale (0-10), which indicates the orientation capacity of the patient,
- IADL_SP value on a 2 point nominal scale, which indicates the ability of the patient to move outside of his home.

Thus there are two different methods of data selection to diagnose dementia pathologies. One method is to use the 28 variables measured by the MMSE, ADL, IADL the Hachinski scale value, the Global Determination scale (GDS) value and three social variables (age, sex and education) to extract the three principal components and apply **T.R.A.C.E.** in the classification mode to obtain the diagnoses. The other method is to use just one of the feature selection techniques which indicate the measures in their natural units, so just a few items need to be measured from the battery of neuropsychological tests available. This latter feature selection procedure, although it may not be as accurate in all cases does provide attributes which are medically meaningful and directly measurable.

The classifications presented in Tables 5 and 6 may be imprecise, so the correction technique was implemented with respect to the data results of the classification of Alzheimer's disease and vascular dementia for the three principal components results and for the data set regarding probable and possible Alzheimer's disease for the three CART variable selection results.

In the first case, one patient classed as Alzheimer's disease resulted imprecisely classified and should be classed as vascular dementia according to this algorithm, while two patients declared afflicted with vascular dementia were according to the correction procedure of **T.R.A.C.E.** to be classed in Alzheimer's disease. In the second case, three patients diagnosed with possible Alzheimer's disease were on the basis of the evidence to be classified as probable Alzheimer's disease patients and none vice versa. On this basis the resulting classification results for just the implementations of a stratified sample designs are indicated in Table 7.

Table 7. Classification results for T.R.A.C.E.: precision in the verification samples for various selection techniques and stratified sampling procedures, after correction procedure (two classes, 28 variables per instance).

	Trace Stratified
3 CART variables for classification of subtypes of Alzheimer's disease	0.9636
3 principal components for classification of the two type of dementia	0.9355

It is important now to dwell on the implications of these results.

4 Discussion

The aim of this section is to examine a number of the results obtained for their implication in diagnostic procedures, current clinical practice and therapeutic practice.

The analysis conducted shows that:

- By whatever method the diagnoses of the patients in this data set have been reached clinically, the diagnostic process is a highly consistent process, for otherwise such accurate classification results by a formal algorithm such as **T.R.A.C.E.** could not have resulted. Ambiguity and spurious assignments would lead to lower precision. In fact, the high precision reached with this algorithm implies that there has been an elevated coherence in the class assignment of the instances.
- There is, thus, a formal classification procedure to effect these diagnoses with high accuracy, which on the basis of the results shown increases the precision of the diagnosis for all types of dementia by about 10%, yielding very accurate results of diagnostic precision with a mean of 95%.
- Many attributes can be measured to diagnose a patient, but
  - Three specific attributes are necessary and sufficient for the diagnosis of Alzheimer's disease and vascular dementia,
  - Four attributes are necessary and sufficient to diagnose possible from probable Alzheimer's disease pathologies.

An important point concerns how this diagnostic method might fare among different clinicians and in different clinics. Since the neuropsychiatric tests used as the basic evaluation have been coded [2], it is unlikely that they will differ from clinic to clinic, in their application so they will tend to quite stable. If the values expressed by different clinics and experts are approximately invariant with regard to a patient, then these barycenters or criteria must be applicable from clinic to clinic.

Instead, the clinics and the schools that they represent may give different importance to different aspects of the scales, so that this would reflect differing degrees of precision with respect to the diagnoses made with the **T.R.A.C.E.**

algorithm from site to site and across clinicians. Thus, the comparative application of this algorithm across sites may be very useful in the analysis of comparative diagnoses.

Eventual inconsistencies can be examined and the eventual incorrect diagnoses corrected, then **T.R.A.C.E.** can be retrained. It is easy to show that the procedure will be consistent with the new data. Thus, an invariant test applicable to different clinics, schools and countries can be used to harmonize diagnoses, should there be existing differences. This could provide a form of a 'golden rule diagnostic' method, in which diagnostic experiments can be conducted with the **T.R.A.C.E.** algorithm, to determine if and how these change the classification results.

The principles given in the literature for the diagnoses of dementia [2], do not envisage this type of normalization, although a form of classification of pathological types is frequently given. Thus, certain representative values for the tests according to pathological types are given, and it is suggested to assign the patient to the profile that appears to be closest [41].

For instance, an often quoted criterion is to be guided by the values of the Hachinski scale of a patient. This is an important measure as it is present in both classifications of the data set by the algorithm **T.R.A.C.E.**.

It is not possible, however, to specify a procedure on the basis of the Hachinski scale measure that allows a preliminary diagnosis to be formulated, which is then refined or changed on the sequential evaluation of a certain number of other measures to yield an accurate diagnosis. In fact, all the three or four relevant measures plus the reference measures (the barycenters) must intervene simultaneously to formulate a diagnosis. This criterion is necessary and sufficient. Efficient diagnoses by clinicians also require the measurement of these aspects in a battery of neuropsychological tests and then the application of a numerical algorithm to determine the most appropriate class.

In Table 8, the measures for a number of patients diagnosed for the Alzheimer's disease or the vascular dementia syndrome are given. These results indicate that any value of the Hachinski scale may be associated with either syndrome [41].

Table 8. Examples of patients' patterns and their diagnosis

hachinski	MMSE_O	IADL_SP	Diagnosis
6.000000	8.000000	1.000000	AD
6.000000	4.000000	0.000000	VD
7.000000	5.000000	0.000000	AD
4.000000	5.000000	0.000000	VD
1.000000	0.000000	0.000000	VD
1.000000	2.000000	0.000000	AD

Moreover, it can be verified from the dataset that exceptions can be found even if the criterion based on the Hachinski scale is extended. Thus it is not true that the diagnosis is correct for those values of the Hachinski scale if the IADL_SP is 0.0 (see the second and fifth item in the table) or that this criterion becomes relevant for the case in which the Hachinski scale is as given, the IADL_SP is 0.0 and the MMSE_O value is above 4.0 (This will discriminate some items in the table, but not consistently).

Thus, all three values are required and the measures must be made from the relative norms according to the procedure indicated. If this is done, then the statistical criteria obtained from the extraction of three principal components are applied the precision is almost the same regardless of the medical criteria.

As it has been indicated [28] in order to predict the clinical course of this disease, it is highly relevant to identify diagnostic criteria. Thus, the increased accuracy of diagnosis by this method should be applied to determine the course of the disease and its treatment.

With the **T.R.A.C.E.** algorithm and this data, about 30 barycenters for the classification between Alzheimer's disease and vascular dementia are determined and about 55 barycenters for the classification of possible and probable Alzheimer's disease are formed [26].

Clinical differences may exist that justify the distinction of Alzheimer's disease into separate subtypes. The recognition of possible subgroups for Alzheimer's disease and vascular dementia syndrome may be important in predicting variable clinical courses associated with differences in therapeutic response [49]. The formation of these subgroups may be linked with cognitive decline in Alzheimer's disease, which is discussed frequently without formulating clear conclusions [47, 40, 19]. The poor results obtained may be due to the attempt to determine specific causation factors, rather than associative factors of general structure, which can be achieved with an algorithm like **T.R.A.C.E.**, which avoids eventual subjective inter-comparison of diagnostic factors.

Consider the classification process. These barycenters are average barycenters obtained from the different groups of patients, which are assigned to a barycenter so that the attribute vector will result closer to a barycenter of its own class than to one of the other classes. Thus each barycenter defines a homogeneous group of patients afflicted by the same pathology.

Each barycenter defines a subtype of the patients with that syndrome and these subclasses can be used to define a finer diagnostic distinction for each type of pathology. Thus the periodic redetermination of the basic pathology through the relevant clinical tests will indicate how the patients move through the groups, or whatever relative movement is manifested.

For instance, although no statistical tests have yet been performed on this data, it has been noticed that within this classification, the groups tend to be homogeneous with age. Thus patients with diagnosed vascular dementia tend to cluster with age and exhibit similar values for some neuropsychiatric tests.

If this can be verified by stringent statistical tests, it would mean that the pathology is such that there occur a similar pathology, related with age, for all the patients as the pathology progresses, which is testified by this movement through the groups. This fact could be used to monitor the progress of the pathology and the effect of particular medication might have on the progression. In short, an accurate comparative diagnostic instrument could be devised to evaluate progress in treating the pathology, which is unavailable at this moment.

Thus by periodic repetition of the battery of neuropsychological tests, the relative movements of each patient can be determined. By differentiating the therapy, it might be found that the relative movements differ, suggesting further experimental designs and tests.

It is of interest to notice that of the six patients which were reclassified, the one that had been assigned to the class Alzheimer's disease, but was reclassified by **T.R.A.C.E.** to the class vascular dementia has recently suffered an ictus, which confirms the machine diagnosis. The other five patients, which the clinician recognized as involved in complex and difficult diagnoses: to confirm or alter their pathology, their destiny must be awaited [35].

5 Conclusions

The suggested methodology seems to be important and applicable in performing a fine sensitivity and specificity analysis of patients in diagnosis and in treatment. The results indicate that the current description of the disease is not quite complete and that some partially duplicate aspects confound.

The clinical analysis, both for diagnosis and treatment, indicates more detailed characterizations are needed and new synergy effects should be studied. It also shows that since all measures may carry systematic indications of the pathology and a noise element, care must be taken regarding increasing the number of attributes considered, which may just have the result of swamping the valuable information and prevent accurate diagnoses.

In this chapter the problem of diagnosing dementia has been examined. It has been found that the process is indistinguishable from good medical diagnostic methods, if the latter are conceived as potentially complex measurement and inference systems.

The advantage of formal machine diagnostic systems is that there will be no subjectivity in operation as it may happen with different experts, while it will provide the same diagnoses from the same data. Thus the principle of coherent replicability, the very essence of scientific method is guaranteed.

Through such a system a "golden rule" for diagnoses with a very high accuracy may be derived. Experts will be able to suggest improvements in the "golden rule" and modify it immediately for the whole community. If this leads to an improvement the modification will be kept, otherwise it is easy to return to the previous methods.

Finer diagnostic distinctions can be experimentally defined, which may better predict the actual state of a patient. Hence, through a careful study of how these patients progress through the pathological states, valuable insights can be obtained on the clinical course of this disease and its treatment.

References

1. L. Amaducci, M. Falcini, and A. Lippi. Descriptive epidemology and risk factors for Alzheimer's disease. *Acta Neurologica Scandinavica (Supplementum)*, 139:21, 1992.

2. American Psychiatric Association. *Diagnostic and Statistical Manual of Mental Disorders*. 4th edition, American Psychiatric Association, Washington, DC, 1994.

3. A. C. Atkinson and A. N. Donev. *Optimum Experimental Design*. Clarendon Press, Oxford, 1992.

4. L. Berg, D. W. McKeel Jr, J. P. Miller, M. Storandt, E. H. Rubin, J. C. Morris, J. Baty, M. Coats, J. Norton, A. M. Goate, J. L. Price, M. Gearing, S. S. Mirra, and A. M. Saunders. Clinicopathologic studies in cognitively healthy aging and Alzheimer's disease: relation of histologic markers to dementia severity, age, sex, and apolipoprotein e genotype. *Archives of Neurology*, 55:326–335, 1998.

5. L. M. Bierer, P. R. Holf, and D. P. Purohit. Neocortical neurofibrillary tangles correlate with dementia severity in Alzheimer's disease. *Archives of Neurology*, 52:81–88, 1995.

6. G. Bonifazi, P. Massacci, L. Nieddu, and G. Patrizi. The classification of industrial sand-ores by image recognition methods. In *Proceedings of 13th International Conference on Pattern Recognition Systems, vol.4: Parallel and Connectionist Systems*, pages 174–179, Los Alamitos, CA, 1996. IEEE Computer Society Press.

7. J. V. Bowler, D. G. Munoz, H. Merksey, and V. Hachinski. Fallacies in the pathological confirmation of the diagnosis of Alzheimer's disease. *Journal of Neurology, Neurosurgery and Psychiatry*, 64:18–24, 1998.

8. H. Braak and E. Braak. Diagnostic criteria for neuropathologic assessment of Alzheimer's disease. *Neurobiology of Aging*, 18:85–88, 1997.

9. W. G. Cochran and G. M. Cox. *Experimental Designs*. 2nd edition, Wiley, New York, 1957.

10. J. L. Cummings. Current perspectives in Alzheimer's disease. *Neurology*, 51:1–8, 1998.

11. D. W. Desmond. Vascular dementia. *Clinical Neuroscience Research*, 3:437–448, 2004.

12. D. W. Desmond, T. Erkinjuntti, M. Sano, J. L. Cummings, J. V. Bowler, F. Pasquier, J. T. Moroney, S. H. Ferris, Y. Stern, P. S. Sachdev, and V. C. Hachinski. The cognitive syndrome of vascular dementia: implications for clinical trials. *Alzheimer Disease and Associated Disorders*, 13(Suppl. 3):s21 – s29, 1999.

13. R. O. Duda and P. E. Hart. *Pattern Recognition and Scene Analysis*. Wiley, New York, 1973.

14. O. Firschlein and M. Fischler. Automatic subclass determination for pattern recognition applications. *IEEE Transactions on Electronic Computers*, 12:137–141, 1963.

15. G. B. Frisoni. *Diagnosi e Terapia della Malattia di Alzheimer: Un Percorso per le Unità di Valutazione Alzheimer*. IRCCS San Giovanni di Dio - Fatebenefratelli, Brescia, Italy, 2001.

16. M. Gearing, S. S. Mirra, J. C. Hedreen, S. M. Sumi, L. A. Hansen, and A. Heyman. The consortium to establish a registry for Alzheimer's disease (CERAD).

Part X: Neuropathology confirmation of the clinical diagnosis of Alzheimer's disease. *Neurology*, 45:461–466, 1995.

17. T. Gomez-Lima, R. Hollister, H. West, S. Mui, and J.H. Growdon. Neuronal loss correlates with but exceeds neurofibrillary tangles in Alzheimer's disease. *Annals of Neurology*, 1:17–24, 1997.

18. R. C. Hamdy. Featured cme topic: Dementia, fact sheet. *Sothern Medical Journal*, 94:673–677, 2001.

19. E. Helmes, J. V. Bowler, H. Merskey, D. D. Munoz, and V. C. Hachinski. Rates of cognitive decline in Alzheimer's disease and dementia with lewy bodies. *Dementia and Geriatric Cognitive Disorders*, 15:67–71, 2003.

20. A. Heyman, G. G. Fillenbaum, K. A. Welsh-Bohmer, M. Gearing, S. S. Mirra, R. C. Mohs, B. L. Peterson, and C. F. Pieper. Cerebral infarcts in patients with autopsy-proven Alzheimer's disease, CERAD Part XVIII. *Neurology*, 51:159–162, 1998.

21. E. Hogervorst, S. Bandelow, M. Combrinck, S. Irani, and A. D. Smith. The validity and reliability of 6 sets of clinical criteria to classify Alzheimer's disease and vascular dementia in cases confirmed post-mortem: added value of a decision tree approach. *Dementia and Geriatric Cognitive Disorders*, 16:170–180, 2003.

22. E. Hogervorst, L. Barnetson, K. A. Jobst, Zs. Nagy, and M. Combrinck an A. D. Smith. Diagnosing dementia: interrater reliability assessment and accuracy of the nincds/adrda criteria versus CERAD histopathological criteria for Alzheimer's disease. *Dementia and Geriatric Cognitive Disorders*, 11:107–113, 2000.

23. S. Holroyd and M. L. Shepherd. Alzheimers's disease: A review for the ophthalmologist. *Survey of Ophthalmology*, 45(6):516–524, 2001.

24. M. Kendall, A. Stuart, and J. K. Ord. *The Advanced Theory of Statistics*. Griffin, London, 1979.

25. H. S. Konjin. *Statistical Theory of Sample Design and Analysis*. North Holland, Amsterdam, 1973.

26. A. Lisi. Metodi della ricerca operativa applicati al problema della classificazione in campo medico: La malattia di Alzheimer. Technical report, Tesi di Laurea, Facoltà di Scienze Statistiche, Università degli Studi di Roma, La Sapienza, a.a. 2000-2001, 2001.

27. A. Lobo, L. J. Launer, L. Fratiglioni, K. Andersen, A. Di Carlo, M. M. Breteler, J. R. Copeland JR, J. F. Dartigues, C. Jagger, J. Martinez-Lage, H. Soininen, and A. Hofman. Prevalence of dementia and major subtypes in europe: A collaborative study of population-based cohorts. neurologic diseases in the elderly research group. *Neurology*, 54(11 Suppl 5):S4–9, 2000.

28. C. Marra, M. C. Silveri, and G. Gainotti. Predictors of cognitive decline in the early stage of probable Alzheimer disease. *Dementia and Geriatric Cognitive Disorders*, 11:212–218, 2000.

29. G. McKhann, D. Drachman, M. Folstein, R. Katzmann, D. Price, and E. M. Stadlam. Clinical diagnosis of Alzheimer's disease: report of the nincds-adrda work group under the auspices of the department of health and human services task force on Alzheimer's disease. *Neurology*, 34:939–944, 1984.

30. S. S. Mirra, M. Gearing, and F. Nash. Neuropathological assessment of Alzheimer's disease. *Neurology*, 49 (Suppl.3):S14–S16, 1997.

31. D. Mitchie, D.J. Spiegelhalter, and C.C. Taylor. *Machine Learning, Neural and Statistical Classification*. Ellis Horwood, New York, 1994.

32. J. T. Moroney, E. Bagiella V. C. Hachinski, P. K. Molsa, L. Gustafson, A. Brun, P. Fischer, T. Erkinjuntti, W. Rosen, M. C. Paik, T. K. Tatemichi, and D. W. Desmond. Misclassification of dementia subtype using the Hachinski Ischemic Score: results of a meta-analysis of patients with pathologically verified dementias. *Annals of the New York Academy of Science*, 826:490–492, 1997.

33. J. C. Morris, A. Heyman, R. C. Mohs, J. P. Hughes, G. van Belle, G. Fillenbaum, E. D. Mellits, and C Clark. The consortium to establish a registry for Alzheimer's disease (CERAD). Part I: Clinical and neuropsychological assessment of Alzheimer's disease. *Neurology*, 39:1159–1165, 1989.

34. J. C. Morris, M. Storandt, and D.W. McKeel. Cerebral amyloid deposition and diffuse plaques in normal aging: evidence for presymptomatic and very mild Alzheimer's disease. *Neurology*, 46:707–719, 1996.

35. L. Nieddu, A. Di Cioccio, A. Lisi, V. Pescosolido, and G. Patrizi. A study on the accuracy of diagnostic dementia levels by an automatic reclassification algorithm. In *La Gerariatria per la Longevita, IV Convegno Geriatrico Dottor Angelico, Aquino (Frosinone), Italy*, pages 26–27, 2002.

36. L. Nieddu and G. Patrizi. Formal properties of pattern recognition algorithms: A review. *European Journal of Operational Research*, 120:459–495, 2000.

37. The Ronald & Nancy Reagan Research Institute of the Alzheimer's Association and the National Institute on Aging Working Group. Consensus report of the working group on: Molecular and biochemical markers of Alzheimer's disease. *Neurobiology of Aging*, 19:109–116, 1998.

38. G. Patrizi. Optimal clustering properties. *Ricerca Operativa*, 10:41–64, 1979.

39. G. Patrizi, G. Addonisio, C. Giannakakis, A.Onetti Muda, Gr. Patrizi, and T. Faraggiana. Diagnosis of Alport syndrome by pattern recognition techniques. In P. M. Pardalos, V. Boginski, and A. Vazacopoulos, editors, *Data Mining in Biomedicine*. Springer, New York, 2006.

40. D. X. Rasmusson, K. A. Carson, R. Brookmeyer, C. Kawas, and J. Brandt. Predicting rate of cognitive decline in probable Alzheimer's disease. *Brain and Cognition*, 31:133–147, 1996.

41. L. Ravizza. Demenze. In G. B. Cassano P. Pancheri, editor, *Trattato Italiano di Psichiatria*, pages 1035–1170. 2nd edition, Masson, Milan, 1999.

42. E. M. Reiman and R. J. Caselli. Alzheimer's disease. *Maturitas*, 31:185–200, 1999.

43. G. C. Roman, T. K. Tatemichi, T. Erkinjuntti, J. L. Cummings, J. C. Masdeu, J. H. Garcia, L. Amaducci, J. M. Orgogozo, A. Brun, and A. Hofman. Vascular dementia: Diagnostic criteria for research studies. report of the NINDS-AIREN workshop. *Neurology*, 43:250–260, 1993.

44. G. A. F. Seber. *Multivariate Observations*. Wiley, New York, 1984.

45. D. A. Snowdon, L.H. Greiner, J.A. Mortimer, K.P.Riley, P.A. Greiner, and W.R.Markesbery. Brain infraction and the clinical expression of Alzheimer's disease. *Journal of the American Medical Association*, 277:813–817, 1997.

46. M. E. Strauss, M.M. Lee, and J.M. Di Filippo. Premorbid personality and behavior symptoms in Alzheimer's disease. *Archives of Neurology*, 54:257–259, 1997.

47. P. Tariska and K. Urbanics. Clinical subtypes of Alzheimer's disease. *Archives of Gerontology and Geriatrics*, 21:13–20, 1995.

48. D. T. Villareal, E. Grant, J. P. Miller, M. Storandt, D. W. McKeel, and J. C. Morris. Clinical outcomes of possible versus probable Alzheimer's disease. *Neurology*, 61:661–667, 2003.

49. D. T. Villareal and J. C. Morris. The diagnosis of Alzheimer's disease. *Alzheimer's Disease Review*, 3:142–152, 1998.

50. S. Watanabe. *Pattern Recognition: Human and Mechanical*. Wiley, New York, 1985.

## Part III

Data Mining Studies in Genomics and Proteomics

A Hybrid Knowledge Based-Clustering Multi-Class SVM Approach for Genes Expression Analysis

Budi Santosa[1], Tyrrell Conway[2], and Theodore Trafalis[1]

[1] School of Industrial Engineering, University of Oklahoma
`bsantosa@gmail.com, ttrafalis@ou.edu`
[2] Department of Botany and Microbiology, University of Oklahoma
`tconway@ou.edu`

Summary. This study utilizes Support Vector Machines (SVM) for multi-class classification of a real data set with more than two classes. The data is a set of E. coli whole-genome gene expression profiles. The problem is how to classify these genes based on their behavior in response to changing pH of the growth medium and mutation of the acid tolerance response gene regulator GadX. In order to apply these techniques, first we have to label the genes. The labels indicate the response of genes to the experimental variables: 1-unchanged, 2-decreased expression level and 3-increased expression level. To label the genes, an unsupervised K-Means clustering technique is applied in a multi-level scheme. Multi-level K-Means clustering is itself an improvement over standard K-Means applications. SVM is used here in two ways. First, labels resulting from multi-level K-Means clustering are confirmed by SVM. To judge the performance of SVM, two other methods, K-nearest neighbor (KNN) and Linear Discriminant Analysis (LDA) are implemented. The Implementation of Multi-class SVM used one-against-one method and one-against-all method. The results show that SVM outperforms KNN and LDA. The advantage of SVM includes the generalization error and the computing time. Second, different from the first application, SVM is used to label the genes after it is trained by a set of training data obtained from K-Means clustering. This alternative SVM strategy offers an improvement over standard SVM applications.

Key words: Distance Measures, Euclidean Distance, Generalization Error, K-Means Algorithm, Kernel Function, KNN, Minimum Distance, Neural Networks, Optimization, RBF, Statistics, Supervised Learning, Support Vector Machine, Unsupervised Learning

1 Introduction

Recently, a number of new microarray and macroarray technologies have been developed for analyzing biological processes such as gene function, cancer, and design of new pharmaceutical. These technologies include DNA gene expression macroarrays that allow biologists to study patterns of gene expression for any cell at any time under a specific set of conditions. These arrays produce large amounts of data that can provide valuable insight for gene function. For example, co-expression of novel genes may provide insight to the functions of several genes for which limited information is available. Gene expression data coming from microarray and macroarray can be analyzed either at the single gene level or at multiple genes level. In the first case each gene's behavior is investigated in a control versus an experimental or treatment situation. In the second case clusters of genes are analyzed in terms of interactions or co-regulation. Another approach is to discover the gene or protein networks that are related to specific patterns. For each gene X the data consist of a finite number of measurements $X_{c1}, X_{c2}, .., X_{cn}$ (control measurements) and X_{t1}, X_{t2} .., X_{tm} (treatment measurements). The measurements are represented through the logarithms of the gene expression levels. Note that treatment refers to any experimental condition different than the control.

The fundamental question in gene analysis is to determine whether the level of gene expression is significantly different in the control and treatment situations, respectively. In the literature empirical or t-test statistical techniques have been used [8]. However a better framework is needed since the replicate measurements are usually small (population size n=m=1,2,..,5), because of the experimental cost and difficulty of the experiments. Macroarray or microarray expression data provides a new method for classifying genes based on their expression profile. Numerous unsupervised and supervised learning methods have been applied to the task of discovering and learning to recognize classes of co-expressed genes. One of the most frequent tools used to solve problems in this area is support vector machine (SVM) [13]. The use of SVM in gene expression analysis is relatively new. Brown et al. [7] applied SVM to classify genes from *S. cerevisiae* based on gene expression. They proved that SVM method outperforms the other techniques including Parzen windows, Fisher's linear discriminant, and two decision tree learners [7].

In this study, we apply several clustering techniques to classify unknown genes into functional categories based on DNA macroarray expression data using a few known genes. The results of the classification step obtained from applying the clustering techniques then are confirmed by applying SVM.

SVM was originally developed for the binary class case. Here SVM is extended to the multi-class case by applying *one- against- all* method and *one-against-one* method. SVM, Linear Discriminant Analysis (LDA) and K-Nearest Neighbor (KNN) are included in the supervised learning technique. To be able to apply these techniques we need a set of data with known labels. Therefore, before applying those three techniques, we apply clustering algorithms to label the data. Multi-level K-Median and K-Means are implemented before running SVM, LDA and KNN. By comparing the error measurements of the three techniques the best technique is obtained.

The chapter is organized as follows. In Section 2, partitioning-clustering algorithm is briefly reviewed. Sections 3 and 4 give illustration and formulation of SVM. An explanation about the DNA data is given in Section 5. In Section 6, the experimental setting is described. Section 7 provides the results. In section 8, the use of SVM as an exploration tool is discussed, and the results are given. Finally, Section 9 concludes the chapter.

2 Clustering Algorithm

Based on computing time and memory reasons, partitioning clustering is preferred to hierarchical clustering especially in large-scale problems. In cases where the number of clusters is known, partitioning clustering is more efficient than hierarchical clustering, although the effectiveness is not guaranteed. Included in partitioning clustering are *K-Means* and *K-Median* [12]. In this study *K-Median* and *K-Means* are used with L_1-norm and L_2-norm. Therefore, we have four different algorithms to run. For each clustering algorithm, the variance within each cluster is calculated. The one with minimum variance within a cluster is chosen to label the data.

The algorithm can be described as follows:

1. Select the number of clusters k.
2. Initialize k cluster centers by binning the data into $3k$ bins.
   K bins with the most data are selected from $3k$ bins and the midpoints of those bins are computed. If the number of nonzero bins is less than k, then the initial center is randomly determined. From empirical observations the choice of $3k$ bins gives good results.
3. Assign each object/data to the closest cluster.
   The closeness is expressed with the L_2-norm distance between each object/data to the centers. Then, the data will be clustered into $k$clusters.
4. Re-compute the centers using current cluster memberships.
   The center of a certain cluster is the *median/mean* of all objects/data in the cluster.
5. Re-assign each object to the new centers.
   If the centers do not change, stop. Otherwise, return to step 3 until the centers do not change.

3 Support Vector Machines (SVMs)

Consider a problem with two classes, as in Figure 1, where a classifier is sought to separate two classes of points. The SVM formulation can be written as follows [9]:

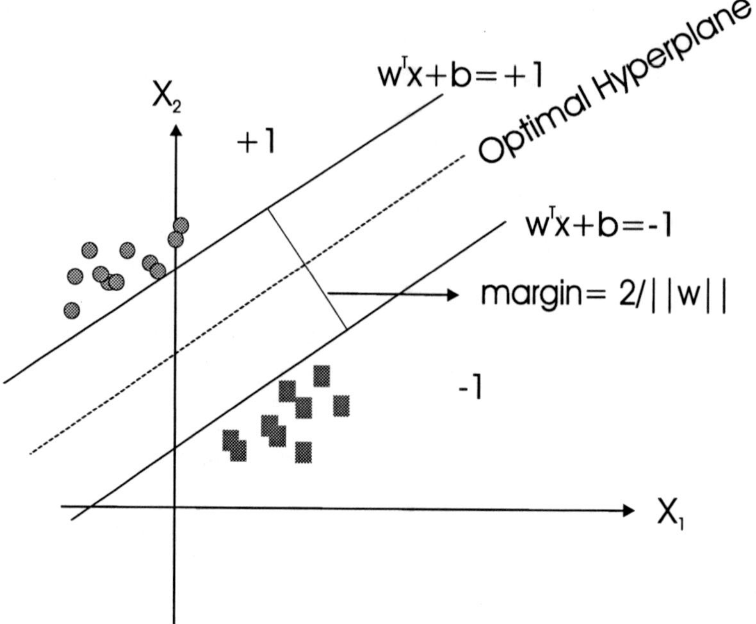

Fig. 1. SVM classifier for a binary classification problem.

$$\min_{w,b,\eta} \frac{1}{2}||w||^2 + C \sum_{i=1}^{\ell} \eta_i \tag{1}$$

$$s.t. \ y_i\,(w.x_i + b) + \eta_i \geq 1, \ \ \eta_i \geq 0, \ i = 1, ..., \ell$$

where C is a parameter to be chosen by the user, w is the vector perpendicular to the separating hyperplane, b is the offset and η_i are referring to the slack variables for possible infeasibilities of the constraints. By this formulation one wants to maximize the margin between two classes by minimizing $||w||^2$. Simultanously, the second term of the objective function is used to minimize the misclassification errors that are described by the slack variables η_i. A larger C corresponds to assigning a larger penalty to slack variables η_i. Introducing positive Lagrange multipliers α_i, to the inequality constraints in model (1), we obtain the following dual formulation [9]:

$$\min_{\alpha} \ \frac{1}{2} \sum_{i=1}^{\ell} \sum_{j=1}^{\ell} y_i y_j \alpha_i \alpha_j x_i x_j - \sum_{i=1}^{\ell} \alpha_i$$

$$s.t. \ \sum_{i=1}^{\ell} y_i \alpha_i = 0 \tag{2}$$

$$0 \leq \alpha_i \leq C, \ i = 1, ..., \ell$$

The solution of the primal problem is then given by $\mathbf{w} = \sum_i \alpha_i y_i x_i$ where $\mathbf{w}$ is the vector that is perpendicular to the separating hyperplane. The free coefficient b can be found from $\alpha_i(y_i\ (\mathbf{w}\cdot\ \mathbf{x}_i + b) - 1) = 0$, for any i such that α_i is not zero.

SVMs map a given set of binary labeled training data into a high-dimensional feature space and separate the two classes of data linearly with a maximum margin hyperplane in the feature space. In the case of nonlinear separability, each data point $\mathbf{x}$ in the input space is mapped into a higher dimensional feature space using a feature map φ. In the new space, the dot product $\langle x, x' \rangle$ becomes $\langle \varphi(x), \varphi(x') \rangle$. A nonlinear kernel function, $k(x, x')$, can be used to substitute the dot product $\langle \varphi(x), \varphi(x') \rangle$. The use of a kernel function allows the SVM to operate efficiently in a nonlinear high-dimensional feature space without being adversely affected by the dimensionality of that space. Indeed, it is possible to work with feature spaces of infinite dimension [1]. Moreover, it is possible to learn in the feature space without even knowing the mapping φ and the feature space F. The matrix $K_{ij} = \langle \varphi(x_i), \varphi(x_j) \rangle$ is called the *kernel matrix*. In general, this hyperplane corresponds to a non-linear decision boundary in the input space. It can be shown that for each continuous positive definite function $K(x, y)$, there exists a mapping, φ, such that $K(x, y) = \langle \varphi(x), \varphi(y) \rangle$ for all $x, y \in R_o$, where R_o is the input space [1].

There are several kernel functions usually used in the SVM [1] such as:

(1) linear: $k(x,y) = x^T y$,

(2) polynomial: $k(x, y) = (x^T y + 1)^p$, and

(3) radial basis function (RBF): $k(x,y) = \exp(-\left[2\sigma^2\right]^{-1} ||x - y||^2)$.

When we use SVM, for each selection of kernel function, there are some parameters for which the values can be altered. The parameters are: trade off cost constant (C), spread σ (for RBF kernel function), and degree of polynomial p (for polynomial function).

4 SVM for Multi-Class Classification

As mentioned before, SVM were originally designed for binary classification. How to effectively extend the SVM approach to multi-class classification is still an ongoing research issue. Currently there are two main approaches for multi-class SVM. One is by constructing and combining several binary classifiers. The other one is by directly considering all data in one optimization formulation [2, 3, 5]. In this work, we used the first approach. There are two methods included in the first approach: One-against-all (OAA) and One-against-one (OAO) [3].

4.1 One-against-all (OAA) Method

By this method, for k-class classification problem, we construct k SVM models where k is the number of classes. The ith SVM is trained with all of the examples in the ith class with positive labels and all other examples with negative labels. Given l training data points $(x_1, y_1),...,(x_l, y_l)$ where $x_i \in R^n$, $i = 1, 2, \ldots, l$ and $y_i \in S = \{1, ..., k\}$ is the class of x_i, the ith SVM solves the following optimization problem [3]:

$$\min_{w^i, b^i, \eta^i} \frac{1}{2}(w^i)^T w^i + C \sum_{j=1}^{\ell} \eta_j^i \tag{3}$$

$$(w^i)^T \varphi(x_j) + b^i \geq 1 - \eta_j^i, \ \ \text{if} \ y_j = i,$$
$$(w^i)^T \varphi(x_j) + b^i \leq -1 + \eta_j^i, \ \ \text{if} \ y_j \neq i,$$
$$\eta_j \geq 0, \ j = 1, ..., \ell$$

where $\varphi(x_j)$ is the map of point x_j, C is a parameter to be chosen by the user, w^i is the vector perpendicular to the separating hyperplane, b^i is the offset and η_j are referring to the slack variables for possible infeasibilities of the constraints. After solving (3) there are k decision functions:

$$(w^1)^T \varphi(x) + b^1, .., (w^k)^T \varphi(x) + b^k$$

Then, the class of a new point x is determined by the largest value of the decision function:

$$j = \arg \max_{i=1...k} ((w^i)^T \varphi(x) + b^i, \ \text{where} \ j \in S \tag{4}$$

Practically, we solve the dual problem of (3) whose number of variables is the same as the number of data in (3). Hence k l-variable quadratic programming problems are solved.

4.2 One-against-one (OAO) Method

This method constructs $k(k-1)/2$ classifiers where each one is trained on data from two classes. To find classifiers between i^{th} and j^{th} classes, we solve the following binary classification problem [3]:

$$\min_{w^{ij}, b^{ji}, \eta^{ji}} \frac{1}{2}(w^{ij})^T w^{ij} + C \sum_t \eta_t^{ij} \tag{5}$$

s.t.

$$(w^{ij})^T \varphi(x_t) + b^{ij} \geq 1 - \eta_t^{ij}, \text{if} \ y_t = i,$$
$$(w^{ij})^T \varphi(x_t) + b^{ij} \leq -1 + \eta_t^{ij}, \text{if} \ y_t \neq j,$$
$$\eta_t^{ij} \geq 0$$

where $\varphi(x_t)$ is the map of point x_t, C is a parameter to be chosen by the user, w^{ij} is the vector perpendicular to the separating hyperplane, b^{ij} is the offset and η_t^{ij} are referring to the slack variables for possible infeasibilities of the constraints. Superscripts ij for each parameter denote the classifiers between class i and class j. There are different methods for doing the future testing after all $k(k-1)/2$ classifiers are constructed. One strategy can be described as follows; if the sign $(w^{ij})^T \varphi(x) + b^{ij}$ of point x is in i^{th} class, then the vote for i^{th} class is added by one. Otherwise, the jth is increased by one. Then we predict x as being in the class with largest vote. In the case where two classes have identical votes; we select the one with smaller index. Practically we solve the dual problem of (4) whose number of variables is the same as the number of data points in two classes. Hence if in average each class has l/k data points, we have to solve $k(k-1)/2$ quadratic programming problems where each of them has about $2l/k$ variables [3].

5 DNA Macroarray Data

The data used for this study is from DNA macroarray experiments [10, 11]. An experiment starts with a commercial macroarray, on which several thousand DNA samples are fixed to a membrane. These immobilized DNAs serve as probes of known sequence, each corresponding to a single gene from the organism under investigation (*E. coli*). The DNA array is used to monitor gene expression in biological samples. The biological samples (i.e., bacterial cultures) are grown under various growth conditions that are designed to ask questions about gene expression under that particular condition. The first step in gene expression of any gene involves the synthesis of messenger RNA; each mRNA corresponds to a specific gene and the amount is proportional to the level of the gene product required by the cell to function properly. Total RNA is extracted from the biological sample and is labeled by making cDNA copies of the mRNAs that contain radioactive phosphate (32-P). The mixture of labeled cDNAs is hybridized to the DNA array, the excess label is washed away, and the array is scanned for radioactive cDNA bound by the DNA array. The amount of radioactivity bound to each probe on the array is proportional to the original level of mRNA in the biological sample. Following hybridization, a scanned densitometry image of the DNA array is made and the the pixel density of each probe is determined. The pixel density is proportional to the amount of cDNA that hybridizes with DNA affixed to the membrane and thus represents the expression level of a particular gene. Each data point produced by a DNA macroarray hybridization experiment represents the expression level of a particular gene under specific conditions.

The number of data points, that is the number of genes, is 4290. Prior knowledge about these genes is extensive; 75% have known function. The dimension of data is 4290 rows by 3 columns. The three columns are the attributes: pH5-5GadX(4)vsWt(4).log ratio, pH5-5GadX(kan-)vsWt.log ratio, and pH5-5GadX(kan+)vsWt.log ratio. Each attribute represents a replicated experiment in which gene expression in the wild type (control) is compared to that in the GadX mutant (experimental). For example, in the pH5-5GadX(4)vsWt(4).log ratio experiment, the data column contains the ratio of gene expression in two samples, representing an experimental condition and a control condition: the GadX+ (normal function in wildtype strain) and the GadX- mutant with a defect in the GadX regulator gene, respectively, both grown in culture medium that was adjusted to pH5.5. The other two data columns are replicates of the same comparison of GadX- vs. GadX+ strains, each with a separately derived mutant strains. Despite extensive knowledge of many *E. coli* genes, little is known about the genes that are regulated by the acid tolerance regulator GadX. GadX is thought to code for regulator that controls the expression level of several genes that are involved in survival in acid environments. The purpose of the biological experiments was to identify the genes whose expression is regulated by GadX and to determine how these target genes function to make *E. coli* tolerant of mild acid.

6 Experiments

6.1 Clustering

Clustering is performed by using two different algorithms: K-Median and K-Means. For each algorithm two different norms such as city block distance (L_1-norm) and Eucledian distance (L_2-norm) are applied. We considered three types of possible responses to mutation of GadX: 1-unchanged, 2-decreased expression level and 3-increased expression level. In these clustering implementations, the number of clusters, k, is set to be equal to 3. Before running the cluster algorithms, there are some genes that the responses to the absence of mutant GadX are already known. Based on this prior knowledge of some genes, the labels of the other genes can be known after implementing clustering algorithms. However, with k = 3, the results of clustering are too general. In order to obtain more limited genes in class 2, the algorithm was rerun in two additional levels. In the first run we are sure that the number of genes in class 2 (decreased expression) is too broad. The next run is applied only for genes in class 2, again with k = 3. In the third run, k = 3, only small number of genes are clustered together with the known genes. By applying this three-level clustering we finally have 11 labels for the whole genes.

6.2 SVM, LDA and KNN

The SVMs, LDA, and KNN are included in the supervised approaches. Hence, to be able to apply these methods we need some data training with known labels. The labels are obtained from experiments in section 6.1. Prior knowledge of some genes is very important to label the genes based on the clustering results. Based on known genes, we label other genes clustered together with these known genes. After labeling the genes, the SVM can be applied. There are 4290 data points (number of genes), 60% of data points for training and 40% for testing. The experiments are run for three sets of training and testing samples.

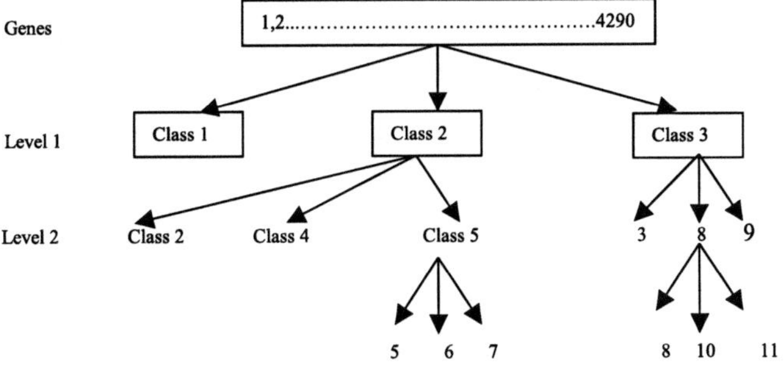

Fig. 2. Labeling Scheme

The one-against-all (OAA) and one-against-one (OAO) methods are used to implement SVMs for multi-class. OSU SVM Classifier Matlab Toolbox by Junshui et al. [4] is used for implementing the Multi-Class SVM. KNN are implemented by using SPIDER software package, which is using a Matlab platform. Likewise, LDA is implemented by using the Statistics toolbox in Matlab. For LDA, a Mahalanobis distance is used. The Mahalanobis distance is defined as the distance between two N dimensional points scaled by the statistical variation in each component of the point. For example, if $\bar{x}$ and $\bar{y}$ are two points from the same distribution which have covariance matrix C, then the Mahalanobis distance is given by $((\bar{x} - \bar{y})'C^{-1}(\bar{x} - \bar{y}))^{\frac{1}{2}}$. The Mahalanobis distance is the same as the Euclidean distance if the covariance matrix is the identity matrix.

7 Results

Clustering implementation results by using some different algorithms are given in Table 1. In the table the values of sum of total variance (SST) and sum of variance within clusters (SSW) are shown.

Table 1. Variances/scatter within cluster for each algorithm

	Kmedian1	Kmedian2	Kmean1	Kmean2
SST	66.2298	66.2298	66.2298	66.2298
SSW	41.9221	41.5656	39.4995	**38.4629**

From Table 1, we observe that the variance within cluster (SSW) value of K-Mean with L_2-norm (Kmean2) has the smallest value. Based on this fact, we can judge that K-Means with L_2-norm is the best one. Therefore, to label the data, K-Means with L_2-norm is used. The results of classification using three levels of K-Means with k = 3 are satisfactory and consistent with biological prediction where the number of genes in class 2 is low. In Table 2, summary of multi-level clustering results is shown. After we run K-Means for 3 levels, there are 5 most decreased expression genes and 40 most increased expression genes.

Table 2. Number of Genes for each level for each Label, $k = 3$

Label	Level 1–2–3
1-unchanged	2392 genes
2-decreased	825 genes–51 genes–5 genes
3-increased	1073 genes –302 genes–40 genes

Table 3 gives the list of 51 genes that resulted from running two levels of the K-Means clustering algorithm. Included in this set are 5 genes with the largest differential expression level. The genes are fliC, gadA, yhiE, gadB and gadX. The fact that gadX is differentially expressed in these DNA arrays is a result of the mutation that inactivated the gene; this serves as an internal control to indicate that the array was working properly. The *fliC* gene encodes a protein that is involved in bacterial cell swimming and its role in acid tolerance is not understood. The *gadA* and *gadB* genes both encode enzymes, glutamate decarboxylase, that are essential for glutamate-dependent acid resistance. It is thought that decarboxylation of glutamate consumes a proton inside of the cell and thereby reduces acidity. The *yhiE* gene encodes another regulator and has recently been identified as being directly involved in regulating *gadA* and *gadB*, along with several other genes involved in acid resistance [6].

Table 4 contains the results of SVM, LDA and KNN implementations. Three samples are selected from the whole data set by considering that each training sample has to contain genes from all labels or classes. There are 11 labels for the whole set of genes obtained from K-Means clustering. Each sample consists of a training and testing data set. For each sample, the SVM method is run for two different kernel functions: polynomial and RBF. For each kernel we apply some parameter values for the same penalty C.

8 Using SVM as an Exploration Tool

Results from K-Means algorithm implementation summarized in Table 2 are again used as an input for SVM implementation. Now, we consider using only 3 classes/labels for the whole genes. With this classification scheme, there are 2392 genes in class 1, 51 genes in class 2 and 40 genes in class 3. In this experiment we try to obtain new classification for the rest of 1807 genes that

Table 3. List of 51 Genes Resulted from K-Means Algorithm Implementation

AceK	pphB	**fliC**	ycgB
Arp	purM	fliD	ycgR
ArtM	rpsQ	**gadA**	**yhiE**
A1016	rpsV	**gadB**	yhiF
A1155	sodA	glgS	yhiM
A1724	yafJ	hdeA	**GadX**
B2352	yafK	hdeB	yhjL
B2353	yahO	hdeD	yifN
B2354	ybaD	himD	yjbH
Dps	ybbL	hofD	yjgG
ElaB	ycaC	moeA	yjgH
FimI	yccJ	pheL	ymgC
FlgD	ycfA	phoE	

Table 4. Misclassification Error for SVM, LDA and KNN

SVM Polynomial C = 10,000	*Degree*		*3*	*5*	*10*	*15*
	OAA	*Misclassification Error(%)*	3.32	2.49	2.14	2.33
	OAO	*Misclassification Error(%)*	1.32	1.18	1.2	1.24
SVM RBF C = 10,000	*Sigma (σ)*		*0.1*	*1*	*10*	*20*
	OAA	*Misclassification Error(%)*	10.53	3.30	2.16	2.16
	OAO	*Misclassification Error(%)*	1.65	1.36	1.37	1.42
LDAs		*Misclassification error(%)*	11.71			
KNN		*Misclassification error(%)*	5.73			

might be different from K-Means results. It is well known that K-Means is using L_1 and L_2 distance measures as a basis to group objects into clusters and is a local approach providing a local minimum. Therefore, the results of K-Means implementation might not be precisely correct. We used 2392 genes from class 1, 13 genes from class 2 and 40 genes from class 3 as a training set. As a validation set, we use 38 genes from class 2. See Figure 3 for the experiment scheme. During the validation process, we assure that our SVM model classifies genes correctly without error. SVM model then is applied to classify the rest of 1807 genes. We are most interested in the genes in class 2 which indicate decreasing expression level in response to the experimental variables. Therefore, only those genes included in class 2 are presented here. Table 5 shows the genes that are classified in class 2 by SVM implementation.

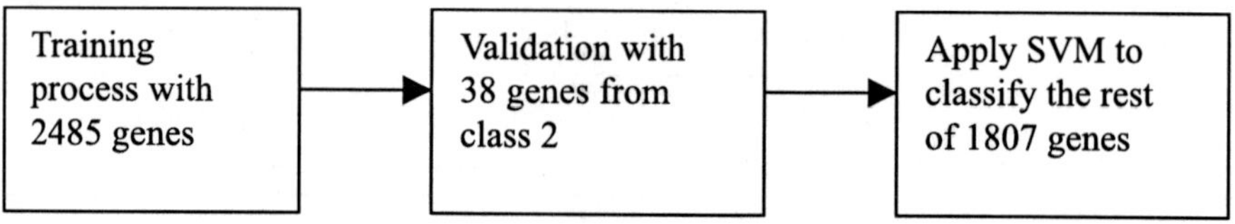

Fig. 3. SVM Implementation

Table 5. List of Additional Genes Classified in Class 2 by SVM

fimA	YabQ	yfjI	ymfI
flgE	YadQ	ygjQ	grxB
ycdF	YhiN	yhiO	pflA
b2760	YieO	yjdI	yhbH
hyfB	b1454	ylcD	yhcN
ybaS	YeaQ	ytfA	yqjC
yneJ	Hha	b1777	yciG
slp	PyrL	rpmH	yieJ
ycjM	YbiM	yccD	b0105
yfjW	b2833	b2073	b3776
tktB	CspE	hsdM	dksA
b1810	Fis	polA	wecD
cheY	NusB	yggH	ydbK
fliZ	RplP	tyrB	ygfE
osmY	RpmG	xasA	yhbO
proP	RpsT	yaiE	yhbP
rbsR	YadF	ycjC	
rfaJ	YajG		

9 Conclusions

Multi-level K-Means clustering produced very good performance in this data set. The prior knowledge about the type of response of the genes to the experimental variables gives benefit in multi-level K-Means implementation in order to obtain a limited amount of genes with largest differential expression level.

From the results shown in Section 7, we see that SVM method produced a better misclassification error than KNN and LDA for confirming the results from K-Means. The performance of SVM is really promising. Considering the kernel function selection can improve the performance of the SVM. In addition, the appropriate selection of parameter values is also very significant to improve the performance of SVM. The other advantage of SVM method is that the computing time is significantly shorter than KNN. The SVM method is also better than KNN since SVM always converges to the same optimal solution while KNN does not. LDA and KNN produce bad results when the distribution of the classes of the training data is not balanced. This occurs in those experiments where the number of data in each class is not equal or almost equal. As we know, in LDA the Mahalanobis distance for each observation in the test sample to the mean of training sample for each class depends on the number of observations in the training sample. This does not occur for SVM. Although the distribution of the training sample is not balanced, SVM still produces good results. In SVM the discriminant function does depend on the geometric mean of the data. In specific for the multi-class SVM, one-against-one (OAO) is better than one-against-all method both for computing time and generalization error.

Using SVM as an exploration tool to label unknown/unlabeled data is a good approach to identify unknown genes of similar function from expression data. SVMs can use similarity functions, defined through a kernel, that operate in high-dimensional feature spaces and allows them to consider correlations between gene expression measurements. The SVM approach is different from clustering approaches and it provides an interesting tool to explore distinctions between different types of genes with some prior information available in terms of training data labels.

References

1. B. Schölkopf and A. Smola. *Learning with Kernels*. MIT Press, Cambridge, MA, 2002.
2. C.C. Chang and C.J. Lin. *LIBSVM: A Library for Support Vector Machines*, 2001, http://www.csie.ntu.edu.tw/~cjlin/libsvm.
3. C.-W. Hsu and C.-J. Lin. A Comparison of Methods for Multi-class Support Vector Machines. *IEEE Transactions on Neural Networks*, 13: 415-425, 2002.
4. J. Ma, Y. Zhao, and S. Ahalt. *OSU SVM Classifier Matlab Toolbox*. Available at http://eewww.eng.ohio-state.edu/~maj/osu_svm/
5. K.P. Bennett and E.J., Bredensteiner. Multicategory Classification by Support Vector Machines. *Computational Optimization and Applications*, 12: 53-79, 1999.
6. Z. Ma, S. Gong, D.L. Tucker, T. Conway, and J.W. Foster. GadE (YhiE) activates glutamate decarboxylase-dependent acid resistance in Escherichia coli K12. *Molecular Microbiology*, 49: 1309-1320, 2003.
7. M.P.S. Brown, W.N., Grundy, D. Lin, N. Cristianini, C., Sugnet, M. Ares, and D. Haussler. Knowledge-based Analysis of Microarray Gene Expression Data Using Support Vector Machines. *Proceedings of the National Academy of Sciences*, 97(1): 262-267, 2000.
8. P. Baldi and S. Brunak. *Bioinformatics: A Machine Learning Approach*. MIT Press, Cambridge, MA, 2002.
9. S. Haykin. *Neural Networks: A Comprehensive Foundation*. 2nd edition, Prentice-Hall, Upper Saddle River, NJ, 1999.
10. T. Conway, B. Kraus, D.L. Tucker, D.J. Smalley, A.F. Dorman, and L. McKibben. DNA Array Analysis in a Microsoft Windows Environment. *Biotechniques*, 32: 110-119, 2002.
11. H. Tao, C. Bausch, C. Richmond, F.R. Blattner, and T. Conway. Functional Genomics: Expression Analysis of Escherichia coli Growing on Minimal and Rich Media. *Journal of Bacteriology*, 181: 6425-6440, 1999.
12. R. O. Duda, P. E. Hart, and D.G. Stork. *Pattern Classification*. John Wiley & Sons, New York, 2001.
13. V.N. Vapnik. *The Nature of Statistical Learning Theory*. John Wiley & Sons, New York, 1995.

Mathematical Programming Formulations for Problems in Genomics and Proteomics

Cláudio N. Meneses[1]*, Carlos A.S. Oliveira[2], and Panos M. Pardalos[1]

[1] Dept. of Industrial and Systems Engineering, University of Florida,
303 Weil Hall, Gainesville, FL, 32611, USA
`{claudio,pardalos}@ufl.edu`
[2] School of Industrial Engineering and Management, Oklahoma State University,
Stillwater, OK, USA
`coliv@okstate.edu`

Summary. Computational biology problems generally involve the determination of discrete structures over biological configurations determined by genomic or proteomic data. Such problems present great opportunities for application of mathematical programming techniques. We give an overview of formulations employed for the solution of problems in genomics and proteomics. In particular, we discuss mathematical programming formulations for string comparison and selection problems, with high applicability in biological data processing.

Key words: Integer programming, mathematical programming, computational biology.

1 Introduction

Problems in genomics and proteomics are among the most difficult in computational biology. They usually arise from the task of determining combinatorial properties of biological material. Researchers in this area are interested in comparing, understanding the structure, finding similarities, and discovering patterns in genomic and proteomic sequences.

Genomic and proteomic data is composed of a sequence of elements, which can be thought of as being part of an alphabet $\mathcal{A}$. A sequence of such elements will be called a *string*. The strings of genetic material (e.g., DNA and RNA) encode "instructions" to produce the proteins that regulate the life of organisms. Proteins themselves can be mathematically modeled as sequences over an alphabet of 20 characters (representing the available amino-acids).

* Supported in part by the Brazilian Federal Agency for Post-Graduate Education (CAPES) - Grant No. 1797-99-9.

In the last few years, advances in biology allowed the study of such sequences of data to move from pure biological science to other areas as well. This was made possible due to the opportunity of analyzing genomic and proteomic material using mathematical models.

The analysis of biological sequences gives rise to interesting and difficult combinatorial problems. As many of these problems are NP-hard, the study of improved techniques is necessary in order to solve problems exactly (whenever possible), or at least with some guarantee of solution quality.

We discuss problems related to configurations of genomic and proteomic sequences. Our interest is in mathematical programming formulations, generally involving integer linear programming (ILP) models. The importance of this study stems from the fact that knowing the mathematical programming properties for a problem usually makes it easier to derive exact as well approximation solution techniques. Another reason is that ILP models can be solved automatically, using standard algorithms and commercial packages for integer programming, and therefore the knowledge of better mathematical models may improve the efficiency of such techniques for the specific problem in hand. More information about applications of optimization in biology can be viewed in the surveys [10, 23].

This chapter is organized as follows. In Section 2, we discuss some string comparison problems, including the closest and farthest string, as well as the closest and farthest substring problems. In Section 3, the protein structure prediction problem is discussed. Sorting by reversals is an interesting problem related to the comparison of gene sequences in different species, presented in Section 4. Section 5 discusses the important task of identifying biological agents in a sample. Integer programming is also very useful in the design of probes for the study of DNA. An example is the minimum cost probe set problem, presented in Section 6. Finally, in Section 7 we present some concluding remarks.

2 String Comparison Problems

Comparison of strings is an important subproblem in many biological applications. Problems involving string comparison appear for example in genetics, when comparing genetic material of similar species [12], designing probes for the identification of specific positions in genetic sequences [24], or determining similarities in functionality among different genes [8].

Comparison problems in strings can be classified according to the type of comparison function employed. The most common type of comparison function is called the *editing distance* measure [19]. The editing distance between two strings s^1 and s^2 is the number of simple operations that must be used to transform s^1 into s^2. The operations used when computing the editing distance normally involve the deletion or addition of characters, as well as the mutation of a character into another. If a cost is given to each of these simple operations, then the *total cost* of the editing transformation is given by the sum of costs of each operation.

Among problems using the editing distance we can cite for example the sequence alignment problem [7, 12], and local sequence alignment [13].

A second type of comparison function, which has many applications in computational biology, is the Hamming distance. In this kind of measure, the distance between two strings with the same size is simply the number of positions where they differ. This type of distance measure gives rise to many interesting combinatorial problems. Some of these problems are discussed in more detail in the remaining of this section: the *closest string problem*, the *closest substring problem*, the *farthest string problem*, and the *farthest substring problem*. For these problems, mathematical programming methods have been developed by some researchers, giving optimum or near optimal solutions.

2.1 Closest String Problem

In the closest string problem, given a set $\mathcal{S}$ of input strings $s^1, \ldots, s^n$, each of length m, the objective is to find a target string s^* such that $\max_i H(s^*, s^i)$ is minimum (where H denotes the the Hamming distance).

In [20], three integer programming formulations for this problem have been discussed and compared. The first formulation can be described using variables z_k^i, where $z_k^i = 1$ if the k-th character in the i-th string and in the target string are the same, and zero otherwise. There are also variables t_k, for $k \in \{1, \ldots, n\}$, storing the index of the k-th character of the target string, as well as constants x_k^i, giving the index of the k-th character in input string s^i.

The first mathematical formulation proposed is this

$$\text{P1:} \qquad \min d \tag{1}$$

$$\text{s.t.} \qquad \sum_{k=1}^{m} z_k^i \leq d \qquad\qquad i = 1, \ldots, n \tag{2}$$

$$t_k - x_k^i \leq K z_k^i \qquad i = 1, \ldots, n; \; k = 1, \ldots, m \tag{3}$$

$$x_k^i - t_k \leq K z_k^i \qquad i = 1, \ldots, n; \; k = 1, \ldots, m \tag{4}$$

$$z_k^i \in \{0, 1\} \qquad i = 1, \ldots, n; \; k = 1, \ldots, m \tag{5}$$

$$d \in \mathbb{Z}_+ \tag{6}$$

$$t_k \in \mathbb{Z}_+ \qquad\qquad k = 1, \ldots, m, \tag{7}$$

where $K = |\mathcal{A}|$ is the size of the alphabet used by instance $\mathcal{S}$.

In the notation used in the above problem, d is the distance that need to be minimized, and according to constraint (2) it is equal to the maximum number of differences between the target string and the input strings. The following two constraints (3) and (4) give upper and lower bounds for the difference between the index of characters appearing in a position, for any strings $s \in \mathcal{S}$.

The mathematical formulation P1 above can be improved in many ways. The first one is by tightening the bounds on constraints (3) and (4) by exploring information about the characters appearing on strings $s^1, \ldots, s^n$. This can be done initially by finding the maximum difference between indices of characters appearing in a specific position. If we call this difference the *diameter* of position k, denoted by K_k, then constraints (3) and (4) become:

$$t_k - x_k^i \leq K_k z_k^i \qquad i = 1, \ldots, n;\ k = 1, \ldots, m \qquad (8)$$

$$x_k^i - t_k \leq K_k z_k^i \qquad i = 1, \ldots, n;\ k = 1, \ldots, m. \qquad (9)$$

Another improvement in formulation P1 comes from the idea of making the index variable t_k to be assigned to one of the characters really appearing in the k-th position, in at least one of the input strings. This avoids the possibility of having a solution with characters which are different from all other characters in one position, and therefore reduces the size of the feasible solution set for this formulation. We call the resulting formulation P2. For example, let $\mathcal{S} = \{\text{``ABC''}, \text{``DEF''}\}$. Then, "BBF" is a feasible solution for P1, but not for P2, since character 'B' does not appear in the first position in any $s \in \mathcal{S}$. This example shows that P2 $\subset$ P1.

The proposed improvement can be implemented by adding an extra variable $v_{j,k} \in \{0,1\}$, for $j \in \{1, \ldots, C_k\}$ and $k \in \{1, \ldots, m\}$, where C_k is the number of distinct characters appearing in position k in the input strings.

The resulting integer programming formulation is the following:

$$\text{P2:} \qquad \min d \qquad\qquad (10)$$

$$\text{s.t.} \qquad \sum_{j=1}^{C_k} v_{j,k} = 1 \qquad\qquad k = 1, \ldots, m \qquad (11)$$

$$\sum_{j=1}^{C_k} j v_{j,k} = t_k \qquad\qquad k = 1, \ldots, m \qquad (12)$$

$$\sum_{k=1}^{m} z_k^i \leq d \qquad\qquad i = 1, \ldots, n \qquad (13)$$

$$t_k - x_k^i \leq K_k z_k^i \qquad i = 1, \ldots, n;\ k = 1, \ldots, m \qquad (14)$$

$$x_k^i - t_k \leq K_k z_k^i \qquad i = 1, \ldots, n;\ k = 1, \ldots, m \qquad (15)$$

$$v_{j,k} \in \{0,1\} \qquad j \in V_k;\ k = 1, \ldots, m \qquad (16)$$

$$z_k^i \in \{0,1\} \qquad i = 1, \ldots, n;\ k = 1, \ldots, m \qquad (17)$$

$$d \in \mathbb{Z}_+, \quad t_k \in \mathbb{Z}_+ \qquad k = 1, \ldots, m \qquad (18)$$

Another formulation for the closest string problem presented in [20] explores the idea of bounding the difference between m and the number of matching characters in the target t and the current string s^i, for $i \in \{1, \ldots, n\}$. This integer program has the advantage that it needs less variables than P2, and therefore it can be solved more efficiently:

$$P3: \quad \min \ d \tag{19}$$

$$\text{s.t.} \quad \sum_{j \in V_k} v_{j,k} = 1 \qquad\qquad k = 1, \ldots, m \tag{20}$$

$$m - \sum_{j=1}^{m} v_{p,j} \le d \qquad p \text{ is the index of } s_j^i \text{ in } V_j \qquad i = 1, \ldots, n \tag{21}$$

$$v_{j,k} \in \{0,1\} \qquad\qquad j \in V_k,\ k = 1, \ldots, m \tag{22}$$

$$d \in \mathbb{Z}_+. \tag{23}$$

A result relating the formulations P1, P2, and P3 is summarized bellow.

Theorem 1 ([20]). *Let RP1 and RP2 be the continuous relaxations of formulations P1 and P2, respectively. If z_1^* is the optimum value of RP1 and z_2^* is the optimum value of RP2, then $z_1^* = z_2^*$.*

Theorem 2 ([20]). *The IP formulations P1 and P3 satisfy $P3 \subseteq P1$.*

Results in [20] suggest that the last formulation P3 is very effective for the closest string problem. In fact, for all instances run in the computational experiments, formulation P3 returned either the optimum solution in a few minutes of computation or a solution very close to the optimum.

2.2 Farthest String Problem

In the farthest string problem, given a set $\mathcal{S}$ of input strings with the same length, it is required to find a string t such that the Hamming distance between t and any $s \in \mathcal{S}$ is maximized. This problem is in practice the opposite of the closest string problem, and it can be solved using similar techniques. In [21], an integer programming model for this problem was proposed. The model can be described in a similar way as the formulation P3 shown above:

$$\max \ d \tag{24}$$

$$\text{subject to} \quad \sum_{j \in \mathcal{A}} v_{j,k} = 1 \qquad\qquad k = 1, \ldots, m \tag{25}$$

$$m - \sum_{j=1}^{m} v_{s_j^i, j} \ge d \qquad\qquad i = 1, \ldots, n \tag{26}$$

$$v_{j,k} \in \{0,1\} \qquad j \in \mathcal{A},\ k = 1, \ldots, m \tag{27}$$

$$d \in \mathbb{Z}_+. \tag{28}$$

The difference between this model and the model P3 is that now we are concerned with maximizing the distance. Consequently, the sign of constraint (26) must also be inverted, in order to correctly bound the variable d occurring in the objective function.

Experimental results with this model have been presented in [21]. Although it cannot be solved in polynomial time (since the problem is NP-hard), the experiments have shown that exact results can be obtained for instances of medium size (around $n = 25$ and $m = 300$). Moreover, the linear relaxation of this integer programming model has proven to yield a very close approximation for larger instances.

2.3 Closest Substring and Farthest Substring Problems

The idea of comparing strings may be useful, even when the input data does not have exactly the same length. However, in this case the problem becomes more complicated, since similarities may occur in different parts of a string. The *closest substring* and *farthest substring* problems are formulated in a similar way to the closest string and farthest string, but the input strings may have variable length [17]. More formally, if the set of input strings is $\mathcal{S}$, we assume that the length of each $s^i \in \mathcal{S}$ is at least m. Then, the closest substring problem requires the determination of a target string t of length m such that the Hamming distance between t and at least one substring s of s^i, for each $s^i \in \mathcal{S}$, is minimum. Similarly, the farthest substring problem is defined as finding a target string t of length m such that the Hamming distance between t and all substrings s of s^i, for $s^i \in \mathcal{S}$, is maximum.

A formulation for the farthest substring problem can be readily given by extending the formulation for the closest string problem (P3 as described above). We do this by defining variables $v_{j,k}$ and d as before and using the following integer program:

$$Q1: \qquad \max d \qquad\qquad\qquad (29)$$

$$\text{s.t.} \qquad \sum_{j \in V_k} v_{j,k} = 1 \qquad\qquad k = 1, \ldots, \max_i |s^i| \qquad (30)$$

$$m - \sum_{j=1}^{m} v_{p,j} \geq d \qquad p \text{ is the index of } s^i_{j+l} \text{ in } V_{j+l} \qquad (31)$$
$$\text{for } l = 1, \ldots, |s^i| - m, \ i = 1, \ldots, n$$

$$v_{j,k} \in \{0, 1\} \qquad\qquad j \in V_k, \ k = 1, \ldots, \max_i |s^i| \qquad (32)$$

$$d \in \mathbb{Z}_+. \qquad\qquad\qquad\qquad (33)$$

In this integer programming formulation, the binary variable $v_{j,k}$ determines if one of the characters appearing at position k, for $k \in \{1, \ldots, \max_i |s^i|\}$, is used in the target string t. The model Q1 has $O(nm^2|\mathcal{A}|)$ constraints.

On the other hand, a similar linear programming model could not be applied for the closest substring problem, since in this case the minimum must be taken over at least one substring of each $s^i \in \mathcal{S}$. In this case, the formulation should be extended with disjunctive constraints as described, e.g., in [22,

Chapter 1], and this would make the resulting model much more difficult to solve.

3 Protein Structure Prediction

One of the difficult problems in the analysis of proteins is related to the determination of its three-dimensional form, given only information about its sequence of amino-acids. This problem has been investigated by many researchers, and although some algorithms have been proposed, the general problem is not considered to be solved [16, 18].

Examples of techniques for protein structure prediction based on integer programming are discussed at length in [10]. We present here two of these techniques: contact maps [5] and protein threading [26].

3.1 The Contact Map Overlap Problem

Among the problems related to the three-dimensional structure of proteins, one of the most interesting from the mathematical point of view is the *contact map overlap*. This problem has the objective of finding a sequence of amino-acids that maximizes the number of common contact points appearing in two input sequences, given the matrix of possible contact points, also called *contact map*.

A contact map is a binary matrix where the elements equal to one represent the fact that two amino-acids are in contact, i.e., the distance between them is less than some threshold. Contact maps can be produced by experimental means and give accurate information about the three-dimensional structure of a protein. It is known that there are techniques allowing one to pass from the data in a contact map to a three-dimensional representation [25]. This possibility justifies the increase of interest in solving problems using the contact map representation. The following formalization of the contact map problem has been introduced recently in [5].

The objective of the contact map overlap problem is to compare two given contact maps (corresponding generally to proteins), and determine the degree of similarity between them. This is done by matching residues from both structures and computing the number of elements in the respective contact maps where both positions represent a contact (i.e., are equal to one).

The problem can also be posed using graphs in the following way. Let G be the graph corresponding to the $A^{n \times n}$ contact map, i.e., G has n nodes, and $(i, j) \in E(G)$ if and only if $A_{ij} = 1$. A *non-crossing matching* m for the contact map problem is one where $m(i, j) = 1$, $m(k, l) = 1$ and $i < j$ implies $k < l$. In the problem, we are given the graphs $G^1 = (V^1, E^1)$ and $G^2 = (V^2, E^2)$. A *common contact* is defined as the matching of a pair $i, j \in V^1$ to a pair $k, l \in V^2$, such that $(i, j) \in E^1$ and $(k, l) \in E^2$. The *contact map* overlap problem can now be formalized as finding a matching m such that the number of common contacts is maximized.

The contact map problem as defined above is known to be NP-hard [25]. Therefore, computational techniques have been studied in order to solve at least specific classes of instances to optimality. For example, the mathematical programming formulation presented in [5] is the following. Let $x \in \{0,1\}^L$ be the incidence vector of a non-crossing matching, where $L = V^1 \times V^2$. Also, if $l = (i_1, i_2)$ and $m = (j_1, j_2)$, define constants $a_{lm} = 1$ if $(i_1, j_1) \in E^1$ and $(i_2, j_2) \in E^2$, $a_{lm} = 0$ otherwise. Then the problem can be formulated as

$$\max \sum_{l \in L, \, m \in L} b_{lm} x_l x_m$$

subject to

$$x \in \{0,1\}^L,$$

where b_{lm} is a value such that $b_{lm} + b_{ml} = a_{lm} = a_{ml}$.

This is a quadratic program, however it can be linearized by defining a new variable

$$y_{ij} \in \{0,1\} \quad \text{s.t.} \quad y_{ij} = 1 \text{ if and only if } x_i = 1 \text{ and } x_j = 1$$

This can be represented using the following linear constraints

$$y_{ij} \leq x_i \qquad \text{for all } i, j \in L$$

$$y_{ij} = y_{ji} \qquad \text{for all } i, j \in L, \, l < m$$

Clearly, the original formulation described above is very hard to solve, due to the exponential number of constraints. Therefore, a better formulation was proposed to allow for computational efficiency. The reformulation is based on Lagrangian relaxation, where some constraints of the problem are added to the objective function with a negative multiplier. This penalizes the objective cost whenever the constraints are not satisfied.

The resulting formulation given in [5] is

$$\max \sum_{l \in L, \, m \in L} b_{lm} y_{lm} + \sum_{l, m \in L, \, l < m} \lambda_{lm} (y_{lm} - y_{ml})$$

where λ_{lm} and λ_{ml} are weights with the objective of penalizing non-feasible solutions. The vector λ of weights can be computed using sub-gradient optimization, an efficient technique used for example in [11]. Results for this approach were reported to be satisfactory for instances with about 1000 residues and 2000 contacts.

3.2 Linear Model for Protein Threading

A mathematical programming framework for protein prediction has also been proposed in the RAPTOR (rapid protein threading by operations research techniques) package [26]. In RAPTOR, the problem of predicting protein structure is modeled using information about contact maps, as well as some additional information derived by locality and fitness of some possible assignments.

Given a set S of known protein structures, it is interesting to predict if a specific sequence of amino-acids will fold into a structure similar to any of the ones in S. This problem is referred to in the literature as the *protein threading* problem. Typically, information is given about the structures in S in the format of contact maps, as discussed in the previous section. The data stored in the contact maps are paired with some additional information about the target sequence t being considered, in order to determine the structure that is most similar to the real structure of t. The motivation for this type of problem is the need of determining the structure of the large number of existing proteins from a smaller set of known structures. This is necessary in computational biology practice, since determining the exact structure of a protein is a long and costly process, which can be done just for a few proteins, compared to the large number that exists in nature.

The integer linear program used in RAPTOR is large and will not be presented here. The objective function tries to minimize the summation of some energy functions used to quantify the fitness of a specific structure for the current target sequence. Constraints of the formulation are concerned with the feasibility of the structure when compared to the target. For example, the alignment of the two structures cannot present any crossings, such as described for contact maps above. Crossings in the candidate alignments are called *conflicts* and solved by the addition of constraints that make the solution infeasible. In [26] a number of different constraints to avoid conflicts are presented, and their computational advantages discussed.

Although the resulting integer linear program used in RAPTOR is more complex than the one used in [5], the relaxed linear program could be solved for much larger problems, giving results provably very close to the optimum. The formulation used there also has the characteristic of using sparse matrices, and therefore the computational time can be reduced by the careful application of techniques for large-scale sparse matrix computation.

4 The Sorting by Reversals Problem

Sorting by reversals is a problem occurring in the study of sequences of genes in chromosomes. It is known that genes may change their position on chromosomes according to single permutations [9]. Therefore, one way of determining the similarity of two chromosomes a and b is determining the number of single permutations of genes, required to transform a into b. Sometimes the similarity of gene permutations in a chromosome yields a better indication of genetic similarity than the more conventional techniques of individual gene comparison.

The sorting by reversals problem is a formalization of the gene reversal problem, where given two sequences, one wants to find the minimum number of permutations that can be used to transform the first sequence into the second. Caprara [3] proved that the sorting by reversals is an NP-hard problem using properties of alternating cycles in bipartite graphs, answering an open question proposed in [14].

The number of single permutations to transform string a into b is called the *reversal distance*. The general problem of transforming a sequence a into b can be clearly reduced to the computation of the reversal distance between a permutation s and the identity permutation $[1, \ldots, n]$ (if we denote $|s| = n$).

We describe an integer program for the sorting by reversals problem proposed in [6]. The main result presented there is a column generation algorithm based on the given formulation. To develop this formulation, we describe the related cycle decomposition problem [1].

Given a permutation $[\pi_1, \ldots, \pi_n]$, construct a graph $G(V, E^1 \cup E^2)$ in the following way. Let $V = \{0, 1, \ldots, n + 1\}$ for a permutation with n elements. Let E^1 be the set of edges $(i, i+1)$ such that $|\pi_i - \pi_{i+1}| \neq 1$. Similarly, let E^2 be the set of edges $(i, i+1)$ such that $|p(i) - p(i + 1)| \neq 1$, where $p(i)$ is the position of element i in the permutation. An *alternating cycle* in the graph G described above is a cycle where edges are taken alternately from E^1 and E^2. A *cycle decomposition* of G is a disjoint set $\mathcal{S}$ of alternating cycles, such that each edge in E appears exactly once in $\mathcal{S}$.

A result from [1] that relates the cycle decomposition problem to the sorting by reversals problem is the following:

Theorem 3. *Given a graph $G(V, E^1 \cup E^2)$ constructed as described above from permutation $\pi = [\pi_1, \ldots, \pi_n]$, let c be the minimum size of a cycle decomposition of G; then $|E^1| - c$ is a lower bound for the optimal solution to the sorting by reversals problem on π.*

The theorem above specifies that solving the alternating cycle decomposition problem may give a good bound for the sorting by reversals problem. In fact, computational experiments and theoretical results [4] have demonstrated that this is a strong bound. Thus, using this bound, an integer program for sorting by reversals is defined in the following way. Given a permutation π, let $\mathcal{S}$ be the set of all alternating cycles in the $G[\pi]$ constructed as shown above. Let $x_{C \in \mathcal{S}}$ be a binary variable equal to 1 if and only if the cycle $\mathcal{C} \in \mathcal{S}$ is selected. Then the integer program is:

$$\min \sum_{\mathcal{C} \in \mathcal{S}} x_{\mathcal{C}}$$

subject to

$$\sum_{C:\, e\in C} x_C \leq 1 \qquad \text{for all } e \in E$$

$$x \in \{0,1\}^{\mathcal{S}}$$

Clearly, the number of variables in this formulation is exponential in $|V|$, and therefore extremely difficult to solve exactly. A first simplification of the problem above consists of relaxing the binary constraints to have a linear problem, with variables $x_C \geq 0$. A second technique suggested in [6] to make the problem tractable is to use a *column generation approach*, where the LP is solved for a small number of variables, and in a second step the optimality of the solution with respect to other variables is tested. If the solution found is optimal with respect to all variables, then the problem is indeed optimal for the problem. Otherwise, one of the variables that can improve the solution is entered into the basis and a new LP is solved.

In fact, determining the optimality of a solution is equivalent to proving that for all alternating cycles $C \in \mathcal{S}$ there is no edge $e \in E$ such that

$$\sum_{e\in C} u_e < 1,$$

where u_e is a value associated to edge e, for all $e \in E$.

The formulation shown above was found to be still very difficult to compute, and therefore the weaker notion of *surrogate alternating cycle* was introduced in [6]. Surrogate alternating cycles have the same requirements of alternating cycles, but they allow for a node to appear more than once in a cycle. This simple modification makes the code to find violating cycles much faster, and therefore the resulting LP can be solved quickly. The authors report the exact solution of instances with size up to 200. Instances with size up to 500 could also be approximately solved with an optimality gap of less than 2%.

5 Non-Unique Probe Selection Problem

Identification of biological agents in a sample is an important problem arising in medicine and bio-threat reduction. In this problem, one seeks to determine the presence or absence of targets – virus or bacteria – in a biological sample. A frequently used approach for making that identification is based on oligonucleotide arrays. This is better explained using an example. Suppose that we would like to identify certain virus types in a sample. By observing if a number of probes – short oligonucleotides – hybridizes to the genome of the virus one can say if a virus is contained in a sample. In case one has more than one virus in the sample, the approach is readily extensible. However, there are drawbacks associated with this approach, since finding unique probes (i.e., probes that hybridize to only one target) is difficult in case of closely related virus subtypes. An alternative approach is using non-unique probes, i.e., probes that hybridize to more than one target.

In this section we describe an integer programming formulation for the case where non-unique probes are used. This model appears in [15], and it is assumed that a target-probe incidence matrix $H = (H_{ij})$ is available, see Table 1, where $H_{ij} = 1$ if and only if probe j hybridizes to target i.

Given a target-probe incidence matrix H the goal is to select a minimal set of probes that helps us to determine the presence or absence of a single target. In Table 1, if only one of the targets $t_1, \ldots, t_4$ is in the sample, then the set with probes p_1, p_2, p_3 resolve the experiment, i.e., p_1, p_2, p_3 detect the presence of a single target. This is seen by observing that for t_1, probes p_1, p_2, p_3 hybridize; for t_2 probes p_1, p_3 hybridize but p_2 does not; for t_3 probes p_2, p_3 hybridize but p_1 does not; for t_4 probes p_2 hydridize but not p_1, p_3. In other words, the logical OR of the row vectors in Table 1 has '1' in the columns corresponding to probes p_1, p_2, p_3.

The problem becomes more difficult when targets t_2 and t_3 are in the sample. In this case, the set p_1, p_2, p_3 hybridize the targets $t_1, \ldots, t_4$, and this situation cannot be distinguished from the one where only t_1 is in the sample. One way to resolve this is selecting probes $p_1, \ldots, p_9$. Note that the hybridization pattern for each subset of two targets is different from the one for every other subset of cardinality one or two. Selecting all probes is often not cost effective since the experimental cost is proportional to the number of probes used in the experiment. We notice that using probes p_1, p_4, p_5, p_6 and p_8 resolve the experiment, since any target taken individually can be identified uniquely.

	p_1	p_2	p_3	p_4	p_5	p_6	p_7	p_8	p_9
t_1	1	1	1	0	1	1	0	0	0
t_2	1	0	1	1	0	0	1	1	0
t_3	0	1	1	1	0	1	1	0	1
t_4	0	1	0	0	1	0	1	1	1

Table 1. Target-probe incidence matrix H [15].

Due to errors in the experiment (i.e., the experiment should report a hybridization but it does not or the experiment reports a hybridization but none should be reported), one may require that two targets must be separated by more than one probe and that each target hybridizes to more than one probe. Many other constraints may arise in practice.

Now we state formally the non-unique probe selection problem. Given a target-probe incidence matrix H with non-unique probes and two parameters minimum coverage c_{min} and minimum Hamming distance h_{min}, find a minimal set of probes such that all targets are covered by at least c_{min} probes and all targets are separated with Hamming distance at least h_{min}.

It can be shown that non-unique probe selection problem is NP-hard using a reduction from the set covering problem [15] . The following formulation is based on the one for the set covering problem. Let $P = \{p_1, \ldots, p_n\}$ denote the set of probes, $T = \{t_1, \ldots, t_m\}$ denote the set of targets, and $M = \{(i, k) \in \mathbb{Z} \times \mathbb{Z} \mid 1 \leq i < k \leq m\}$. Let x_j, $j \in P$, be 1 if probe p_j is chosen and 0 otherwise. Then we have the following integer program.

$$\min \sum_{j=1}^{n} x_j$$

subject to

$$\sum_{j=1}^{n} H_{ij}x_j \geq c_{min} \qquad \text{for all } i \in T \qquad \text{[Coverage]}$$

$$\sum_{j=1}^{n} |H_{ij} - H_{kj}|x_j \geq h_{min} \quad \text{for all } (i, k) \in M \quad \text{[Hamming distance]}$$

$$x_j \in \{0, 1\} \qquad\qquad j = 1, \ldots, n,$$

where $|x - y|$ in the Hamming distance constraints stands for the absolute value of the difference between the real numbers x and y. Note that H_{ij} and H_{kj} are constants.

In [15], this formulation is used for solving real and artificial instances of the non-unique probe selection problem.

6 Minimum Cost Probe Set Problem

The analysis of microbial communities gives rise to interesting combinatorial problems. One such problem is that of minimizing the number of oligonucleotide probes needed to analyze a given population of clones. Clones and probes are represented as sequences over the alphabet $\{A,C,G,T\}$. This problem is relevant since the cost of analysis is proportional to the number of probes used in an experiment.

A probe p is said to distinguish a pair of clones c and d, if p is a substring of exactly one of c or d. In some applications clones have length approximately 1500 and probes have length between 6 and 10.

The probe set problem can be defined as follows. Let $C = \{c_1, \ldots, c_m\}$ denote a set of clones and $\mathcal{P} = \{p_1, \ldots, p_m\}$ denote a set of probes. Let $C^2 = \{(c, d) \mid c, d \in C, c < d\}$, where "$<$" is an arbitrary (e.g., lexicographic) ordering of C. We denote by $\Delta_S \subseteq C^2$ the set of pairs of clones that are distinguished by $S \subseteq \mathcal{P}$. In order to analyze C at a low cost one needs to find a smallest set of probes from $\mathcal{P}$ such that $\Delta_S = C^2$. This problem is called minimum cost probe set (MCPS). It can be shown, via a reduction from the vertex cover problem, that the MCPS is NP-hard [2].

Next we describe an integer programming formulation for the MCPS that appears in [2]. Let $x = (x_p \mid p \in \mathcal{P})$ be a binary vector satisfying $x_p = 1$ if probe $p \in S$ and $x_p = 0$ if $p \notin S$. Then we can model the MCPS as follows.

$$\min |S| = \sum_{p \in \mathcal{P}} x_p$$

subject to

$$\sum_{p \in \mathcal{P}} \delta_{p,c,d} x_p \geq 1 \qquad \text{for all } (c,d) \in \mathcal{C}^2 \qquad \text{[Distinguish constraint]}$$

$$x_p \in \{0,1\} \qquad \text{for all } p \in \mathcal{P},$$

where $\delta_{p,c,d} = 1$ if and only if $(c,d) \in \Delta_p$.

We note that due to the number of probes, this formulation may have a large number of constraints and the resulting coefficient matrix may be dense. Some instances tested in [2] have about 1,200,000 constraints and 5,000 variables. In [2], an algorithm based on Lagrangian relaxation is used to find near-optimal solutions for MCPS instances.

7 Conclusion

In this chapter we presented several examples of mathematical programming formulations for problems in genomics and proteomics. These formulations show that integer and linear programs constitute important solution tools for many problems appearing in computational biology.

We have seen that although the effectiveness of most of these techniques is remarkable, they are not straightforward to implement and require detailed knowledge of the structure of the problem. This demonstrates the importance of research concerning with the combinatorial structure of problems in computational biology. There remains a large number of open questions about efficient mathematical programming models and solution techniques for many of the problems discussed here. These will surely stay during the next years as challenging issues for researchers in computational genomics and proteomics.

References

1. V. Bafna and P.A. Pevzner. Genome rearrangements and sorting by reversals. *SIAM Journal on Computing*, 25(2):272–289, 1996.
2. J. Borneman, M. Chroback, G. D. Vedova, A. Figueroa, and T. Jiang. Probe selection algorithms with applications in the analysis of microbial communities. *Bioinformatics*, 17:S39–S48, 2001.
3. A. Caprara. Sorting by reversals is difficult. In *Proceedings of the First Annual International Conference on Computational Colecular Biology*, pages 75–83. ACM Press, 1997.
4. A. Caprara. On the tightness of the alternating-cycle lower bound for sorting by reversals. *Journal of Combinatorial Optimization*, 3:149–182, 1999.
5. A. Caprara and G. Lancia. Structural alignment of large-size proteins via lagrangian relaxation. In *Proceedings of the Sixth Annual International Conference on Computational Biology*, pages 100–108. ACM Press, 2002.
6. A. Caprara, G. Lancia, and S. K. Ng. A column-generation based branch-and-bound algorithm for sorting by reversals. In M. Farach, S. Roberts, M. Vingron, and M. Waterman, editors, *Mathematical Support for Molecular Biology*, volume 47 of *DIMACS series in Discrete Mathematics and Theoretical Computer Science*, pages 213–226. The American Mathematical Society, 1999.
7. M.A. Dayhoff, R.M. Schwartz, and B.C. Orcutt. A model of evolutionary change in proteins. In *Atlas of Protein Sequence and Structure*, chapter 5, pages 345–352. National Biomedical Research Foundation, Washington, DC, 1978.
8. R.F. Doolittle, M.W. Hunkapiller, L.E. Hood, S.G. Devare, K.C. Robbins, S.A. Aaronson, and H.N. Antoniades. Simian sarcoma virus onc gene, v-sis, is derived from the gene (or genes) encoding a platelet-derived growth factor. *Science*, 221:275–277, 1983.
9. N. Franklin. Conservation of genome form but not sequence in the transcription antitermination determinants of bacteriophages λ, $\phi21$ and P22. *Journal of Molecular Evolution*, 181:75–84, 1985.
10. H.J. Greenberg, W.E. Hart, and G. Lancia. Opportunities for combinatorial optimization in computational biology. *INFORMS Journal on Computing*, 16(3):211–231, 2004.
11. M. Held and R.M. Karp. The traveling salesman problem and minimum spanning trees: Part II. *Mathematical Programming*, 1:6–25, 1971.
12. S. Henikoff and J.G. Henikoff. Amino acid substitution matrices from protein blocks. In *Proceedings of the National Academy of Sciences*, volume 89, pages 10915–10919, 1992.
13. X. Huang and W. Miller. A time efficient, linear space local similarity algorithm. *Advances in Applied Mathematics*, 12:337–357, 1991.
14. J. Kececioglu and D. Sankoff. Exact and approximation algorithms for sorting by reversals, with application to genome rearrangement. *Algorithmica*, 13:180–210, 1995.
15. G. W. Klau, S. Rahmann, A. Schliep, M. Vingron, and K. Reinert. Optimal robust non-unique probe selection using integer linear programming. *Bioinformatics*, 20:i186–i193, 2004.
16. G. Lancia, R. Carr, B. Walenz, and S. Istrail. Optimal PDB structure alignments: a branch-and-cut algorithm for the maximum contact map overlap problem. In *Proceedings of the 5th RECOMB*, pages 193–201, 2001.

17. K. Lanctot, M. Li, B. Ma, S. Wang, and L. Zhang. Distinguishing string selection problems. *Information and Computation*, 185(1):41–55, 2003.

18. C. Lemmen and T. Lengauer. Computational methods for the structural alignment of molecules. *Journal of Computer-Aided Molecular Design*, 14:215–232, 2000.

19. V.I. Levenshtein. Binary codes capable of correcting deletions, insertions and reversals. *Soviet Physics Doklady*, 6:707–710, 1966.

20. C.N. Meneses, Z. Lu, C.A.S. Oliveira, and P.M. Pardalos. Optimal solutions for the closest string problem via integer programming. *INFORMS Journal on Computing*, 16(4):419–429, 2004.

21. C.N. Meneses, C.A.S. Oliveira, and P.M. Pardalos. Optimization techniques for string selection and comparison problems in genomics. *IEEE Engineering in Biology and Medicine Magazine*, 24(3):81–87, 2005.

22. G.L. Nemhauser and L.A. Wolsey. *Integer and Combinatorial Optimization*. Wiley Interscience Series in Discrete Mathematics and Optimization. Wiley and Sons, 1988.

23. P.M. Pardalos. Applications of global optimization in molecular biology. In C. Carlsson and I. Eriksson, editors, *Global & Multiple Criteria Optimization and Information Systems Quality*, pages 91–102. Abo Akademis Tryckeri, Finland, 1998.

24. A.D. Sharrocks. The design of primers for PCR. In H.G. Griffin and A.M Griffin, editors, *PCR Technology, Current Innovations*, pages 5–11. CRC Press, London, 1994.

25. M. Vendruscolo, E. Kussell, and E. Domany. Recovery of protein structure from contact maps. *Folding and Design*, 2:295–306, 1997.

26. J. Xu, M. Li, D. Kim, and Y. Xu. RAPTOR: Optimal protein threading by linear programming. *Journal of Bioinformatics and Computational Biology*, 1(1):95–117, 2003.

Inferring the Origin of the Genetic Code

Maria Luisa Chiusano[1]*, Luigi Frusciante[2], and Gerardo Toraldo[3]

[1] Department of Genetics, General and Molecular Biology,
University of Naples "Federico II", via Mezzocannone 8, 80134 Naples, Italy
`chiusano@unina.it`
[2] Department of Soil, Plant and Environmental Sciences (DISSPA)
University of Naples "Federico II", via Università 100, 80055 Portici, Italy
[3] Department of Agricultural Engineering,
University of Naples "Federico II", via Università 100, 80055 Portici, Italy

Summary. The extensive production of data concerning structural and functional aspects of molecules of fundamental biological interest during the last 30 years, mainly due to the rapid evolving of biotechnologies as well as to the accomplishment of the Genome Projects, has led to the need to adopt appropriate computational approaches for data storage, manipulation and analyses, giving space to fast evolving areas of biology: Computational Biology and Bioinformatics. The design of suitable computational methods and adequate models is nowadays fundamental for the management and mining of the data. Indeed, such approaches and their results might have strong impact on our knowledge of biological systems. Here we discuss the advantages of novel methodologies to building data warehouses where data collections on different aspects of biological molecules are integrated. Indeed, when considered as a whole, biological data can reveal hidden features which may provide further information in open discussions of general interest in biology.

1 Introduction

Nowadays, genome, transcriptome and proteome projects aim to determine the structural organization of the genomes and of the expressed molecules, such as RNAs and proteins, for all the biotic world, from viruses to *Homo sapiens*. The main goal of these international projects is to characterize the genetic material of an organism, its expression and its final products, to collect data concerning the functionalities of the molecules in each cell, during life stages and/or under specific stimuli. These efforts give rise to the exponential growth of biomolecular data. The data are today available mainly in the form of sequences, representing the primary level of molecular structure, or in the form of higher order organizations, to define the three-dimensional distribution of molecule atoms in the space.

---

* Corresponding author

The biological data are collected in molecular databases (see Database issue 32 of Nucleic Acids Research (2004) for a description of the currently available biological databases). The organization, collection and management of these large datasets represent only the starting point of their analysis. Data need to be investigated to understand molecular structure organization and functions; to derive peculiarities and affinities of specific cellular systems in specific conditions; to understand the evolution of living beings. This is the key to enhancing our knowledge of the determinants of the organization and physiology of the organisms, and of the mechanisms that, in a finely tuned network of interactions, control the expression of a genome, in time, i.e. during development, and in space, i.e. in the single compartments, from the cellular locations to the tissue and to the apparatus, when dealing with a multicellular life form [76, 67, 75, 52, 89, 85].

Bioinformatics and Computational Biology support the analysis of the structure and function of biological molecules, with applications of major impact in medicine and biotechnologies. This is mainly due to the suitability of the computational approaches for large scale data analysis and to the higher efficiency of simulations with predictive models.

Though many different experimental and computational programs are being developed to characterize the structure and function of genomes and their products, only 10% of genome functionalities is known.,This means that we are still far from achieving the favorable ambitious goal of simulation in silico of complex living systems. Indeed, this would require the knowledge of the functionalities of the genome, transcriptome and proteome, and of their links with cellular physiology and pathophysiology. This accounts for the increasing need for efficient computational methods which could be suitable to extract the still hidden information from the large amount of biological data already available and to be determined [14, 87, 17, 56, 93, 7, 41].

Integrated collections of data in flexible, easily accessible data banks [8, 80, 42, 60, 57, 59] and the design of algorithms and software for their efficient management and analysis [11, 22, 25, 26, 83, 84, 71] are fundamental prerequisites for fruitful mining of biological data. Therefore, one of the main topics stressed in this chapter is the need of a biologically meaningful integration of the different data sources available. This requires a challenging effort which is acquiring wide interest in Bioinformatics and in Computational Biology [14, 8, 15, 80, 2, 17, 27, 42, 59, 85, 88].

In this chapter, we present a method for comprehensive mining on nucleic acid and protein structures. Furthermore, we report some of the results we obtained by applying the proposed method. Data concerning sequence and higher order structure information of both nucleic acid and protein molecules are integrated to produce a comprehensive data warehouse, which could support datamining methodology based on graphical approaches and on suitable statistics in order to extract biologically interesting information.

To point out the usefulness of the proposed method, we report a specific application to the study of relationships between coding region compositions and the corresponding encoded protein structures. Among others [18, 20, 21, 34, 81], one of the most interesting results obtained by applying our approach to the analysis of different datasets from different organisms was that the nucleotide frequencies in the second codon position of a coding sequence are remarkably different when the encoded amino acid subsequences correspond to different secondary structures in a protein, namely helix, β-strand and aperiodic structures [19]. This evidence, clearly showing a correlation between nucleotide frequencies and the corresponding protein structures, was discussed in terms of the relationship between the physicochemical properties of the secondary structures of proteins and the organization of the genetic code. This result is considered an interesting contribution in the context of discussions about the origin of the genetic code, suggesting that the genetic code organization evolved to preserve the secondary structures of proteins. Indeed, our results show that the genetic code is organized so as to disfavor nucleotide mutations that could be deleterious as they modify amino acid composition and hence the average physicochemical properties required to determine a given secondary structure in a protein [19].

1.1 Theories on the origin of the genetic code

Amino acid coding at nucleotide level obeys the following scheme: four distinct symbols {A,C,G,U}, representing the alphabet F of letters encoding each of the four nucleotides in a nucleic acid molecule, are grouped as three consecutive, non-overlapping, elements, called triplets, or codons. Each nucleotide in the triplet/codon can be numbered as the first, the second and the third codon position ($X_1X_2X_3$, where $X \in F$). When considering all the possible combinations of nucleotides, 64 triplets (i.e. 4^3) are possible. Each triplet corresponds to one of the twenty different amino acids that can encode a protein, or to a STOP signal which marks the end of a nascent protein during protein synthesis, i.e. the biological process that produces an amino acid chain [13]. There are more triplets encoding the same amino acid. From a mathematical point of view, the genetic code is a surjective mapping among the set of the 64 possible three-nucleotide codons and the set of 21 elements composed of the 20 amino acids plus the STOP signal [44]. The rules of encoding, i.e. those that determine which amino acid will be present in the protein at a given triplet in the nucleic acids, are termed the genetic code. The organization of the genetic code is illustrated in Table 1, where each nucleotide of a triplet/codon is read using respectively the first (first column), the second (central columns) and the third (last columns) position in the table, and the encoded amino acid is obtained as the element localized by three nucleotides in the table. Therefore, redundancy and degeneracy follow, because one amino acid can be encoded by more than one triplet. Codons encoding the same amino acid are named synonymous codons.

The genetic code (Table 1) establishes a mapping between the world of nucleic acids and the world of proteins in the sense that specific regions, the coding region, of a specific nucleic acid polymer, the messenger RNA (mRNA), include a succession of non-overlapping triplets, i.e. of three nucleotides, that determine one and only one sequence of amino acids, i.e. a specific protein. Therefore, understanding the origin of the organization of the genetic code entails comprehending the logic according to which a certain amino acid was assigned to a certain triplet and, hence, understanding the meaning of the relative positions of the amino acids within the genetic code (Table 1), as well as understanding why amino acids are encoded by different numbers of codons.

Table 1. Genetic Code organization. The amino acids are represented using the three-letter abbreviation, while the 1st, the 2nd and the 3rd position represent the nucleotide content at the three codon positions using the four letter alphabet for nucleotides.

	2nd base in the codon				
1st base in the codon	U	C	A	G	3rd base in the codon
U	Phe	Ser	Tyr	Cys	U
U	Phe	Ser	Tyr	Cys	C
U	Leu	Ser	Stop	Stop	A
U	Leu	Ser	Stop	Trp	G
C	Leu	Pro	His	Arg	U
C	Leu	Pro	His	Arg	C
C	Leu	Pro	Gln	Arg	A
C	Leu	Pro	Gln	Arg	G
A	Ile	Thr	His	Ser	U
A	Ile	Thr	His	Ser	C
A	Ile	Thr	Gln	Arg	A
A	Met	Thr	Gln	Arg	G
G	Val	Ala	Asp	Gly	U
G	Val	Ala	Asp	Gly	C
G	Val	Ala	Glu	Gly	A
G	Val	Ala	Glu	Gly	G

Speculation about the origin of the genetic code began even before the code was deciphered [77, 37, 69], and this is because the origin of the nucleic acid directed protein biosynthesis is one of the main problems in studying the origin of living systems. Although the mechanisms of protein biosynthesis have been elucidated, the origin of this process remains unknown. Therefore, different theories are still discussed to understand the forces and events that determined the actual specific association between one (or more) triplet(s) and the corresponding amino acid.

The *stereochemical hypothesis* suggests that the origin of the genetic code must lie in the stereochemical interactions between codons (or anti-codons[4]) and amino acids. In other words, this hypothesis establishes that, for instance, lysine must have been codified by the codons AAA or AAG (Table 1) because lysine is somehow stereochemically correlated with these codons. Several models have been proposed to support this theory [37, 79, 99, 32, 102, 9, 10, 66, 68, 4, 51, 90, 106].

The *physicochemical hypothesis* suggests that the driving force behind the origin of the genetic code structure was the one that tended to reduce the physicochemical distances between amino acids codified by codons differing in one position [92, 101]. In particular, Sonneborn [92] identified the selective pressure tending to reduce the deleterious effects of mutations in one of the codon positions as the driving force behind the definition of amino acid allocations in the genetic code table. By contrast, Woese et al. [101] suggested that the driving force behind the definition of the genetic code organization must lie in selective pressure tending to reduce the translation errors of the primitive genetic message. In both cases, amino acid that should be preserved in their position are encoded by similar codons.

Another essentially similar hypothesis is the *ambiguity reduction hypothesis*. This hypothesis [100, 35, 36] suggests that groups of related codons were assigned to groups of structurally similar amino acids and that the genetic code, therefore, reached its current structuring through the reduction of the ambiguity in the coding between and within groups of amino acids. A point for discussion in the proposed hypotheses concerns whether there was physicochemical similarity between amino acids and the triplets coding for them, which might have promoted the origin of the genetic code. Some studies indicate that this might be the case [98, 53, 62].

The earliest traces of another hypothesis of the origin of the genetic code, the *coevolution theory*, were found by Nirenberg et al. [70], who recognized the existence of contiguity between codons that codify for amino acids synthesized by a common precursor. However, it was Pelc [78] and, above all, Dillon [31] who recognized that the distribution of codons among amino acids might have been guided by the biosynthetic relationships between amino acids. But it was not until later that the genetic code coevolution hypothesis was clearly formulated [104]. This hypothesis suggests that the codon system structure is primarily an imprinting of the prebiotic pathways that formed amino acids. Consequently, the origin of the genetic code could be clarified on the basis of the precursor-product relationships between amino acids visible in the current biosynthetic pathways. In other words, the hypothesis suggests that early on in the genetic code, only precursor amino acids were codified and that, as these

---

[4] The complementary sequence of the codon present on a transfer RNA, the molecule which transports a specific amino acid according to its anticodon, and recognizes the exact codon on the mRNA to insert the required amino acid in the forming protein chain.

gradually formed product amino acids, some of the codons in the precursor domain were conceded to the products [104, 105].

Since the genetic code mediates the translation of mRNA into proteins, a simple conjecture suggests that some fundamental themes of protein structure are reflected in the genetic code table, as these themes might have characterized the main selective pressure promoting code structuring. There are some indications that support this hypothesis. Jurka and Smith [54, 55] suggest that the β-turns of proteins became objects for selection in the prebiotic environment and influenced the origin of the genetic code and the biosynthetic pathways of amino acids, as precursor amino acids are also the most abundant ones in β-turns. Furthermore, a study aiming to clarify how the physicochemical properties of amino acids are distributed among the pairs of amino acid in precursor-product relationships and those that are not, but which are nevertheless defined in the genetic code, found that the pairs in precursor-product relationships reflect the β-sheets of proteins through the bulkiness or, more generally, the "size" of amino acids [29]. These two studies, therefore, seem to point out that β-turns and β-sheets were the main adaptive themes promoting the origin of genetic code organization. In favor of this view are the observations and suggestions of other authors who state that these structural motifs characterized primitive proteins [72, 73, 74, 12, 97, 65, 49]. The presumed identification of β-turns and β-sheets as the fundamental themes of primitive proteins might give some information as to what type of message characterized ancestral mRNAs.

In conclusion, although several hypotheses have been proposed as possible explanations for the origin of the genetic code organization, this topic is still an open question among researchers.

1.2 Protein secondary structure features

Different secondary structures of proteins exhibit considerable differences in amino acid frequencies [94, 48, 50, 82, 24, 43, 23, 64]. Some amino acids are, indeed, more prone to be present in specific secondary structures while others tend to disrupt them [3]. Propensities of amino acids for a secondary structure correlate with their physicochemical properties. These properties have provided the basic information used in prediction methods.

Protein secondary structures reflect the physicochemical properties of the most frequent amino acids in those structures. For example, the β-strand structure is strongly hydrophobic, while aperiodic structures contain more hydrophilic amino acids. Therefore, constraints on the secondary and tertiary structures tend to limit accepted mutations to those in which an amino acid is replaced by another amino acid with similar properties [33, 46, 16].

Other investigations have addressed the possible correlation between the nucleotides at each codon position and the properties of amino acids [45, 103, 91, 6, 96, 95]. In particular, hydrophobic amino acids are encoded by codons having U in the second position, while hydrophilic amino acids are encoded by triplets with A in the second position. However, previous attempts to link protein secondary structures to the organization of the genetic code [29] have been unsuccessful [86, 45], except in the case of β-turns [55] and β-strands [29]. Recently Gupta et al. [47] reported that the average frequencies of U and A at the second codon position are remarkably different between α-helix and β-strand, although they report no further results or discussion on this topic.

The present chapter shows that the nucleotide distributions in second codon positions are strongly related to the average physicochemical properties of protein secondary structure, and this relationship sheds further light on the origin of the genetic code.

2 Methodology

2.1 One coding region analysis

Our methodology is implemented in C language [20]. It is based on a comprehensive analysis of the structural information available "from the gene to the encoded protein". In particular, we considered all the information that can be related to a coding region, i.e. the region of the mRNA that, if read from the first to the last nucleotide by triplets, gives rise to a sequence of amino acids according to the rules embedded in the genetic code. Therefore, we considered the coding region codon composition, the encoded protein sequence and the information concerning different levels of protein organization from a structural and functional point of view. Protein information can be derived from specific features of the involved molecules that can be obtained using specific software and/or can be retrieved from related databases. All the information available can be collected and organized in a suitable graphical approach, reporting all the data "aligned" in a biologically significant way.

Graphics and basic statistics can be derived to inspect relationships and correlations between the different levels of structural and functional organization of the considered molecules as a basis to perform specific analysis and carry out oriented mining on data concerning different aspects.

The software is able to determine the compositional features of the nucleotide sequence as well as of the encoded amino acid chain in a straightforward way. Therefore, profiles related to the physicochemical properties of the amino acid can be reported too. Moreover, the software is designed to consider external sources of information, derived from data banks or exploitation of already available software:

i)   the structural and the functional information derived from the Swissprot data bank;

ii) the prediction of the protein secondary structures, described in terms of β-strand, α-helix, turn and aperiodic structures, from a consensus of five different predictive methods [40, 63, 28, 38, 39];

iii) the three-dimensional information derived from the DSSP program [58], when the experimentally determined three-dimensional structure of the protein is available.

All the information the user needs to consider in his/her analysis is aligned versus the nucleic acid sequence, because of the linear correspondence that relates this sequence to the amino acid sequence and to the protein structure information. The aligned format supports the investigation into the structural features of a protein from its nucleotide coding sequence to its three-dimensional organization. In particular, this software can be useful:

- to compare different structural aspects and levels of information. For example, the information coming from secondary structure predictions can be compared with different structure information determined by predictive approaches and/or from experimental data (presence of specific functional domains, crystallographic data, presence of disulfide bonds);
- to derive information on compositional features of the coding sequence and the corresponding protein structure;
- to check data contained in the data bank that often, due to lack of control, can include mistakes. For example, the comparison of the hydrophobicity profiles, of the structure predictions, of the domains or of other structural information contained in the protein data banks, can support the inspection of discrepancies that could be further investigated.

Typical graphical outputs of the software are shown in Figures 1 and 2 and described in the corresponding figure legends.

2.2 Multiple coding region analysis

The proposed approach can be extended to allow the analysis of multiple aligned sequences.

An example of output for a set of multiple aligned nucleotide sequences is shown in Figure 3. The alignment is made of three sequences of the Adenosine 1 receptor from different mammalian orders. The sequences share a high similarity. The Adenosine 1 receptor is a transmembrane protein whose functional domains are shown in the figure as they are reported in the Swissprot data bank (extracellular (light lines), cytoplasmic (left-right lines), transmembrane (dark lines) domains).

The software compares the aligned coding region sequences reporting the nucleotide synonymous (those leaving the amino acid unchanged) and nonsynonymous (those changing the amino acid) substitutions. The superimposition of profiles with different colors (non-overlapping profiles) highlights the compositional differences. The aligned information also shows the localization and possible effects of the substitutions at the amino acid level. As an example, it is possible to detect from Figure 3 amino acid substitutions that cause dramatic

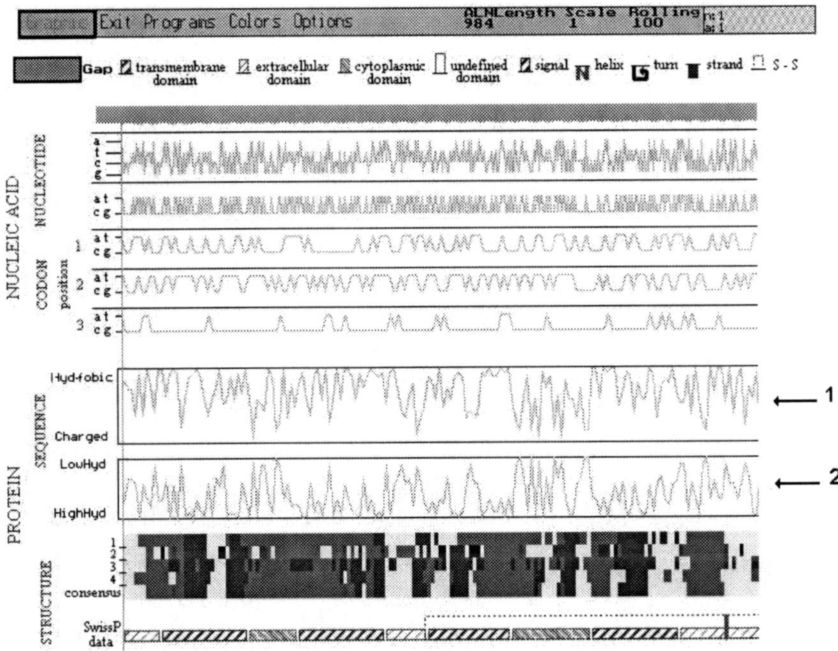

Fig. 1. Example of output in the graphical analysis of a coding region. Nucleic acid sequence information is reported as a profile describing the composition along the entire sequence (in this example each nucleotide is represented, as well as the AT/GC content of the sequence) and in terms of the composition in the three codon positions (in this example AT/GC content is reported for each codon position). Two similar but not equivalent profiles are reported for describing the protein sequence: in the first profile (1) the amino acids are ordered as hydrophobic, polar, and charged ones, while in the second profile (2) the amino acids are ordered according to the Kyte and Doolittle scale [61]. The protein structure information reported in this examples corresponds to the predictions (in terms of helix, β-strand, turn and aperiodic structure) of four different methods, while in the last line a consensus of these analyses is reported too. The protein sequence information and the functional domains reported in the Swissprot data bank are included. In this example, a disulfide bridge (- - -) is reported for the sequence under analysis.

changes in hydrophobicity. In the example reported in Figure 3, the contemporary display of different data sources also indicates regions that correspond to transmembrane domains, and confirms that these functional domains are predicted to assume an α-helix conformation. Moreover, it is evident that the transmembrane domains are highly conserved both at nucleotide and amino acid level, and that the rare nonsynonymous substitutions detectable tend to be conserved substitutions in terms of physicochemical properties. The perfect correspondence between the functional domains as reported in the Swissprot

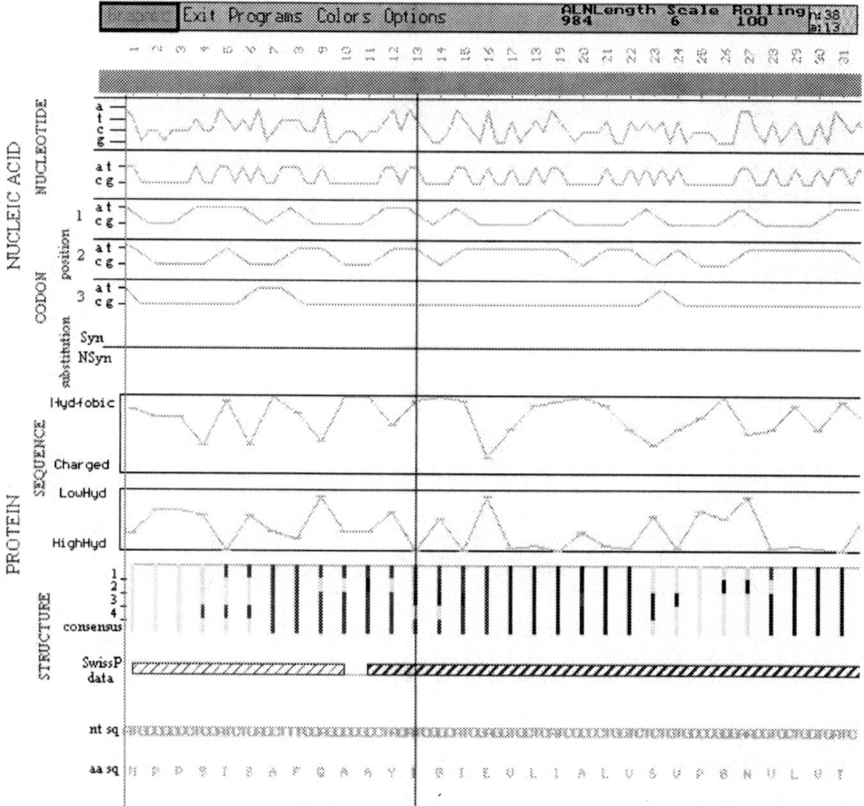

Fig. 2. Example of output in the graphical analysis of a coding region. The image on the computer screen can be scaled to investigate more detailed information. Through this option, it is possible to read both the nucleic acid (nt sq) and the amino acid (aa sq) sequences. The vertical bar on the screen can be moved by the user to scan the sequences and the aligned information, while selected positions are indicated on the menu bar (nucleotide 38 and amino acid 13, in this example).

data bank for all the three different sequences in Figure 3 confirms the functional similarity due to the high sequence similarity, and the correctness of the alignment under consideration.

The proposed approach can be useful not only to identify conserved regions in a sequence-structure alignment, but also to identify regions which are susceptible of variation, either for functional reasons (for example, a specific binding site for a ligand in one of the aligned protein may differ from that of another protein though the general structure similarity) or due to a lack of involvement in a specific function and, therefore, no presence of selective constraints that maintain structural conservation.

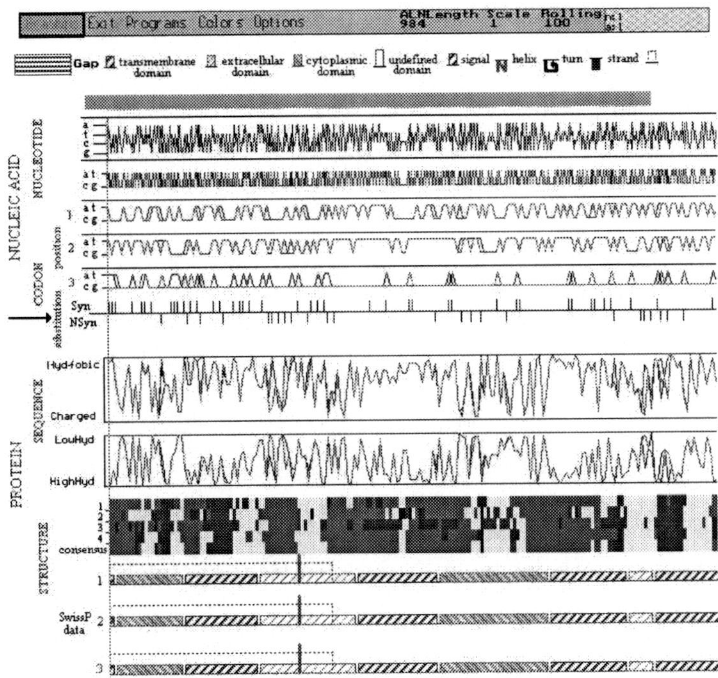

Fig. 3. Example of output in the graphical analysis of multiple aligned sequences. Three sequences are aligned in this example. Nucleotide and amino acid sequences profiles are superimposed so that nucleotide and amino acid substitutions could be identified. In this example, predictions of protein structures are reported only for one of the sequence in the alignment (user defined choice). The Swissprot information is reported for all the three aligned sequences (1,2,3). Positions of synonymous and nonsynonymous substitutions are reported along the alignment ($\rightarrow$).

Graphical approaches permit a straightforward inspection of the data information content, which is really useful in the phase of "knowledge detection" that could drive successive automated analyses.

The software can also produce simple text-based tables to report data integrating different levels of molecular information, to apply suitable statistics to determine significant relationships.

An example of data text-based report is found in Figure 4, where three groups of aligned information are visible: i) a multiple alignment of 48 amino acid sequences; ii) the corresponding predicted structures derived from a consensus of specific software [40, 63, 28, 38, 39]; the experimental structure information available for two of the aligned amino acid sequences, as well as a predicted model of the three-dimensional structure based on the amino acid sequence. This sample output allows the analysis of relationships between aligned amino acid sequences sharing sequence similarity and the higher or-

der structure information available from other source data. All the sequence
and structure information available, including experimental data, were then
used in a comprehensive approach to evaluate, in the reported example, the
reliability of the predicted model [34].

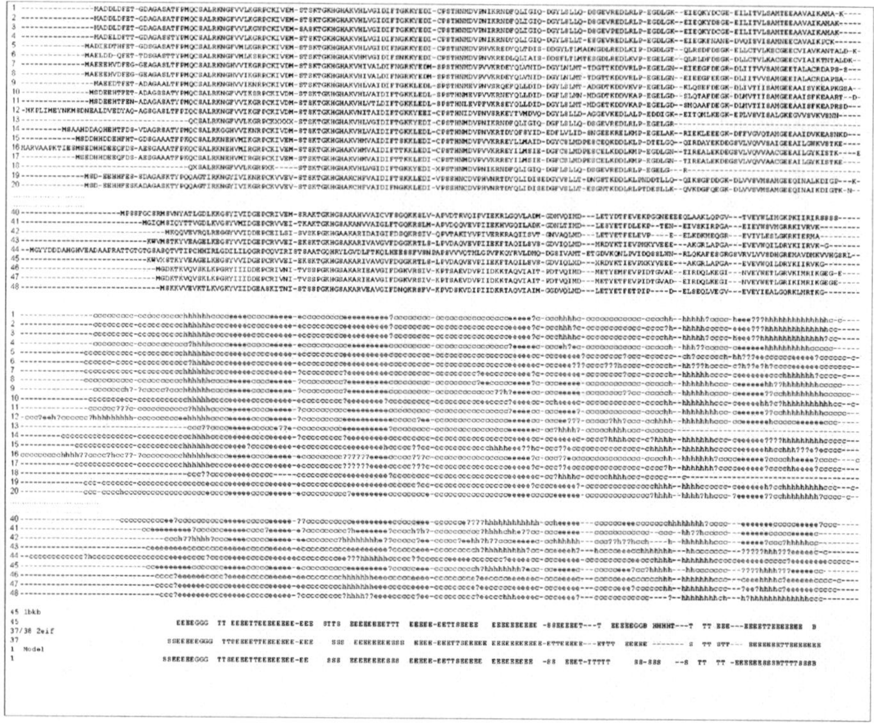

Fig. 4. Example of text output of aligned data. In this example, a partial view of
an alignment of 48 homologous protein sequences (first group) compared with the
corresponding predicted protein secondary structures (second group). 1bkb and 2eif
are the experimentally determined structures of sequence 45 and sequences 37 and
38, respectively. Model is the secondary structure of a predicted three-dimensional
model [34].

Focusing on selected features or selected regions, the software can analyse
subsets of data, extracting information at different structure levels. For example, when the secondary structures of a protein (helices, β-strands, or coils)
or a domain in a set of predefined possibilities (for instance a transmembrane
domain, a signal sequence or other) are selected as "reference regions", a set of
related information is retrieved and stored in different user-requested formats
so as to become available for analyses in greater depth.

In conclusion, the software supports a comprehensive analysis of functional and structural related sequences in the form of multiple alignment to obtain further information. It may be used to check the information available in the databases, to extend the known information to similar sequences for which no information is yet available and to evaluate the quality of sequence multiple alignments, which is no minor problem in the computational analysis of biological sequences.

The software description and some examples of output have been reported in this chapter to show the integrated approach we propose. Integrated data sources highlight information which is otherwise undetectable and support a better comprehension of functional, structural and evolutionary relationships of biological molecules.

To stress the usefulness of the method proposed and the keen interest in managing integrated data to perform comprehensive datamining, we report, in the following sections, our application of the described method to the study of the relationships between the intragenic variability of the nucleotide composition at the coding sequence level and the secondary structure of the encoded proteins.

3 Database construction

Two data sets of experimentally determined structures comprising 77 human and 232 prokaryotic proteins were used in order to investigate the relationships between the nucleotide composition in the second position of triplet/codon of coding sequences and the secondary structures of the encoded proteins.

The human protein set was obtained from ISSD [1], a database of 77 protein structures aligned with their coding regions, while the prokaryotic proteins where obtained from the PDB database [5].

Nucleotide coding region information aligned with the secondary structures of experimentally determined protein structures was used to derive the frequency of the four nucleotides (A,C,G,U) in the codon second position for each type of secondary structure.

The protein secondary structures, assigned by the DSSP program [58], were described in terms of β-strand, helix (including 3_{10} helices and α-helices), and aperiodic structure (including the turn structure and the protein segments that are not defined and/or lack periodicity). The average hydrophobicity levels, based on the Gravy scale [61], and molecular weights of the amino acids in each of the three structures were also calculated.

Basic statistics were used to mine the data and the T-test for dependent samples was used to evaluate the significance of the pairwise differences in nucleotide composition in the three secondary structures.

4 Results

In Table 2, we report the mean values of the nucleotide frequencies (A_2, C_2, G_2, U_2) at codon second positions in human and prokaryotic proteins calculated in the coding regions corresponding to different secondary structures.

The three structures show marked differences in the frequency of U in the codon second position (U_2) in both groups of organisms: the aperiodic structure shows the lowest values; the β-strand structure shows the highest ones, and the helix structure has an intermediate behavior. A_2 also differs among the three structures, with higher values in helix and aperiodic structures and lower ones in the β-strand structure. G_2 and C_2 have consistently lower values in all three structures compared to A_2 and U_2, with higher figures in the aperiodic structure in comparison to helix and β-strand structures.

Table 2. Mean, standard deviation, minimum and maximum values of the nucleotide frequencies determined at second codon positions in the coding regions corresponding to the three secondary structures.

		aperiodic				helix				b-strand			
		U2	A2	C2	G2	U2	A2	C2	G2	U2	A2	C2	G2
human	mean	0.17	0.36	0.23	0.24	0.26	0.38	0.19	0.17	0.42	0.25	0.18	0.15
	SD	0.05	0.07	0.05	0.06	0.09	0.09	0.07	0.08	0.08	0.07	0.06	0.06
	min	0.04	0.15	0.12	0.11	0	0.22	0	0	0.24	0.1	0	0
	max	0.28	0.53	0.35	0.46	0.5	0.67	0.4	0.43	0.66	0.53	0.3	0.3
prokaryotic	mean	0.2	0.38	0.22	0.2	0.3	0.39	0.19	0.12	0.44	0.27	0.16	0.13
	SD	0.05	0.09	0.06	0.06	0.07	0.11	0.08	0.06	0.09	0.09	0.06	0.05
	min	0.05	0.19	0.05	0	0	0.1	0	0	0.2	0	0	0
	max	0.32	0.63	0.46	0.47	0.5	0.78	0.67	0.28	0.8	0.61	0.36	0.32

The differences in nucleotide frequency in the codon second position can be explained by the different amino acid composition in the three structures (Table 3). As expected, the amino acids have different propensities for each structure. Interestingly, all amino acids with U in the second position exhibited the highest frequencies in the β-strand structure, while helix structures exhibited higher frequencies for these amino acids than aperiodic structures. The amino acids that contribute most to these differences are phenylalanine, isoleucine and valine, while leucine is strongly differentiated only between the aperiodic structure and the remaining two structures. With the sole exception of tyrosine, amino acids that have A in the second position are more frequent in aperiodic and helix structures than in the β-strand structure. These results indicate that the differences of U_2 and A_2 observed among structures are not due to any particular amino acid that is very frequent in one structure and seldom present in the others. Rather, they demonstrate that there is a coherent behavior. Amino acids with G_2 in their codons do not exhibit preferences for any particular structure, with the sole exception of glycine, which is very frequent in the aperiodic structure. Amino acids with C_2 also exhibit no common behavior, although alanine is most frequent in the α-helix and proline is very frequent in the aperiodic structure.

Table 3. Amino acid composition in the three secondary structures. The nucleotide at the second codon position is reported for each amino acid.

	Nucleotide in the 2nd position	Total	b-strand	helix	aperiodic
Ile	U	0.064	0.084	0.052	0.027
Phe	U	0.068	0.067	0.044	0.031
Val	U	0.041	0.122	0.062	0.047
Leu	U	0.06	0.107	0.105	0.063
Met	U	0.046	0.027	0.027	0.011
Lys	A	0.067	0.055	0.076	0.071
Asp	A	0.041	0.022	0.037	0.064
Tyr	A	0.068	0.044	0.032	0.027
Glu	A	0.049	0.056	0.089	0.059
Asp	A	0.039	0.025	0.05	0.074
Gln	A	0.023	0.032	0.046	0.039
His	A	0.034	0.022	0.024	0.023
Thr	C	0.021	0.07	0.047	0.062
Ala	C	0.047	0.054	0.098	0.052
Pro	C	0.054	0.023	0.026	0.073
Trp	G	0.077	0.019	0.016	0.011
Cys	G	0.092	0.032	0.023	0.026
Gly	G	0.015	0.047	0.04	0.112
Arg	G	0.027	0.037	0.054	0.047
Ser	G/C	0.066	0.056	0.055	0.08

A further step in examining nucleotide preferences in the secondary structures of proteins was to analyse their distributions by considering intergenic variability. For this purpose all amino acids belonging to the same type of secondary structure were pooled for each gene (Figures 5 and 6). The distinct distributions of the base frequencies in the three structures is clearly visible in both human (Figure 5) and prokaryotic (Figure 6) data sets: three separate *"clouds"* can be distinguished, each corresponding to a given structure. U_2 (abscissa) is the frequency that best separates the *"clouds"*. The β-strand structure shows a U_2 distribution that is never lower than 0.24 (hu-

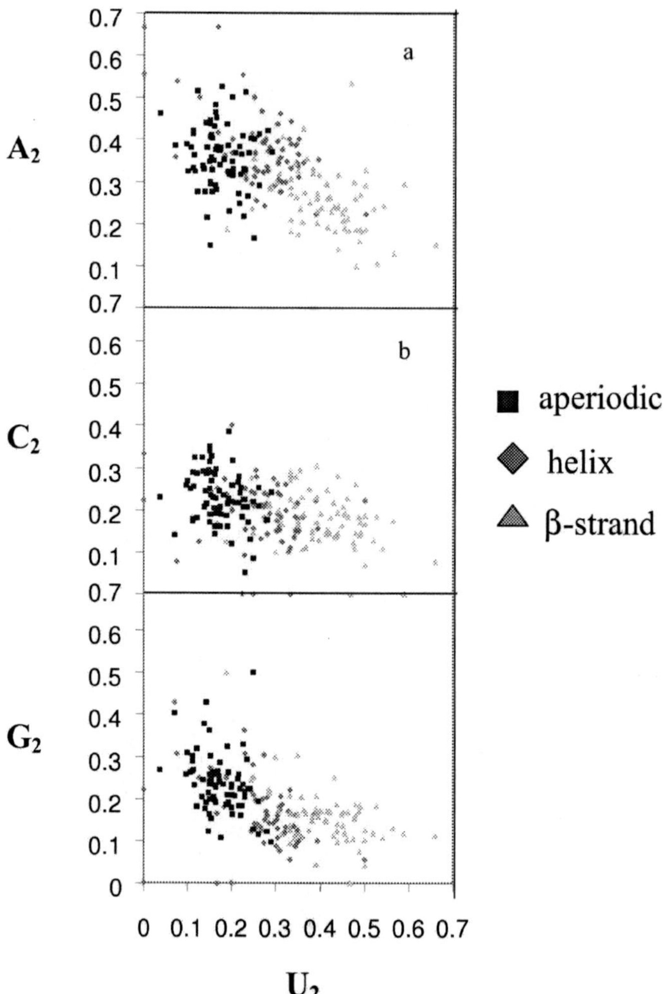

Fig. 5. Scatterplots of A2, C2 and G2 versus U2 in human data.

man data) or 0.2 (prokaryotic data), and that overlaps only marginally with the U_2 distribution for the aperiodic structure. Although the distributions of U_2 frequencies in the helix structure have some overlapping with both aperiodic and β-strand structures, it occupies a clearly differentiated, intermediate position.

Both Figures 5a and 6a show a separation of the A_2 distributions (ordinate) between aperiodic and helix structures on the one hand, with higher A_2 values, and β-strand on the other, with lower A_2 values.

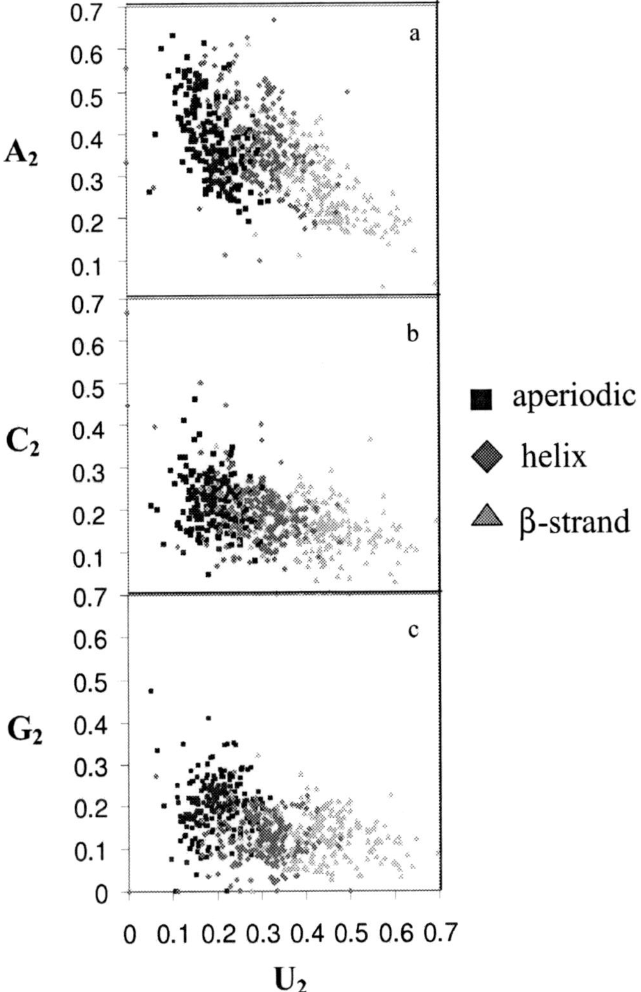

Fig. 6. Scatterplots of A2, C2 and G2 versus U2 in prokaryotic data. Symbols are as in Fig. 1.

Although G_2 and C_2 distributions do not distinguish the three structures as clearly as A_2 and U_2, their values tend to be higher for the aperiodic structure compared to helix and β-strand structures, especially in the case of G_2. In this latter case, the higher values of G_2 in the aperiodic structure are due to the contribution of turn structure. This is shown in Figure 3 where G_2 is plotted versus U_2 in human proteins, separating the contribution of turns from the remaining types of aperiodic structure. It is evident that G_2 frequency is higher in the turn structure than in the other types of aperiodic struc-

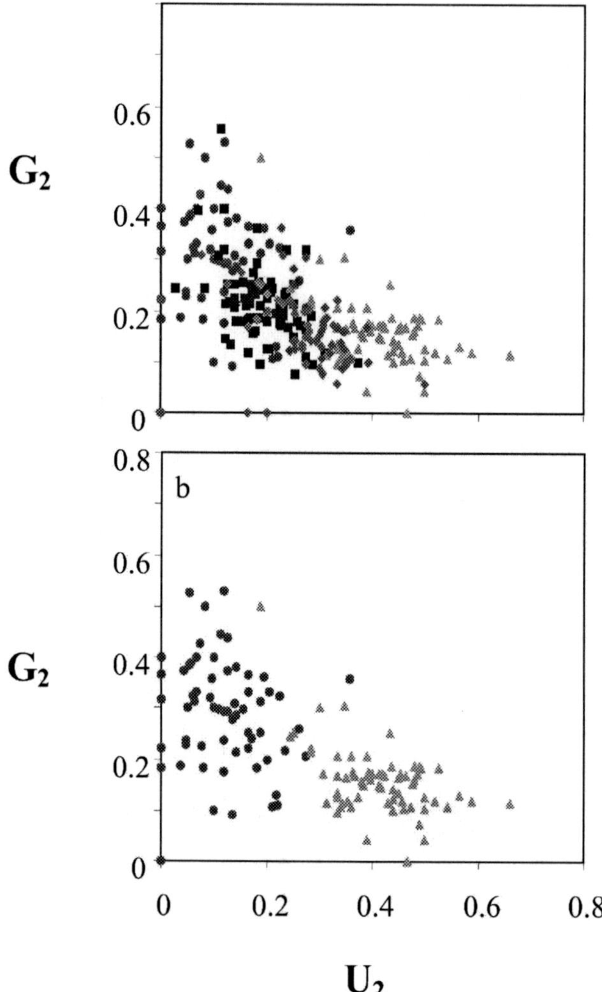

Fig. 7. Scatterplots of G2 versus U2 in the secondary structures (a) and in turn and b-strand structures (b). In (b) the contribution of the turn structure is separated from the remaining aperiodic structure.

ture, while U_2 is lower, making the separation between turn and β-strands even stronger (Figure 7b) than that obtained between aperiodic and β-strand structure (Figure 5c, 6c). The same comparison for prokaryotic proteins yields similar results (not shown).

Analysis at the "intragenic level", made in order to understand the contribution of single elements of structure to nucleotide frequencies in a given protein secondary structure, suggests that each individual element follows the general behavior described above for intergenic analysis, though the larger

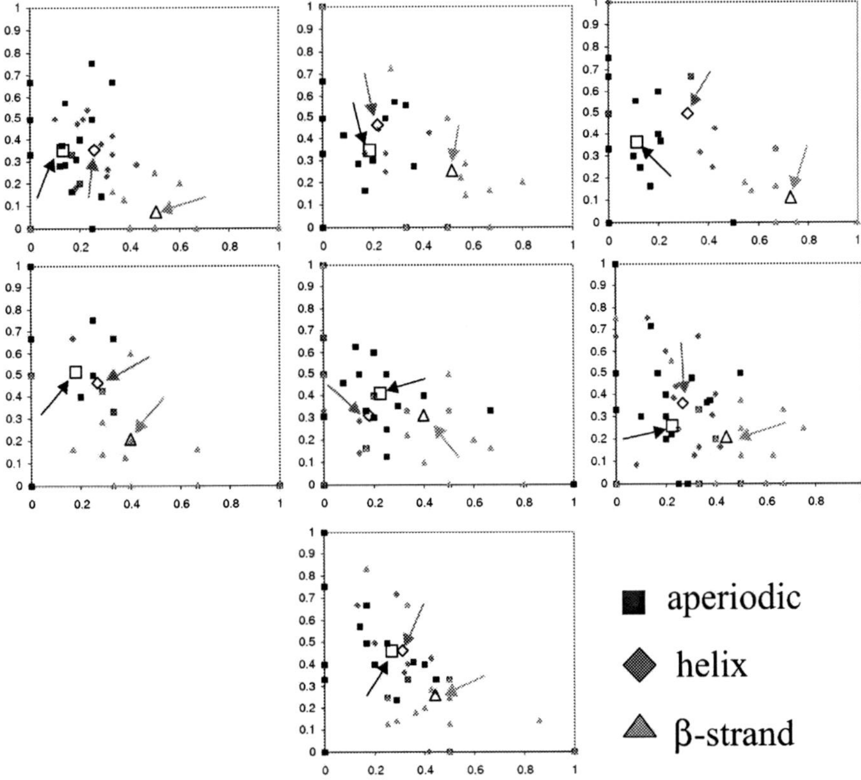

Fig. 8. Intragenic analyses of A2 versus U2 for some sample proteins. Colored symbols represent average nucleotide frequencies in individual element of structure. Open symbols represent average nucleotide frequencies of all the elements of structure of the same type in the protein (corresponding to dots in Fig. 5 and 6).

variability, which is very likely due to the smaller sample size. In Figure 8, examples for some genes are shown.

4.1 Physicochemical properties of structures

To further understand the reasons for the nucleotide preferences in the three structures in the context of the relationship between the physicochemical properties of amino acids and the genetic code, we investigated the average hydrophobicity values and molecular weights of the amino acids in the protein secondary structures.

By plotting the molecular weights versus hydrophobicity, in both human and prokaryotic proteins (Figure 9, panels a and b, respectively), we can observe that the distributions of the values differ among the three types of secondary structure. The β-strand structure has higher hydrophobicity when

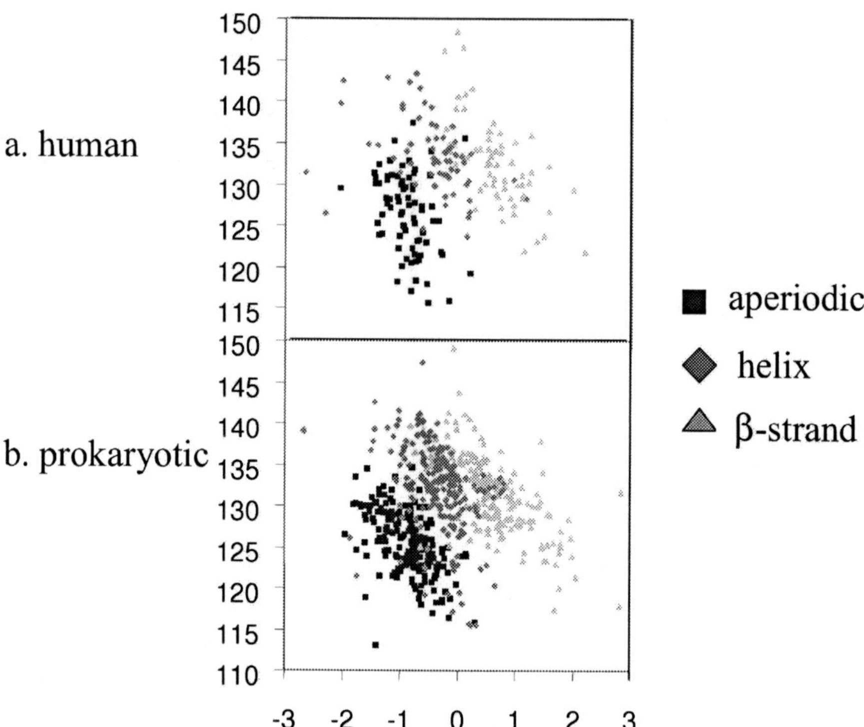

Fig. 9. Molecular weights versus hydrophobicity

compared to aperiodic and helix structures. In particular, the hydrophobicity interval of aperiodic and β-strand structures shows only a small overlap. Molecular weights show different, yet widely overlapping distributions. β-strands and helix structures exhibit molecular weights that are higher than those of the aperiodic structure.

From this analysis general information may be derived on universal rules concerning physicochemical properties of the structures. β-strand and aperiodic structures have diametrically opposite tendencies. The β-strand has higher hydrophobicity and, on average, amino acids with higher molecular weights, while the aperiodic structure is less hydrophobic and consists of amino acids with lower molecular weight on average. Moreover, the helix structure is intermediate, sharing a similar distribution with the β-strand structure in terms of molecular weight, while it follows the behavior of the aperiodic structure in its hydrophobicity patterns.

More important though is the fact that molecular weight and hydrophobicity are negatively correlated when each kind of secondary structure is considered separately, the correlations being statistically highly significant. This correlation disappears completely, if not becoming positive, when all the structures are considered together. Moreover, for the same hydrophobicity the difference among the structures is given by the molecular weight; likewise, if the molecular weight is kept constant, the difference among the structures is given by the hydrophobicity values (Figure 9). In other words, if one of these two variables is kept constant, changing the other would necessarily imply a change in the secondary structure.

Finally, we would like to mention that while each of these two physicochemical variables is not powerful enough for discriminating the secondary structures when considered separately (due to considerable overlapping), the combination of the two is indeed a strong predictor.

5 Discussion

In Section 5, we reported the analysis of nucleotide frequencies in the codon second position corresponding to different secondary structures of proteins. We noted that the average nucleotide composition in the three secondary structures represents a valid source of information for alternative predictive methods and also that our results reveal strong implications of protein secondary structures on the origin of the genetic code.

As far as the first point is concerned, we showed in this chapter that the nucleotide frequencies in the codon second position are so strikingly different that one could imagine straightforward secondary structure predictions from a nucleic acid sequence data. So far, attempts in this direction have only been made by Lesnik and Reiss (1996-1998), who used sequence information to predict putative transmembrane α-helix domains. The advantages of using nucleotide frequencies (or molecular weights and hydrophobicity) is that this approach would imply estimating 4 (or 2) parameters (the nucleotides) instead of 20 (the amino acids), with the resulting increase in the reliability of the estimates.

However, the main contribution of the present analyses is that the physicochemical requirements of protein structures, strongly dependent on amino acid composition, are indeed related to well defined choices of second codon position. This link between protein structure, amino acid preferences and the second position of the codons suggests that the organization of the genetic code must somehow reflect the secondary structures of proteins, and hence one of the main adaptive forces driving the organization of the genetic code should be the protein secondary structure. Even though this could be in part expected, there are still no clear indications that the link between the amino acid properties and the organization of the code evolved because these properties are structural determinants, as we are stating, or are involved in the codon-anticodon interactions that, according to the stereochemical hypothesis [30], could have promoted the organization of the genetic code. Indeed, earlier attempts to link secondary protein structures to the genetic code have been unsuccessful [86, 45]. More recent papers, that analyze the relationship between the genetic code and the putative primitive protein structures, have led to the conclusion that β-turn [55] and β-strand structures [29] are linked to the structure of the genetic code, suggesting that these structures could have molded the code. However, none of these previous approaches defined the link between the organization of the genetic code and protein structure as clearly as was possible here.

The stereochemical theory suggests that the genetic code originated from the interactions between codons or anticodons and amino acids, thus establishing a direct correlation between the amino acids and their synonymous codons. If this is true, i.e., if these interactions between amino acids and their codons were indeed relevant for the organization of the genetic code, it would imply the existence of a highly sophisticated physicochemical determinism. In other words, the genetic code would have been organized via intervening forces related to the interactions between amino acids and their codons/anticodons, the same amino acidic determinants of this interaction being also linked to properties involved in the organization of the secondary structures of the proteins. The implication of this proposal is that two effects participated in the organization of the code, namely the amino acid-anticodon (or -codon) interactions on the one hand, and their role in determining the secondary structure of proteins on the other. If one considers this implication too deterministic for the putative conditions in which the organization of the genetic code took place, the stereochemical hypothesis may be weakened by our results. However, the possibility that the same physicochemical properties of amino acids (such as hydrophobicity and molecular weight) that were definitely relevant to determining the secondary structures of proteins, could be important for the interaction of amino acids and their anticodons, is also likely.

We expect that natural selection acts on protein structures, while it is not immediately obvious why the organization of the code should reflect the secondary structures of proteins. The answer is probably related to putative selective pressures operating during the origin of the genetic code, in such a way that an organization like the one we observe tends to reduce the rate of very deleterious translation errors that could disrupt protein structure through deleterious effects on secondary structure requirements. This means that the secondary structures were the three-dimensional elements of the proteins that had to be preserved at the time the genetic code was organized. To reduce translation errors while assembling the amino acid sequence required for the determination of a given structure, amino acids with similar physicochemical properties were grouped by identical second positions in their codons, leading to their present organization in the columns of the standard representation of the code (Table 1).

Based on this hypothesis, further details can be discussed, such as the clear separation of aperiodic structure and β-strand structure, shown by the second codon position analysis. The separation reveals that these two structures were the fundamental ones when the genetic code was organized. The intermediary role of the helix structure demonstrates its flexibility in terms of structural requirements, but also stresses that the helix structure is not clearly distinguished by the genetic code. One hypothesis, which could explain why the helix structure was not preserved by the genetic code, is that the rules for helix formation are not related to amino acid side chains, but are the results of backbone hydrogen bonds. It would not, therefore, have been necessary to select amino acids through common choices in their second codon positions, in order to guarantee the folding of a region of a protein into a helix structure. Moreover, analysis of the preferred second codon position usage on turn structure, reported here, has shown that this structure is better separated from the β-strand structure than are the remaining kinds of aperiodic structures. This separation could be viewed as supporting either the hypothesis that the primitive structures could have been the β-turn [55] and/or the β-strand structure [29].

6 Concluding Remarks

Building a data warehouse through specific and suitable computational analyses that allow us to collect information on relationships between nucleic acid and protein structures can be considered a productive approach to exploiting the wide source of information today available in both biological sequence and structure data banks and to derive still hidden information from the data.

The integration of data is one of the best approaches to characterize known biological features and reveal new ones. Integrated computational and graphical approaches applied to data analysis help expand our knowledge on these molecules, provide the information required for reliable structural predictions, for successful interpretation of the whole data, with the end goal of obtaining computer simulations of the biological systems which could indeed reflect reality.

The main message of this chapter is the need to design integrated datasets through computational bench work that, overcoming the difficulties of reliable computational analyses, related both to incomplete knowledge of biological systems and computational limits, could help in-depth investigation into nucleic acid and protein data.

The implementation of computational methods that not only allow the analyses of multiple related results obtained by specialized software but that also integrate different sources of information related to the data under consideration is fundamental in the era of *systems biology*. The main concept underlying this approach is that information that can be derived from the study of each of the different aspects involved in the organization of complex biological systems separately cannot achieve the vision that can be obtained from considering a comprehensive collection of all these aspects together, in a holistic picture from where laws governing the integrated collective properties can be derived.

The examples reported here, discussing the various aspects related to nucleic acid and protein structures, in particular, the features of protein structures at coding sequence level, helps to stress the usefulness of such methods. The elements of protein structure on which our attention was more deeply focused are the secondary structures, described in terms of helix, β-strand, and aperiodic structure. Indeed, through the analysis of information that is related, but is often taken into consideration separately, some interesting, non-evident relationships between nucleic acid and protein molecules were derived. In particular, relationships between protein secondary structures and the corresponding coding regions were considered in the reported work, showing unexpected peculiarities in the average nucleotide frequencies in the second codon position of experimentally determined structures. Further investigations on the physicochemical aspects of the structures, on both the human and prokaryotic datasets also added the possibility to determine average properties of protein secondary structures in terms of average hydrophobicity and molecular weight (Figure 9). The representation of these properties with an original graphical approach was useful to describe the general organization of protein structures. Another example of useful information derived from this method is that, there are possible interesting starting points for the design of new algorithms for protein secondary structure predictions.

However, another interesting observation concerns the limited variability of each protein secondary structure properties, as this observation strongly contributed to the discussions about the origin of the genetic code organization [30, 19].

Acknowledgments

This research is supported by the "Agronanotech" Project funded by MIPAF.

The authors wish to thank Dr. Massimo Di Giulio for his contribution to the current work, and Dr. Giorgio Bernardi, Prof. Giovanni Colonna, Prof. Fernando Alvarez-Valin for suggesting and supporting applications of the methodology proposed.

References

1. I.A. Adzhubei, A.A. Adzhubei, and S. Neidle. An Integrated Sequence-Structure Database incorporating matching mRNA sequence, amino acid sequence and protein three-dimensional structure data. *Nucleic Acids Research*, 26: 327-331, 1998.

2. R. Apweiler, M.J. Martin, C. O'Donovan, and M. Pruess. Managing core resources for genomics and proteomics. *Pharmacogenomics*, 4(3): 343-350, 2003.

3. I. Bahar, M. Kaplan, and R.L. Jernigan. Short-range conformational energies, secondary structure propensities, and recognition of correct sequence-structure matches. *Proteins*, 29: 292-308, 1997.

4. R. Balasubramanian, P. Seetharamulu, and G. Raghunathan. A conformational rational for the origin of the mechanism of nucleic acid-directed protein synthesis of living organisms. *Origins Life*, 10: 15-30, 1980.

5. H.M. Berman, J. Westbrook, Z. Feng, G. Gilliland, T.N. Bhat, H. Weissig, I.N. Shindyalov, and P.E. Bourne. The Protein Data Bank. *Nucleic Acids Research*, 28: 235-242, 2000.

6. G. Bernardi and G. Bernardi. Compositional constraints and genome evolution. *Journal of Molecular Evolution*, 24(1-2): 1-11, 1986.

7. P. Bertone and M.Gerstein. Integrative data mining: the new direction in bioinformatics. *IEEE Engineering in Medicine and Biology Magazine : The Quarterly Magazine of the Engineering in Medicine & Biology Society*, 20(4): 33-40, 2001.

8. E. Birney, D. Andrews, P. Bevan, M. Caccamo, G. Cameron, Y. Chen, L. Clarke, G. Coates, T. Cox, J. Cuff, V. Curwen, T. Cutts, T. Down, R. Durbin, E. Eyras, X.M. Fernandez-Suarez, P. Gane, B. Gibbins, J. Gilbert, M. Hammond, H. Hotz, V. Iyer, A. Kahari, K. Jekosch, A. Kasprzyk, D. Keefe, S. Keenan, H. Lehvaslaiho, G. McVicker, C. Melsopp, P. Meidl, E. Mongin, R. Pettett, S. Potter, G. Proctor, M. Rae, S. Searle, G. Slater, D. Smedley, J. Smith, W. Spooner, A. Stabenau, J. Stalker, R. Storey, A. Ureta-Vidal, C. Woodwark, M. Clamp, and T. Hubbard. Ensembl 2004. *Nucleic Acids Research*, Database Issue: 468-470, 2004.

9. S. Black. A theory on the origin of life. *Advances in Enzymology*, 38: 193-234, 1973.

10. S. Black. Prebiotic 5-substituted uracils and a primitive genetic code. *Science*, 268: 1832, 1995.

11. S. Bottomley. Bioinformatics: smartest software is still just a tool. *Nature*, 429: 241, 2004.

12. A. Brack, L.E. Orgel. Beta structures of alternating polypeptides and their possible prebiotic significance. *Nature*, 256(5516): 383-387, 1975.

13. T.A. Brown. *Genomes*. Second Edition, BIOS Scientific Publishers, Oxford, 2002.

14. S. Buckingham. Bioinformatics: Data's future shock. *Nature*, 428: 774-777, 2004.

15. J. Chen, P. Zhao, D. Massaro, L.B. Clerch, R.R. Almon, D.C. DuBois, W.J. Jusko, and E.P. Hoffman. The PEPR GeneChip data warehouse, and implementation of a dynamic time series query tool (SGQT) with graphical interface. *Nucleic Acids Research*, 32, Database Issue: 578-581, 2004.

16. T.P. Chirpich. Rates of protein evolution: A function of amino acid composition. *Science*, 188: 1022-1023, 1975.

17. M. Chicurel. Bioinformatics: bringing it all together. Nature, 419: 751-755, 2002.

18. M.L. Chiusano, G. D'Onofrio, F. Alvarez-Valin, K. Jabbari, G. Colonna, and G. Bernardi. Correlations of nucleotide substitution rates and base composition of mammalian coding sequences with protein structure. *Gene*, 238(1): 23-31, 1999.

19. M.L. Chiusano, F. Alvarez-Valin, M. Di Giulio, G. D'Onofrio, G. Ammirato, G. Colonna, and G. Bernardi. Second codon positions of genes and the secondary structures of proteins. Relationships and implications for the origin of the genetic code. *Gene*, 261(1): 63-69, 2000.

20. M.L. Chiusano. *Implementation and Application of Computational Methods for the Analysis of Nucleic Acids and Proteins*. Ph.D. thesis, 2000.

21. M.L. Chiusano, L. Frappat, P. Sorba, and A. Sciarrino. Codon usage correlations and Crystal Basis Model of the Genetic Code. *Europhysics Letters*, 55(2): 287-293, 2001.

22. M.L. Chiusano, T. Gojobori, and G. Toraldo. A C++ Computational Environment for Biomolecular Sequence Management. *Computational Management Science*, 2(3): 165-180, 2005.

23. P.Y. Chou and G.D. Fasman. Conformational parameters for amino acids in helical, beta-sheet, and random coil regions calculated from proteins. *Biochemistry*, 13(2): 211-222, 1974.

24. D.A. Cook. The relation between amino acid sequence and protein conformation. *Journal of Molecular Biology*, 29: 167-71, 1967.

25. A.J. Cuticchia and G.W. Silk. Bioinformatics needs a software archive. *Nature*, 429: 241, 2004.

26. G. Della Vedova and R. Dondi. A library of efficient bioinformatics algorithms. *Applied Bioinformatics*, 2(2): 117-121, 2003.

27. G. Dennis Jr., B.T. Sherman, D.A. Hosack, J. Yang, W. Gao, H.C. Lane, and R.A. Lempicki. DAVID: Database for Annotation, Visualization, and Integrated Discovery. *Genome Biology*, 4(5): 3, 2003.

28. G. Deleage and B. Roux. An algorithm for protein secondary structure prediction based on class prediction. *Protein Engineering*, 1(4): 289-294, 1987.

29. M. Di Giulio. The beta-sheets of proteins, the biosynthetic relationships between amino acids, and the origin of the genetic code. *Origins of Life and Evolution of the Biosphere: The Journal of the International Society for the Study of the Origin of Life*, 26: 589-609, 1996.

30. M. Di Giulio. On the origin of the genetic code. *Journal of Theoretical Biology*, 187: 573-581, 1997.

31. L.S. Dillon. Origins of genetic code. *The Botanical Review*, 39: 301-345, 1973.

32. P. Dunnill. Triplet nucleotide-amino-acid pairing; a stereochemical basis for the division between protein and non-protein amino-acids. *Nature*, 215: 355-359, 1966.

33. C.J. Epstein. Role of the amino-acid "code" and of selection for conformation in the evolution of proteins. *Nature*, 210(31): 25-28, 1966.

34. A. Facchiano, P. Stiuso, M.L. Chiusano, M. Caraglia, G. Giuberti, M. Marra, A. Abbruzzese, and G. Colonna. Homology modelling of the human eukaryotic initiation factor 5A (eIF-5A). *Protein Engineering*, 14: 11-12, 2001.

35. W.M. Fitch. An improved method of testing for evolutionary homology *Journal of Molecular Biology*, 16: 9-16, 1966.

36. W.M. Fitch and K. Upper. The phylogeny of tRNA sequences provides evidence for ambiguity reduction in the origin of the genetic code. *Cold Spring Harbor Symposia on Quantitative Biology*, 52: 759-767, 1987.

37. G. Gamow. Possible Relation between Deoxyribonucleic Acid and Protein Structures. *Nature*, 173: 318, 1954.

38. C. Geourjon and G. Deleage. SOPM: A self-optimized method for protein secondary structure prediction. *Protein Engineering*, 7(2): 157-164, 1994.

39. C. Geourjon and G. Deleage. SOPMA: Significant improvements in protein secondary structure prediction by consensus prediction from multiple alignments. *Computer Applications in the Biosciences*, 11(6): 681-684, 1995.

40. J.F. Gibrat, J. Garnier, and B. Robson. Further developments of protein secondary structure prediction using information theory. New parameters and consideration of residue pairs. *Journal of Molecular Biology*, 198(3): 425-443, 1987.

41. J. Glasgow, I. Jurisica, and R. Ng. Data mining and knowledge discovery in molecular databases. *Pacific Symposium on Biocomputing*, 12: 365-366, 2000.

42. A. Goesmann, B. Linke, O. Rupp, L. Krause, D. Bartels, M. Dondrup, A.C. McHardy, A. Wilke, A. Puhler, and F. Meyer. Building a BRIDGE for the integration of heterogeneous data from functional genomics into a platform for systems biology. *Journal of Biotechnology*, 106(2-3): 157-167, 2003.

43. D.E. Goldsack. Relation of amino acid composition and the Moffitt parameters to the secondary structure of proteins. *Biopolymers*, 7: 299-313, 1969.

44. D.L. Gonzalez. Can the genetic code be mathematically described? *Medical Science Monitor*, 10(4): HY11-17, 2004.

45. M. Goodman and G.W. Moore. Use of Chou-Fasman amino acid conformational parameters to analyze the organization of the genetic code and to construct protein genealogies. *Journal of Molecular Evolution*, 10: 7-47, 1977.

46. R. Grantham. Composition drift in the cytochrome c cistron. *Nature*, 248(5451): 791-793, 1974.

47. S.K. Gupta, S. Majumdar, T.K. Bhattacharya, and T.C. Ghosh. Studies on the relationships between the synonymous codon usage and protein secondary structural units. *Biochemical and Biophysical Research Communications*, 269(3): 692-696, 2000.

48. A.V. Guzzo. The influence of amino-acid sequence on protein structure. *Biophysical Journal*, 5: 809-822, 1965.

49. H. Hartman. Speculations on the origin of the genetic code. *Journal of Molecular Evolution*, 40: 541-544, 1995.

50. B.H. Havsteen. Time-dependent control of metabolic systems by external effectors. *Journal of Theoretical Biology*, 10: 1-10, 1996.

51. L.B. Hendry, E.D. Bransome Jr., M.S. Hutson, and L.K. Campbell. First approximation of a stereochemical rationale for the genetic code based on the topography and physicochemical properties of "cavities" constructed from models of DNA. *Proceedings of the National Academy of Sciences*, 78: 7440-7444, 1981.

52. L. Huminiecki, A.T. Lloyd, K.H. Wolfe. Congruence of tissue expression profiles from Gene Expression Atlas, SAGEmap and TissueInfo databases. *BMC Genomics*, 4(1): 31, 2003.

53. J.R. Jungck. The genetic code as a periodic table. *Journal of Molecular Evolution*, 11(3): 211-224, 1978.

54. J. Jurka and T.F. Smith. Beta turns in early evolution: chirality, genetic code, and biosynthetic pathways. *Cold Spring Harbor Symposia on Quantitative Biology*, 52: 407-410, 1987.

55. J. Jurka and T.F. Smith. Beta-turn-driven early evolution: the genetic code and biosynthetic pathways. *Journal of Molecular Evolution*, 25(1): 15-19, 1987.

56. I. Jurisica and D.A. Wigle. Understanding biology through intelligent systems. *Genome Biology*, 3(11): 4036, 2002.

57. P. Janssen, A.J. Enright, B. Audit, I. Cases, L. Goldovsky, N. Harte, V. Kunin, and C.A. Ouzounis. COmplete GENome Tracking (COGENT): A flexible data environment for computational genomics. *Bioinformatics*, 19(11): 1451-2, 2003.

58. W. Kabsch and C. Sander. Dictionary of protein secondary structure: pattern recognition of hydrogen-bonded and geometrical features. *Biopolymers*, 22: 2577-637, 1983.

59. P. Kemmeren and F.C. Holstege. Integrating functional genomics data. *Biochemical Society Transactions*, 31: 1484-1487, 2003.

60. J.Kohler, S. Philippi, and M. Lange. SEMEDA: Ontology based semantic integration of biological databases. *Bioinformatics*, 19(18): 2420-2427, 2003.

61. J. Kyte and R.F. Doolittle. A simple method for displaying the hydropathic character of a protein. *Journal of Molecular Biology*, 157: 105-32, 1982.

62. J.C. Lacey Jr., N.S. Wickramasinghe, and G.W. Cook. Experimental studies on the origin of the genetic code and the process of protein synthesis: a review update. *Origins of Life and Evolution of the Biosphere: The Journal of the International Society for the Study of the Origin of Life*, 22(5): 243-275, 1992.

63. J.M. Levin, B. Robson, and J. Garnier. An algorithm for secondary structure determination in proteins based on sequence similarity. *FEBS Letters*, 205(2): 303-308, 1986.

64. M. Levitt. Conformational preferences of amino acids in globular proteins. *Biochemistry*, 17: 4277-85, 1978.

65. D.I. Marlborough. Early Assignments of the Genetic Code Dependent upon Protein Structure. *Origins Life*, 10: 3-14, 1980.

66. G. Melcher. Stereospecificity of the genetic code. *Journal of Molecular Evolution*, 3: 121-141, 1974.

67. A. Nantel. Visualizing biological complexity. *Pharmacogenomics*, 4(6): 697-700, 2003.

68. G.L. Nelsestuen. Amino acid-directed nucleic acid synthesis. A possible mechanism in the origin of life. *Journal of Molecular Evolution*, 11: 109-120, 1978.

69. M.W. Nirenberg and J.H. Matthei. The dependence of cell-free protein synthesis in E. coli upon naturally occurring or synthetic polyribonucleotides. *Proceedings of the National Academy of Sciences*, 47: 1588, 1961.

70. M.W. Nirenberg, O.W. Jones, P. Leder, B.F.C. Clark, W.S. Sly, and S. Petska. On the coding of genetic information. *Cold Spring Harbor Symposia on Quantitative Biology*, 28: 549-557, 1963.

71. T. Okayama, T. Tamura, T. Gojobori, Y. Tateno, K. Ikeo, S. Miyazaki, K. Fukami-Kobayashi, and H. Sugawara. Formal design and implementation of an improved DDBJ DNA database with a new schema and object-oriented library. *Bioinformatics*, 14(6): 472-478, 1998.

72. L.E. Orgel. A possible step in the origin of the genetic code. *Israel Journal of Chemistry*, 10: 287-292, 1972.

73. L.E. Orgel. Prebiotic Polynucleotides and Polypeptides. *Israel Journal of Chemistry*, 14: 11-16, 1975.

74. L.E. Orgel. The Organization and Expression of the Eukaryotic Genome. *Proceedings of the International Symposium*, Academic Press, London, 1977.

75. Z.M. Ozsoyoglu, J.H. Nadeau, G. Ozsoyoglu. Pathways database system. *OMICS: A Journal of Integrative Biology*, 7(1): 123-125, 2003.

76. J. Papin and S. Subramaniam. Bioinformatics and cellular signaling. *Current Opinion in Biotechnology*, 15(1): 78-81, 2004.

77. L. Pauling and M. Delbruk. The Nature of the intermolecular forces operative in biological processes. *Science*, 92: 77-79, 1950.

78. S.R. Pelc. Correlation between coding-triplets and amino-acids. *Nature*, 207: 597-599, 1965.

79. S.R. Pelc and M.G.E. Welton. Stereochemical relationship between coding triplets and amino-acids. *Nature*, 209: 868-870, 1966.

80. S. Philippi. Light-weight integration of molecular biological databases. *Bioinformatics*, 20(1): 51-57, 2004.

81. N. Potenza, R. Del Gaudio, M.L. Chiusano, G.M.R. Russo, and G. Geraci. Cloning and molecular characterization of the first innexin of the phylum annelida-expression of the gene during development. *Journal of Molecular Evolution*, 57(1): 165-173, 2002.

82. J.W. Prothero. Correlation between the distribution of amino acids and alpha helices. *Biophysical Journal*, 6: 367-70, 1966.

83. S. Rajasekaran, H. Nick, P.M. Pardalos, S. Sahni, and G. Shaw. Efficient Algorithms for Local Alignment Search. *Journal of Combinatorial Optimization*, 5: 117-124, 2001.

84. S. Rajasekaran, Y. Hu, J. Luo, H. Nick, P.M. Pardalos, S. Sahni, and S. Shaw. Efficient Algorithms for Similarity Search. *Journal of Combinatorial Optimization*, 5: 125-132, 2001.

85. R.B. Russell. Genomics, proteomics and bioinformatics: all in the same boat. *Genome Biology*, 3(10): reports 4034.1-4034.2, 2002.

86. F.R. Salemme, M.D. Miller, and S.R. Jordan. Structural convergence during protein evolution. *Proceedings of the National Academy of Sciences*, 74: 2820-2824, 1977.

87. C.W. Schmidt. Data explosion: bringing order to chaos with bioinformatics. *Environmental Health Perspectives*, 111(6): A340-5, 2003.

88. P. Shannon, A. Markiel, O. Ozier, N.S. Baliga, J.T. Wang, D. Ramage, N. Amin, B. Schwikowski, and T. Ideker. Cytoscape: a software environment for integrated models of biomolecular interaction networks. *Genome Research*, 13(11): 2498-504, 2003.

89. G. Sherlock and C.A. Ball. Microarray databases: storage and retrieval of microarray data. *Methods in Molecular Biology*, 224: 235-48, 2003.

90. M. Shimizu. Specific aminoacylation of C4N hairpin RNAs with the cognate aminoacyl-adenylates in the presence of a dipeptide: origin of the genetic code. *The Journal of Biochemistry*, 117: 23-26, 1995.

91. M. Sjostrom and S. Wold. A multivariate study of the relationship between the genetic code and the physical-chemical properties of amino acids. *Journal of Molecular Evolution*, 22: 272-277, 1985.

92. T.M. Sonneborn. Degeneracy of the genetic code: Extent, nature and genetic implication. In V. Bryson and H. Vogel, editors, *Evolving Genes and Proteins*, pages 377-397. Academic Press, New York, 1965.

93. L. Stein. Creating a bioinformatics nation. *Nature*, 417: 119-120, 2002.

94. A.G. Szent-Gyotgyi and C. Cohen. Role of proline in polypeptide chain configuration of proteins. *Science*, 126: 697, 1957.

95. N. Tolstrup, J. Toftgard, J. Engelbrecht, and S. Brunak. Neural network model of the genetic code is strongly correlated to the GES scale of amino acid transfer free energies. *Journal of Molecular Biology*, 243: 816-820, 1994.

96. F.J. Taylor and D. Coates. The code within the codons. *Biosystems*, 22: 177-187, 1989.

97. G. Von Heijne, C. Blomberg, and H. Baltscheffsky. Early Evolution of Cellular Electron Transport: Molecular Models for the Ferredoxin-Rubredoxin-Flavodoxin Region. *Origins Life*, 9: 27-37, 1978.

98. A.L. Weber and J.C. Jr. Lacey. Genetic code correlations: Amino acids and their anticodon nucleotides. *Journal of Molecular Evolution*, 11(3): 199-210, 1978.

99. M.G.E. Welton and S.R. Pelc. Specificity of the stereochemical relationship between ribonucleic acid-triplets and amino-acids. *Nature*, 209: 870-872, 1966.

100. C.R. Woese. On the evolution of the genetic code. *Proceedings of the National Academy of Sciences*, 54: 1546-1552, 1965.

101. C.R. Woese, D.H. Dugre, S.A. Dugre, M. Kondo, and W.C. Saxinger. On the fundamental nature and evolution of the genetic code. *Cold Spring Harbor Symposia on Quantitative Biology*, 31: 723-736, 1966

102. C.R. Woese. *The Genetic Code*. Harper and Row, New York, 1967.

103. R.V. Wolfenden, P.M. Cullis, and C.C. Southgate. Water, protein folding, and the genetic code. *Science*, 206: 575-7, 1979.

104. J.T. Wong. A co-evolution theory of the genetic code. *Proceedings of the National Academy of Sciences*, 72: 1909-1912, 1975.

105. J.T. Wong. Evolution of the genetic code. *Microbiological Sciences*, 5: 174-181, 1988.

106. M. Yarus. A specific amino acid binding site composed of RNA. *Science*, 240: 1751-1758, 1988.

Deciphering the Structures of Genomic DNA Sequences Using Recurrence Time Statistics

Jian-Bo Gao[1], Yinhe Cao[2], and Wen-wen Tung[3]

[1] Department of Electrical and Computer Engineering
University of Florida, Gainesville, FL 32611, USA
`gao@ece.ufl.edu`
[2] 1026 Springfield Drive, Campbell, CA 95008, USA
`contact@biosieve.com`
[3] National Center for Atmospheric Research
P.O. BOX 3000, Boulder, CO 80307-3000, USA
`wwtung@ucar.edu`

Summary. The completion of the human genome and genomes of many other organisms calls for the development of faster computational tools which are capable of easily identifying the structures and extracting features from DNA sequences. Such tools are even more important for sequencing uncompleted genomes of many other organisms, such as floro- and neuro- genomes. One of the more important structures in a DNA sequence is repeat-related. Often they have to be masked before protein coding regions along a DNA sequence are to be identified or redundant expressed sequence tags are to be sequenced. Here we report a novel recurrence time based method for sequence analysis. The method can conveniently study all kinds of periodicity and exhaustively find all repeat-related features from a genomic DNA sequence. An efficient codon index can also be derived from the recurrence time statistics, which has two salient features of being largely species-independent and working well on very short sequences. Efficient codon indices are key elements of successful gene finding algorithms, and are particularly useful for determining whether a suspected expressed sequence tag belongs to a coding or non-coding region. We illustrate the power of the method by studying the genomes of *E. coli*, the yeast *S. cervisivae*, the nematode worm *C. elegans*, and the human, *Homo sapiens*. Our method only requires approximately $6 \cdot N$ byte memory and a computational time of $N \log N$ to extract all the repeat-related and periodic or quasi-periodic features from a sequence of length N without any prior knowledge about the consensus sequence of those features, therefore enables us to carry out analysis of genomes on the whole genomic scale.

Key words: Genomic DNA sequence, repeated-related structures, coding region identification, recurrence time statistics

1 Introduction

Although DNA is relatively simple, the structure of human genome and genomes of other organisms is very complicated. With the completion of many different types of genomes, especially the human genome, one of the grand challenges for the future genomics research is to comprehensively identify the structural and functional components encoded in a genome [1]. Among the outstanding structural components are repeat-related structures [2, 3], periodicity and quasi-periodicity, such as period-3, which is considered to reflect codon usage [4], and period 10-11, which may be due to the alternation of hydrophobic and hydrophilic amino acids [5] and DNA bending [6]. Extracting and understanding these structural components will greatly facilitate the identification of functional components encoded in a genome, and the study of the evolutionary variations across species and the mechanisms underlying those variations. Equally or even more important, repeat-related features often have to be masked before protein coding regions along a DNA sequence are to be identified or redundant expressed sequence tags (ESTs) are to be sequenced.

More important than finding repeat-related structures in a genome is the identification of genes and other functional units along a DNA sequence. In order to be successful, a gene finding algorithm has to incorporate good indices for the protein coding regions. A few representative indices are the Codon Bias Index (CBI) [7], the Codon Adaptation Index (CAI) [8, 9], the period-3 feature of nucleotide sequence in the coding regions [10, 11, 12, 13] and the recently proposed YZ score [14]. Each index captures certain but not all features of a DNA sequence. The strongest signal can only be obtained when one combines multiple different sources of information [15]. In order to improve the accuracy and simplify the training of existing coding-region or gene identification algorithms (see the recent review articles [16, 17] and the references therein), and to facilitate the development of new gene recognition algorithms, it would be highly desirable to find new codon indices.

Over the past decades, sequence alignment and database search [18, 19, 20, 21, 22, 23, 24, 25, 26, 27] have played a significant role in molecular biology, and extensive research in algorithms has resulted in a few basic software tools such as FASTA [28, 29] and BLAST [30, 31, 32, 33]. Although these tools have been routinely used in many different types of researches, finding biologically significant information with these tools is far from trivial, for the following reasons: i) The results of sequence alignment and database search strongly depend on some model-based score function [19, 20, 21]. However, the model may not be general enough to be appropriate for the biological problem under study. For example, a widely used affine gap cost model [21] assumes that insertions and deletions are exponentially less likely to occur as their length increases. Nevertheless, long insertions and deletions may occur as a single event, such as insertion of a transposon element. ii) The dynamic programming approach to the maximization of the score function, although mathematically sound, requires a computational time at least proportional to the product of the length of the two sequences being compared. This makes the approach less feasible for sequence comparison on the whole genomic scale. iii) Assessment of the statistical significance of an alignment is hard to make [34, 35, 36, 37, 38, 39, 40]. All theoretically tractable score statistics are based upon certain probability models about the sequence under study. However, those models may not capture interesting sequence segments such as repeat structures inherent in natural DNA sequences. For example, it is a common experience for a user of BLAST that for some input sequence, the output from BLAST may be very noisy: many matches with very high score may only hit on low complexity regions and are not biologically interesting, while biologically significant units such as binding sites, promoter regions, or expression regulation signals do not have a chance to show up as the output due to their low scores [26, 27].

Here, we propose a simple recurrence time based method, which has a number of distinct features: i) It does not utilize a score function in general and does not penalize gaps in particular. This makes it most convenient to find out large blocks of insertions or deletions. ii) Computationally it is very efficient: with a computational time proportional to $N \log N$, where N is the size of the sequence, and a memory of $6N$, it can exhaust all repeat-related and periodic or quasi-periodic structures. This feature allows us to carry out genome analysis on the entire genomic scale by a PC. iii) It is model-free in the sense that it does not make any assumption about the sequences under study. Instead, it builds a representation of the sequence in such a way that all interesting sequence units are automatically extracted. (iv) The method defines an efficient codon index, which is largely species-independent and works well on very short sequences. This feature makes the method especially appealing for the study of short ESTs. Below, we shall illustrate the power of the method by extracting outstanding structures including insertion sequences (ISs), rRNA clusters, repeat genes, simple sequence repeats (SSRs), transposons, and gene and genome segmental duplications such as inter-chromosomal duplication from genome sequences. We shall also discuss the usefulness of the method for the study of the evolutionary variations across species by carefully studying mutations, insertions and deletions. Finally, we shall evaluate the effectiveness of the recurrence time based codon index by studying all of the 16 yeast chromosomes.

2 Databases and Methods

Recurrence time based method for sequence analysis can be viewed as a 2-layer approach: first organize structures/patterns embedded in a sequence by the values of their recurrence times, then find where those sequences are. Before we explain the basic idea of recurrence time statistics, let us briefly describe the sequence data analyzed here.

2.1 Databases

We have studied the DNA sequence data from the following four species:
(a) *E. coli*[4] [42] , (b) the yeast *S. cervisivae*[5] [43], (c) the nematode worm
C. elegans[6] [44], (d) and the human, *Homo sapiens*[7] [2, 45]. Except the E.
coli genome, the other three contain gaps, since they are not yet completely
sequenced. Those gaps are deleted before we compute the recurrence times
from them. For the yeast *S. cervisivae*, the sequences of chromosome 1 to
chromosome 16 are joined together into a single sequence in ascending order.

2.2 Basic Idea of Recurrence Time Statistics

Notations

Let us denote a sequence we want to study by $S = b_1 b_2 b_3 \cdots b_N$, where N is
the length of the sequence, b_i, $i = 1, \cdots, N$, are nucleotide bases. For instance,
if we take

$$S_1 = \text{ACGAAAAACGATTTTAAA},$$

then $N = 18$, $b_1 = \text{A}$, $b_2 = \text{C}, \cdots, b_{18} = \text{A}$. Next, we group consecutive
nucleotide bases of window size w together and call that a word of size w.
Using maximal overlapping sliding window, we then obtain $n = N - w + 1$
such words. We associate these words with the positions of the original DNA
sequence from 1 to n, i.e., $W_i = b_i b_{i+1} \cdots b_{i+w-1}$ is a word of size w associated
with the position i along the DNA sequence. Two words are considered equal
if all of their corresponding bases match. That is, $W_i = W_j$, if and only
if $b_{i+k} = b_{j+k}$, $k = 0, \cdots, w - 1$. $S[u \rightarrow v] = b_u b_{u+1} \ldots b_v$ will denote a
subsequence of S from position u to v.

Recurrence time

The recurrence time, $T(i)$, of position i for a DNA sequence S is a discretized
version of the recurrence times of the second type for dynamical systems
introduced recently by Gao [46, 47, 48]. It is defined as follows.

> **Definition:** The recurrence time $T(i)$ for a position i along the DNA
> sequence is the smallest $j - i$ such that $j > i$ and $W_j = W_i$. If no such
> j exists, then there is no repeat for the word W_i after position i in the
> sequence S, and we indicate such a situation by $T(i) = -1$.

To analyze the $T(i)$ sequence, we first filter out all those $T(i) = -1$, then denote the remaining positive integer sequence by $R(k)$, $k = 1, \cdots, m$, and
finally estimate the probability distribution functions for both $R(k)$ and

---

[4] http://www.genome.wisc.edu/sequencing/k12.htm
[5] ftp://genome-ftp.standford.edu/pub/yeast/data_download/
[6] http://www.sanger.ac.uk/Projects/C_elegans/
[7] http://www.ncbi.nlm.nih.gov/genome/guide/human/

$\log_{10} R(k)$ sequence. These two probability distribution functions are what we mean by the recurrence time statistics. The reason that we also work on $\log_{10} R(k)$ is that the largest $R(k)$ computed from a genomic DNA sequence can be extremely long, hence, it may be difficult to visualize the distribution for $R(k)$ in linear scale.

Let us take S_1 as an example. If $w = 3$, then $n = 16$, and its recurrence time series $T(i)$ is:

$$7, 7, -1, 1, 1, 10, -1, -1, -1, -1, -1, 1, -1, -1, -1, -1$$

Discarding all the -1 terms from the $T(i)$ sequence, we then get the following recurrence time $R(i)$ series:

$$7, 7, 1, 1, 10, 1$$

where $m = 6$. The motivation for introducing the above definition is that the recurrence time sequence $T(i)$, $i = 1, \cdots, n$, for a DNA sequence and a completely random sequence will be very different, and that by exploiting this difference, we would be able to exhaustively identify most of the interesting features contained in a DNA sequence.

2.3 Recurrence Time Statistics for Completely Random (Pseudo-DNA) Sequences

To find the difference between a DNA sequence and a completely random sequence in terms of the recurrence times, we study a completely random sequence first. We have the following interesting theorem.

Theorem: Given a sequence of independent words W_i, $i = 1, \cdots, n$, where there are a total of K distinct words, each occurring with probability $p = 1/K$, the probability that the recurrence time $T(i)$ being $T \geq 1$ is given by

$$P\{T(i) = T\} \propto [n - T] \cdot p \cdot [1 - p]^{(T-1)} \qquad (1 \leq T < n). \qquad (1)$$

Proof: It suffices to note that the probability for an arbitrary word W_i, where i is from the positions 1 to $n - T$, to repeat exactly after $T \geq 1$ positions is given by the geometrical distribution, $p \cdot [1 - p]^{(T-1)}$. Since there are a total of $n - T$ such positions or words, while each position along the sequence from 1 to $n - T$ has the same role, the total probability is then proportional to the summation of $n - T$ terms of $p \cdot [1 - p]^{(T-1)}$. This completes the the proof.

If we assume the four chemical bases A, C, T and G to occur completely randomly along a (pseudo) DNA sequence, then there are a total of 4^w words of length w, each occurring with probability $p = 4^{-w}$. Hence, the probability for a word to repeat exactly after $T \geq w$ locations is given by Eq. (1), while the distribution for the log-recurrence time $\log_{10} R(k)$ is given by

$$f(t) = C \cdot T \cdot [n - T] \cdot p \cdot [1 - p]^{(T-1)}, \qquad (0 \le t < \log_{10} n), \qquad (2)$$

where $T = 10^t$, and C is a normalization constant. To prove Eq. (2), it suffices to note that $p(T)dT = f(t)dt$.

2.4 Recurrence Time Statistics for DNA Sequences

(A) Recurrence time statistics and a novel codon index

In Fig. 1, we have plotted the probability density functions (pdfs) of log recurrence time, i.e., $\log_{10} R(i)$, for the DNA sequence data from the four species. The red curves in Fig. 1 are computed according to Eq. (2) and represent those of completely random sequences with their length and the word size chosen to analyze them the same as those of the DNA sequences. The word sizes used are 12, 15, 16, 15 for Fig. 1 (a) to Fig. 1 (d) respectively. We observe two interesting features: (i) the pdfs for the genome sequences are very different from those for the random sequences, as signaled by the many sharp peaks in the curves of the pdfs for the genome sequences; (ii) The degree of this difference varies vastly among the four genomes studied. In fact, Eq. (2) describes the background distribution for the $\log_{10} R(i)$ sequence for *E. coli* fairly well, but very poorly describes the same distribution for the chromosome 16 of the human. This suggests that the longer a genome sequence has evolved, the more it deviates from the completely random sequence. Each sharp peak in Fig. 1 may actually represents many sharp peaks if we plot the pdf for the $R(i)$ sequence instead of that for the $\log_{10} R(i)$ sequence. This is because with logarithmic scale, a whole interval of $R(i)$ will be lumped together. It is important to emphasize that all the sharp peaks indicate distinct features of a genome sequence. To better understand this, let us take an example. A sequence of $(A)_l$, which represents a consecutive sequence of A's of length l, contributes to a peak at $R = 1$, if l is larger than the word length w. In fact, when $l > w$, $(A)_l$ contributes a total of $l - w$ counts to $R = 1$. Other single base repeats similarly contribute to $R = 1$. As another example, we note that a sequence such as $(AC)_l$ contributes to $R = 2$ a total of $2l - w$ counts.

We are now ready to propose a novel recurrence time based codon index. This index is based on the period-3 feature. To appreciate the idea, in Fig. 2, we have shown the probability distributions for the recurrence times not greater than 40, for the genome sequences of four species, E.Coli, Yeast, C. elegans, and the Human. The black curves are for the coding regions. The red curves in Fig. 2(b-d) are for the non-coding regions. Due to the low percentage of non-coding regions in the E.Coli genome, such a curve is not computed. We observe that the black curves all have very well defined peaks at recurrence times of 3, 6, 9, etc. Also note that the black curves are very similar among the four different species. Such period-3 feature can be conveniently used to define a codon index, which we shall denote by RT_{p3}:

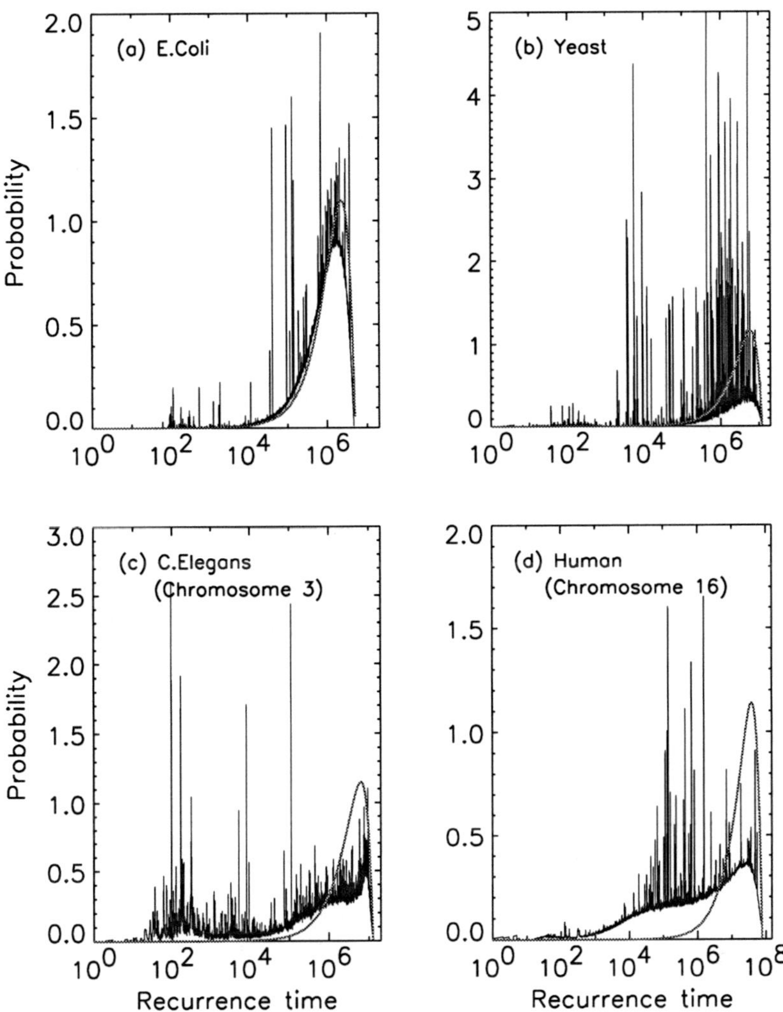

Fig. 1. The probability density function (pdf) for the recurrence time $\log_{10} R(i)$ sequence computed from the DNA sequence of (a) *E. coli*, (b) the yeast *S. cervisivae*, (c) chromosome 3 of the nematode worm *C. elegans*, and (d) chromosome 16 of the human. Red curves are computed from Eq. (2) and represent the situation where the four bases A, C, T, and G occur completely randomly with equal probability.

328   Jian-Bo Gao, Yinhe Cao, and Wen-wen Tung

$$RT_{p3} = \sum_{i=1}^{m} [2p(3i) - p(3i+1) - p(3i+2)] \tag{3}$$

where $p(i)$ is the probability for the recurrence time $T = i$ calculated for a coding or non-coding sequence, n is the number of bases of the coding/non-coding sequence, and m is a cutoff parameter typically chosen not to be larger than 20 so that very short sequences can be studied.

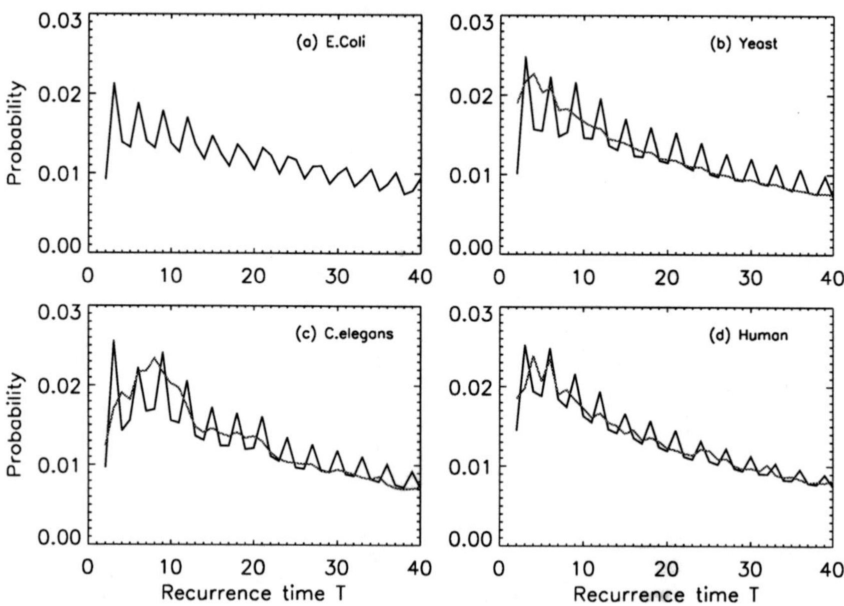

Fig. 2. The probability distribution curves computed from the genomes of four organisms studied. The red and black curves are for non-coding and coding sequences, respectively. A window size of $w = 3$ was used.

Before we move to evaluate the efficiency of RT_{p3} as a codon index, let us focus on how we can exhaustively find all the repeat-related structures by tracing the peaks in Fig. 1 back to the DNA sequence. This can be easily done.

(B) Computation of exact repeat elements from recurrence times

Let $T(i)$ be the recurrence time for position i of a DNA sequence S, where $i = 1, 2, \cdots, n$. For each particular value $T(1 \leq T < n)$ of the recurrence time, we build a list of indices $L(T : S)$ by linearly scanning $T(i)$ from $i = 1$ to $i = n$ and adding i to $L(T : S)$ whenever $T(i) = T$. Denote the index set $L(T : S)$ by

$$L(T : S) = \{i_1, i_2, \cdots, i_C\},$$

where $T(i_k) = T$, $k = 1, 2, \cdots, C$, $i_k < i_{k+1}$ for $k = 1, 2, \cdots, C-1$, and C is the count of the occurrence of T in the recurrence time series $T(i)$. If we take S_1 as an example, then

$$L(1 : S_1) = \{4, 5, 12\}, \ L(7 : S_1) = \{1, 2\}, \ L(10 : S_1) = \{6\}.$$

When the count C is larger than 1, we define the gap between two consecutive indices of $L(T : S)$ as:

$$g_k = i_{k+1} - i_k \quad (k = 1, 2, \ldots, C-1)$$

Let $g^* = w$. What happens when all $g_k \leq g^*$? In this situation, the sequence segment $S_r = S[i_1 \to i_C - i_1 + w - 1]$ is an exact repeat with period T. To see this, we note that for each term i_k in $L(T : S)$, we have $W_{i_k} = W_{i_k + T}$, or more explicitly, $b_{i_k + j} = b_{i_k + T + j}$ for $j = 0, 1, \cdots, w - 1$, and $k = 1, 2, \ldots, C$. Hence, we can concatenate b_{i_2} at $b_{i_1 + g_1}$ to combine the repeats starting from b_{i_1} and b_{i_2}. More concretely, for $k = 1$, we have $b_{i_1 + j} = b_{i_1 + T + j}$, $j = 0, 1, \cdots, w - 1$; similarly, for $k = 2$, we have $b_{i_2 + j} = b_{i_2 + T + j}$, $j = 0, 1, \cdots, w - 1$. Noting $g_1 \leq w$ means $i_2 - i_1 \leq w$, or $i_2 \leq i_1 + w$, we have $b_{i_1 + j} = b_{i_1 + T + j}$ for $j = 0, 1, \cdots, i_2 - i_1 + w - 1$. Continuing this procedure till $k = C$, we have $b_{i_1 + j} = b_{i_1 + j + T}$ for $j = 0, 1, \cdots, i_C - i_1 + w - 1$. Hence, the sequence $S_r = S[i_1 \to i_C - i_1 + w - 1]$ is an exact repeat with period T with its length $l = i_C - i_1 + w$. If $T < l$, S_r is then a tandem repeat.

In general, some gaps g_k may be larger than $g^* = w$. When there are P such gaps, we can decompose $L(T : S)$ into $P + 1$ subsets, such that within each subset all of the gaps are not larger than g^*. Then following the procedure detailed in the last paragraph, we see that each subset represents an exact repeat of period T. Taking S_1 as an example again, then from $L(7 : S_1) = \{1, 2\}$ we get an exact repeat ACGA of period 7, and from $L(1 : S_1) = \{4, 5, 12\}$, we get two exact repeats of period 1: the first one is AAAAA which is a simple sequence repeat, the other is TTTT, which is also a simple sequence repeat.

Before we move on to evaluating the efficiency of RT_{p3} as a codon index, we emphasize that the procedure outlined here makes the recurrence time method largely independent of the word size w, which implies that any feature with length longer than w can be re-combined. For simplicity, we shall call this a **recombination algorithm**. This algorithm, together with the features related to mutations, deletions, and insertions, which are to be discussed shortly, makes the recurrence time based method especially convenient for identifying horizontally transfered genes.

(C) Single Nucleotide Mutation and Single Nucleotide Polymorphism (SNP)

Suppose we have two exactly repeating sequence segments, S_{lead} and S_{lag}, where the subscript lead and lag mean S_{lead} appears earlier than S_{lag} in a genome. In the simplest case, each word constructed from segments of S_{lead}

has the same period T. In general, however, a few words constructed from segments of S_{lead} may have smaller recurrence times, due to the possibility that those words may find their copies in between S_{lead} and S_{lag}. Now suppose one nucleotide somewhere within S_{lead} is mutated. Since the mutated nucleotide appears in a consecutive w words, each of length w, we see that for those words, the period T will have to be different than T. This means if we plot out the recurrence time vs. the sequence position curve, then we should observe a gap of length w in an otherwise almost constant (T) curve. When this is the case, we can suspect that there may be a single nucleotide mutation at the end of the gap. If the gap corresponds to recurrence times larger than T or equal to -1 (meaning no repeats), then we can conclude that there is a single nucleotide mutation at the end of the gap. This is actually a sufficient condition, since it excludes the possibility that a few words may have copies in between S_{lead} and S_{lag}.

The study of single nucleotide mutation is most relevant to the study of Single Nucleotide Polymorphism (SNP), where DNA sequence variations occur when a single nucleotide (A, T, C, or G) in the genome sequence among different populations is changed, possibly due to evolution. It is clear that if we concatenate two genome sequences for different subjects together, then we can treat SNP as a special type of single nucleotide mutation.

(D) Insertion/Deletion and relations between repeat sequences of different periods

Suppose we have a sequence segment starting from the position i_a. What happens if we insert a sequence of length, say, a few thousand bases, in the middle of that segment, then let the segment with insertion to repeat somewhere in the genome? Equivalently, the original sequence segment can be considered as a result of deletion from the longer (i.e., with insertion) sequence segment. This interesting situation is revealed by a jump in the recurrence time vs. sequence position plot, with the height of the jump being the length of the insertion sequence, as we explain below.

Let $T(i)$ denote the recurrence time for the word at the position i of the sequence S. Suppose $i_a < i_b$, $\quad T(i_a) = T(i_{a+1}) = \cdots = T(i_b) = T_1 > 0$, and

$i_b < i_c \leq i_b + w < i_d, \quad T(i_c) = T(i_{c+1}) = \cdots = T(i_d) = T_2 > T_1.$
Then the sequence segment

$$S_a = S[i_a \to (i_d + w - 1)]$$

can be considered the result of deleting the sequence

$$S_{deletion} = S[(i_c + T_1 - 1) \to (i_c + T_2 - 1)]$$

from the sequence segment

$$S_b = S[(i_a + T_1) \to (i_d + w - 1 + T_2)].$$

Equivalently, S_b is the result of inserting the sequence $S_{deletion}$ into S_a right before the position i_c. Note that the condition of $i_c \leq i_b + w$ comes from the fact that the boundary always affects a consecutive w words, each of length w. When the first w bases of the deleted sequence segment do not have their copies at the positions starting from i_c, we have $i_c = i_b + w$. Otherwise, we have in-equalities.

3 Results and Discussion

In this section, we present examples of repeat-related structures extracted by the proposed method and evaluate the efficiency of the codon index.

3.1 Extraction of Repeat-Related Structures

We now present examples of structures which can be found by tracing the peaks in Fig. 1 back to the genome sequences. These structures include insertion sequences (ISs), rRNA clusters, repeat genes, simple sequence repeats (SSRs), transposons, and gene and genome segmental duplications such as inter-chromosomal duplication. We shall illustrate most of these structures using the yeast *S. cervisivae* as an example.

We first study SSRs. SSRs are perfect or slightly imperfect tandem repeats of particular k-mers. They have been extremely important in human genetic studies, because they show a high degree of length polymorphism in human population owing to frequent slippage by DNA polymerase during replication [2]. Any tandem repeat of k-mers, disregarding its exact content, will contribute to the count of occurrence of period $T = k$ in recurrence time statistics, hence can be easily found by following the peak of $T = k$ in Fig. 1. As an example, we note that there are 39 sequence segments contributing to $k = 13$. Three of them are CCACACCCACACA, GGTGTGTGGGTGT, and TACCGACGAGGCT. Note that by Fig. 1(a) we can conclude that E . *coli* has very few SSRs.

One of the more striking features of the yeast *S. cervisivae* genome is that it contains many copies of transposon yeast (**Ty**) elements. Each **Ty** element is about 6.3 kb long, with the last 330 bp at each end constituting direct repeats, called δ. Those direct end repeats are responsible for the peaks around 5500 in Fig. 1(b), which enable us to find all of those **Ty** elements on both strands of the genome. As two examples, we mention that the transposon Ty3-1 on the Watson strand of chromosome 7 starts at the position 707196 and ends at 712546, and has a period of 5011. The transposon Ty1-1 on the Crick strand of chromosome 1 starts at the position 166162 and ends at 160238, and has a period of 5588.

Gene duplication is an important source of evolutionary novelty. Many duplicate genes have been found in the yeast *S. cerevisiae* genome, and they often seem to be phenotypically redundant [49, 50, 51]. Any gene duplication will contribute to one of the sharp peaks in Fig. 1(b). As an example, we note that a gene (standard name MCH2, systematic name YKL221W), which is on chromosome 6 starting from the position 6931, is repeated on chromosome 13, starting from the position 7749.

Genome segmental duplications consist of large blocks that have been copied from one region of the genome to another. They have been found among genomes of many species including the yeast *S. cervisivae* [49], and the *Homo sapiens* [2, 52]. In fact, they contribute to some of the the sharpest peaks in Fig. 1. An example of such segmental duplications is the inter-chromosomal duplication corresponding to the peak at $T = 5150433$ in Fig. 1(b).

3.2 Evaluation of the Novel Codon Index

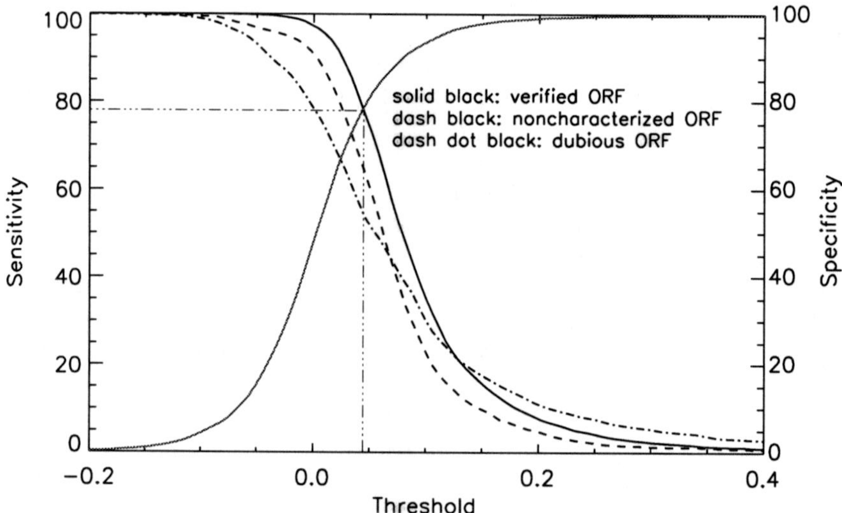

Fig. 3. The specificity and sensitivity curves for the RT_{p3} index evaluated on all of the 16 yeast chromosomes.

In order to evaluate the effectiveness of the RT_{p3} as a codon index, we study all of the 16 yeast chromosomes. Our sample pool is comprised of two sets of DNA segments: the coding set (fully coding regions or exons), which contains 4125 verified ORFs, 1626 uncharacterized ORFs, and 812 dubious ORFs, and the non-coding set, which contains 5993 segments (fully non-coding regions or introns). Some of these coding and non-coding segments are very short. Regardless of their length, each segment is counted as one when calculating the sensitivity and specificity curves. Fig. 3 shows the specificity and sensitivity curves for all of the 16 yeast chromosomes, where the red curve is the cumulative distribution function for RT_{p3} for the non-coding regions, and the black curves are the complementary cumulative distribution function for the coding regions. For clarity, we have computed such distributions for verified ORFs, uncharacterized ORFs, and dubious ORFs, separately. To understand the meaning of such curves, let us focus on the intersection of the solid black curve and the red curve. When we chose RT_{p3_0} as a threshold value, then with 78% probability a coding sequence is characterized as coding sequence, while with 78% probability a non-coding sequence is also taken as a non-coding sequence. As expected, this percentage is lower for uncharacterized and dubious ORFs. It is interesting to note that the percentage of accuracy calculated on Human genomes is around 74%, close to 78%. Because of this (see also Fig. 2), we conclude that the method is largely species-independent.

It is interesting to note that the period-3 feature is often quantified by performing the Fourier spectral analysis on fairly long DNA sequences. In order to make such analysis applicable to sequences as short as 162 bases, recently a lengthen-shuffle algorithm is proposed [11]. Fourier spectral analysis together with the lengthen-shuffle algorithm gives about 69% of sensitivity and specificity when evaluated on a prokaryote genome, the V.*cholerae* chromosome I, and about 61% when evaluated on eukaryotic genomes [12]. It is clear that the RT_{p3} index is more accurate. Other features of the recurrence time based method are: (i) DNA sequences as short as 40 bases can be very well studied. Noting that an expressed sequence tag (EST) is usually very short and that little may be known about the genome to which the EST belongs, this feature, together with the species-independent one, makes the method particularly useful for determining whether a suspected EST belongs to a coding or non-coding region. (ii) The method directly works on the DNA sequence. In contrast, numerical sequences have to be obtained by certain mapping rules in order to use the Fourier spectral analysis based methods.

3.3 Discussion

In this chapter, we have proposed a simple recurrence time based method for DNA sequence analysis, and shown that the method can conveniently exhaust all repeat-related structures of length greater than an arbitrarily chosen small word of size w in a genome. We have also shown that the method is very convenient for the study of mutations, insertions and deletions, hence, it holds great potential for the study of evolutionary variations across species and the mechanisms underlying it. By characterizing the peaks at multiples of 3, we have defined a very efficient codon index which is largely species independent and works well on very short sequences. We emphasize that one of the more appealing features of RT_{p3} as a codon index is that no priori knowledge about the sequence is used. Hence, the method will be especially convenient for the study of genome sequences that very little is known. This is the case, for example, when a genome sequence is to be sequenced by a few small research groups by studying expressed sequence tags (ESTs).

While the accuracy of 78% for the yeast genome is already satisfactory, we note that it is possible to improve this percentage by designing other indices from the recurrence times. Readers interested in this issue are encouraged to contact the authors for the raw recurrence time data.

4 Acknowledgments

The Authors have benefited considerably by participating workshops organized by the Institute of Pure and Applied Mathematics at UCLA.

References

1. F.S. Collins, E.D. Green, A.E. Guttmacher, and M.S. Guyer. A vision for the future of genomics research. *Nature* 422 (6934): 835-847, 2003.
2. International Human Genome Sequencing Consortium, Initial sequencing and analysis of the human genome *Nature* 409: 860-921, 2001.
3. J. Jurka. Repeats in genomic DNA: mining and meaning. *Current Opinion in Structural Biology*, 8: 333-337, 1998.
4. R. Guigo. DNA Composition, Codon Usage and Exon Prediction. In M.J. Bishop, editor, *Genetics Databases*, pages 53-80. Academic Press, San Diego, CA, 1999.
5. H. Herzel, D. Weiss, and E.N. Trifonov. 10-11 bp periodicities in complete genomes reflect protein structure and DNA folding. *Bioinformatics*, 15 (3): 187-193, 1999.
6. A. Fukushima, T. Ikemura, M. Kinouchi, et al. Periodicity in prokaryotic and eukaryotic genomes identified by power spectrum analysis. *Gene*, 300 (1-2): 203-211, 2002.
7. J.L. Bennetzen and B.D. Hall. Codon selection in yeast. *Journal of Biological Chemistry*, 257: 3026-3031, 1982.
8. P.M. Sharp and W.-H. Li. The codon Adaptation Index–a measure of directional synonymous codon usage bias, and its potential applications. *Nucleic Acids Research*, 15: 1281-1295, 1987.
9. R. Jansen, H.J. Bussemaker, and M. Gerstein. Revisiting the codon adaptation index from a whole-genome perspective: analyzing the relationship between gene expression and codon occurrence in yeast using a variety of models. *Nucleic Acids Research*, 31: 2242-2251, 2003.
10. S. Tiwari, S. Ramachandran, A. Bhattacharya, S. Bhattacharya, and R. Ramaswamy. Prediction of probable genes by Fourier analysis of genomic sequences. *Computer Applications in the Biosciences*, 13: 263-270, 1997.
11. M. Yan, Z.S. Lin, and C.T. Zhang. A new Fourier transform approach for protein coding measure based on the format of the Z curve. *Bioinformatics*, 14: 685-690, 1998.
12. B. Issac, H. Singh, and H. Kaur. Locating probable genes using Fourier transform approach. *Bioinformatics*, 18: 196-197, 2002.
13. D. Kotlar and Y. Lavner. Gene prediction by spectral rotation measure: A new method for identifying protein-coding regions. *Genome Research*, 13: 1930-1937, 2003.
14. C.T. Zhang and J. Wang. Recognition of protein coding genes in the yeast genome at better than 95% accuracy based an the Z curve. *Nucleic Acids Research*, 28: 2804-2814, 2000.
15. M. Snyder and M. Gerstein. Genomics - Defining genes in the genomics era. *Science*, 300: 258-260, 2003.
16. J.W. Fickett and R. Guig'o. Computational gene identification In S. Swindell, R. Miller, and G. Myers, editors, *Internet for the Molecular Biologist*, pages 73-100. Horizon Scientific Press, Wymondham, UK, 1996.
17. M.Q. Zhang. Computational prediction of eukaryotic protein-coding genes. *Nature Reviews Genetics*, 3: 698-709, 2002.
18. S.B. Needleman and C. Wunsch. A general method applicable to the search for similarities in the amino acid sequence of two proteins. *Journal of Molecular Biology*, 48: 443-453, 1970.

19. T.F. Smith and M.S. Waterman. Identification of common molecular subsequences. *Journal of Molecular Biology*, 147: 195-197, 1981.

20. W.M. Fitch and T.F. Smith. Optimal sequence alignments. *Proceedings of the National Academy of Sciences*, 80: 1382-1386, 1983.

21. S.F. Altschul and B.W. Erickson. Optimal sequence alignment using affine gap costs. *Bulletin of Mathematical Biology*, 48: 603-616, 1986.

22. W.R. Pearson. Comparison of methods for searching protein sequence databases. *Protein Science*, 4: 1145-1160, 1995.

23. A.L. Delcher, S. Kasif, R.D. Fleischmann, J. Peterson, O. White, and S.L. Salzberg. Alignment of whole genomes. *Nucleic Acids Research*, 27: 2369-2376, 1999.

24. A.L. Delcher, A. Phillippy, J. Carlton, and S.L. Salzberg. Fast algorithems for large-scale genome alignment and comparison. *Nucleic Acids Research*, 30: 2478-2483, 2002.

25. S. Henikoff and J.G. Henikoff. Performance evaluation of amino acid substitution matrices. *Proteins*, 17: 49-61, 1993.

26. J. Jurka, P. Klonowski, V. Dagman, and P. Pelton. CENSOR-A program for identification and elimination of repetitive elements from DNA sequences. *Computers and Chemistry*, 20: 119-122, 1996.

27. A.F.A. Smit. The origin of interspersed repeats in the human genome. *Current Opinion in Genetics & Development*, 6: 743-748, 1996.

28. D.J. Lipman and W.R. Pearson. Rapid and sensitive protein similarity searches. *Science*, 227: 1435-1441, 1985.

29. W.R. Pearson and D.J. Lipman. Improved tools for biological sequence comparison. *Proceedings of the National Academy of Sciences*, 85: 2444-2448, 1988.

30. S.F. Altschul, W. Gish, W. Miller, E.W. Myers, and D.J. Lipman. Basic local alignment search tool. *Journal of Molecular Biology*, 215: 403-410, 1990.

31. S.F. Altschul, M.S. Boguski, W. Gish, and J.C. Wootton. Issues in searching molecular sequence databases. *Nature Genetics* 6 119-129, 1994.

32. S.F. Altschul, T.L. Madden, A.A. Schäffer, J. Zhang, Z. Zhang, W. Miller, and D.J. Lipman. Gapped BLAST and PSI-BLAST: a new generation of protein database search programs. *Nucleic Acids Research*, 25: 3389-3402, 1997.

33. A.A. Schäffer, L. Aravind, T. L. Madden, S. Shavirin, J. L. Spouge, Y. I. Wolf, E. V. Koonin, and S. F. Altschul. Improving the accuracy of PSI-BLAST protein database searches with composition-based statistics and other refinements. *Nucleic Acids Research*, 29: 2994-3005, 2001.

34. R.A. Lippert, H.Y. Huang, and M.S. Waterman. Distributional regimes for the number of k-word matches between two random sequences. *Proceedings of the National Academy of Sciences*, 99: 13980-13989, 2002.

35. S. Karlin and S.F. Altschul. Methods for assessing the statistical significance of molecular sequence features by using general scoring schemes. *Proceedings of the National Academy of Sciences*, 87: 2264-2268, 1990.

36. M.S. Waterman and M. Vingron. Rapid and accurate estimates of statistical significance for sequence database searches. *Proceedings of the National Academy of Sciences*, 91: 4625-4628, 1994.

37. M.S. Waterman and M. Vingron. Sequence comparison significance and Poisson approximation. *Statistical Science*, 9: 367-381, 1994.

38. T.F. Smith, M.S. Waterman, and C. Burks. The statistical distribution of nucleic acid similarities. *Nucleic Acids Research*, 13: 645-656, 1985.

39. S.F. Altschul and W. Gish. Local alignment statistics. *Methods in Enzymology*, 266: 460-480, 1996.

40. J.G. Reich, H. Drabsch, and A. Daumler. On the statistical assessment of similarities in DNA sequences. *Nucleic Acids Research*, 12: 5529-5543, 1984.

41. F. Takens. Detecting strange attractors in turbulence. In D.A. Rand and L.S. Young, editors, *Dynamical Systems and Turbulence, Lecture Notes in Mathematics*, Vol. 898, p. 366, Springer-Verlag, Berlin, 1981.

42. F.R. Blattner, G. Plunkett III, C.A. Bloch, N.T. Perna, V. Burland, M. Riley, J. Collado-Vides, J.D. Glasner, C.K. Rode, G.F. Mayhew, J. Gregor, N.W. Davis, H.A. Kirkpatrick, M.A. Goeden, D.J. Rose, B. Mau, and Y. Shao. The complete genome sequence of Escherichia coli K-12. *Science*, 277: 1453-1474 (1997).

43. H.W. Mewes, K. Albermann, M. Bhr, D. Frishman, A. Gleissner, J. Hani, K. Heumann, K. Kleine, A. Maierl, S.G. Oliver, F. Pfeiffer, and A. Zollner. Overview of the yeast genome. *Nature*, 387: 7-8 (1997).

44. The C. elegans Sequencing Consortium, Genome Sequence of the Nematode Caenorhabditis elegans-A Platform for Investigating Biology. *Science*, 282: 2012-2018, 1998.

45. The Celera Genomics Sequencing Team, The sequence of the human genome. *Science*, 291: 1304-1351, 2001.

46. J.B. Gao. Recurrence Time Statistics for Chaotic Systems and Their Applicaitons. *Physical Review Letters*, 83: 3178-3181, 1999.

47. J.B. Gao and H.Q. Cai. On the structures and quantification of recurrence plots. *Physics Letters A*, 270: 75-87, 2000.

48. J.B. Gao. Detecting nonstationarity and state transitions in a time series. *Physical Review E*, 63, 066202, 2001.

49. K.H. Wolfe and D.C.Shields. Molecular evidence for an ancient duplication of the entire yeast genome. *Nature*, 387: 708-13, 1997.

50. C. Seoighe and K.H. Wolfe. Updated map of duplicated regions in the yeast genome. *Gene*, 1: 253-261, 1999.

51. G. Glaever, et al. Functional profiling of the *Saccharomyces cerevisiae* genome. *Nature*, 418: 387-391, 2002.

52. J. Brendan, et al. Genome duplications and other features in 12 Mb of DNA sequence from human chromosome 16p and 16q. *Genomics*, 60: 295-308, 1999.

53. R.D. Kornberg and Y. Lorch. Twenty-five years of nucleosome, fundamental particle of the eukaryote chromosome. *Cell*, 98: 285-294, 1999.

54. K. Luger, A.W. Mader, R.K. Richmond, et al. Crystal structure of the nucleosome core particle at 2.8 Å resolution. *Nature*, 389: 251-260, 1997.

55. A. Stein and M. Bina. A Signal encoded in vertebrate DNA that influences nucleosome positioning and alignment. *Nucleic Acids Research*, 27: 848-853, 1999.

56. M.A. El Hassan and C.R. Calladine. Two Distinct Modes of Protein-induced Bending. *Journal of Molecular Biology*, 282: 331-343, 1998.

Clustering Proteomics Data Using Bayesian Principal Component Analysis

Halima Bensmail[1], O. John Semmes[2], and Abdelali Haoudi[3]

[1] University of Tennessee
Statistics Department
Knoxville, TN 37996-0532
bensmail@utk.edu
[2] Eastern Virginia Medical School
Department of Microbiology and Molecular Cell Biology
Norfolk, VA 23507 semmesoj@evms.edu
[3] Eastern Virginia Medical School
Department of Microbiology and Molecular Cell Biology
Norfolk, VA 23507
haoudia@evms.edu

Summary. Bioinformatics clustering tools are useful at all levels of proteomic data analysis. Proteomics studies can provide a wealth of information and rapidly generate large quantities of data from the analysis of biological specimens from healthy and diseased individuals. The high dimensionality of data generated from these studies requires the development of improved bioinformatics tools for efficient and accurate data analysis. For proteome profiling of a particular system or organism, specialized software tools are necessary. However, there have not been significant advances in the informatics and software tools necessary to support the analysis and management of the massive amounts of data generated in the process. Clustering algorithms based on probabilistic and Bayesian models provide an alternative to heuristic algorithms. The number of diseased and non-diseased groups (number of clusters) is reduced to the choice of the number of component of a mixture of underlying probability. Bayesian approach is a tool for including information from the data to the analysis. It offers an estimation of the uncertainties of the data and the parameters involved. We present novel algorithms that cluster and derive meaningful patterns of expression from large scaled proteomics experiments. We processed raw data using principal component analysis to reduce the number of peaks. Bayesian model-based clustering algorithm was then used on the transformed data. The Bayesian model-based approach has shown a superior performance, consistently selecting the correct model and the number of clusters, thus providing a novel approach for accurate diagnosis of the disease.

Key words: Clustering, Principal component analysis, Proteomics, Bayesian analysis

1 An Approach to Proteome Analysis Using SELDI-Time of Flight-Mass Spectrometry

There is a variety of new methods for proteome analysis. Unique ionization techniques, such as electrospray ionization and matrix-assisted laser-desorption ionization (MALDI), have facilitated the characterization of proteins by Mass Spectrometry (MS) [26, 23]. Surface-enhanced laser desorption-ionization time of flight mass spectrometry (SELDI-TOF-MS), originally described in [24] overcomes many of the problems associated with sample preparations inherent with MALDI-MS. The underlying principle in SELDI is surface-enhanced affinity capture through the use of specific probe surfaces or chips. This protein biochip is the counterpart of the array technology in the genomic field and also forms the platform for Ciphergen's ProteinChip array SELDI MS system [34]. A 2-DE analysis separation is not necessary for SELDI analysis because it can bind protein molecules on the basis of its defined chip surfaces. Chips with broad binding properties, including immobilized metal affinity capture, and with biochemically characterized surfaces, such as antibodies and receptors, form the core of SELDI. This MS technology enables both biomarker discovery and protein profiling directly from the sample source without preprocessing. Sample volumes can be scaled down to as low as 0.5 μl, an advantage in cases when sample volume is limiting. Once captured on the SELDI protein biochip array, proteins are detected through the ionization-desorption, TOF-MS process. A retentate (proteins retained on the chip) map is generated in which the individual proteins are displayed as separate peaks on the basis of their mass and charge (m/z). Wright et al. [48] demonstrated the utility of the ProteinChip SELDI-MS in identifying known markers of prostate cancer and in discovering potential markers either over- or underexpressed in prostate cancer cells and body fluids. SELDI analyses of cell lysates prepared from pure populations from microdissected surgical tissue specimens revealed differentially expressed proteins in the cancer cell lysate when compared with healthy cell lysates and with benign prostatic hyperplasia (BPH) and prostate intraepithelial neoplasia cell lysates [48]. SELDI is a method that provides protein profiles or patterns in a short period of time from a small starting sample, suggesting that molecular fingerprints may provide insights into changing protein expression levels according to health status, i.e. healthy, benign, premalignant or malignant lesions. This appears to be the case because distinct SELDI protein profiles for each cell and cancer type evaluated, including prostate, lung, ovarian and breast cancer, have been described recently [1, 12, 30, 39, 38]. In these studies, protein profiling data is generated by SELDI ProteinChip Array technology followed by the analysis utilizing numerous types of software algorithms.

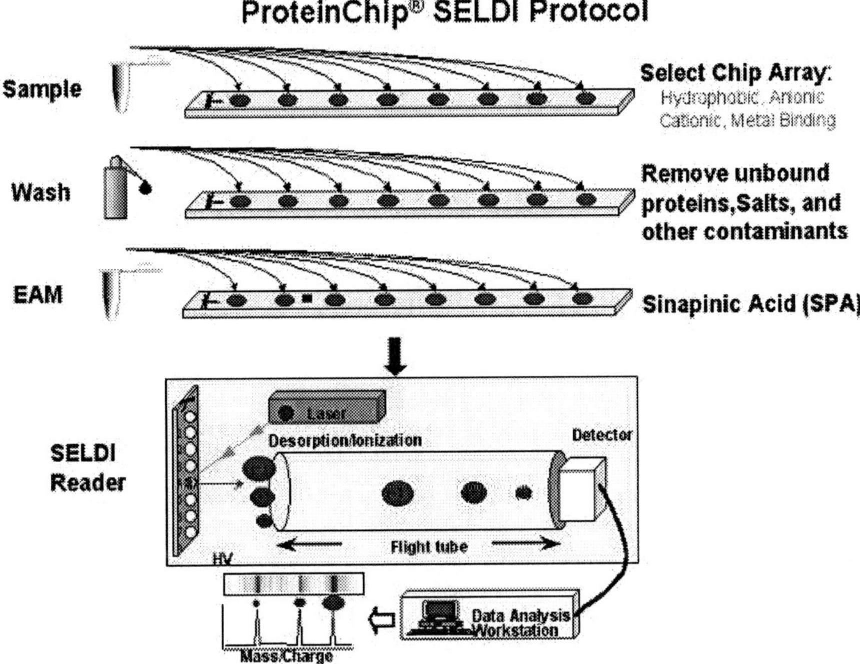

Fig. 1. The proteinChip SELDI-TOF-MS process. sample proteins are applied to the chip array of the desired chemistry. The samples are analyzed by the mass reader and the time of flight recorded. This information is converted to protein mass. The process is highly automated and designed for high throughput analysis.

2 Clustering and Classification Methods for Large Proteomics Datasets

Due to the large array of data that is generated from a single proteomic data analysis, it is essential to implement the algorithms that can detect expression patterns from such large volumes of data correlating to a given biological/pathological phenotype from multiple samples [6]. Under normality assumption, covariance matrices play an important role in statistical analysis including clustering. Particularly, a covariance matrix provides information on the structure of the data to cluster. The geometrical structure of the dataset is expressed through the eigenvalues and eigenvectors of the covariance matrix. Eigenvectors represents the orientation of the data, whereas eigenvalues represents its shape (dispersion). K-means [31] uses a minimum distance criteria to find the clusters which are not supposed to overlap (which is not the case here), and where the number of clusters are given as an input. Self-Organizing maps [27], Neural Networks [8] and other clustering algorithms pay less attention to this structure. K-means puts weight on the cluster mean. Kohonen Mapping

(SOM) uses a topological structure classification using a priori known weight vector to initialize the topology of the clusters. Neural Network, a method borrowed from computer science, uses a black box non-linear transformation to cluster data. All the previous methods do not account for the choice of the number of clusters; it has to be specified a priori. A probabilistic model, particularly, Bayesian model, offers this possibility. It also proposes probabilistic criteria for finding the number of clusters, their geometry, and the uncertainty involved in the calculation.

3 Serum Samples from HTLV-1-Infected Patients

Protein expression profiles generated through SELDI analysis of sera from HTLV-1 (Human T-cell Leukemia virus type 1)-infected individuals were used to determine the changes in the cell proteome that characterize ATL (Adult T-cell leukemia), an aggressive lymphoproliferative disease from HAM/TSP (HTLV-1-Associated Myelopathy/Tropical Spastic Paraparesis), a chronic progressive neurodegenerative disease clinically similar to Multiple Sclerosis. Both diseases are associated with the infection of the CD4+ T-cell population by HTLV-1, which is estimated to infect approximately 20 million people worldwide. The HTLV-1 virally encoded oncoprotein Tax has been implicated in the retrovirus mediated cellular transformation and is believed to contribute to the oncogenic process through induction of genomic instability affecting both DNA repair integrity and cell cycle progression [18, 19]. Serum samples were obtained from the Virginia Prostate Center Tissue and Body Fluid Bank. All samples had been procured from consenting patients according to protocols approved by the Institutional Review Board and stored frozen. None of the samples had been thawed more than twice. Triplicate serum samples ($n = 68$) from healthy or normal ($n_1 = 37$), ATL ($n_2 = 20$) and HAM ($n_3 = 11$) patients were processed. A bioprocessor, which holds 12 chips in place, was used to process 96 samples. Each chip can hold up to 8 samples. Each chip contained two Quality Control (QC) samples consisting of a serum sample from a known cancer patient and a serum sample from a healthy patient. The QC samples were applied to each chip along with the unknown test samples in a random fashion. The QC spots served as quality control for assay and chip variability. The samples were blinded for the technicians who processed the samples. The reproducibility of the SELDI spectra, i.e., mass and intensity from array to array on a single chip (intraassay) and between chips (interassay), was determined with the pooled normal serum QC sample.

4 Principal Component and Model Selection

Principal Component Analysis (PCA) was first developed by Pearson [37] and Hotelling [22]. It is a multivariate procedure to reduce the dimensionality of

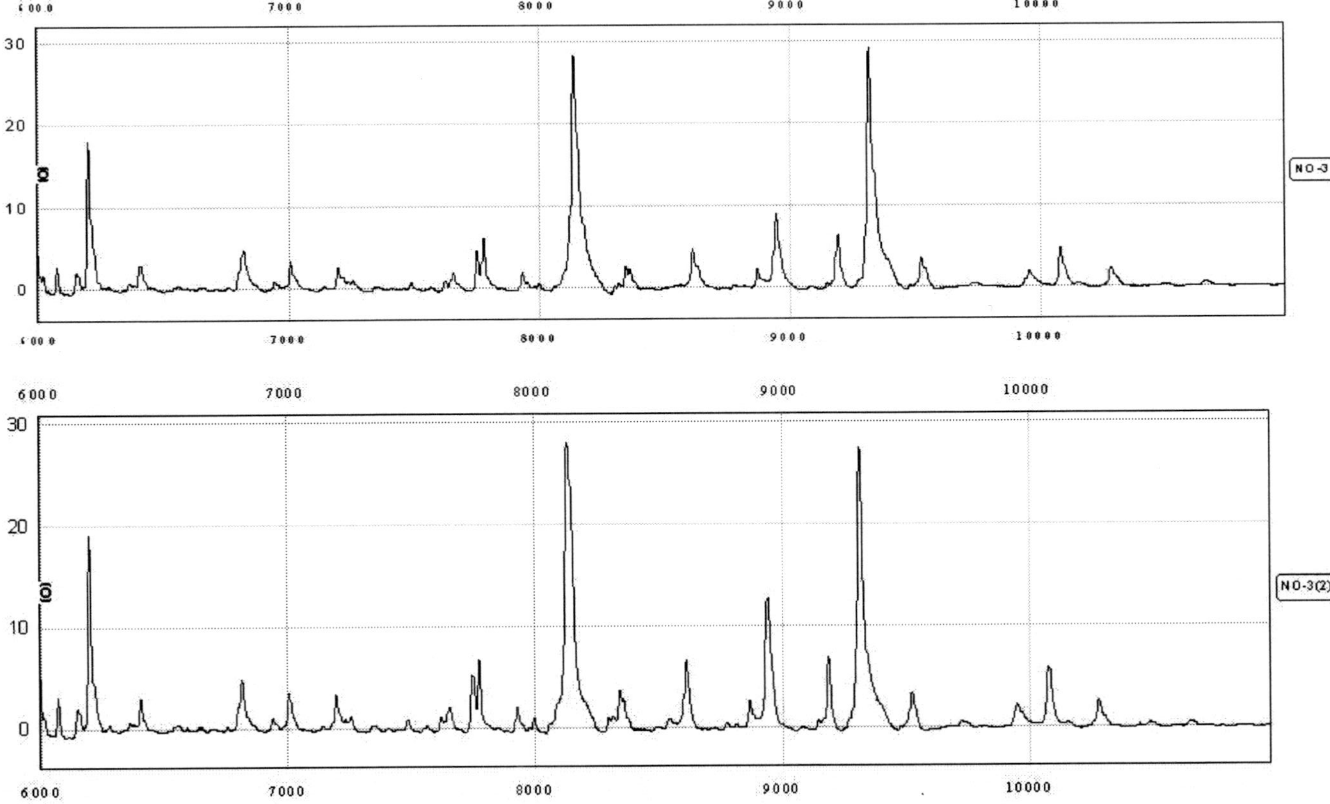

Fig. 2. Reproducibility of a representative spectra from Normal samples

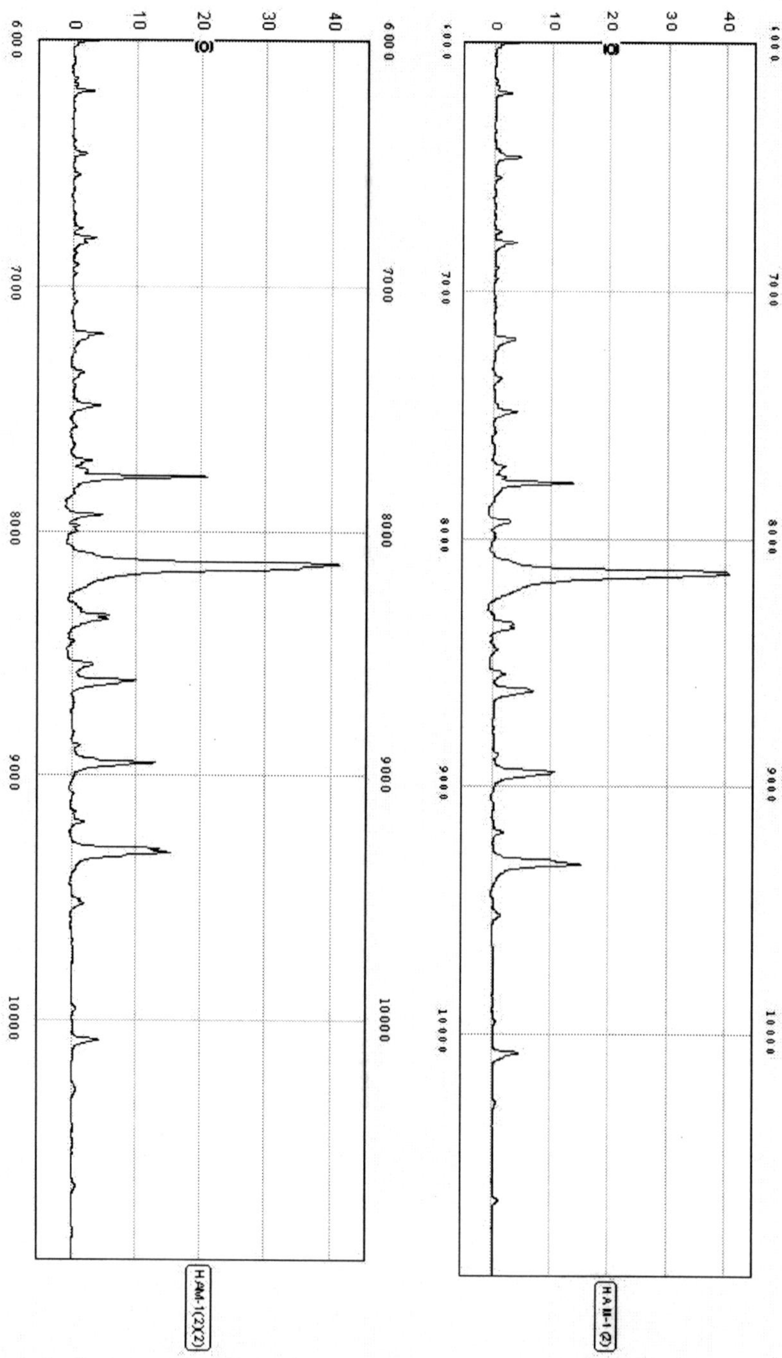

Fig. 3. Reproducibility of a representative spectra from HAM samples

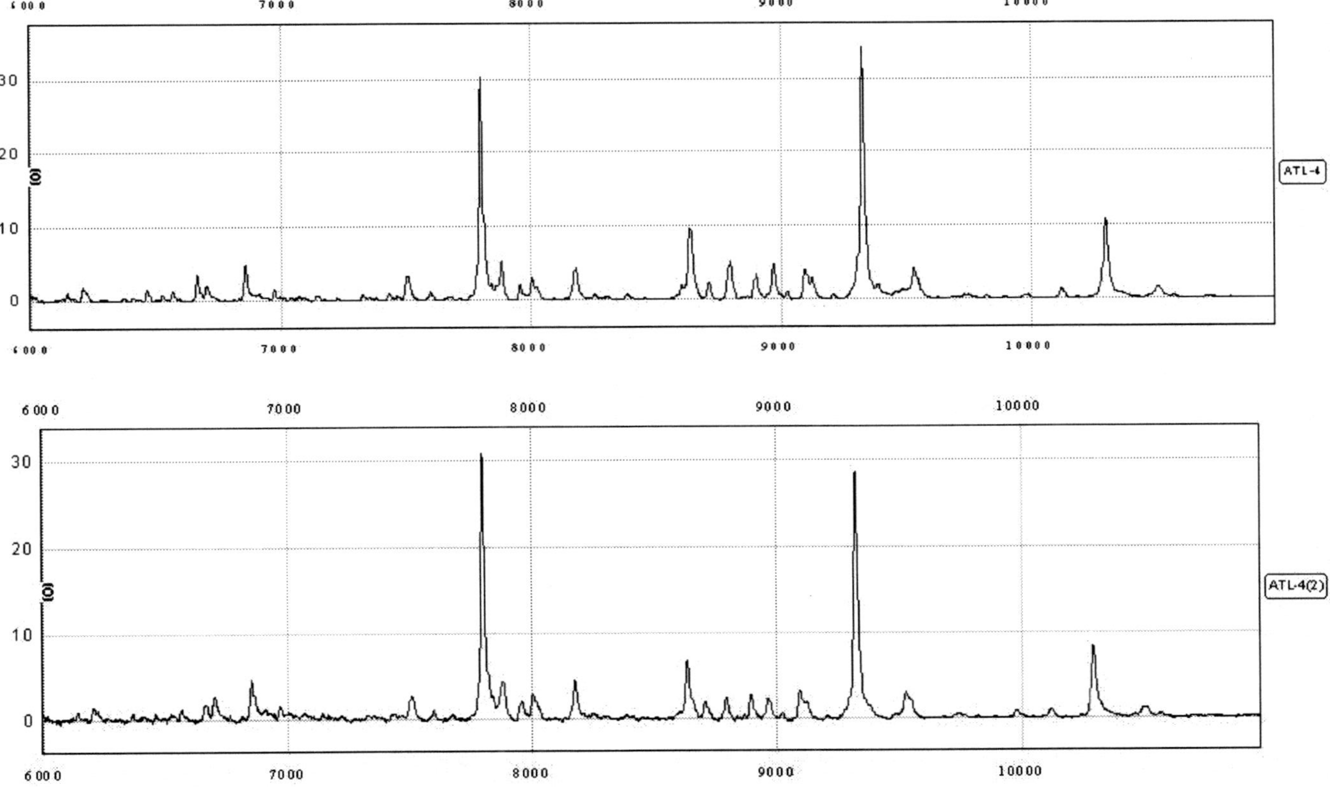

Fig. 4. Reproducibility of a representative spectra for ATL samples

the data. The idea is to transform a large set of correlated variables into a small set of uncorrelated variables which are ordered by reducing variability. The derived uncorrelated variables are linear combinations of the original variables. The uncorrelated variables that have small variability can be removed or deemed as residuals without losing much information of the original data. There are several algorithms to calculate the principal components that can generate the same results from the same dataset. If there is more than one possible transformation for the same maximum variation, then the results may differ. Suppose X is an $n \times m$ matrix with n rows and m columns. It can be decomposed into its principal components as follows:

$$X = t_1 p_1^t + t_2 p_2^t + \ldots\ldots + t_k p_k^t + E \tag{1}$$

where k is the number of components retained in the model with $k <$ $\min(n, m)$, t_i is the score vector, and p_i is the loading vector, which is the eigenvector of the covariance matrix if X $(i = 1, 2, \ldots, k)$. In this expression, X is written as a sum of k products of vectors t_i and p_i and the residual matrix (E). The residual matrix contains all the information of the insignificant components which are not included in the model. The p_i's define the principal component coordinate system and are mutually orthogonal. The t_i's are the scores of the data on that system. The data can be projected onto the principal component space, and these projections are the scores:

$$XP = T \tag{2}$$

where X is an $n \times k$ matrix, P is a $k \times k$ matrix of loading vectors, T is an $n \times k$ score matrix. The pairs (t_i, p_i) are arranged in descending order so that the first pair explains the largest amount of variation of the data, the second pair explains the next to the largest amount of variation and the kth explains the least. Usually only a few components are necessary to be kept in the model which explain most of the variations in the data. Consequently, the PCA can be used to reduce the dimensionality of the data and at the same time retain the most information of the original data. One of the effective methods to compute the PCA is the singular value decomposition (SVD) method. Suppose X is an $n \times m$ matrix with n rows and m columns. It can be decomposed as follows: $X = USV^t$ where U is an $n \times m$ matrix whose columns are orthogonal; S is an $m \times m$ diagonal matrix S with positive or zero elements, which are termed the singular values; V is an $m \times m$ orthogonal matrix. The squares of singular values correspond to the eigenvalues of covariance matrix of X by descending orders. The columns of V are the eigenvectors associated with the eigenvalues. We use the first k eigenvectors of V to define an $m \times k$ matrix P, where $P = [V_1, ..., V_k]$. Using P, we obtain k principal components as $T = XP$, where T is an $n \times m$ matrix, X is an $n \times m$ matrix and P is an $m \times k$ matrix. The transformed PC's are defined as $Z = XB$ with covariance matrix $\Sigma_Z = Cov(Z)$. We can use Bayesian Information Criteria (BIC) [42] to choose the number of principal components. For this information criterion, small values of BIC are preferred. The BIC values calculated by the software R for different number of principal component are given in Figure 5. We can see that the smallest BIC values correspond to two PC's. Therefore, these two PC's will be further used in the clustering algorithm. The scatter plot of

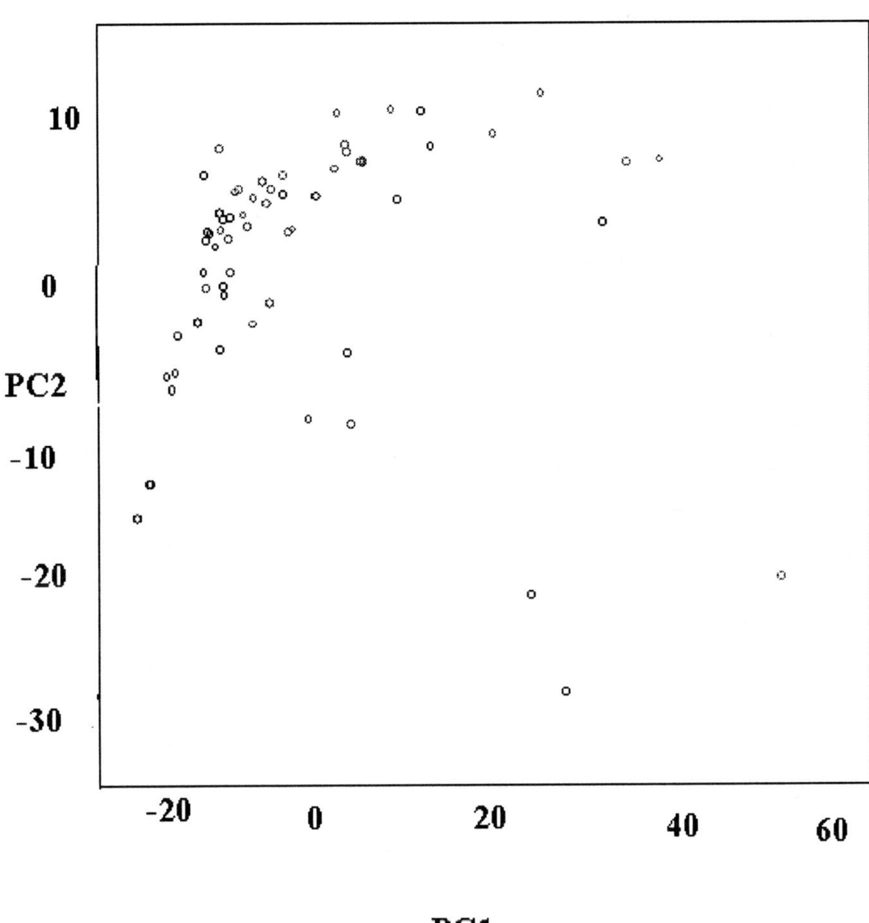

Fig. 5. Projection of all spectral data (Normal, Ham and ATL) in the space spanned by the first two principal component vectors. X-axis is the first principal component. Y-axis is the second principal component.

The information explained by the first two PCs are much higher. However, there is no obvious cut point by looking at the cumulative percentage curve; the first two PCs just explained about 30% of the total variation.

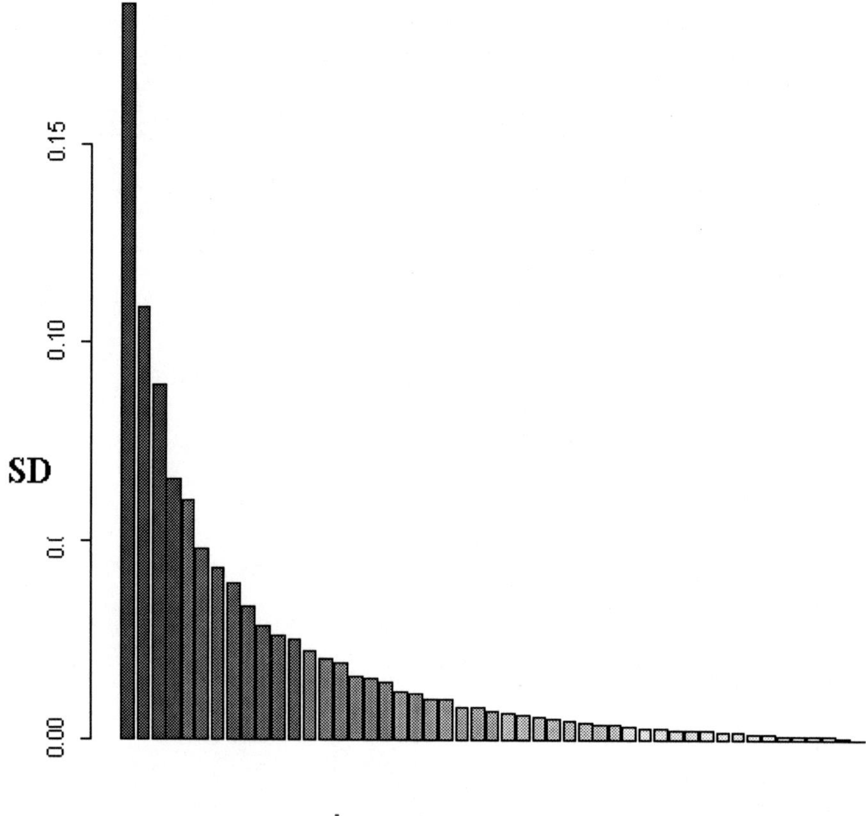

Fig. 6. Level of standard deviation of each Principal Component. SD: standard deviation, PCs: principal components

According to BIC's criterion, the smallest ICOMP value comes from the two-PC model. The two-PC model has the smallest BIC values. The first two PCs will be used for clustering the data.

$$X = T \times P + residual \tag{3}$$

$$= t_1 pc_1 + t_2 pc_2 + residual \tag{4}$$

The transformed data is given in Table 1 that mentions the new coordinates of the first four observations.

5 Bayesian Cluster Analysis

In cluster analysis, we consider the problem of determining the structure of the data with respect to clusters when no information other than the observed

Table 1. Transformed data

Observation	$PC1$	$PC2$
1	-8.49687	1.590751
2	26.40243	12.00707
3	-5.85782	-4.34584
4	768.59	776.42

values is available [20, 17, 28, 31, 47, 43, 9]. Various strategies for simultaneous determination of the number of clusters and the cluster membership have been proposed [15, 11, 10].

Mixture models provide a useful statistical frame of reference for cluster analysis. In the theory of finite mixture models, the data to be classified are viewed as coming from a mixture of probability distributions, each representing a different cluster, so the likelihood is expressed as

$$p(\theta_1, \ldots, \theta_K; \pi_1, \ldots, \pi_K | \mathbf{y}) = \prod_{i=1}^{n} \sum_{k=1}^{K} \pi_k f_k(\mathbf{y}_i | \theta_k) \qquad (5)$$

where π_k is the probability that an observation belongs to the kth component or cluster ($\pi_k \geq 0; \sum_{k=1}^{K} \pi_k = 1$), f_k is the density function of each component distribution, and $\theta_k = (\mu_k, \Sigma_k)$ is the underlying parameter involved.

The Bayesian approach is promising for a variety of mixture models, both Gaussian and non-Gaussian. Here, our approach uses a Bayesian mixture model based on a variant of the standard spectral decomposition of Σ_k, namely

$$\Sigma_k = \lambda_k \mathbf{D}_k \mathbf{A}_k \mathbf{D}_k^t \qquad (6)$$

where λ_k is a scalar, $\mathbf{A}_k = diag(1, a_{k2}, \ldots, a_{kp})$ where $1 \geq a_{k2} \geq \ldots a_{kp} > 0$, and $\mathbf{D}_k$ is an orthogonal matrix for each $k = 1, \ldots, K$.

We assume that the data are generated by a mixture of underlying probability distributions; each component of the mixture represents a different cluster so that the observations $\mathbf{x}_i$, ($i = 1, \ldots, n; \mathbf{x}_i \in R^p$) to be classified arise from a random vector x with density $p(\theta, \pi | X = \mathbf{x})$ as in (5), where $f_k(\cdot | (\theta_k = \mu_k, \Sigma_k))$ is the multivariate normal density function, μ_k is the mean, and Σ_k is the covariance matrix for the kth group. The vector $\pi = (\pi_1, \ldots, \pi_K)$ is the mixing proportion ($\pi_k \geq 0$, $\sum_{k=1}^{K} \pi_k = 1$).

We are concerned with Bayesian inference about the model parameters θ, π and the classification indicators ν. Markov Chain Monte Carlo methods (MCMC) provide an efficient and general recipe for Bayesian analysis of mixtures. Given a classification vector $\nu = (\nu_1, \ldots, \nu_n)$, we use the notation $n_k = \#\{i : \nu_i = k\}$ for the number of observations in cluster k, $\bar{\mathbf{x}}_k = \sum_{i:\nu_i=k} \mathbf{x}_i / n_k$ for the sample mean vector of all observations in the cluster k, and $\mathbf{W}_k = \sum_{i:\nu_i=k}(\mathbf{x}_i - \bar{\mathbf{x}}_k)(\mathbf{x}_i - \bar{\mathbf{x}}_k)^t$ for the sample covariance matrix.

We use conjugate priors for the parameters π, μ and Σ of the mixture model. The prior distribution of the mixing proportions is a Dirichlet distribution

$$(\pi_1, \ldots, \pi_K) \sim Dirichlet(\beta_1, \ldots, \beta_K), \tag{7}$$

The prior distributions of the means μ_k of the mixture components conditionally on the covariance matrices Σ_k are Gaussian

$$\mu_k|\Sigma_k \sim N_p(\xi_k, \Sigma_k/\tau_k) \tag{8}$$

with known scale factors $\tau_1, \ldots, \tau_K > 0$ and locations $\xi_1, \ldots, \xi_K \in R^p$. The conjugate prior distribution of the covariance matrices depends on the model. It will be given for each model as detailed in Table 2.

Table 2. Different geometrical characteristics of the four models for the covariance matrix

Model	Volume	Shape	Orientation
λI	same	same(spherical)	undefined
$\lambda_k I$	different	same(spherical)	undefined
$\lambda D A D^t$	same	same	same
$\lambda_k D A D^t$	different	same	same
$\lambda_k D_k A_k D_k^t$	different	different	different

We estimate the parameters of the models by simulating from the joint posterior distribution of π, and ν using the Gibbs sampler, and at iteration $(t+1)$ the Gibbs sampler steps go as follows:

1. Simulate the classification variables $\nu_i^{(t+1)}$, $i = 1, \ldots, n$, independently according to the posterior probabilities $p_{ik} = P(\nu_i^{(t)} = k|\pi, \theta)$, $k = 1, \ldots, K$ conditional on the current values for $\pi^{(t)}$ and $\theta^{(t)}$,

$$p_{ik} = \pi_k f_k(\mathbf{x}_i|\mu_k^{(t)}, \Sigma_k^{(t)}) / \sum_{k=1}^{K} \pi_k^{(t)} f_k(\mathbf{x}_i|\mu_k^{(t)}, \Sigma_k^{(t)}) \qquad (i = 1, \ldots, n) \tag{9}$$

2. Simulate the vector $\pi^{(t+1)} = (\pi_1^{(t+1)}, \ldots, \pi_K^{(t+1)})$ of mixing proportions from its posterior distribution, namely

$$\pi^{(t+1)} \sim Dirichlet(\beta_1 + \sum_{i=1}^{n} \#\{\nu_i^{(t+1)} = 1\}, \ldots, \beta_K + \sum_{i=1}^{n} \#\{\nu_i^{(t+1)} = K\}) \tag{10}$$

where β_k are the known prior parameters of the Dirichlet distribution.

3. Simulate the parameter $\theta^{(t+1)}$ of the model from the posterior distribution $\theta|\nu^{(t+1)}$.

4. Iterate the steps 1 to 3.

The validity of this procedure, namely the fact that the Markov chain associated with the algorithm converges in distribution to the true posterior distribution of θ, was shown. The proof is based on a *duality principle*, which uses the finite space nature of the chain associated with the ν_i's. This chain is ergodic with state space $\{1, \ldots, K\}$, and is thus geometrically convergent. These properties transfer automatically to the sequence of values of θ and π, and important properties as the central limit theorem or the law of the iterated logarithm are then satisfied. Next, we describe and estimate parameters of the four proposed models we used in this chapter:

(1) Model $[\lambda I]$: similar spherical clustering

This model assumes that the clusters are spherical with the same volume (λ). As shown above, the prior distribution of the parameter μ_k is given in (8). The posterior distribution of $(\mu_k|\lambda, \nu)$ is a multivariate normal distribution with mean $\bar{\xi}_k = (n_k\bar{\mathbf{y}}_k + \tau_k\xi_k)/(n_k + \tau_k)$ and covariance matrix $\lambda/(n_k + \tau_k)I$. As λ is a scale measure, we use an inverse Gamma distribution $G_a^{-1}(m_0/2, \rho_0/2)$ as a prior with scale and shape parameters m_0 and ρ_0. The posterior distribution of $\lambda|\mu, \mathbf{y}$ is an inverse Gamma distribution

$$G_a^{-1}\left(\{n + m_0 p\}/2, \left\{\rho_0 + tr(\mathbf{W} + \sum_k \frac{n_k\tau_k}{n_k + \tau_k}(\bar{\mathbf{x}}_k - \xi_k)(\bar{\mathbf{x}}_k - \xi_k)^t)\right\}/2\right) \quad (11)$$

(2) Model $[\lambda_k I]$ different spherical clustering

This model assumes that the clusters are spherical with different volume (λ_k). If the prior distribution of λ_k is an inverse Gamma $G_a^{-1}(m_k/2, \rho_k/2)$ with scale and shape parameters m_k and ρ_k for each cluster, then the posterior distribution of $\lambda_k|\mu, x$ is an inverse Gamma distribution

$$G_a^{-1}\left(\frac{1}{2}(n_k + p m_k), \left\{\rho_k + tr(\mathbf{W}_k + \frac{n_k\tau_k}{n_k + \tau_k}(\bar{\mathbf{x}}_k - \xi_k)(\bar{\mathbf{x}}_k - \xi_k)^t)\right\}/2\right) \quad (12)$$

(3) Model $[\Sigma]$: similar ellipsoidal clustering

Using this model, we suppose that all clusters have the same volume, same shape and same orientation ($\Sigma = \lambda\mathbf{D}\mathbf{A}\mathbf{D}^t$). The prior distribution of Σ is an inverse Wishart distribution with degrees of freedom m_0 and sample variance $\mathbf{\Psi}_0$. The posterior distribution of $\Sigma|\mu, \mathbf{x}$, has the following inverse Wishart distribution

$$W_p^{-1}\left(n_k + m_0, \mathbf{\Psi}_0 + \mathbf{W} + \sum_k \frac{n_k\tau_k}{n_k + \tau_k}(\bar{\mathbf{x}}_k - \xi_k)(\bar{\mathbf{x}}_k - \xi_k)^t\right) \qquad (13)$$

(4) Model $[\lambda_k\Sigma]$: proportional ellipsoidal clustering

Using this model, we suppose that all clusters have different volume, same shape and same orientation ($\Sigma = \lambda_k\mathbf{D}\mathbf{A}\mathbf{D}^t$). The prior distribution of Σ is an inverse Wishart distribution with degrees of freedom m_0 and sample variance $\mathbf{\Psi}_0$. The prior distribution of λ_k is an inverse Gamma $G_a^{-1}(m_k/2, \rho_k/2)$ with scale and shape parameters m_k and ρ_k for each cluster, then the posterior distribution of $\lambda_k|\Sigma, \mu, x$ is an inverse Gamma distribution

$$G_a^{-1}(\frac{1}{2}(n_k + pm_k), \qquad (14)$$
$$\{\rho_k + tr(\Sigma(\mathbf{W} + \Psi_0 + \sum_k \frac{n_k\tau_k}{n_k + \tau_k}(\bar{\mathbf{x}}_k - \xi_k)(\bar{\mathbf{x}}_k - \xi_k)^t))\}/2)$$

The posterior distribution of $\Sigma|\lambda_k, \mu, \mathbf{x}$, has the following inverse Wishart distribution

$$W_p^{-1}\left(n + m_0, \mathbf{\Psi}_0 + \mathbf{W} + \sum_k \frac{n_k\tau_k}{n_k + \tau_k}(\bar{\mathbf{x}}_k - \xi_k)(\bar{\mathbf{x}}_k - \xi_k)^t\right) \qquad (15)$$

(5) Model $[\Sigma_k]$: different ellipsoidal clustering)

In this case, we suppose that clusters have different shape, different volume and different orientation ($\Sigma = \lambda_k\mathbf{D}_k\mathbf{A}_k\mathbf{D}_k^t$). If the prior distribution of Σ_k is an inverse Wishart distribution with degrees of freedom m_k and a sample variance $\mathbf{\Psi}_k$, then the posterior distribution of $\Sigma_k|\mu, \mathbf{x}$, has the following inverse Wishart distribution

$$W_p^{-1}\left(n_k + m_k, \mathbf{\Psi}_k + \mathbf{W}_k + \frac{n_k\tau_k}{n_k + \tau_k}(\bar{\mathbf{x}}_k - \xi_k)(\bar{\mathbf{x}}_k - \xi_k)^t\right) \qquad (16)$$

We propose a Bayesian approach for selecting the number of clusters. We also need to specify which one of the models described above offers the best classification. Therefore, we provide an adaptive and flexible clustering algorithm which considers not only the number of clusters, but also the shape of the proposed clusters. The models described above use the geometrical specification of the clusters depending upon the data to be analyzed.

6 Model selection

The deviance information criterion (DIC) has been recently introduced as a means of comparing models. DIC uses the posterior expectation of the log likelihood as a measure of model fit. For a particular model M, the DIC is defined as

$$DIC = 2\overline{D} - D\left(\bar{\theta}_M\right), \tag{17}$$

where

$$\bar{D} = -2 \int \left[\log p\left(\mathbf{x}|\theta_M\right)\right] p\left(\theta_M|\mathbf{x}, M\right) d\theta_M \tag{18}$$

$$= E_{\theta_M|x}\left[D\left(\theta_M\right)\right], \tag{19}$$

is the posterior expectation of the so-called deviance $D\left(\theta_M\right) = -2 \log p\left(\mathbf{x}|\theta_M\right)$. The second term in the right hand side of (17) is the deviance evaluated at the posterior mean of the parameter vector θ_M. In order to motivate (17), an expansion of the deviance around the posterior mean $\bar{\theta}_M$, and taking expectations with respect to the posterior distribution of θ gives the expected deviance [45]

$$\bar{D} \approx D\left(\bar{\theta}_M\right) + tr\left\{\left[-\frac{\partial^2 \log p\left(\mathbf{y}|\theta_M\right)}{\partial\theta_M\partial\theta'_M}\right]_{\theta_M=\bar{\theta}} Var\left(\theta_M|\mathbf{y}\right)\right\} \tag{20}$$

where $D\left(\bar{\theta}_M\right) = -2 \log p\left(\mathbf{y}|\bar{\theta}_M\right)$, and $Var\left(\theta_M|\mathbf{y}\right)$ is the variance–covariance matrix of the posterior distribution of θ_M. The second term on the right hand side of (20) is measuring model complexity and is called the "effective number of parameters". Expression (17) is a result of combining the posterior expectation of the deviance, given by (18), with the effective number of parameters, which from (20), is given approximately by $\bar{D} - D\left(\bar{\theta}_M\right)$. Thus, DIC has a term that reflects the fit of the model and another term which introduces a penalty due to the complexity of the model. The penalty inherent in DIC is stronger than other measures, such as the posterior Bayes Factor (apart from the second term in (17), $\bar{D}$ already includes a penalty factor). Models having a smaller DIC should be favored as this indicates a better fit and a lower degree of model complexity. Spiegelhalter et al. [46] show that DIC is related to other model comparison criteria and has an approximate decision-theoretic justification.

DIC is very easily calculated using the MCMC output. The first term in the DIC is estimated using twice the average of the simulated values of $-\log p\left(\mathbf{y}|\theta_M\right)$, and the second term is the plug-in estimate of the deviance using the average of the MCMC simulated values of θ_M.

7 Results

We applied the Bayesian clustering algorithm and succeeded to detect the clusters and their geometrical representation and also provided the uncertainty of the classification for each sample. First, the comparison of the Deviance Criteria for different number of clusters and different models is summarized in Table 3.

Table 3. DIC scores for the number of clusters and different models

Clusters/Models	1	2	3	4	5
1	828.36	817.95	818.20	832.63	814.10
2	761.36	780.63	753.92	742.34	659.02
3	703.76	660.07	668.43	652.63	676.68
4	766.56	752.62	770.37	765.73	771.02
5	768.59	776.42	782.43	780.55	782.19
6	764.02	765.92	770.53	795.09	774.38
7	817.38	790.34	789.99	815.82	819.77

Using 1000 iterations of the Gibbs sampler was necessary for the stability of the Markov chain. Convergence was immediate as shown in Figures 7 and 9. The correct model $[\lambda_k \Sigma]$ (model 4) and the correct number of groups (3 clusters), are strongly favored. DIC scored best for the model $[\lambda_k \Sigma]$ with three components, which means that data proposes three proportional groups with different volume. When we wanted to investigate the performance of this choice, we compared the proposed clusters with the ones proposed by experts. The error rate of misclassification obtained by the optimal model is equal to 8%. Due to the limited space, here we show the time series plots for μ only. Figure 7 shows the convergence of each component of μ to the mode or expected parameter estimates.

The expected values of the parameter estimates μ, λ, Σ and π are $\mu_1 = (18.03, 3.06)$, $\mu_2 = (-7.33, 3.47)$, $\mu_3 = (-8.86, -7.048)$, $\lambda = (2.35, 1.23, 4.43)$,

$$\Sigma_1 = \begin{pmatrix} 17.60 & -4.51 \\ -4.51 & 7.44 \end{pmatrix}, \ \Sigma_2 = \begin{pmatrix} 2.37 & -0.67 \\ -0.67 & 1.63 \end{pmatrix}, \ \Sigma_3 = \begin{pmatrix} 11.42 & -1.34 \\ -1.34 & 3.11 \end{pmatrix}, \text{ and}$$

$\pi = (0.30, 0.38, 0.31)$.

The second best choice proposed by DIC is three ellipsoidal groups with different volume and different orientation. The expected values of the parameter estimates μ, Σ and π are $\mu_1 = (15.77, 1.45)$, $\mu_2 = (-8.82, -1.41)$,

$$\pi = (0.33, 0.66), \ \Sigma_1 = \begin{pmatrix} 6.65 & -2.95 \\ -2.95 & 5.29 \end{pmatrix}, \text{ and } \Sigma_2 = \begin{pmatrix} 3.46 & 0.017 \\ 0.017 & 1.54 \end{pmatrix}.$$

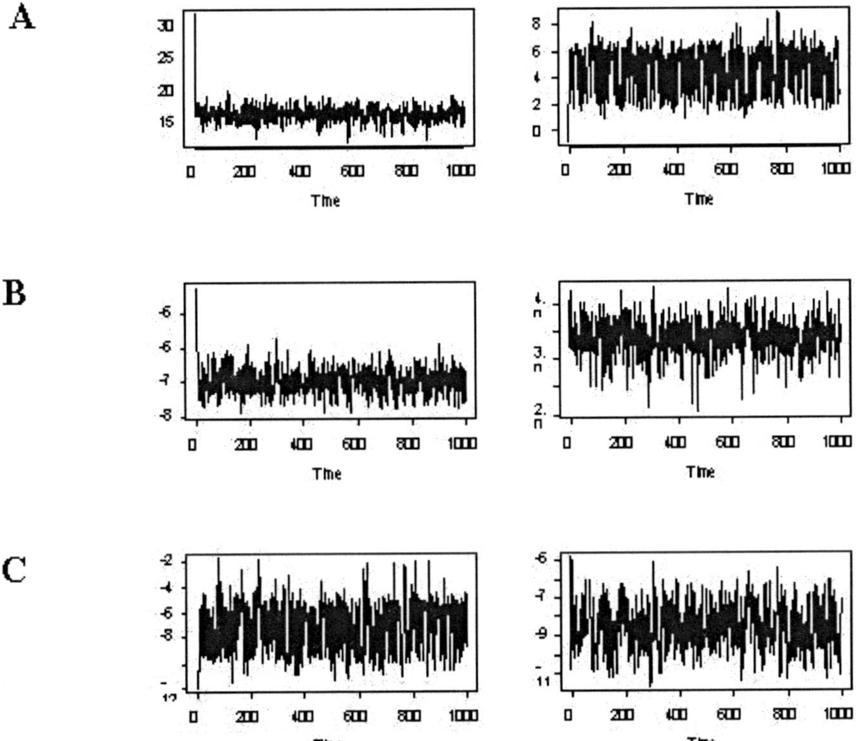

Fig. 7. Simulation history of three cluster mean vectors and its convergence. (a): the two panels represent the mean vector of the first cluster. (b): the two panels represent the mean vector of the second cluster. (c): the two panels represent the mean of the third cluster.

8 Summary

Cancer proteomics encompasses the identification and quantitative analysis of differentially expressed proteins relative to healthy tissue counterparts at different stages of disease, from preneoplasia to neoplasia. The high dimensionality of the data obtained through SELDI-TOF-MS and other recent mass spectrometry-based proteomics technologies underscores the need for developing an artificial-intelligence algorithm capable of analyzing such high volume data to develop an efficient and reproducible classifier. In order to achieve this goal, we have utilized the Bayesian-based principal component algorithm to efficiently cluster the data. With our data set, the PCA Bayesian approach not only showed an extremely good performance, but it also proposed the correct number of clusters and their geometrical configuration and the correct model using DIC. The diagonal model proposed by DIC performed well. We may be misled by the the second best choice (two clusters), which might be surprising

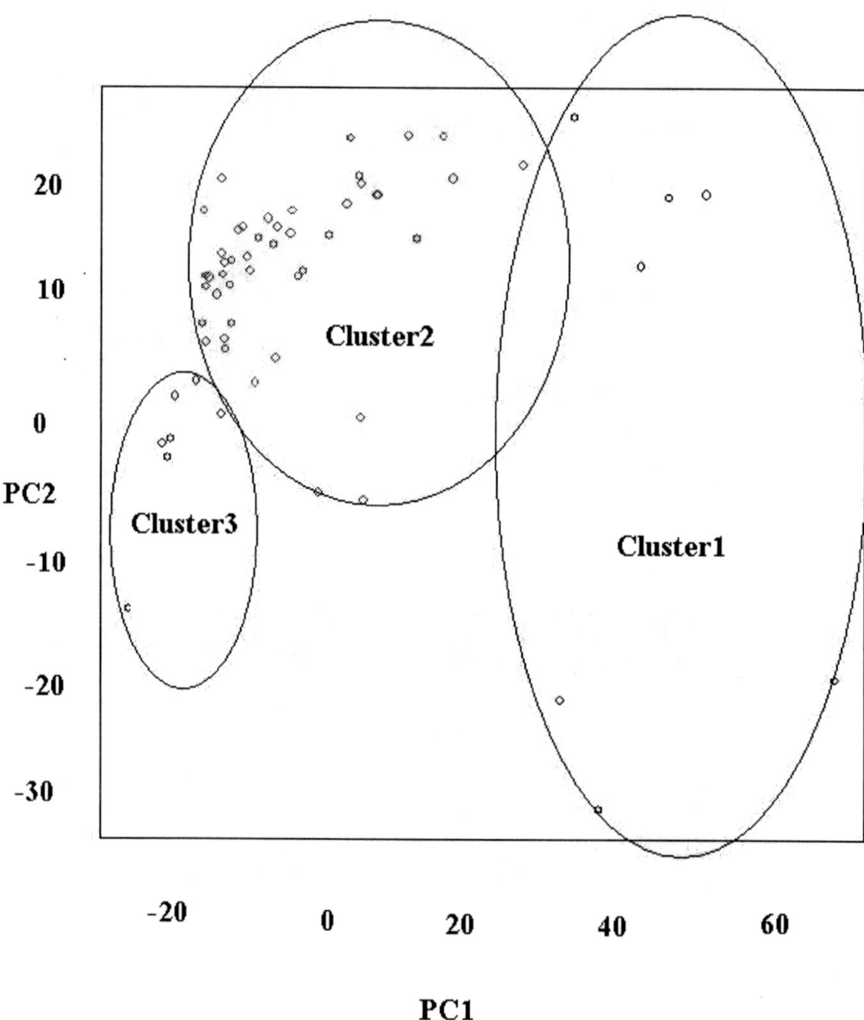

Fig. 8. 2D projection of the Bayesian algorithm. Shown are the three clusters found by the Bayesian algorithm.

but not contradictory. The DIC recognizes the normal group but puts HAM and ATL together in a second group. This is probably due to the fact that HAM and ATL patients were infected by the same virus that might be targeting the same group of genes, and therefore proteins, in the two diseases. The first and the best choice proposed by DIC generated an ellipsoidal model, with different covariance matrices, and proposed three clusters. The three groups have a superior similarity to the real data sets (Normal, ATL and HAM) with an error rate of 9%. This means that the intensity profile of the proteome can distinguish between the normal patients and the diseased patients. Within the

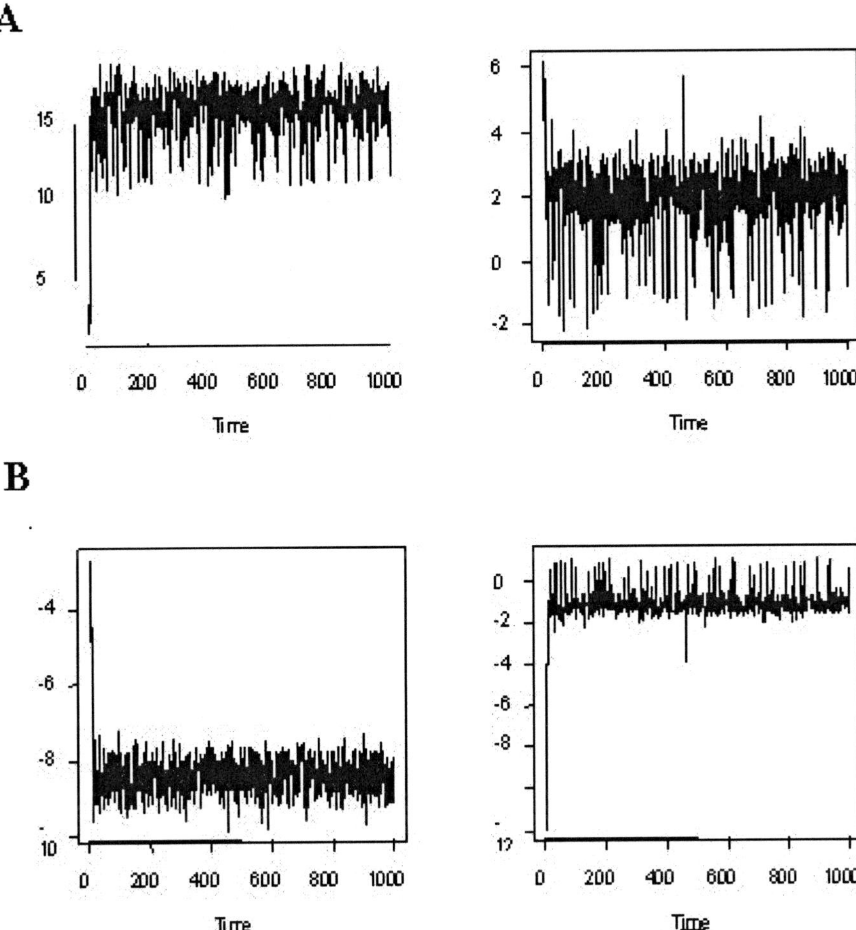

Fig. 9. Simulation history of two cluster mean vectors and its convergence. (a): the two panels represent the mean vector of the first cluster. (b): the two panels represent the mean vector of the second cluster.

two diseased patients it recognizes the Adult T cell leukemia (ATL) and the HTLV-Associated Myelopathy/Tropical Spastic Paraparesis (HAM) and separates them in two different clusters. HTLV-1 might be targeting a similar set of proteins; however, these proteins can be expressed differentially in the two diseases, which is reflected by different intensity profiles. The high specificity obtained using this approach represents a significant advancement in the clustering of high dimensional data especially when more than two patient groups are considered. Classifying proteomics data generated from studies containing more than two patient groups represents a serious challenge, and the outcome of such classification usually has low specificity and sensitivity rates. To boost

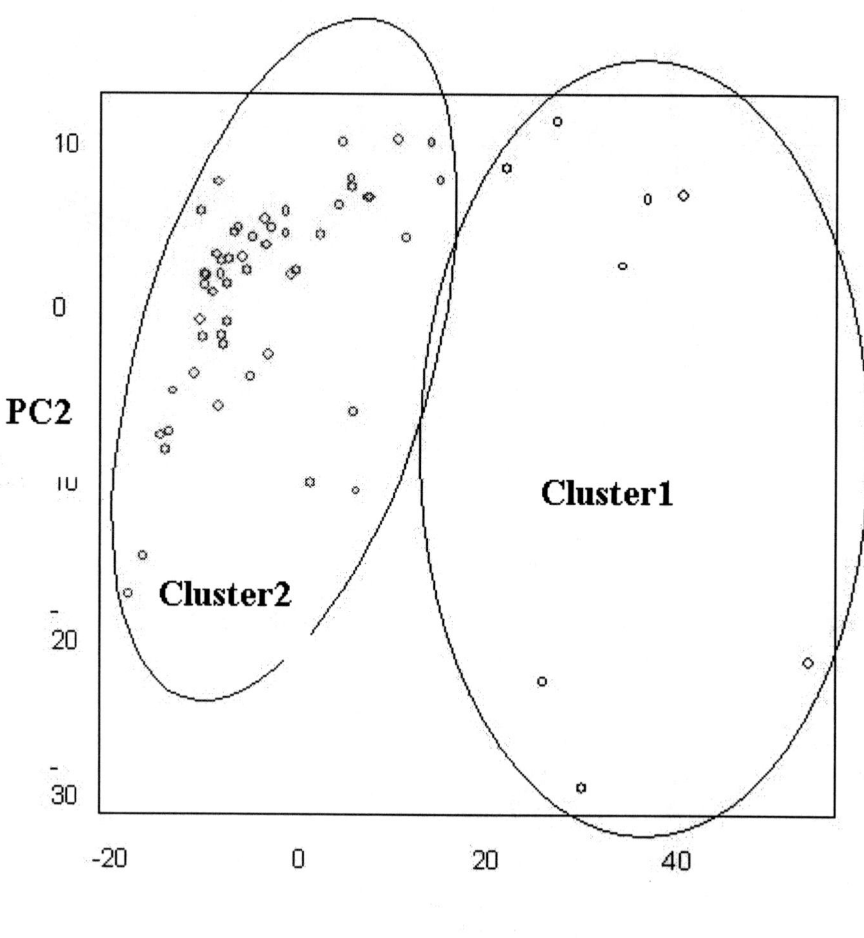

PC1

Fig. 10. 2D projection of the Bayesian algorithm. Shown are the two clusters found by the Bayesian algorithm.

the capabilities of this algorithm and enhance the rate of correct classification, one may instead use a non-parametric approach to cluster the proteomics data. The use of the normality assumption was strongly emphasized and pushed on the transformed data, which causes a satisfactory but still not perfect score especially when the mass is used instead of the intensity.

Acknowledgments

This work was supported by grant from the Leukemia & Lymphoma Society to OJS and by a grant from the Scholarly Research Grant Program of the College of Business Administration and the University of Tennessee to HB.

References

1. B.L. Adam, Y. Qu, J.W. Davis, M.D. Ward, M.A. Clement, L.H. Cazares, O.J. Semmes, P.F. Schelhammer, Y. Yasui, F. Ziding, and G.L. Wright. Serum protein fingerprinting coupled with a pattern-matching algorithm distinguishes prostate cancer from benign prostate hyperplasia and healthy men. *Cancer Research*, 62: 3609-3614, 2002.

2. H. Akaike. Factor analysis and AIC. *Psychometrika*, 52: 317-332, 1987.

3. J.D. Banfield and A.E. Raftery. Model-based Gaussian and non-Gaussian clustering. *Biometrics*, 49: 803-821, 1993.

4. R.E. Banks, M.J. Dunn, D.F. Hochstrasser, J.C. Sanchez, W. Blackstock, and D.J. Pappin. Proteomics: new perspectives, new biomedical opportunities. *Lancet*, 356: 1749-1756, 2000.

5. H. Bensmail, G. Celeux, A.E. Raftery, and C. Robert. Inference in model-based cluster analysis. *Computing and Statistics*, 7: 1-10, 1997.

6. H. Bensmail and A. Haoudi. Postgenomics: proteomics and bioinformatics in cancer research. *Journal of Biomedicine and Biotechnology*, 4: 217-230, 2003.

7. D.H. Binder. Approximations to Bayesian clustering rules. *Biometrika*, 68: 275-285, 1981.

8. C.M. Bishop. *Neural Networks for Pattern Recognition,*. Oxford, University Press, New York, 1995.

9. H.H. Bock. On some significance tests in cluster analysis. *Journal of Classification*, 2: 77-108, 1985.

10. H.H. Bock. Probability models in partitional cluster analysis. *Computational Statistics and Data Analysis*, 23: 5-28, 1996.

11. H. Bozdogan. Choosing the number of component clusters in the mixture model using a new informational complexity criterion of the inverse Fisher information matrix. In O. Opitz, B. Lausen, and R. Klar, editors, *Information and Classification*, pages 40-54, Springer-Verlag, 1993.

12. L.H. Cazares, B.L. Adam, M.D. Ward, S. Nasim, P.F. Schellhammer, O.J. Semmes, and G.L. Wright, Jr. Normal, benign, preneoplastic, and malignant prostate cells have distinct protein expression profiles resolved by surface enhanced laser desorption/ionization mass spectrometry. *Clinical Cancer Research*, 8: 2541-52, 2002.

13. A. Dasgupta and A.E. Raftery. Detecting Features in Spatial Point Processes with Clutter via Model-Based Clustering. *Journal of the American Statistical Association*, 93: 294-302, 1998.

14. J. Diebolt and C.P. Robert. Bayesian Estimation of Finite Mixture Distributions. *Journal of of the Royal Statistical Society*, Series B, 56: 363-375, 1994.

15. L. Engelman and J.A. Hartigan. Percentage Points of a Test for Clusters. *Journal of the American Statistical Association*, 64: 1647-1648, 1969.

16. A.E. Gelfand and D.K. Dey. Bayesian Model Choice: Asymptotics and Exact Calculations. *Journal of the Royal Statistical Society*, 56: 501-514, 1994.

17. A.D. Gordon. *Classification: Methods for the Exploratory Analysis of Multivariate Data*. 2nd edition, Chapman and Hall, New York, 1999.

18. A. Haoudi and O.J. Semmes. The HTLV-1 tax oncoprotein attenuates DNA damage induced G1 arrest and enhances apoptosis in p53 null cells. *Virology*, 305: 229–239, 2003.

19. A. Haoudi, R.C. Daniels, E .Wong, G. Kupfer, and O.J. Semmes. Human T-cell Leukemia Virus-I Tax Oncoprotein Functionally Targets a Subnuclear Complex Involved in Cellular DNA Damage-Response. *Journal of Biological Chemistry*, 278: 37736-37744, 2003.

20. J.A. Hartigan. *Clustering Algorithms*. Wiley, New York, 1975.

21. T. Hastie and R. Tibshirani. *Generalized Additive Models*. Chapman and Hall, London, 1990.

22. H. Hotelling. Analysis of a complex of statistical variables into principle components. *Journal of Educational Psychology*, 24: 417-520, 1933.

23. F. Hillenkamp, M. Karas, R.C. Beavis, and B.T. Chait. Matrix-assisted laser desorption/ionization mass spectrometry of biopolymers. *Analytical Chemistry*, 63: 1193A-1203A, 1991.

24. T.W. Hutchens and T.T. Yip. New desorption strategies for the mass spectrometric analysis of micromolecules. *Rapid Communications in Mass Spectrometry*, 7: 576-580, 1993.

25. H. Jeffreys. *Theory of Probability*. Clarendon, 1961.

26. M. Karas and F. Hillenkamp. Laser desorption ionization of proteins with molecular masses exceeding 10 000 daltons. *Analytical Chemistry*, 60: 2299-2301, 1988.

27. T. Kohonen. *Self-Organizing Maps*. 2nd edition, Springer, Berlin, 1997.

28. L. Kaufman and P.J. Rousseeuw. *Finding Groups in Data. An Introduction to Cluster Analysis*. Wiley, New York, 1990.

29. S.M. Lewis and A. Raftery. Estimating Bayes factor via posterior simulation with the Laplace-Metropolis estimator. *Journal of the American Statistical Association*, 92: 648-655, 1997.

30. J. Li, Z. Zhang, J. Rosenzweig, Y.Y. Wang, and D.W. Chan. Proteomics and bioinformatics approaches for identification of serum biomarkers to detect breast cancer. *Clinical Chemistry*, 48: 1296-1304, 2002.

31. J.B. MacQueen. Some methods for classification and analysis of multivariate observations. *Proceedings of the Fifth Berkeley Symposium on Mathematical Statistics and Probability*, 1: 281-297, 1967.

32. G. Mclachlan and D. Peel. *Finite Mixture Models*, John Wiley and Sons, New York, 2000.

33. U. Menzefricke. Bayesian clustering of data sets. *Communication in Statistics-Theory and Methods*, 10: 65-77, 1981.

34. M. Merchant and S.R. Weinberger. Recent advancements in surface-enhanced laser desorption/ionization-time of flight-mass spectrometry. *Electrophoresis*, 21: 1164-1167, 2000.

35. F. Murtagh and A. Raftery. Fitting straight lines to point patterns. *Pattern Recognition*, 17: 479-483, 1984.

36. M. Mukherjee, E.D. Feigelson, G.J. Babu, F. Murtagh, C. Fraley and A. Raftery. Three types of gamma ray bursts. *Astrophysical Journal*, 508: 314-327, 1998.

37. K. Pearson. On lines and planes of closest fit to systems of points in space. *Philosophical Magazine*, 6: 559-572, 1901.

38. E.F. Petricoin, A.M. Ardekani, B.A. Hitt, P.J. Levine, V.A. Fusaro, S.M. Steinberg, G.B. Mills, C. Simone, D.A. Fishman, E.C. Kohn, and L.A. Liotta. Use of proteomic patterns in serum to identify ovarian cancer. *The Lancet*, 359: 572-577, 2002.

39. Y. Qu, B.L. Adam, Y. Yasui, M.D. Ward, L.H. Cazares, P.F. Schelhammer, Z. Feng, O.J. Semmes, G.L. Wright, Jr. Boosted decision tree analysis of surface enhanced laser desorption/ionization mass spectral serum profiles discriminates prostate cancer from noncancer patients. *Clinical Chemistry*, 10: 1835-1843, 2002.

40. S. Richardson and P.J. Green. On Bayesian analysis of mixtures with an unknown number of components. *Journal of the Royal Statistical Society*, Series B, 59: 731-792, 1997.

41. K. Roeder and L. Wasserman. Practical bayesian density estimation using mixture of normals. *Journal of the American Statistical Association*, 92: 894-902, 1997.

42. G. Schwartz. Estimating the dimension of a model. *The Annals of Statsitics*, 6: 461-464, 1978.

43. A.J. Scott and M.J. Symons. Clustering methods based on likelihood ratio criteria. *Biometrics*, 27: 387–397, 1971.

44. B. Silverman. Some aspects of the spline smoothing approach to nonparametric regression curve fitting. *Journal of the Royal Statistical Society B*, 47: 1-52, 1985.

45. D. Sorensen, and D. Gianola. *Likelihood, Bayesian, and MCMC Methods in Quantitative Genetics*. John Wiley and Sons, 2003.

46. D.J. Spiegelhalter, J.P. Myles, D.R. Jones, K.R. Abrams. Methods in health service research: An introduction to Bayesian methods in health technology assessment. *British Medical Journal*, 319: 508-512, 1999.

47. J.H. Wolfe. Comparative cluster analysis of patterns of vocational interest. *Multivariate Behavioral Research*, 13: 33-44, 1978.

48. G.L. Wright, L.H. Cazares, S.M. Leung, S. Nasim, B.L. Adam, T.T. Yip, P.F. Schelhammer, L. Gong, and A. Vlahou. Proteinchip surface enhanced laser desorption/Ionization (SELDI) mass spectrometry: A novel protein biochip technology for detection of prostate cancer biomarkers in complex protein mixtures. *Prostate Cancer and Prostatic Diseases*, 2: 264-276, 1999.

49. K.Y. Yeung, A. Fraley, A. Murua, A. Raftery, and W.L. Ruzzo. Model-based clustering and data transformations for gene expression data. *Bioinformatics*, 17: 977-987, 2001.

Bioinformatics for Traumatic Brain Injury: Proteomic Data Mining

Su-Shing Chen[1,6], William E. Haskins[1,2,4], Andrew K. Ottens[1,2,4], Ronald L. Hayes[2,3,4], Nancy Denslow[1,5], and Kevin K.W. Wang[1,2,3,4]

[1] Center of Neuroproteomics and Biomarkers Research,
University of Florida, Gainesville, FL 32610
[2] Center for Traumatic Brain Injury Studies
University of Florida, Gainesville, FL 32610
[3] Department of Psychiatry,
University of Florida, Gainesville, FL 32610
[4] Department of Neuroscience,
University of Florida, Gainesville, FL 32610
[5] Interdisciplinary Center of Biomedical Research,
University of Florida, Gainesville, FL 32610
[6] Computer and Information Science Engineering,
University of Florida, Gainesville, FL 32610

Summary. The importance of neuroproteomic studies is that they will help elucidate the currently poorly understood biochemical mechanisms or pathways underlying various psychiatric, neurological and neurodegenerative diseases. In this chapter, we focus on traumatic brain injury (TBI), a neurological disorder currently with no FDA approved therapeutic treatment. This chapter describes data mining strategies for proteomic analysis in traumatic brain injury research so that the diagnosis and treatment of TBI can be developed. We should note that brain imaging provides only coarse resolutions and proteomic analysis yields much finer resolutions to these two problems. Our data mining approach is not only at the collected data level, but rather an integrated scheme of animal modeling, instrumentation and data analysis.

1 Introduction

With the complete mapping of the human genome, we are now armed with a finite number of possible human gene products (human proteome). There are approximately 30,000 to 40,000 hypothetical protein products transcribable from the human genome [2, 25, 26, 31, 32, 60, 61]. The study of the proteome is also aided by recent advances of protein separation, identification and quantification technologies not available even 3-5 years ago. Yet, the proteome is still extremely complex because by definition, proteome is organ-, cell type-, cell state- and time-specific. Proteins are also subjected to various posttranslational modifications. In addition, cellular proteins are almost constantly subjected to various forms of posttranslational modifications (PTM), including phosphorylation/dephosphorylation by different kinases and phosphatases, proteolysis or processing acetylation, glycosylation and crosslinking by transglutaminases or protein conjugation to small protein tags such as ubiquitin or SUMO [35, 58, 59]. It has been proposed that a more feasible approach is to focus on a subproteome, such as that of single tissue or a subcellular organelle [33]. On the other hand, we proposed that focusing on the study of the proteome of the central and peripheral nervous systems (CNS and PNS) maybe more manageable and productive [17]. We further submitted that although the applications of proteomic technologies to nervous system disorders (e.g. neural injury, neurodegeneration, substance abuse and drug addiction) is still in its infancy, the potential insights one would gain from such endeavors are tremendous. The importance of neuroproteomic studies is that they will help elucidate the currently poorly understood biochemical mechanisms or pathways underlying various psychiatric, neurological and neurodegenerative diseases. The example we will focus on here is traumatic brain injury (TBI), a neurological disorder currently with no FDA approved therapeutic treatment. In general, proteomic studies of TBI create a huge amount of data and the bioinformatic challenge is two-fold: (i) to organize and archive such data into a useful and retrievable database format and (ii) to data-mine such database in order to extract the most useful information that can be used to advance our understanding of the protein pathways relevant to TBI.

This chapter reports the bioinformatics component of the TBI research at the Center of Neuroproteomics and Biomarkers Research and Center for Traumatic Brain Injury Studiesat the University of Florida. In particular, we describe data mining strategies for proteomic analysis in TBI research so that the diagnosis and treatment of TBI can be developed. We should note that brain imaging provides only coarse resolutions and proteomic analysis yields much finer resolutions to these two problems. Our data mining approach is not only at the collected data level, but rather an integrated scheme of animal modeling, instrumentation and data analysis. Thus computing infrastructure is essential at all the protein separation, protein identification/quantification and bioinformatics levels.

The organization of the chapter is as follows. In Section 2, we describe traumatic brain injury (TBI). Section 3 considers animal models, while Section 4 deals with the source of biological materials. Sections 5 and 6 address samples collection and pooling. Proteomic analysis is overviewed in Section 7. In Section 8, we present bioinformatics for TBI proteomics. Finally in conclusion we consider our future work and the prospect of systems biology in TBI research.

2 Traumatic Brain Injury (TBI)

Traumatic brain injury or traumatic head injury is characterized as a direct physical impact or trauma to the head, causing brain injury [17]. Annually there are 2 million traumatic brain injury (TBI) cases in the U.S. alone. They result in 500,000 hospitalizations, 100,000 deaths, 70,000-90,000 people with long-term disabilities and 2,000 survive in permanent vegetative state. Medical costs of TBI are estimated to be over $48 billion annually in U.S. The cause of TBI can be broken down into the following catalogues: motor vehicle accidents (50%), falls (21%), assault & violence (12%), sports & recreation (10%) and all others (7%) Importantly, 30-40% of all battlefield injuries have a head injury component.

Due to intensive research in both clinical setting and experimental animal models of TBI, there is now a general understanding of the pathology of TBI. It all starts with the impact zone, where there is mechanical compression-induced direct tissue injury and often associated with hemorrhage. Significant amount of cell death will occur very rapidly in this zone. More distal to the injury zone, due to the impact of the force, contusion injury also result, long fiber tracts (axons) are especially at risk to this type of injury. Usually after the first phase of cell injury/cell occurs, there is also the secondary injury which is believed to be mediated by neurotoxic glutamate release (neurotoxicity). Other significant alterations include inflammation responses by microglia cells, astroglia activation and proliferation and stem cells differentiation. Over time, if the TBI patient survives, these events lead to long-lasting brain tissue remodeling. Therefore, the spatial and temporal levels of biochemical and proteomic changes of TBI can be investigated.

3 Animal Models of TBI

Over the past decades, basic science researchers have developed several animal models for TBI [19, 54]. There are several well characterized models of TBI, including controlled cortical impact (CCI) with compressed gas control, fluid percussion model that transduce a contusion force due to the movement of fluid in the chamber, and the vertical weight drop model with which a weight is dropped from a certain height within a hollow chamber for guidance. Thus it creates an acceleration force which direct on the top of the skull (either unilateral or bilateral injury [19]). In our work, we employ the rat CCI model of TBI. We have argued that the use of proteomic will greatly facilitate the biochemical mechanisms underlying the various phases of TBI pathology [17].

4 Source of Biological Materials

Proteomic studies for traumatic brain injury can be generally categorized into human studies, animal and cell culture-based studies. For the purposes of this review, cell culture-based studies will not be discussed further. When comparing human vs. animal studies, there are pro and con in each scenarios. Regarding the sample types that can be exploited for proteomic analysis, they will include brain tissues, cerebrospinal fluid, blood (serum and plasma). For human TBI studies, samples that are the easiest to obtain would be blood samples (which are further fractionated into plasma or serum). Interestingly, there is increasing interests now focus on using cerebrospinal fluid (CSF) as its status will reflect the status of the central nervous system itself. Following severe traumatic brain injury, spinal shun or spinal tap are routinely done thus obtaining the CSF is not an issue. One of the major challenges of using clinical samples-based proteomic studies is that it is extremely difficult to control individual (biological) and environmental variables

(I) Brain Samples

Human brain materials from TBI would inevitably come from deceased TBI patients. These brain samples will be subjected to postmortem artifacts, compounded by various and significant time delay before samples can be obtained.

The biggest advantage of animal neuroproteomic studies over human counterparts is the ability to obtain brain tissues in a controlled laboratory environment. Furthermore, it is possible to harvest samples from defined anatomic regions. For example, for traumatic brain injury studies, we often focus on cortical and hippocampal samples. This is important as different brain regions might be selectively more vulnerable to traumatic or ischemic insults.

(II) CSF

CSF can be collected from the cisterna magna from lab animals, such as rats and mice. CSF contains rich brain proteome information that is relevant to disease diagnosis [16]. However, only about 50-100 ul can be withdrawn from a rat and 25-30 ul from a mouse. Care must also be taken not to contaminate samples with blood due to puncture. While more than one CSF draw might be possible, in our laboratory, we generally withdraw only one CSF sample followed by sacrifice. In the case of human TBI, CSF can also be collected routinely from ventriculotomy or from spinal tap.

(III) Blood Samples (Serum and Plasma)

In both human and animal traumatic brain injury studies, blood can be routinely collected and usually further processed into either serum or plasma fractions before subjecting to proteomic analysis. Like CSF, most proteomic researchers believe there is significant proteomic information in the blood that would reflect the status of the brain, particularly after TBI with possible blood-brain barrier compromise [53, 57].

5 Samples Collection and Processing Consistency

It needs to be emphasized here that for proteomic to be consistent and reproducible, one needs to take extra attentions to ensure the variables can be kept to minimal. All sample collection procedure should be discussed and finalized and the operators made familiar with the procedures. Some practice runs are highly desirable. For human studies, detailed record keeping is extremely important for future analysis or trouble-shooting purposes. For human studies, for example, CSF or blood samples should be taken at consistent intervals and ideally, food consumption might significantly affect blood proteomic profile. For animal studies which are conducted in controlled environment, it should be possible to keep brained and befouled sample collection time and routine as standardized as possible. Also, for animal studies, the animal subjects should be tagged and observed carefully and regularly; with any out-of-the norm observations recorded. They might become very helpful in enhancing proteomic analysis. Both tissue and biofluid samples, once obtained and processed, should be snap frozen and store at -85C until use.

6 Sample Pooling Considerations

There is also an important decision to be made before the proteomic analysis, i.e. whether to pool samples for analysis or analyze individual samples. Pooling samples significantly reduce minor individual variability and reduce the amount of workload. Yet, at the same time, its disadvantage is that it might miss certain proteomic changes that are present in only a subset of samples. On the other hand, analysis of individual samples has the advantage of being an exhaustive analysis of individual proteomic profile but it can be highly time-consuming and cost-prohibiting. If the protein amount in the samples are limiting factor, it would be useful to pool samples. Additionally, if there is a biochemical marker that correlates with TBI (such as alphaII-spectrin breakdown products), it can be used as positive controls for quality assurance and might even be used to guide inclusion criteria for sample pooling [51, 52]. It is also possible to incorporate both pooling and individual proteomic analysis in the same studies. For example, for pilot studies or initial proteomic profiling of TBI, pooled samples can be used while the final detailed analysis can be done with individual samples.

7 Proteomic Analysis Overview

Regardless whether we are dealing with human or animal samples or whether they are tissue lysate or biofluid (CSF, serum or plasma). The strategy we developed can be organized into three interacting scientific disciplines or phases:

protein separation, protein identification and quantification and bioinformatics analysis. By design, any proteomic center should spend two-thirds of its scientific and financial resources to establish robust readily usable proteomic platforms. However, it is equally important for the center to develop new or improve existing neuroproteomics technologies on all fronts.

7.1 Protein Separation Methods

In TBI neuroproteomic studies, we are less interested in descriptive and exhaustive characterization of the whole neuroproteomic, but rather we will focus on protein level or posttranslational changes that occur in TBI. With this in mind, it is important to devise methods in comparing and contrasting the two proteomic data sets: "control" versus "TBI". In order to productively identify all the proteins in a specific system of interest (subproteome) or a subset of proteins that are differentially expressed in TBI, it is essential that complex protein mixtures (such as brain sample or biofluid) be first subjected to multi-dimensional protein separation. Since proteins differ in size, hydrophobicity, surface charges, abundance and other properties, to date there is no single protein separation method that can satisfactorily resolve all proteins in a proteome.

Currently, there are two main stream protein separation methods used for proteomic analysis: (i) 2D-gel isoelectrofocusing/electrophoresis and (ii) multi-dimensional liquid chromatography.

(i) 2-dimensional gel electrophoresis approach

Two-Dimensional gel electrophoresis (2D-gel) is the most established protein separation method for the analysis of a proteome or subproteome [7]. It is achieved by subjecting protein mixtures to two protein separation methods under denaturing condition, in the presence of 6-8 M urea and cationic detergent such as SDS. Traditionally, proteins are first separated based on their PI value with a tube gel (polyacrylamide) by isoelectrofocusing with the aid of mobile ampholytes with different PI values. After IEF, the tube gel is placed atop a polyacrylamide gradient gel within which the SDS-bound proteins are separated by size. Due to poor gel-consistency, the IEF step (the first dimension) is most variable; however, a recent breakthrough in IEF technology utilizing immobilized pH gradient strips (IPG) for 2D-gel analysis provides improved reproducibility [6, 24, 31, 36]. Another disadvantage with 2D-gels is the inevitable gel-to-gel variability in exact location and patterns of protein spots. This proves problematic when comparing two samples directly (such as control vs. substance abuse brain). The recent advance of 2D-differential-in-gel-electrophoresis (2D-DIGE) has resolved this [49, 64]. Two protein mixtures are labeled with the fluorescent cyanine dye pairs Cy3/Cy5 that match in molecular weight and charge but matched have distinct excitation and emission wavelengths. These advantages are incorporated into our approach.

They include in particular the high resolving power for complex mixtures of proteins, and the capability of resolving post-translationally modified proteins, including acetylation, phosphorylation, and glycosylation and protein crosslinking [35, 58]. It is possible to annotate each protein of a proteome by PI and molecular weight values as X-Y coordinates to form a 2D protein map of which there is already a wealth of 2D-brain protein coordinates in publicly accessible and searchable databases [3, 20, 21, 42, 44]. There are however, several persistent weaknesses of 2D-gels. Proteins of extreme PI or minute quantity and proteins that are either very small or very large may be missed. Also, integral membrane proteins of which many are CNS disorder drug targets (membrane-bound receptors or neurotransmitter transporters) are lost due to their extreme hydrophobicity.

Regarding protein separation, there are also research in the direction of microfluidic 2D- protein separation with miniaturized IEF and electrophoresis. This approach is the advantage of reducing waste and sample usage without compromising detection sensitivity [18, 56].

(ii) 2-dimensional liquid chromatography approach

Alternative protein separation methods are needed to overcome some of the shortcoming of 2D-gels. Recently, there is significant movement toward multidimensional liquid chromatography methods to resolve complex protein mixtures [50]. The general idea draws on classic chromatographic principles including size chromatography (SEC) (gel filtration), ionic interaction (strong cation exchange (SCX) and strong anion exchange (SAX), hydrophobic interaction (C4- or phenyl-agarose chromatography), and isoelectrofocusing chromatography. One can envision combining multiple chromatographic approaches in series to achieve multidimensional separations. When selecting chromatographic separation methods, considerations must also been given to take advantage of the size, pI and hydrophobicity differences of the proteins of interests. IN addition, when dealing with membrane-bound proteins, the chromatographic method must be compatible with the use of proper neutral detergent (such as Triton X-100 or CHAPS). Importantly, minute proteins can be further concentrated to enhance their detectability. One weakness of this approach is that even with 2D LC separation, it is often not possible to separate all proteins individually. This problem will be addressed under the "Protein identification and Quantification" section. In summary, when compared to the 2D-gel electrophoresis method, the tandem liquid chromatography method described here is more compatible with membrane-bound proteins as well as can enrich proteins in minute quantity.

7.2 Protein Identification and Quantification Methods

The approach we are taking represents an effort to apply systematically the most contemporary proteomics approaches to identify and develop clinically useful biomarkers for brain injury from traumatic causes, disease or drugs. Classical methods of protein identification involving protein separation by gels or liquid chromatography coupled to mass spectrometry to provide a potent and novel methodological array never applied systematically before to the detection of biomarkers of CNS injury, either alone or in combination. This integrated strategy makes possible both "targeted" analyses of known potential biomarker candidates as well as "untargeted" searches for novel proteins and protein fragments that could prove even more useful. Each of these technologies has advantages and disadvantages that together are complementary to each other. Thus, multiple proteomic strategies optimize opportunities for successful brain injury proteomic studies. Lastly, protein identification research also benefited from improved bioinformatics tools for protein database searching [9]. Thus, importantly, research designs must incorporate appropriate bioinformatics support.

(i) Mass spectrometry approach

(a) *MALDI-TOF* (matrix-assisted laser desorption ionization mass spectrometry) - time-of-flight (TOF) approach: the most classical method for protein identification in a given protein mixture is to perform 2D gel electrophoresis followed by in-gel digestion of gel band(s) of interest followed by identification of proteins by mass spectrometry. The 2D-gel method has been improved by the use of immobilized pH gradient strips for the first dimension and the ability to label protein samples from control and experimental tissues with Cy dyes (Cy3 and Cy5) that form co-migrating labeled samples that are compared in the same gel. Differentially expressed proteins are easily found, cut from the gel, digested in the gel spot by trypsin, and then identified by MALDI-TOF [5]. It is important to understand that MALDI identifies peptides based on accurate determination of peptide masses since each amino acid has a unique mass and thus any given peptide which is composed of a unique combination of sequence will have a unique mass. However, this method of protein identification is not infallible. Although rare, peptides can have identical amino acid composition with which the order of these amino acid residues could be different. Thus, it is common practice that in order to positively identify the presence of a specific protein, at least two peptide fragments from the protein must be independently identified based on their mass. In addition, any posttranslational modifications when occurs at significantly high tachometric ratio, will make this type of mass prediction extremely difficult. This method is useful for distinguishing proteins that are either up-regulated or down-regulated due to injury, but it is also sub-optimal for finding small peptides from basic, very acidic, or hydrophobic proteins. Complementary to this method are direct mass spectrometry procedures that capture the entire range of proteins and peptides, but may not distinguish proteins that are post-translationally modified, also the maximal protein size is limited to about 25,000 to 30,000 Delton. This approach is taken advantaged of by a modified MALDI approach called SELDI (invented by Ciphergen) which combines a protein separation phase with the MALDI using an affinity matrix based "Protein Chips" [65].

(b) *LC-MS/MS approach.* There are several 1D- and 2D-chromatography techniques [1] that can substitute for the 2D-gels that give reasonable resolution and include proteins that could be missed by the gel methods. These chromatography techniques can now be coupled to protein fragmentation (trypsinization) and reverse-phase chromatographic peptide separation, which is then coupled in-line with mass spectrometers. The main advantage of the in-line techniques is better recovery of peptides and thus, greater sensitivity. It is now possible to identify proteins that are present in tissues at the pM range. High-powered mass spectrometers including the quadrupole ion-trap (LCQ-Deca), the quadrupole time-of-flight (QSTAR), and the FT-ICR (Bruker BioApex 4.7) mass spectrometers can be used for identifying proteins. These methods work extremely well, especially when coupled with database searching and bioinformatics. Importantly, some of these MS can be configured to become tandem MS. The advantage of tandem MS (MS/MS) is that it can provide peptide sequence information while single MS can also provide peptide mass (see above) [29]. Briefly, in MS/MS, when peptides are ionized at the ion source in the first mass analyzer, selected peptide ions were further ionized in the collision cell. Due to the high energy of ionization inside the collision cells, peptides are actually fragmented randomly along the peptide backbone. Depending on whether the fragmentation site is at the N-terminal or the C-terminal, for each residue site, pair of a- b- and y-daughter ions will be generated. The exact mass of all the b-and y-daughter ions are then determined in the second mass analyzer. Thus, by analyzing this mass information using now available bioinformatic software, the sequence of peptide of interest can be reconstructed without ambiguity.

(ii) Protein and peptide quantification by MS

There are now no less than half a dozen MS-based protein/peptide quantification methods, which are reviewed recently [17]. In this section, we will focus on two most validated quantification methods that are applicable to TBI proteomics.

(a) *ICAT*: A direct chromatographic approach to evaluate differential expression is the use of isotope-coded affinity tags (ICAT) [28]. These tags can be used to label the protein samples on cysteine residues that are then compared mixed together following digested by trypsin. Fragments that are labeled by the tags can then be selectively isolated and analyzed by mass spectrometry. Differential expression is determined by relative peak heights of the two samples, and MS/MS sequencing and database searching directly identify the differentially expressed proteins [50, 66]. This method is very powerful and quick. We already have experience with this approach and it works well.

(b) *AQUA*: Another innovative method to quantify differential expression is through the use of *Absolute QUAntitation* (AQUA) probes [22], which involves creating synthetic peptides containing heavy isotopes that can be spiked into the trypsin digest to act as exogenous calibrants for quantitation. For this method to work one must first identify by other means the protein that is differentially expressed, for example 2D-gel electrophoresis coupled to mass spectrometry or ICAT. The calibrant peptide is then synthesized and used within tryptic digests. This is a much quicker way of evaluating the effectiveness of a biomarkers and validating differential expression than waiting a specific antibody to be developed. Our preliminary experience with this method suggests that it will be a powerful way to proceed.

(iii) Antibody panel /array approach

Protein identification is also assisted by the availability of various platforms of antibody arrays or panels (Zyomyx protein Biochips, BD Powerblot and BD antibody arrays] [27, 34, 41, 48]. These methods all rely on antibody-based capturing of protein of interest. The quantification of the captured protein is either achieved by pre-labeling (including differential labeling) of protein with fluorsencent dye (dye-labeling detection), such as BD antibody arrays, similar to the gene chip mRNA quantification method. Alternatively quantitative detection with a second primary antibody specific to the same protein antigen (sandwich detection), similar to the sandwich enzyme-linked immunoabsorbant assay (sandwich ELISA) method (such as the Zyomyx protein chips). Thirdly, the BD Powerblot, as a variant, is in fact a high-throughput western blotting (immunoblotting) system with two distinct protein samples differentially subjected to a set of 5 blots. Each blot has 39 usable lanes with the use of a manifold system. Each lane is developed with 5-6 different fluorophores-linked monoclonal antibodies (toward antigen with non-overlapping molecular weight) Thus with this method, the samples will be probed with a total of 1,000 monoclonal antibodies. We have actually conducted several Powerblot experiments with animal TBI studies.

The major advantage of the antibody panel or array approach is that proteins of interest can be readily identified since all antibodies used have known antigens and their positional assignment on the antibody chip or panel is known Also, quantification is already built-into this antibody-based approach, without any additional effort. On the other hand, the major disadvantage of this approach is that it is practically impossible to be exhaustive as one would only have high fidelity antibodies to a subset of proteins. Furthermore, if antibodies are collected from many different sources it will likely results in uneven detection sensitivity. As in other immunoassay methods (Western blotting, immunostaining or ELISA). It is a given that antibody based method will likely detect specially bound protein as well as non-specifically bound proteins or other substances. This will likely give rise to high background or false positive reactions or both. The authors believe that despite its shortcoming, the antibody-based protein identification approach is a perfect complement to the MS-based approach discussed above.

8 TBI Proteomic Bioinformatics

The current advance in databases and web portals has a natural convergence for knowledge and data sharing among local and remote scientists in any NIH domain. Large databases will be networked, while web portals will "federate and access" large databases. Such efforts need to develop for the neuroproteomic domain. Neuroscience has one of the most complex information structures - concepts, data types and algorithms - among scientific disciplines. Its richness in organisms, species, cells, genes, proteins and their signal transduction pathways provides many challenging issues for biological sciences, computational sciences, and information technology. The advances in neuroscience need urgently developing portal services to access databases for analyzing and managing information: sequences, structures, and functions arising from genes, proteins, and peptides (e.g. protein segments and biomarkers) [9].

In this bioinformatics component, two interlinked mandates are: (i) to build a local user-friendly proteomic databases, and (ii), to develop interoperable proteomic tools and architecture for multiple data integration and to integrate user and public domain-based databases. Data analysis applications should be interoperable with database operations and portal access. The TBI proteomics core technologies will provide an integrative approach to genomic and proteomic information by developing a common portal architecture, the TBI proteomics portal, at the University of Florida for data archiving and retrieval among core researchers and end users, and data linking and sharing to national and international neuroproteomic websites (e.g., Human Proteome Organization (HUPO, USA) [30] and Human Brain Proteome Project (Germany) [47]). (iii) Lastly, bioinformatics tools and software are also needed to enhance our ability to mine data, as well as to study protein-protein interaction, protein pathways and networks and complex post-translational modification such as (protein phosphorylation, processing, crosslinking and conjugation). This will help us develop knowledge bases about neuroproteomic functions and signal transduction pathways in terms of dynamic objects and processes [45, 46, 55, 62]. In addition, clinical information should be integrated with genomic and proteomic databases. The following diagram depicts the neuroproteomics bioinformatics core:

The three major functions of the bioinformatics component of TBI proteomic research can be further explained as follows:

Neuroproteomics Bioinformatics Core

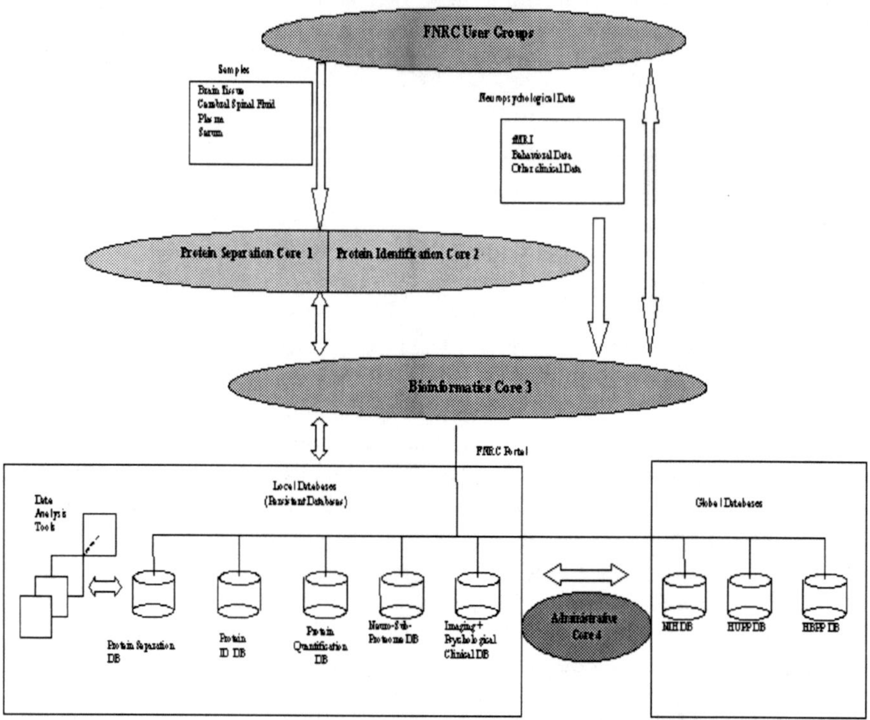

Fig. 1. Neuroproteomics Bioinformatics Core

(i) Permanence

Permanence is defined here as developing local databases for proteomics separation and identification, and link with national and international data sources. Local databases will include chromatograms, mass spectra, gel images, peptide and protein sequences, and fMRI images for control and diseased samples. Data modeling and semantics will be developed by proteomics and computer scientists together so that semantic equivalence of search attributes and semantic associations can be established.

Our Bioinformatics Core is in the process of combining different data semantics and knowledge trees in separate genomic, proteomic, and clinical databases. Our main contribution will be the development of data modeling and semantics by proteomics and computer scientists together so that semantic equivalence of search attributes and semantic associations can be established. A key requirement is the development of semantics (or ontology) of biological information, which are then captured in two components - semantic indexing and meta-information - of the intelligent search engines. A recent book of S. Chen [15] has described these two important methods. Semantic indexing extends the existing full-text, hypertext, and database indexing schemes to include semantics or ontology of information content. Meta-information (or metadata) means "information about information" concerning content, context, and archival description. Both semantic indexing and meta-information are necessary to the semantic equivalence in intelligent search engines. Meta-information contains information about not only individual objects but also whole data collections and even resources. It provides collection-wide semantics to organize a widely distributed collection of information resources better. Furthermore both semantic indexing and meta-information complement each other, reduce the complexity of neural taxonomy and classification, and correlate semantically the proteomic types and phenotypes (e.g. behavior in drug abuse) at various (subcellular, cellular, and tissue or fluid) levels of neural activities. Dissemination to national and international data sources (e.g. HUPO-USA and HBPP-Germany) will be consistently maintained through our intelligent search engines.

(ii) Interoperability

Interoperability is defined here as integrating existing data analysis tools with local databases. A proteomic problem-solving environment will be established to provide users with rapid access to TBI neuroproteome center databases and analysis tools. This will include existing tools for proteomics research and drug abuse research. The range of these tools is very broad, from peptide sequencing and protein identification to image processing for fMRI images and data analysis for neuropsychological tests and diagnosis.

A critical component of our Bioinformatics Core will be distributed search at widely distributed resources of data analysis and multiple levels of proteomic clinical and behavior information. The distributed collections of heterogeneous information resources will be large-scale. The intelligent search engines are beyond the capability of current web search engines and protocols. A distributed information retrieval system, Emerge, has implemented some aspects of semantic indexing and meta-information of NIH's PubMed and Entrez databases, in a collaboration with NCSA of UIUC. The TBI neuroproteome center distributed information retrieval component is a set of search engines extending Emerge. Such an intelligent search engine should allow nomenclature, syntactic, and semantic differences in queries, data, and meta-information. It should permit type, format, representation, and model differences as well in databases. In our TBI neuroproteome research, we have to compare information among proteomic and clinical data, such as chromatograms, mass spectra, gel images, peptide and protein sequences, and fMRI images. This intelligent search engine must go to different databases to retrieve various data of potentially different types, formats, representations, and models. In an asynchronous way, data are compared to an abstract and conceptual schema for neuroscience domains. The object-oriented data modeling helps us to establish these mappings between the abstract and conceptual schema and different database schemas. Due to the diverse nature of neuroscience information, we will need a set of interoperable search engines to guide users finding information of various domains, formats, types, and levels of granularity (e.g. peptide, protein, cell, and system levels). Since some abstract and conceptual schema has been developed for neuroscience domains, we will need a set of interoperable search engines for a wider set of analysis tools and databases.

Interoperability with analysis tools will be an important component. The starting point will provide a point and click interface for rapid access of neuroscience databases and analysis tools. The interoperability of databases and analysis tools will establish a proteomic problem-solving environment. Thus users of the problem-solving environment will also be factored into the interoperability. Whatever users need - small vs. large data sets, interactive vs. batch computation - will require design and implementation of data and event services. For the current research, we intend to develop a neuroproteomic workbench to gather a collection of data analysis tools for neuroproteomics as well as TBI neuroproteomic data sets (see data samples below) :

(1) Peptide sequencing and protein identification by MALDI-TOF-MS and capillary LC-MS/MS [43, 63].
(2) Protein peak patterns and single protein retention time from 1D or 2D-ion exchange or size exclusion chromatograms.
(3) Protein database searching algorithms such as SEQUEST [67].

The integration of databases with proteomic computational algorithms will be based on the object-oriented data modeling and data semantics discussed earlier. The ODMG compliant data analysis and databases are highly relevant to the Common Component Architecture [8]. In high throughput computing, in terms of parallel or multi-threaded objects, components (data and algorithms alike) may be distributed over a wide area grid of resources and distributed services.

(iii) Data Mining

Our neuroproteomic initiative has placed significant effort in new data mining and analysis tools for differential protein expression, protein network and modification analysis and validation. A unique data-mining workbench will be created to explore protein network and pathways underlying the pathobiology of TBI from a neuroproteomic perspective. Novel data-mining tools will include a differential analysis tool for research on proteins and protein fragments involved in TBI and construction of cognitive maps [4, 40, 68, 69], a graphical network method to represent knowledge and information. Furthermore, the cognitive maps will be used for TBI-induced Differential Neuroproteome Validation and possible brain injury diagnosis and severity monitoring. These data mining steps are described in the following:

Capillary LC separation of tryptic peptides

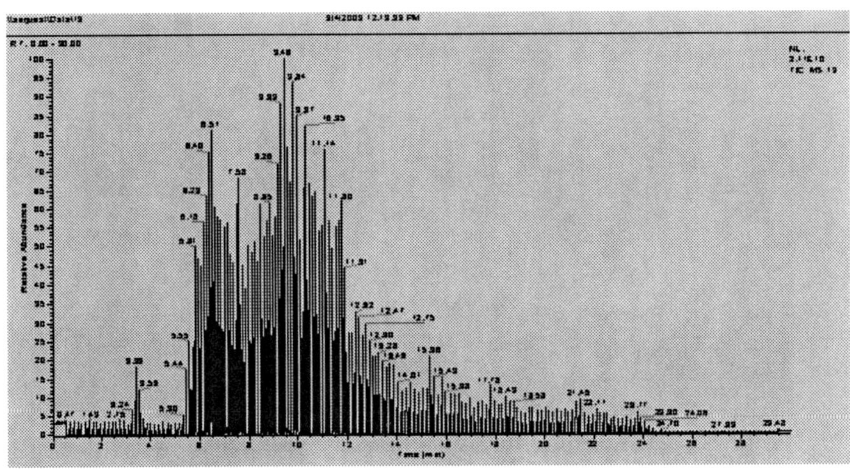

Representative CLC chromatogram of tryptic peptides from
the unlabeled injured (I) sample in gel slice 19.

Tryptic Peptides Observed in the 1st- ranked protein (Synaptojanin) in Gel Slice 22: Injured-Cy3 (I) HC

MAFSKGFRIY HKLDPPPFSL IVETRHKEEC LMFESGAVAV LSSAEKEAIK GTYAKVLDAY GLLGVLRLNL GDTMLHYLVL
VTGCMSVGKI QESEVFRVTS TEFISLRVDA SDEDRISEVR KVLNSGNFYF AWSASGVSLD LSLNAHRSMQ EHTTDNRFFW
NQSLHLHLKH YGVNCDDWLL RLMCGGVEIR TIYAAHKQAK ACLISRLSCE RAGTRFNVRG TNDDGHVANF VETEQVIYLD
DCVSSFIQIR GSVPLFWEQP GLQVGSHRVR MSRGFEANAP AFDRHFRTLK DLYGKQIVVN LLGSKEGEHM LSKAFQSHLK
ASEHASDIHM VSFDYHQMVK GGKAEKLHSV LKPQVQKFLD YGFFYFDGSA VQRCQSGTVR TNCLDCLDRT NSVQAFLGLE
MLAKQLEALG LAEKPQLVTR FQEVFRSMWS VNGDSISKIY AGTGALEGKA KLKDGARSVT RTIQNNFFDS SKQEAIDVLL
LGNTLNSDLA DKARALLTTG SLRVSEQTLQ SASSKVLKNM CENFYKYSKP KKIRVCVGTW NVNGGKQFRS IAFKNQTLTD
WLLDAPKLAG IQEFQDKRSK PTDIFAIDFE EMVELNAGNI VNASTTNQKL WAVELQKTIS RDNKYVLLAS EQLVGVCLFV
FIRPQHAPFI RDVAVDTVKT GMGGATGNKG AVAIRMLFHT TSLCFVCSHF AAGQSQVKER NEDFVEIARK LSFPMGRMLF
SHDYVFWCGD FNYRIDLPNE EVKELIRQ QN WDSLIAGDQL INQKNAGQIF RGFLEGKVTF APTYKYDLFS EDYDTSEKCR
TPAWTDRVLW RRRKWPFDRS AEDLDLLNAS FQDESKILYT WTPGTLLHYG RAELKTSDHR PVVALIDIDI FEVEAEERQK
IYKEVIAVQG PPDGTVLVSI KSSAQENTFF DDALIDELLQ QFAHFGEVIL IRFVEDKMWV TFLEGSSALN VLSLNGKELL
NRTITITLKS PDWIKTLEEE MSLEKISVTL PSSTSSTLLG EDAEVSADED MEGDVDDYSA EVEELLPQHL QPSSSSGLGT
SPSSSSPRTSP CQSPTAPEYS APSLPIRPSR APSRTPGPLS SQGAPVDTQP AAQKESSQTI EPKRPPPPRP VAPPARPAPP
QRPPPPSGAR SPAPARKEFG APKSPGTARK DNGRVQPSPS QAGLRGPSPS GTGAARPTIP ARAGVISAPQ SQARVSAGRL
TPESQSKPLE TSKGPAVLPE PLKPQAAFPP QPSLPTPAQK LQDPLVPIAA PMPPSIPQSN LETPPLPPPR SRSSQSLPSD
SSPQLQQEQP TG

Mass (mono): 142869.8   **Identifier:** gi|1166575   **Database:** C:/Xcalibur/database/rat.fasta
Protein Coverage: 262/1292 = **20.3 %** by amino acid count, 28224.7/142869.8 = **19.8%** by mass

1ˢᵗ-ranked Tryptic Peptide MS-MS analysis in Gel Slice 22 (Synaptojanin): Injured-Cy3 (I) HC

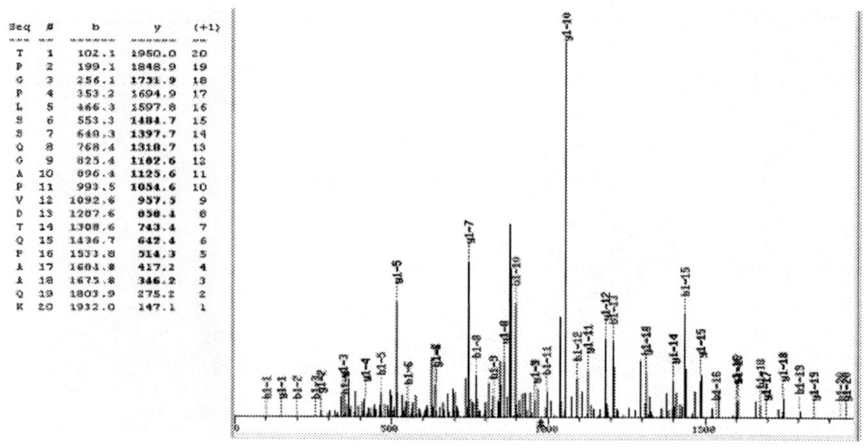

Tryptic peptide ID: TPGPLSSQGAPVDTQPAAQK

Fig. 2. Observed Data of Tryptic Peptides

Dataming Steps

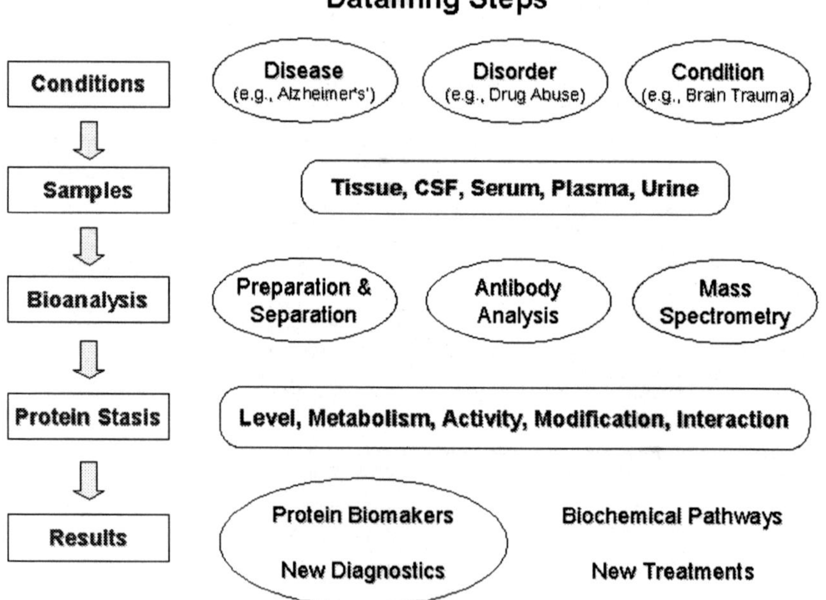

Fig. 3. Data Mining Steps

a) Creating Cognitive Maps for TBI-induced differential proteome

New data mining tools for TBI-induced differential proteome analysis and validation are being developed at our center. There are three major zones of neuroproteomics information (i) pathophysiological stasis (including TBI, other CNS injuries, such as ischemic stroke, aging, environmental toxin or substance abuse-induced brain injury, neurodegenerative diseases such as Alzheimer's disease or Parkinsonism), (ii) neuroproteome stasis (such as differential protein expression, protein synthesis and metabolism, alternative mRNA splicing and RNA editing, protein-protein interaction, enzymatic activity or protein functions) post-translational modifications (such as protein crosslinking, acetylation, glycosylation protein proteolysis and processing, phosphorylation) and protein-protein interaction networks and signal transduction pathways and (iii) sources of neuroproteomic data (brain tissue from different areas or anatomical regions of the brain, such as hippocampus), biological fluids such as the cerebrospinal fluid (CSF), blood samples (including plasma and serum) where brain proteins stasis might be reflected upon via diffusion-based equilibrium or blood brain barrier compromise (e.g. from brain to CSF to blood).

Collection of data from these three components will enable the construction of multiple cognitive maps [4, 40, 68, 69]. For instance, cognitive maps can be constructed for the TBI-induced differential proteome in the following figure. Automated reasoning and knowledge discovery algorithms on the cognitive maps [10, 11, 12, 13, 14, 15, 39] will distill the information and present the knowledge gained from a systems biology perspective. Thus, cognitive maps will enable the brain trauma researchers to gain a greater understanding of the entire TBI-induced differential neuroproteome and hopefully the mechanistic protein-pathways of TBI.

b) Using Cognitive Maps for TBI-induced Differential Neuroproteome Validation

A statistical analysis tool is also being developed for TBI-induced differential neuroproteome validation and possible TBI protein-pathways elucidation. For example, up- or down-regulation of multiple proteins and protein fragments in control and injured samples will be quantified by ICAT, AQUA, or ELISA to validate differential TBI neuroproteome. Linear discriminant analysis (LDA) will be used to calculate the probability of a correct diagnosis given the number of injury-specific biomarkers measured the number of samples, etc. Thus, statistical analysis tools are expected to provide an important component for all the neuroproteomics research conducted at our neuroproteomic center. These statistical analysis data will be fed into the cognitive maps to reach decision on

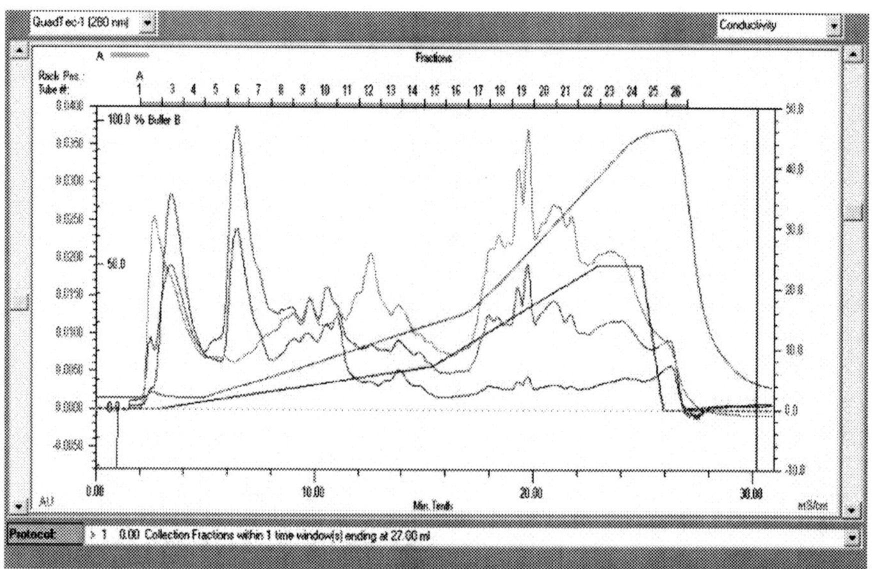

Fig. 4. Input Data for Neuroproteomics Cognitive Maps

diagnosis, monitoring and treatment. We have both statistical/probabilistic and fuzzy reasoning capabilities in our cognitive maps [40, 68, 69].

Cognitive maps are directed graphs representing relations (by links) among concepts/attributes (by nodes). Cognitive maps include several knowledge representation schemes. Semantic networks or frames form a special class of cognitive maps. Inference networks and causal networks form other classes of cognitive maps. In cognitive maps, link weights may be assigned to relations representing their compatibility degrees, and node values may be assigned to concepts and attributes representing relevance factors. A hierarchical cognitive map consists of several cognitive maps, each of which represents gene network interaction or metabolic pathway. The knowledge bases of hierarchical cognitive maps will effectively capture the complex behavior of biological systems. A hierarchical cognitive map is alternatively represented as a large cognitive map combining several individual ones in the following diagram.

Cognitive maps can extend to probabilistic, or fuzzy cognitive maps, and further to neural network learning maps. These numerically enabled cognitive maps can be interfaced with other numerical simulation packages in biology.

Now we briefly describe the relaxation computation in a cognitive map. Let Σ be a collection of biological objects $\{x_1,...,x_n\}$ (e.g., gene sequences, protein structures, metabolites, genotypes, and phenotypes), and let Λ be a collection of labels $\{\lambda_1,...,\lambda_m\}$ with any mathematical structure (e.g., concentrations and intensities). The labeling problem is to find a consistent labeling of biological objects in Σ by Λ, given a set of relations among objects and

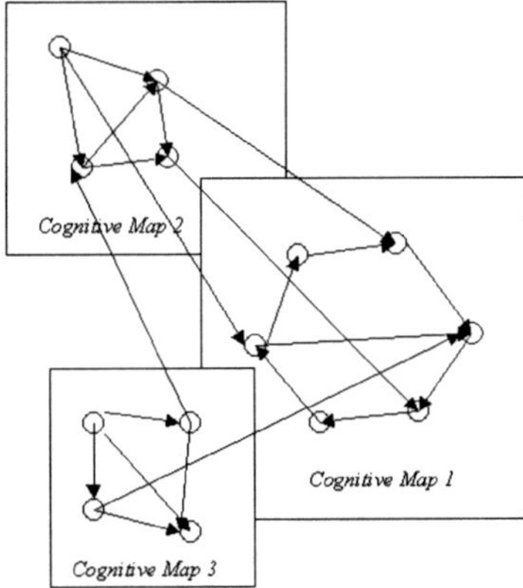

Fig. 5. Hierarchical Cognitive Maps

a set of constraints among objects and their labels. For each x_i, let Λ_i be a subset of Λ that is compatible with x_i. For any pair $\{x_i, x_j\}$ of objects $(i, j$ distinct), let Λ_{ij} be a subset of compatible pairs of labels in $\Lambda_i \times \Lambda_j$. A labeling $L = \{L_1, ..., L_n\}$ is an assignment of a set of labels Λ_i in Λ to each x_i. L is consistent if for each i, j and all λ in Λ_i, $(\{\lambda\} \times \Lambda_j)$ intersects with Λ_{ij}. L is unambiguous if it is consistent and assigns only a single label to each object. The semantic labeling of cognitive maps determines the results of TBI-induced differential neuroproteome validation. The semantic labeling is to assign a measure $m_i(\lambda)$ to the statement "λ is the correct label of x_i". An arbitrary labeling of a neuroproteome may not be consistent and unambiguous, because the constraint satisfaction is required among either objects or a combination of new input evidences. The interaction with external users and systems is through a query system. At the initial stage, the $m_i(\lambda)$ is either estimated by the user or is provided by another cognitive map or simulation tool. Now the initial measures go through a constraint satisfaction checking by the label relaxation, which iterates the process until the convergence to final measures is reached. The final measures are sent back to the query subsystem for either clinical decision or further data analysis.

The relaxation scheme is mathematically described as follows. An initial assignment of measures $\{m_i(0)(\lambda)\}$ to $\{x_i\}$ is given at time 0. A relaxation operator R is defined to transform one set $\{m_i(k)(\lambda)\}$ of measures to another set $\{m_i(k+1)(\lambda)\}$. The limit $\{m_i^*(\lambda)\}$ of $\{m_i(k)(\lambda)\}$ gives the unambiguous labeling under compatibility constraints, as k approaches to infinity. In reality, we expect the limit to be attained after a finite number of iterations. In practice, the limit $\{m_i^*(\lambda)\}$ may not be unique (we are not always getting an unambiguous labeling). The multiple labelings are sent back to the users so that they can select an appropriate result for further analysis.

There are several ways to define the relaxation operator R. A relaxation operator R should produce $m_i(k+1)(\lambda)$ from the combination of

$m_i(k)(\lambda)$ and support $s_i(k)(\lambda)$ by some update equations, where $s_i(k)(\lambda) = \Sigma r_{ij}(\lambda s \lambda')m_j(k)(\lambda')$, where $r_{ij}(\lambda, \lambda')$ is the compatibility function of "label λ is assigned to x_i and label λ' is assigned to x_j", and j-indices are indices of all source nodes leading to the i-th node. A relaxation operator R is defined by the following update equations:

$$m_i(k+1)(\lambda) = min[1, max(0, m_i(k)(\lambda) + s_i(k)(\lambda))],$$

$$s_i(k)(\lambda) = \Sigma(r_{ij}(k)(\lambda, \lambda') + \Delta r_{ij}(k)(\lambda, \lambda'))m_j(k)(\lambda'),$$

$$\Delta r_{ij}(k+1)(\lambda, \lambda') = a_{ij}\Delta r_{ij}(k)(\lambda, \lambda') + b_{ij}m_i(k+1)(\lambda)m_j(k)(\lambda'),$$

where a_{ij} and b_{ij} are learning parameters. The first equation makes sure that $m_i(k+1)(\lambda)$ stays between 0 and 1. The second equation provides the network input to the (i, λ)-th node. The third equation includes the Hebbian learning rule.

9 Conclusion

In summary, proteomic studies of both human and rat traumatic brain injury, if approached systemically, is a very fruitful and powerful analytic technology. In order to obtain a comprehensive TBI neuroproteome data set, it is important to integrate multiple protein separation and protein identification technologies. Equally important is the optimization of individual protein separation identification methods. Bioinformatics platform then becomes the critical adhesive component by serving two purposes: (i) integrating all proteomic data sets and other relevant biological or clinical information, and (ii) inferring and elucidating the protein-based pathways and biochemical mechanisms underlying the pathobiology of TBI and identifying and validating biomarkers for the diagnosis and monitoring of TBI [23]. Ultimately, if we are to be successful in doing these, the TBI proteomic approach outlined here must be further integrated with genomic, cytomics as well as systems biology approaches [37, 38].

References

1. J.N. Adkins, S.M. Varnum, K.J. Auberry, R.J. Moore, N.H. Angell, R.D. Smith, D.L. Springer, and J.G. Pounds. Toward a human blood serum proteome: analysis by multidimensional separation coupled with mass spectrometry. *Molecular & Cellular Proteomics*, 1: 947-955, 2002.

2. R. Aebersold and J.D. Watts. The need for national centers for proteomics. *Nature Biotechnology*, 20(7): 651-651, 2002.

3. R.D. Appel, A. Bairoch, and D.F. Hochstrasser. 2-D Databases on the World Wide Web in Methods in Molecular Biology. In A.J. Link, editor, *2-D Proteome Analysis Protocols*, pages 383-391. Humana Press, Totowa, NJ, 1999.

4. R. Axelrod. *Structure of Decision*. Princeton University Press, 1976.

5. W.V. Bienvenut, C. Deon, C. Pasquarello, J.M. Campbell, J.C. Sanchez, M.L. Vestal, and D.F. Hochstrasser. Matrix-assisted laser desorption/ionization-tandem mass spectrometry with high resolution and sensitivity for identification and characterization of proteins. *Proteomics*, 2(7): 868-876, 2002.

6. B. Bjellqvist, K. Ek, P.G. Righetti, E. Gianazza, A. Gorg, and R. Westermeier. Isoelectric focusing in immobilized pH gradients: Principle, methodology and some applications. *Journal of Biochemical and Biophysical Methods*, 6: 317-339, 1982.

7. J. Boguslavsky. Resolving the proteome by relying on 2DE methods. *Drug Discovery & Development*, 6(7): 57-60, 2003.

8. R.G.G. Cattell, editor. *The Object Database Standard: ODMG-93*, Morgan Kaufmann, 1996.

9. D.N. Chakravarti, B. Chakrarti, and I. Moutsatsos. Informatic tools for proteome profiling. *BioTechniques, Computational Proteomics Supplement*, 32: S4-S15, 2002.

10. S. Chen. Knowledge acquisition on neural networks. In B. Bouchon, L. Saitta and R. R. Yager, editors, *Uncertainty and Intelligent Systems*, pages 281-289. Lecture Notes in Computer Science, Springer-Verlag, Vol. 313, 1988.

11. S. Chen. Some extensions of probabilistic logic. *Proceedings of the AAAI Workshop on Uncertainty in Artificial Intelligence*, Philadelphia, PA, August 8-10, pages 43-48, 1986; An extended version appeared in *Uncertainty in Artificial Intelligence*, Vol. 2, edited by L. N. Kanal and J. F. Lemmer, North-Holland, 1986.

12. S. Chen. Automated reasoning on neural networks: A probabilistic approach. IEEE First International Conference on Neural Networks, San Diego CA, June 21-24, 1987.

13. S. Chen. Knowledge discovery of gene functions and metabolic pathways. IEEE BioInformatic and Biomedical Engineering Conference, Washington, DC, November 2000.

14. S. Chen. Knowledge representation for systems biology. First International Conference on Systems Biology, Tokyo Japan, November 14-16, 2000.

15. S. Chen. *Digital Libraries: The Life Cycle of Information*. Better Earth Publisher, 1998.

384   Su-Shing Chen et al.

16. P. Davidsson, A. Westman-Brinkmalm, C.L. Nilsson, M. Lindbjer, L. Paulson, N. Andreasen, M. Sjogren, and K. Blennow. Proteome analysis of cerebrospinal fluid proteins in Alzheimer patients. *Neuroreport*, 13(5):611-5, 2002.

17. N.D. Denslow, M.E. Michel, M.D. Temple, C. Hsu, K. Saatman, and R.L. Hayes. Application of Proteomics Technology to the Field of Neurotrauma. *Journal of Neurotrauma*, 20: 401-407, (2003).

18. S. Derra. Lab-on-a-chip technologies emerging from infancy. *Drug Discovery & Development*, 6(5): 40-45, 2003.

19. J. Finnie Animal models of traumatic brain injury: a review. *Australian Veterinary Journal*, 79(9): 628-633, 2001.

20. M. Fountoulakis, R. Hardmaier, E. Schuller, and G. Lubec. Differences in protein level between neonatal and adult brain. *Electrophoresis*, 21(3): 673-678, 2000.

21. M. Fountoulakis, E. Schuller, R. Hardmeier, P. Berndt, and G. Lubec. Rat brain proteins: two-dimensional protein database and variations in the expression level. *Electrophoresis*, 20(18): 3572-3579, 1999.

22. S.A. Gerber, J. Rush, O. Stemman, M.W. Kirschner, and S.P. Gygi. Absolute quantification of proteins and phosphoproteins from cell lysates by tandem MS. *Proceedings of the National Academy of Sciences*, 100: 6940-6945, 2003.

23. I. Goldknopf, H.R. Park, and H.M. Kuerer. Merging diagnostics with therapeutic proteomics. *IVD Technology*, 9(1): 1-6, 2003.

24. A. Gorg, W. Postel, and S. Gunther. The current state of two-dimensional electrophoresis with immobilized pH gradients. *Electrophoresis*, 9: 531-546, 1988.

25. S.G. Grant, W.P. Blackstock. Proteomics in neuroscience: from protein to network. *Journal of Neuroscience*, 21(21): 8315-8, 2001.

26. S.G.N. Grant and H. Husi. Proteomics of multiprotein complexes: answering fundamental questions in neuroscience. *Trends in Biotechnology*, 19(10 Suppl):S49-54, 2001.

27. S. Graslund, R. Falk, E. Brundell, C. Hoog, and S. Stahl. A high-stringency proteomics concept aimed for generation of antibodies specific for cDNA-encoded proteins. *Biotechnology and Applied Biochemistry*, 35(Pt 2): 75-82, 2002.

28. S.P. Gygi, B. Rist, S.A. Gerber, F. Turecek, M.H. Gelb, and R. Aebersold. Quantitative analysis of complex protein mixtures using isotope-coded affinity tags. *Nature Biotechnology*, 17: 994-9, 1999.

29. S.P. Gygi and R. Aebersold. Mass spectrometry and proteomics. *Current Opinion in Chemical Biology*, 4: 489494, 2000.

30. S. Hanash. HUPO initiatives relevant to clinical proteomics. *Molecular and Cellular Proteomics*, 3: 298-301, 2004.

31. S. Hanash. Disease proteomics. *Nature*, 422(6928): 226-232, 2003.

32. D.F. Hochstrasser, J.C. Sanchez, and R.D. Appel. Proteomics and its trends facing nature's complexity. *Proteomics*, 2(7): 807-812, 2002.

33. H. Husi and S.G. Grant. Proteomics in neuroscience: from protein to network. *Journal of Neuroscience*, 21(21): 8315-8318, 2001.

34. P. James. Chips for proteomics; a new tool or just hype? *Biotechniques*, Suppl:4-10, 2002

35. D. Janssen. Major approaches to identifying key PTMs. *Genomics and Proteomics*, 3(1): 38-41, 2003.

36. P. Jungblut, B. Thiede, U. Zimny-Arndt, E.C. Muller, C. Scheler, and B. Wittmann-Liebold. Resolution power of two-dimensional electrophoresis and identification of proteins from gels. *Electrophoresis*, 17: 839-847, 1996.

37. H. Kitano. Systems biology: a brief overview. *Science*, 295(5560): 1662-4, 2002.

38. H. Kitano. Computational systems biology. *Nature*, 420(6912): 206-10, 2002.

39. H. Kitano. Perspectives on systems biology. *New Generation Computing*, 18(3): 199-216, 2000.

40. B. Kosko. Fuzzy cognitive maps. *International Journal of Man-Machine Studies*, 24: 65-75, 1986.

41. W. Kusnezow and J.D. Hoheisel. Antibody microarrays: promises and problems. *BioTechniques*, 33(suppl.): 1423, 2002.

42. P.F. Lemkin. Comparing Two-dimensional Electrophoretic Gel Images Across the Internet. *Electrophoresis*, 18: 2759-2773, 1997.

43. B. Lu and T. Chen. A suffix tree approach to the interpretation of tandem mass spectra: applications to peptides of non-specific digestion and post-translational modifications. *Bioinformatics*, 19(Suppl 2): II113-II121, 2003.

44. G. Lubec, K. Krapfenbauer, and M. Fountoulakis. Proteomics in brain research: potentials and limitations. *Progress in Neurobiology*, 69(3): 193-211, 2003.

45. P. Mendes. GEPASI: a software for modeling the dynamics, steady states and control of biochemical and other systems. *Computer Applications in the Biosciences*, 9(5): 563-571, 1993.

46. P. Mendes. Biochemistry by numbers: simulation of biochemical pathways with Gepasi 3. *Trends in Biochemical Sciences*, 22: 361-363, 1997.

47. H.E. Meyer, J. Klose, and M. Hamacher. HBPP and the pursuit of standardisation. *Lancet Neurology*, 2(11): 657-658, 2003.

48. M.D. Moody. Array-based ELISAs for high-throughput analysis of human cytokines. *Biotechniques*, 31: 186-194, 2001.

49. W.F. Patton. Detection technologies in proteome analysis. *Journal of Chromatography. B, Analytical Technologies in the Biomedical and Life Sciences*, 771(1-2): 3-31, 2002.

50. J. Peng, J.E. Elias, C.C. Thoreen, L.J. Licklider, and S.P. Gygi. Evaluation of multidimensional chromatography coupled with tandem mass spectrometry (LC/LC-MS/MS) for large scale protein analysis: The yeast proteome. *Journal of Proteome Research*, 2: 43-50, 2003.

51. B.R. Pike, J. Flint, J.R. Dave, X.-C. Lu, K.K.W. Wang, F.C. Tortella, and R.L. Hayes. Accumulation of calpain and caspase-3 proteolytic fragments of brain-derived αII-spectrin in CSF after middle cerebral artery occlusion in rats. *Journal of Cerebral Blood Flow & Metabolism*, 24(1): 98-106, 2004.

52. B.R. Pike, J. Flint, S. Dutta, E. Johnson, K.K.W. Wang, and R.L. Hayes. Accumulation of non-erythroid αII-spectrin and calpain-cleaved αII-spectrin breakdown products in cerebrospinal fluid after traumatic brain injury in rats. *Journal of Neurochemistry*, 78: 1297-1306, 2001.

53. A. Raabe, C. Grolms, and V. Seifert. Serum markers of brain damage and outcome prediction in patients after severe head injury. *British Journal of Neurosurgery*, 13: 56-59, 1999.

54. R. Raghupathi, D.I. Graham, and T.K. McIntosh. Apoptosis after traumatic brain injury. *Journal of Neurotrauma*, 17(10): 927-938, 2000.

55. V.N. Reddy, M.L. Mavrovouniotis, and M. N. Liebman. Petri net representations in metabolic pathways. *Proceedings of ISMB*, pp. 328-336, 1993.

56. D.R. Reyes, D. Iossifidis, P.A. Auroux, and A. Manz. Micro total analysis systems. 1. Introduction, theory, and technology. *Analytical Chemistry*, 74(12): 2623-2636, 2002.

57. B. Romner, T. Ingebrigtsen, P. Kongstad, S.E. Borgesen. Traumatic brain damage: serum S-100 protein measurements related to neuroradiological findings. *Journal of Neurotrauma*, 17(8): 641-647, 2000.

58. H. Schäfer, K. Marcus, A. Sickmann, M. Herrmann, J. Klose, and H.E. Meyer. Identification of phosphorylation and acetylation sites in alphaA-crystallin of the eye lens (mus musculus) after two-dimensional gel electrophoresis. *Analytical and Bioanalytical Chemistry*, 376(7): 966-972, 2003.

59. D.C. Schwartz and M. Hochstrasser. A superfamily of protein tags: ubiquitin, SUMO and related modifiers. *Trends in Biochemical Sciences*, 28(6): 321-328, 2003.

60. R.F. Service. Gold rush - High-speed biologists search for gold in proteins. *Science*, 294(5549): 2074-2077, 2001.

61. R.D. Smith. Probing proteomes–seeing the whole picture? *Nature Biotechnology*, 18: 1041-1042, 2000.

62. R. Somogyi and C.A. Sniegoski. Modeling the complexity of genetic networks: understanding multigenic and pleiotropic regulation. *Complexity*, 1: 45-63, 1996.

63. D.L. Tabb, W.H. McDonald, and J.R. Yates. DTASelect and contrast: Tools for assembling and comparing protein identifications from shotgun proteomics. *Journal of Proteome Research*, 1: 21-26, 2002.

64. M. Unlu, M. E. Morgan, and J.S. Minden. Difference gel electrophoresis: a single gel method for detecting changes in protein extracts. *Electrophoresis*, 18(11): 2071-2077, 1997.

65. A. Wiesner. Detection of Tumor Markers with ProteinChip(R) Technology. *Current Pharmaceutical Biotechnology*, 5(1): 45-67, 2004.

66. J.R. Yates III, E. Carmack, L. Hays, A.J. Link, and J.K. Eng. Automated Protein Identification using Microcolumn Liquid Chromatography-Tandem Mass Spectrometry. In A.J. Link, editor, *2-D Proteome Analysis Protocols*, pages 553-569. Humana Press, Totowa, NJ, 1999.

67. J.R. Yates, S.F. Morgan, C.L. Gatlin, P.R. Griffin, and J.K. Eng. Method to compare collision-induced dissociation spectra of peptides: Potential for library searching and subtractive analysis. *Analytical Chemistry*, 70: 3557-3565, 1998.

68. W.R. Zhang, S. Chen, W. Wang, and R.S. King. A cognitive map based approach to the coordination of distributed cooperative agents. *IEEE Transactions on Systems, Man, and Cybernetics*, 22: 103-114, 1992.

69. W.R. Zhang, S. Chen, and J.C. Bezdek. Pool2: A generic system for cognitive map development and decision analysis. *IEEE Transactions on Systems, Man, and Cybernetics*, 19: 31-39, 1989.

## Part IV

Characterization and Prediction of Protein Structure

Computational Methods for Protein Fold Prediction: an Ab-initio Topological Approach

G. Ceci[1,5], A. Mucherino[1,5], M. D'Apuzzo[1,4,5], D. Di Serafino[1,4,5*], S. Costantini[2,3,5], A. Facchiano[3,4,5], and G. Colonna[2,4,5]

[1] Department of Mathematics, Second University of Naples
via Vivaldi 43, I-81100 Caserta, Italy
[2] Department of Biochemistry and Biophysics, Second University of Naples,
via Costantinopoli 16, I-80138 Naples, Italy
[3] Institute of Food Science, CNR, via Roma 52 A/C, I-83100 Avellino, Italy
[4] Research Center of Computational and Biotechnological Sciences (CRISCEB),
Second University of Naples, via Costantinopoli 16, I-80138 Naples, Italy
[5] Computational Biology Doctorate, Second University of Naples, Italy

Summary. The prediction of protein native conformations is still a big challenge in science, although a strong research activity has been carried out on this topic in the last decades. In this chapter we focus on ab-initio computational methods for protein fold predictions that do not rely heavily on comparisons with known protein structures and hence appear to be the most promising methods for determining conformations not yet been observed experimentally. To identify main trends in the research concerning protein fold predictions, we briefly review several ab-initio methods, including a recent topological approach that models the protein conformation as a tube having maximum thickness without any self-contacts. This representation leads to a constrained global optimization problem. We introduce a modification in the tube model to increase the compactness of the computed conformations, and present results of computational experiments devoted to simulating α-helices and all-α proteins. A Metropolis Monte Carlo Simulated Annealing algorithm is used to search the protein conformational space.

Key words: Protein fold prediction, Ab-initio methods, Native state topology, Tube thickness, Global optimization, Simulated annealing

1 Introduction

Proteins are heteropolymers that control and regulate many vital functions [66, 67, 68], hence they are considered the building blocks of living organisms. A protein is made of a sequence of amino acid residues connected by peptide

---

* Corresponding author. Email: `daniela.diserafino@unina2.it`.

bonds, called *primary structure*, which folds into a unique three-dimensional conformation, called *tertiary structure* or *native state*. The biological function of a protein is largely determined by its native state; the knowledge of the native state is therefore critical in understanding the role of the protein in the cell and the related molecular mechanisms. Levinthal's paradox [48] and Anfinsen's experiment [5] suggest that the Nature applies an "algorithm" to drive a protein from its primary structure to its own tertiary structure, and that the information needed to perform this algorithm is contained in the primary structure. Understanding the *protein folding problem* means understanding and reproducing this algorithm.

Many scientists have been working on the protein folding problem for nearly half a century. A growing interest in its solution has been observed during the years, because of its impact in several research fields, such as genetic disease treatment, drug design, and the emerging structural and functional genomics. However, despite the research has been very active, we are still far from a clear and full explanation of the protein folding mechanisms and this problem is still considered a big challenge in science.

Different computational approaches to the protein fold prediction have been developed. We focus our attention on the so-called *ab-initio* methods that do not rely heavily on comparisons with known protein structures and appear to be the most promising for determining three-dimensional conformations that have not yet observed experimentally. These methods are usually based on suitable representations of the polypeptide chain and on suitable energy functions reproducing physicochemical interactions among protein atoms. According to Anfinsen's hypotesis, the native state corresponds to the minimum energy of the system and its determination requires the solution of a (computationally demanding) global constrained optimization problem.

Recent studies have emphasized the role of the *topology of the native state* in the protein folding process [11, 42, 69, 80]. In this context, an ab-initio method has been developed that takes into account mainly topological rather than physicochemical features of the protein [7, 8, 9, 10, 54]. This method is based on a very simplified model that represents the polymer chain as a tube of nonzero thickness, without self-contacts. As in other approaches, this formulation leads to a constrained global optimization problem. In this chapter we present a modified version of this model, discuss the choice of model parameters and show results of computational experiments devoted to simulating α-helices and all-α proteins.

The chapter is organized as follows. In Section 2 we provide a very short description of the chemical structure of a protein to better understand the terminology used in the remainder of the chapter. In Section 3 we introduce the three main computational approaches to the protein fold prediction problem: *homology modeling, fold recognition* and *ab-initio methods*. In Sections 4 and 5 we provide a brief description of energy functions and global optimization techniques, that characterize a variety of ab-initio approaches. Following Klepeis and Floudas [37], ab-initio methods can be further classified as ab-initio methods that require database information and "true" ab-initio methods, that are based only on information obtained from physicochemical principles. A survey of methods falling into both classes is provided in Sections 6 and 7. Among the true ab-initio methods, we present also recent approaches based on topological features of the proteins. This survey is not meant to be exhaustive; it rather gives an idea of the evolution of main trends in the ab-initio protein folding research, along with successes and limitations. In Section 8 we focus on a specific topological model and give the mathematical description of the corresponding constrained global optimization problem, while in Section 9 we discuss how the values of the model parameters have been chosen. In Section 10, after a short presentation of the Simulated Annealing algorithm used to solve the optimization problem, we report results of our computational experiments. A few concluding remarks are given in Section 11.

2 The Chemical Structure of a Protein

A protein is a polymer composed by a sequence of genetically driven amino acid residues. Proteins in living cells are built from a set of only 20 different amino acids, all having two main substructures: a common basic substructure composed by an amide group (NH_2), a carboxyl group (COOH) and a hydrogen atom (H), all linked to a central carbon atom called C_α, and a substructure that differentiates each amino acid, called *side chain* or *R-group*, composed by chemically different residues. A schematic representation of an amino acid is given in Figure 2. The carbon atom of the carboxyl group is usually called C'. Consecutive amino acids are connected by a *peptide bond*, i.e. the carboxyl group of the *i*-th amino acid of the sequence is linked, through a covalent bond, to the amide group of the $(i + 1)$-th amino acid and a H_2O molecule is released, as shown in Figure 2. Therefore, the whole structure of the protein consists of a "main chain" of atoms, made of the linked $NC_\alpha C'O$ components of amino acids, and a number of side chains, with a shape similar to a fishbone. For this similarity, the main chain is also called *backbone*. The sequence of amino acids specific of each different protein is called *primary structure*.

As previously observed, the information contained into the chain of amino acid residues determines the unique three-dimensional conformation of a protein, i.e. its own *native state* or *tertiary structure*. Folded proteins usually contain one or more local, repetitive spatial arrangements of amino acid residues, with characteristic conformations, called *secondary structures*. The most common secondary structures found in proteins are α-helices, β-sheets and loop/turns. Examples of α-helices and β-sheets are given in Figure 2.

Protein tertiary structures can be described in terms of bond lengths (i.e. distances between two atoms connected with a covalent bond), bond angles (i.e. angles between two adjacent bond vectors, where a bond vector is identified by two atoms connected with a covalent bond) and dihedral angles (i.e. angles between the normals to the planes defined by suitable consecutive triplets of atoms). When the protein is at its equilibrium state, the bond lengths and bond angles can be considered approximately fixed, so that the three-dimensional conformation is determined by the dihedral angles. These angles are conventionally denoted with the letters Φ, Ψ, ω and χ. The former three angles characterize the protein backbone, while the latter is related to the side chains. A representation of Φ, Ψ and ω is given in Figure 2, where the indices $i-1$, i and $i+1$ identify three consecutive amino acid residues. For more details the reader is referred, for example, to [53].

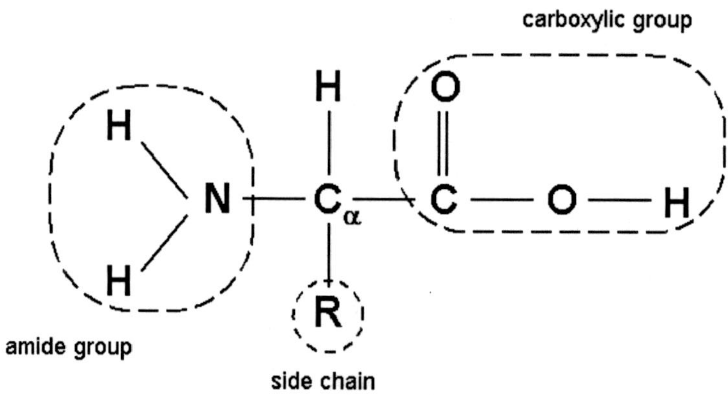

Fig. 1. Schematic representation of an amino acid.

3 Computational Approaches to Protein Fold Prediction

Computational approaches to predict protein three-dimensional conformations are usually classified as *homology modeling* (or *comparative modeling*), *fold recognition* (or *threading*) and *folding ab initio* (see, for example, [18]).

Homology modeling is based on the idea that proteins having strong sequence similarity have also strong structure and function similarity. Given a sequence of amino acid residues, homology modeling methods essentially try to align the target sequence to suitable structure templates, stored in protein databases, and build a three-dimensional conformation by using alignment information (see, for example, [14, 17, 79]). Different alignment methods have been developed, such as BLAST [3], PSI-BLAST [4] and the profile-profile method [41]. The main limitation of the homology modeling methods is that they work effectively only for sequences with at least 30-40% identity. For smaller identity percentages, they have a low reliability (see, for example,

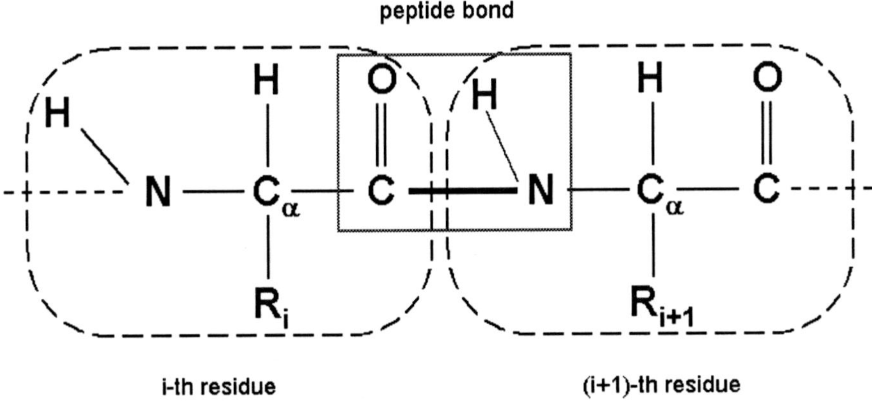

Fig. 2. A peptide bond between two amino acid residues.

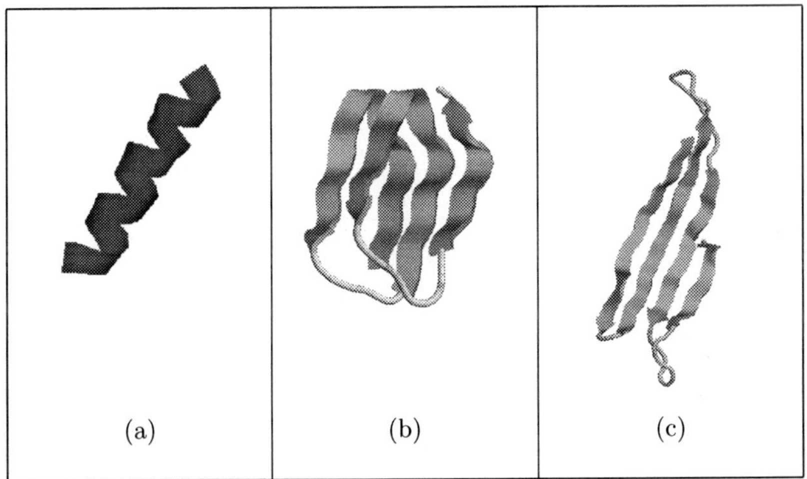

Fig. 3. Examples of protein secondary structures: α-helix (a), parallel β-sheet (b), anti-parallel β-sheet (c).

[22]). A further limitation is that only 15-25% of sequences have homologous proteins with known three-dimensional conformation in a given genome.

Fold recognition methods are based on the idea that there may be only a limited number of different protein folds. Therefore, they try to predict the protein conformation from known three-dimensional structures that do not have homologous characteristics. To this aim, a library of structure templates is defined, then the target sequence is fitted to each library entry and an

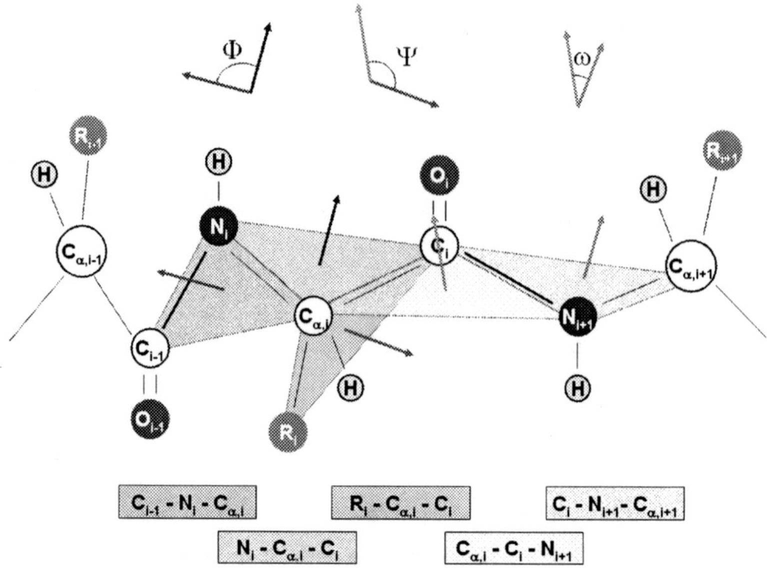

Fig. 4. The protein dihedral angles Φ, Ψ and ω.

energy function is used to evaluate the fit and hence to determine the most suitable template. Obviously, the quality of the obtained model is limited by the actual presence of the correct template into the database and by the actual similarity of the selected templates.

Ab-initio methods are potentially able to predict three-dimensional conformations not yet been observed experimentally. The basic idea behind these methods is that, according to thermodynamic principles, a protein spontaneously folds into its native state, which corresponds to a global minimum of free energy. As already observed, ab-initio techniques can be divided into two categories, one including the methods that use knowledge-based information, such as secondary structure information stored in databases, the other including the methods that do not exploit structural databases during folding predictions.

As discussed in [27, 61], ab-initio methods are generally characterized by suitable protein representations, by energy functions that take into account physicochemical interactions, and by efficient algorithms to search the feasible conformational space. Computational models of proteins explicitly treating all degrees of freedom are currently impractical because of the huge size of the conformational space, of the large number of intramolecular/intermolecular interactions and of the protein complex topology. Both high-resolution and low-resolution models introduce simplifications. High-resolution models taking into account detailed information about the protein conformation are more rigorous, but lead to problems that are more difficult to be solved. On the other hand, low-resolution models, based on simplified molecular descriptions or structural restraints, can provide only simplified fold descriptions, but are able to give insights into thermodynamic and kinetic properties of the protein folding process.

4 Energy Functions

Energy modeling plays a critical role in protein folding simulations. A large number of energy functions, also called force fields, has therefore been developed to represent the interactions among protein atoms. To better understand the ab-initio methods presented in Sections 6 and 7 we give a short description of energy functions. This description follows [21]; for more details the reader is referred there and to references therein.

Over the years, a large number of energy models has been empirically developed for the protein folding problem, such as AMBER [93], CHARMM [15], ECEPP [57, 58], ECEPP/2 [59], ECEPP/3 [60], MM2 [1] and MM3 [2]. These models are typically expressed as the sum of potential energy terms representing *bonded interactions*, i.e. related to bonds, bond angles and dihedral angles, and *nonbonded interactions*, such as van der Waals and electrostatic ones. These potentials are usually described in terms of relative distances of atoms or atom aggregates.

A simple model of bond potential energy is

$$E^{bond} = k^{bond}(r - r_0)^2,$$

which measures how much the bond length r is far from its ideal value r_0. The constant $k^{bond} > 0$ is called "spring constant", in analogy with Hooke's law. This model provides a good approximation of the bond potential just on small motions around the equilibrium configuration. A more detailed representation of bond stretching is obtained by considering the so-called Morse potential:

$$\tilde{E}^{bond} = \tilde{k}^{bond}(1 - e^{a(r-r_0)})^2,$$

with $\tilde{k}^{bond}, a > 0$. However, the first potential is usually considered because it is simpler to evaluate than Morse potential. Small protein structures obtained by X-ray crystallography are typically used to compute r_0.

Angle bending energy is associated with vibrations around the equilibrium bond angle θ_0, therefore its potential can be modeled by Hooke's law too:

$$E^{angle} = k^{angle}(\theta - \theta_0)^2.$$

The value of θ_0 depends on the triplet of atoms defining the bond angle θ and k^{angle} and controls the angle stiffness.

Tortional energy potentials are used to describe the internal rotation energy of dihedral angles. These potentials are usually modeled as

$$E^{dihedral} = \sum_{n=1}^{3} \frac{V_n}{2}[1 + cos(n\psi - \gamma)]$$

where the V_i's are rotation energy barriers, ψ is the torsion angle and γ is the angular offset. Note that some force fields neglect bond stretching and angle bending energies, thus taking into account only torsonal energy.

Nonbonded interactions involve atoms that are not linked by covalent bonds. Usually, non bonded energy terms account for the electrostatic energy and the van der Waals energy.

On each peptide bond between two amino acid residues there is a dipole which is orthogonal to the $N - C$ bond. The energy of this dipole is described by the Coulomb law:

$$E^{elect}_{ij} = \frac{q_i q_j}{4\pi\epsilon_0 r_{ij}}$$

where q_i and q_j are the magnitudes of the two charges of the dipole, r_{ij} is the distance between the charges and ϵ_0 is the dielectric constant.

The main energy involved in the protein stabilization is the non-bonded van der Waals energy, arising from a balance between attractive and repulsive subatomic forces. Attractive forces are longer range than repulsive forces, but, if the distance among atoms is short, they become dominant. This leads to an equilibrium distance in which repulsive and attractive forces are balanced. The van der Waals interaction between two atoms i and j is often modeled through a Lennard-Jones potential, which includes attraction and repulsion terms:

$$E^{vdW}_{ij} = \frac{a_{ij}}{(r_{ij})^{12}} - \frac{b_{ij}}{(r_{ij})^6}.$$

The constants a_{ij} and b_{ij} control the depth and the position of the potential energy well.

The solvent, usually water, has a fundamental influence on the structure, dynamics and thermodynamics of biological molecules, both locally and globally. One of the most important solvent effects is the screening of electrostatic interactions. This can be taken into account implicitly, by including a further dielectric constant ϵ_r in the electrostatic energy potential:

$$E^{elect+solv} = \frac{q_i q_j}{4\pi\epsilon_0\epsilon_r r_{ij}}$$

A more rigorous treatment of solvent effects can be obtained by considering the Poisson-Boltzmann equations. As an alternative, the solvent is explicitly taken into account, by using models based on the assumption that solvation energy is proportional to the protein surface area exposed to the solvent, or to the solvent accessible volume of a hydration layer. These models account also for cavity formations [39].

From a thermodynamic point of view, the difference between two molecular conformations is determined by their difference in free energy, that is defined in terms of enthalpy, H, entropy, S, and absolute temperature, T, of the molecular system:

$$E^{free} = H - T \cdot S.$$

A direct computation of the free energy requires detailed molecular dynamics simulations and hence is too costly. A generally accepted alternative approach based on statistical mechanics describes the free energy contributions by using harmonic approximations [31].

5 Optimization Solvers

As previously observed, ab-initio methods search for conformations corresponding to the global minimum of some energy function, under suitable constraints, i.e. lead to constrained global optimization problems. Hence, it is useful to give a brief description of optimization solvers applied in this context. We focus here on the solvers that are used in the protein fold prediction methods described in the next sections. For more details the reader is referred to [21, 33, 62, 63, 64].

The global optimization solvers can be divided into two main classes: *heuristic* and *deterministic*. The former includes methods based on probabilistic descriptions, for which convergence to a solution is not ensured, or only a convergence in probability is demonstrated. The latter contains methods that, under suitable hypotheses, provide convergence to a solution of the global optimization problem.

Monte Carlo (MC) methods [19] are heuristic methods that simulate the evolution of a system in terms of probability distribution functions. They generate many approximate solutions by random sampling from a probability distribution and get the target solution as an average over the generated samples. In many applications, the variance corresponding to the average solution can be predicted, obtaining an estimate of the number of samples needed to achieve a given error. Enhancements of the basic MC strategy have been developed to reduce the possibility of getting trapped into local minima. They include *Replica Exchange Monte Carlo* (REM) [91], *Parallel Hyperbolic Sampling* [96] and *Electrostatically-Driven Monte Carlo* (EDMC) [72] methods.

A further improvement over MC methods is provided by *Simulated Annealing* (SA) methods [40, 52]. They are based on an analogy with the annealing physical process that consists in decreasing slowly the temperature of a given system (e.g. a liquid metal) in order to obtain a crystalline structure. SA methods are iterative procedures that, at each step, execute a *Metropolis Monte Carlo* algorithm that generates a new candidate approximation of the solution, by applying a random perturbation to the previous one. Through a random mechanism controlled by a parameter called temperature, it is decided whether to move to the candidate approximation or to stay in the current one at the next iteration. The acceptance/rejection of the new approximation is usually based on the evaluation of the so-called Metropolis acceptance function, that is a probability function based on the Boltzmann distribution [55]. Higher temperatures correspond to a larger number of accepted conformations. The temperature parameter plays a crucial role in the whole process; it must be decreased very slowly, to avoid the simulation gets trapped in a local minimum close to the initial state. A modification of the SA strategy called *Monte Carlo Minimization* (MCM) has been also developed, that applies a local Monte Carlo minimization to the current conformation, before checking if the Metropolis acceptance criterion is satisfied [73, 74]. We come back to Metropolis Monte Carlo Simulated Annealing in Section 10.1, since we used this method in our experiments.

Genetic Algorithms (GAs) are heuristic methods based on principles from the evolution theory. Indeed, they represent each feasible point in the conformational space as a chromosome and mimic the evolution of a population of chromosomes. Two chromosomes can generate child chromosomes (crossover operation) and a chromosome can undergo mutations. Furthermore, chromosomes are selected depending on their fitness value, which is defined taking into account the objective function to be minimized. Starting from an initial population, GAs set up an iterative process, where a child population is generated at each step from a parent one, by applying the above evolutionary mechanisms, until suitable termination criteria are satisfied. GAs differ by the mechanisms used to simulate mutation and crossover and by the fitness function. As noted in [21], the choice of these mechanisms greatly influences the ability of finding global minimum energy configurations. A review on GAs is given in [87].

Conformational Space Annealing (CSA) methods work with typical concepts of SA, GAs and MCM. As in GAs, an initial population of variables called first bank is generated and then a subset of bank conformations called seeds are selected. The seeds are perturbed, by replacing (typically small) seed portions with the corresponding portions of bank conformations, and are used as trial conformations, to obtain a new bank. As in MCM, a local minimization is applied to all conformations to work only with the space of local minima. The diversity of sampling is controlled by comparing a suitable distance measure between two conformations with a cutoff value, D_{cut}. A trial conformation is compared with the closest one in the current bank. If their distance is smaller than D_{cut}, they are considered similar and the one with lower energy is chosen. Otherwise, the highest energy conformation in the bank plus the trial one is discarded. The cutoff value is slowly decreased during the simulation process and hence acts as the temperature parameter in SA. The algorithm usually stops when all the bank conformations have been used as seeds and the cutoff parameter has reached a suitably small value. More details can be found in [32, 47].

An example of deterministic global optimization strategy is provided by *Molecular Dynamics* (MD) simulations. MD methods simulate the evolution of a molecular system by applying the equation of motion to the atoms of the system. They have been able to provide detailed information about heteropolymers and to give insights into complex dynamic processes occurring in biological systems, such as protein folding [16].

Branch and Bound (BB) methods fall into the class of deterministic global optimization methods too. These are iterative methods that, at each step, find lower and upper bounds on the global minimum objective value. The iterations are stopped when the difference between the bounds is smaller than a given tolerance. Recently, a deterministic BB algorithm named αBB has been developed by Floudas et al. and applied to molecular conformation problems [6, 35, 36]. αBB determines the upper bounds by function evaluation or local minimization of the original objective function, while the lower bounds are computed by minimizing convex lower-bounding functions obtained by adding a convex term to the original one. The lower-bounding functions depend on a parameter that controls their shape and must be properly chosen to guarantee convexity. Lower-bounding functions are built in such a way that they have properties ensuring the convergence of the algorithm to a global minimum.

6 Ab-initio Methods Using Knowledge-Based Information

Ab-initio methods with knowledge-based information usually build template models by extracting from databases fragments with sequence or structural similarity to fragments of the target sequence. Therefore, there is no clearly defined separation between these methods and the homology modeling or fold recognition ones. Ab-initio methods exploiting both approches are discussed in the next two Sections.

6.1 Lattice models

To reduce the degrees of freedom of the conformational space, models have been developed that are based on a simplified representation of the protein chain over a lattice. These *lattice models* use secondary structure predictions and threading techniques to derive some constraints; then, they search the conformational space by applying Monte Carlo procedures to the lattice. Because of these semplifications, lattice models are generally two orders of magnitude faster than high-resolution models [45]. On the other hand, simplified models of proteins lead to a loss of dynamic mechanisms, so that often predicted conformations do not fit native structures suitably. First lattice studies did not focus on protein structure prediction, but rather on understanding thermodynamic and kinetic properties of protein folding. Indeed lattice models have a long history in modeling polymers, due to their analytical and computational simplicity.

Early in the '90s, Levitt et al. [49] developed a low-resolution method, based on a simple representation of the protein backbone as a self-avoiding chain of connected vertices on a tetrahedral lattice, with several amino acid residues assigned to each lattice vertex. To reduce the space of feasible lattice structures, this model requires the final conformations to be compact and globular. Effects of solvent interactions are not considered, because the lattice model did not represent accurately the exposed surface of a conformation. Starting from observed contact frequencies in X-ray structures, the energy of contact between two lattice vertices is defined and a dynamic programming strategy is applied to find the best conformational energy. This model was validated on real proteins with 52-68 amino acid residues and correct low-resolution structures were found [49]. A drawback is that it can be applied only to proteins with a small number of residues; furthermore, it does not consider interatomic interactions.

Lattice models have undergone an evolution over the years. In [77, 94] Levitt and co-workers presented a lattice-based hierarchical approach. In this case, starting from the sequence of amino acid residues, all feasible compact conformations are identified by using a highly simplified tetrahedral lattice model; a lattice-based scoring function is used to select a subset of these conformations and to build high-resolution (all-atom) models. Then, by using a knowledge-based scoring function, three small subsets are extracted from the set of all-atom models and a procedure based on distance geometry is used to generate the best conformations from each of the subsets. Using this approach, structures of proteins with at most 80 residues were predicted, obtaining RMSD values ranging from 4.1 to 7.4 Å [77]. Unfortunately, the method failed for proteins with complex supersecondary structures.

Lattice models have been also studied by Skolnick and co-workers [43, 44]. They developed a lattice model of the protein structure and dynamics in which the polypeptide chain is represented with a simple cubic lattice. The emphasis is on the side chain role, rather than on geometry of the backbone. The backbone is treated implicitly, since the C_α coordinates are computed by considering the positions of three consecutive side chains. The energy function takes into account sequence independent properties, such as interactions between the i-th and the $(i+4)$-th residues in the α-helix side chains or long distance interactions in the β-sheets, and sequence dependent properties, such as long-range pairwise and multibody interactions that simulate hydrofobic effects. The lowest energy conformation corresponding to the native state is searched by a Replica Exchange Monte Carlo (REM) procedure [91]. The model was tested on small and structurally simple single-domain proteins considering two sets of sequences, one corresponding to single fragments of known structures, the other to known protein tertiary structures. The best results, evaluated by using the RMSD values of the predicted versus the original conformations, were obtained for the set of single fragments. The method evolved into a hierarchical ab-initio lattice approach that uses a combination of multiple sequence comparison, threading, clustering and refinement [83]. In this approach, the starting fragmentary templates for the lattice model are provided by a threading algorithm and a reduced representation of the protein conformational space is used, where the center of mass of the C_α and side-chain atoms are the interaction centers. The energy function is defined through a statistical analysis of known protein structures, leading to statistical potentials for pairwise and multiboby side-chain interactions. The conformational space is sampled by the REM procedure. This method is called SICHO (SIde CHain Only). Results presented at CASP4 meeting [70] showed that it is able to obtain good results on small proteins of not too complex topology [83].

Another structure prediction lattice-method that combines homology and ab-initio modeling is TOUCHSTONE, developed by Skolnick et al. [84]. A first version of this method is based on the SICHO lattice model, with force field including short-range structural correlations, hydrogen-bonding interactions and long-range pair-wise potential. Two threading restraints are used to reduce the conformational search space, concerning side-chain contacts and local distances. The former restraint is obtained by using the PROSPECTOR threading algorithm [71], while the latter is derived from sequence alignments and threading of short sequence fragments. REM is used to search the conformational space. To generate another set of indipendent trajectories, a Monte Carlo sampling scheme, called Parallel Hyperolic Sampling (PHS) [96], is used. Then the structures generated by the simulations are rebuilt at an atomic detail. This method was applied to the genome of *Mycoplasma genitalium* bacterium, that has one of the smallest known genomes among living organisms [85]. 85 proteins with at most 150 amino acid residues were examined, obtaining a correct prediction of the topology of 63% of the proteins.

As discussed in [85], the potential function used in TOUCHSTONE is not suitable for predicting multiple-domain structures. To overcome this limitation, both the lattice representation and the force field have been modified [86, 97]. The SICHO model has been replaced by the CABS one, in which the C_α trace is confined to a lattice system, while the group made by the side chain and the C_β carbon are off-lattice, with positions determined from three adjacent C_α atoms. The energy function takes into account pairwise and multiboby side-chain interactions, short- and long-range hydrogen-bond interactions, contact and local distance restraints obtained through PROSPEC-TOR, burial and electrostatic interactions, global propensities to predicted contact orders and contact numbers, and local stiffness of global proteins. The conformational space search method is PHS, as in the previous TOUCH-STONE version.

Experiments were carried out on a set of 125 proteins (43 all-α proteins, 41 all-β proteins and 51 α/β-proteins, according to Kabsch and Saunder classification [30]), with lengths ranging from 36 to 174 amino acid residues. By using PROSPECTOR restraints, 83 proteins were successfully folded. Comparisons with the previous TOUCHSTONE version showed the efficiency of CABS versus SICHO. Furthermore, it was observed that short-range restraints considerably speedup local structure formations.

Recently, a high-resolution lattice model has been developed by Kolinski [45] that is based on a representation of the protein backbone over a lattice and on the REM searching procedure. For each residue, this model takes into account the C_α and C_β carbons, the side-chain and an additional atom located along the $C_\alpha - C_\alpha$ virtual bound. Only the C_α coordinates are explicitly computed and are used, together with amino acid properties, to calculate the coordinates of off-lattice elements. The force field is based on the CABS model and the potential used takes into account short- and long-range interactions. The simulation process is based on Metropolis Monte Carlo scheme, subject to a simulated annealing procedure or controlled by REM. This lattice model can be applied to perform ab-initio structure predictions as well as in multi-template comparative modeling [45].

6.2 Methods Based on Fragment Assembly

The idea behind these methods is to build protein tertiary structures from small protein segments or secondary structures, obtained through sequence alignment or threading.

Such an approach is implemented, for example, in FRAGFOLD, developed by Jones et al. [26, 28]. In FRAGFOLD simulations, the first step is the selection from a library of protein structures of suitable supersecondary structural fragments at the position of each residue of the target sequence, and hence the prediction of secondary structures by using PSIPRED [25], which applies neural-network techniques and PSI-BLAST sequence alignments. The predicted secondary structures are used as input to FRAGFOLD. Random conformations are then generated until a conformation with no steric clashes is obtained. Starting from this one, a Simulated Annealing algorithm is applied to minimize an energy function, which is a weighted sum of terms expressing short- and long-distance pair potentials, single-residue solvation energy, steric interactions (such as the van der Waals energy), and hydrogen-bond interactions. Results presented at CASP4 and CASP5 [26, 28] showed that FRAGFOLD can correctly predict local domains, but fails in predicting entire three-dimensional structures. In particular, there are problems with the prediction of β-structures, since the formation of these structures is a cooperative process requiring the convergence of many substructures.

Another method which exploits sequence alignment and fragment assembly is Rosetta, developed by Baker et al. [12, 13, 81, 82]. This method is based on the assumption that the distribution of conformations of each three- and nine-residue segment can be reasonably approximated by the distribution of structures adopted by the corresponding sequence (or closely related ones) in known protein conformations. Therefore, Rosetta breaks the target sequence into three- and nine-residue segments and applies a profile-profile comparison procedure to extract fragment libraries from protein structure databases. The fragments are assembled to build three-dimensional structures by using a fragment insertion Metropolis Monte Carlo procedure. Many of such template-based models are generated and then clustered. For sequences with less than 100 residues, an all-atom refinement is used instead of clustering. The energy function used in searching the conformational space describes sequence-dependent properties, such as non-local interactions (e.g. disulfide bonding, backbone hydrogen bonding, electrostatics) and sequence-independent properties, connected to the formation of α-helices, β-strands and to the assembly of β-strands into β-sheets. Only the backbone atoms are considered explicitly, while the side chains are represented as centroids.

Rosetta underwent a significant evolution since its development. The improvements concern the application of filters to reject non-protein-like conformations (local low-order contact conformations and β-strands not properly assembled into β-sheets) [76], the modifications of the methodology for picking up fragments from the structure database, in order to ensure a remarkable diversity of secondary structures when dealing with segments with a weak propensity to fold into a single secondary structure, the use of a new prediction method, JUFO [29], and the exploitation of quantum chemistry calculations, traditional molecular mechanics approaches and protein structural analysis to compute parameters in the energy function [12, 13]. A neural network method is under development with the aim of identifying strand-loop-strand motifs starting from the protein primary structure [46].

Rosetta was applied to CASP5 targets. In particular, for α- or α/β-proteins Rosetta generated models with a correct overall topology and RMSD values ranging from 2.8 to 4.2 Å. Rosetta method failed for proteins having more than 280 residues and a complex topology; furthermore, it sometimes generated models being too globular or having β-strands less exposed than in the native conformation.

7 Ab-initio Methods Without Knowledge-Based Information

Knowledge-based ab-initio methods are dependent on the information stored in structural databases and on statistical analysis of this information; hence they can produce inaccurate predictions of new folds. A way to overcome this problem is offered by "true" ab-initio methods which simulate the folding process by using only protein models based on physicochemical principles. These methods are obviously more challenging, since they require "realistic" representations of atomic interactions and powerful algorithms and computational resources to search the feasible conformational space. A few examples of ab-initio methods without database information are discussed in the next sections.

7.1 Hierarchical Approaches

Hierarchical approaches start from a reduced representation of protein atoms and their interactions and then refine computed reduced conformations to obtain all-atom structures to be optimized.

A simple hierarchical approach to protein folding is given by LINUS (Local Independently Nucleated Units of Structure), developed by Srinivasan and Rose [89, 90]. This procedure has been used to predict secondary structures and to capture a physical interpretation of protein secondary elements. Indeed, Srinivasan and Rose used LINUS to support the physical theory that secondary structure propensities are mainly determined by competing local effects, involving conformational entropy and hydrogen bonding.

A Metropolis Monte Carlo procedure is applied to search the conformational space. The amino acid sequence is considered as an extended chain, where the backbone atoms are represented as points, while the side chains are modeled as different nonoverlapping spheres, according to amino acid type and size. The degrees of freedom are the dihedral angles, Φ, Ψ and χ. A Metropolis Monte Carlo procedure is used to search the conformational space. More precisely, the extended chain is subdivided into subsequences of three consecutive residues, proceeding from the N-terminus to the C-terminus, that are perturbed by using a predefined set of random moves to obtain a new configuration. This configuration is accepted or rejected according to a Metropolis acceptance criterion based on attractive and repulsive contributions [90]. This cycle is completed when all the chain residues have been processed.

LINUS was used also by Maritan and co-workers in order to estimate the rate of successful secondary structure predictions as a function of the temperature [24]. In particular, they showed that at low temperatures local interactions are facilitated and stabilized, leading to α-helices and turns; consequently, β-strands are favoured at high temperatures. At intermediate temperatures some protein subsequences tend to fold into β-strands, while others into α-helices and turns. They also found that α-helices and β-strands can be predicted with an accuracy greater than 40% [24].

A different hierarchical approach has been developed by Scheraga and co-workers [78] to capture pairwise and multibody interactions during the folding process. In this approach, a set of low-energy structures is computed first, by using a reduced model based only on the C_α trace and on the so-called UNRES (UNited-RESidue) potential force field [50, 51], to describe intra-protein interactions and hydrogen bonding. The conformational space is searched by a Conformational Space Annealing (CSA) algorithm [47]. The virtual-bond chains of these low-energy structures are converted to an all-atom backbone, by using the dipole-path method based on alignment of peptide-group dipoles [50]. The backbone conformation is optimized by using EDMC [72], a procedure that iteratively looks for low-energy structures in the conformational space and takes into account electrostatic interactions and thermal effects. All-atom side chains are added to the previous model under constraints of non-overlap; loop and disulfide-bonds are then treated explicitly. The final conformation is obtained by using the ECEPP/3 all-atom energy function [60], with gradual reduction of the C_α-C_α distance of the parent united-residue structure. The ECEPP/3 energy function is the sum of electrostatic, hydrogen-bonded, torsional and non-bonded terms.

The method above described was successfully applied to single-chain proteins as well as to multiple-chain ones. In the latter case, in order to obtain correct predictions, interchain interactions were taken into account by suitably modyfing UNRES and CSA [78].

7.2 A Combinatorial and Global Optimization Approach

A novel true ab-initio approach for the prediction of three-dimensional structures of proteins is implemented in ASTRO-FOLD, developed by Floudas and co-workers [33, 35, 36, 38, 39]. ASTRO-FOLD combines the classical hierarchical view of protein folding, in which the folding process starts from rapid formation of secondary structures and then proceeds to the slower tertiary structure arrangement, with the hydrophobic-collapse view, in which secondary and tertiary structures are formed concurrently. The prediction of a protein conformation is performed into four steps. First, initiation and termination sites of α-helices are identified, then β-strands are identified and β-sheet topologies are predicted, and, later, constraints on the protein structure and information on loop segments are derived. Based on the previous information, the overall protein tertiary structure is predicted by using a model that combines both the above views of the protein folding process and by applying deterministic global optimization, stochastic optimization and torsion-angle dynamics. Therefore, ASTRO-FOLD can be defined as combinatorial and global optimization framework based on a four-step approach.

The main idea behind α-helix determination is that the fold of such secondary structure is based on local interactions. Hence, in order to identify local sites of helix formation, the amino acid sequence is segmented into overlapping oligopeptides and ensembles of low potential states are computed, along with a global minimum energy state, using a detailed atomic level model based on the ECEPP/3 force field [60]. The determination of these state is performed by applying the deterministic branch-and-bound algorithm α-BB [20] and the stochastic CSA algorithm [32, 47]. Free energy calculations are then performed, with a force field which is the sum of potential, entropic, solvation, ionization and internal cavity contributions. The energy values are used to compute the probability that each oligopeptide folds into a helix and to define a helix propensity for each residue.

Once α-helices have been identified, the remaining residues are analyzed to identify the locations of β-strands and β-sheets, and to predict β-sheet topologies as well as disulfide bridges. Since the formation of such structures is driven by long-distance interactions, a different approach is used. The key assumption is that β-structure formation depends on hydrophobic forces [37];

to model them, the prediction of hydrofobic residue contacts is required. To predict a β-sheet, β-strand superstructures are postulated that encompass all the β-strand substructures that may constitute the β-sheet topology. The mathematical model of the superstructures is formulated as a global optimization problem, whose solution maximizes contacts between hydrophobic residues, subject to constraints enforcing physically meaningful configurations for β-strands and disulfide bridges. This approach is used to identify a rank-ordered list of possible β-sheet structures.

Once α-helices and β-sheets have been identified, secondary structure restraints are defined. Dihedral angles, atomic distance and $C_\alpha - C_\alpha$ distance bounds are defined according to the main properties of corresponding secondary structures. Restraints for unassigned residues are also defined either through an analysis of overlapping oligopeptides, such as for α-helices identification, or through predictions of entire loop fragments. Both approaches are implemented by exploiting deterministic and stochastic optimization solvers.

The final stage of ASTRO-FOLD is the prediction of the protein tertiary structure. This problem is formulated as the global minimization of a suitable potential energy, subject to the restraints above discussed. This problem is solved by a combination of α-BB and torsion angle dynamics [35].

As reported in [36], ASTRO-FOLD was tested on CASP5 targets of at least 150 residues, obtaining accurate α-helix and β-strand and impressive β-sheet predictions. Indeed, RMSD values ranging between 4.1 Å and 6.9 Å, and SOV [95] values corresponding to more than 80% accuracy have been obtained for the computed conformations.

As noted in [36], the application of ASTRO-FOLD to medium-size proteins was made possible by using distributed computing environments. The framework was parallelized by taking into account the different problems and solvers at each stage the prediction process.

7.3 Topological Approaches

Experimental and theoretical studies have shown that the folding process is widely influenced by topological properties of the native state. For example, by analyzing a small set of non homologous simple single domain proteins, Baker and co-workers revealed that a statistically significant correlation exists between folding kinetics and native state topological complexity [69]. Starting from their results, Koga and Takada studied the relationships between native topology and folding pathways [42]. By using a simple representation of the polypeptide chain through its C_α trace and a free-energy functional approach, that takes into account chain connectivity, contact interactions and entropy, they were able to correctly describe folding pathways of small single-domain proteins. The correlation between the topology of the native state and the folding pathways was confirmed by Maritan et al. [80], by performing molecular dynamics simulations of the immunoglobulin, using a model that represents only the C_α carbons and an energy function that includes bonding and non-bonding terms. Other studies suggest that folding rates are correlated to topological parameters such as contact order and cliquishness [56].

An interesting topological approach to the protein folding problem has been proposed by a research group led by Banavar and Maritan [7, 8, 9, 10, 54]. In this approach, a protein is modeled as a tube of nonzero thickness without any self-contacts (see Figure 7.3). The axis of the tube is a suitable curve interpolating the C_α carbons and the thickness is expressed in terms of a metric that measures the "distance" among any three points on the curve, x_i, x_j, x_k, as the radius $r(x_i, x_j, x_k)$ of the circle passing through them (r is assumed to be infinity if the points are aligned). Note that $1/(r(x_i, x_j, x_k)^p$, $p > 0$, can be regarded as a three-body potential and hence the tube thickness is related to a certain interaction energy among chain particles [23]. Indeed, the modeled structure is energetically stable, i.e. its conformation corresponds to a minimum of free energy, when it achieves a maximum thickness under constrains preventing self-intersection and aligned triplets of amino acids. As pointed out in [7], despite its simplicity, this model is able to capture the physical thickness of the protein chain, that is due to the presence of the R-groups. Furthermore, a nonzero thickness implies that the interactions between two spatially close tube segments do not depend only on their distance, but also on their relative orientation, so the tube model is able to represent the inherent anisotropy associated with the local directionality of the chain.

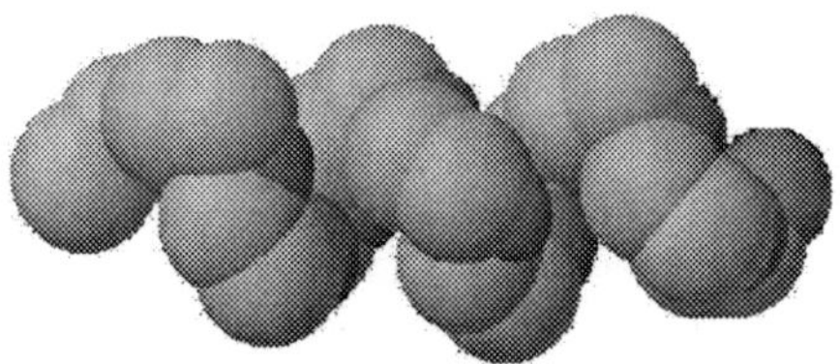

Fig. 5. The sequence of N-C_α-C' units of the crambin helix composed of the amino acids $7 \div 17$. The picture is similar to a thickened tube.

Numerical simulations based on the above model are reported in [7, 10, 92]. Different constraints have been considered to take into account the compactness of a polymer chain, such as a pairwise attractive potential with a given range [7], or suitable bounds on the global and the local gyration radius or on the contact distance and the number of allowed contacts [92]. A Metropolis Monte Carlo procedure has been used to search the conformational space, obtaining helix- and hairpin-like structures.

We have focused our attention on the tube model, because it appears both simple and capable of representing significant features of the protein chain.

Next sections are devoted to describe a modified version of it and related computational experiments.

8 A Modification of the Tube Model

Following [10, 23], we provide a more detailed description of the tube model, which is the basis of our computational approach. Let $X = (x_1, x_2, \ldots, x_n)$ be a n-ple of different points called *conformation*, where each $x_i \in \Re^3$ represents the position of the C_α atom of the i-th amino acid residue of the polypeptide chain. The interaction among any three non-aligned points x_i, x_j, x_k can be measured by the radius of the unique circle among them, which has the following expression:

$$r(x_i, x_j, x_k) = \frac{||x_i - x_j||\,||x_i - x_k||\,||x_j - x_k||}{4A(x_i, x_j, x_k)} = \frac{||x_i - x_j||}{2|\sin\theta|}$$

where $||\cdot||$ is the Euclidean norm, $A(x_i, x_j, x_k)$ is the area of the triangle with vertices x_i, x_j and x_k, and θ is the angle between the vectors $x_i - x_k$ and $x_j - x_k$. If the three points are aligned, $A(x_i, x_j, x_k)$ and $\sin\theta$ are null, hence the above definition can be extended to these points by setting $r(x_i, x_j, x_k) = \infty$. Note that $r(x_i, x_j, x_k)$ can be viewed as an approximation of the standard radius of curvature. Indeed, if the three points vary over a simple (i.e. without knots) and smooth curve C, then

$$\lim_{\substack{x_j, x_k \to x_i \\ x_j, x_k \in C}} r(x_i, x_j, x_k) = \rho(x_i),$$

where $\rho(x_i)$ is the radius of curvature of C at x_i. In the following, the radius $r(x_i, x_j, x_k)$ is referred to as three-body radius.

The thickness of the conformation X can be defined as:

$$D(X) = \min_{\substack{1 \leq i,j,k \leq n \\ i \neq j, j \neq k, k \neq i}} r(x_i, x_j, x_k). \tag{1}$$

$D(X)$ is a "discrete version" of the thickness $\Delta(C)$ of a simple and smooth curve C, which is defined as the maximum thickness of a tube with axis C and circular section, that does not exhibit any self-contacts. $\Delta(C)$ has the following expression:

$$\Delta(C) = \min\left\{\min_{x \in C} \rho(x), \frac{1}{2} \min_{(x,y) \in \Omega} ||x - y||\right\},$$

where Ω is the set of all pairs of points of C such that $x \neq y$ and the vector $x - y$ is orthogonal to the tangents to C at both x and y. In other words, in the continuous case, the tube thickness is the smallest value between the minimum radius of curvature of C and half the minimum distance of closest approach over C. It can be proved that

$$\Delta(C) = \min_{x,y,z \in C} r(x,y,z),$$

where the definition of r is extended by continuity to coinciding points [23].

As pointed out in [10], the three-body radius is able to distinguish among local and non-local interactions along the protein chain. When three consecutive particles are considered, a discrete version of the radius of curvature is used to measure their interaction; when the particles are non-consecutive, the distance of approach between two parts of the chain is taken into account (see Figure 8). The thickness takes into account that the protein backbone cannot have self-contacts and that the side chains cannot overlap; furthermore, it provides a global measure of the free space in the protein conformation.

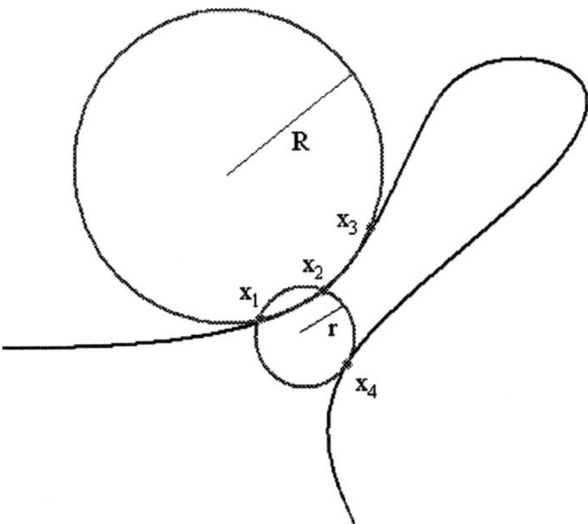

Fig. 6. Three-body radii of consecutive and non-consecutive points.

As observed in Section 7.3, finding an energetically stable conformation can be achieved by maximizing the thickness under suitable constraints. On the other hand, the tube model can be used to predict and analyze compact tube-shaped conformations of given thickness. The latter conformations can be obtained by maximizing a function counting the number of triplets having a three-body radius close to a given thickness value $\bar{D}$:

$$f(X) \equiv f(x_1, x_2, \ldots, x_n) = \sum_{i=1}^{n-2} \sum_{j=i+1}^{n-1} \sum_{k=j+1}^{n} f_{\bar{D}}(r(x_i, x_j, x_k)), \qquad (2)$$

where

$$f_{\bar{D}}(r(x_i, x_j, x_k)) = \begin{cases} 1 \text{ if } & r(x_i, x_j, x_k) \in [\bar{D} - \epsilon, \bar{D} + \epsilon] \\ 0 \text{ otherwise} \end{cases} \qquad (3)$$

and ϵ is a real positive constant. As shown in Section 9, typical values of thickness, characterizing protein structures, can be obtained by analyzing existing protein structure data sets; therefore, maximizing $f(X)$ under suitable constraints, using these typical values of thickness, can provide a means to predict meaningful protein-like three dimensional conformations.

To increase global protein compactness, we have modified f by adding a term forcing the points x_i to be inside an ellipsoid, whose surface is thought as a rough approximation of the protein surface shape. By changing the lengths of the ellipsoid axes, different shapes can be approximated. The added term has the following form:

$$g(X) \equiv g(x_1, x_2, \ldots, x_n) = \sum_{i=1}^{n} g_{(a,b,c)}(x_i) \qquad (4)$$

where

$$g_{(a,b,c)}(x_i) = \begin{cases} 1 \text{ if} & \dfrac{(x_i^1 - x_G^1)^2}{a^2} + \dfrac{(x_i^2 - x_G^2)^2}{b^2} + \dfrac{(x_i^3 - x_G^3)^2}{c^2} \leq 1 \\ 0 \text{ otherwise} \end{cases}, \qquad (5)$$

$x_G = (x_G^1, x_G^2, x_G^3)$ is the barycenter of X, a, b and c are the lengths of the ellipsoid semiaxes, and the superscripts are used to denote the Cartesian coordinates of a point.

Constraints have been imposed to explicitly take into account that two consecutive α-carbons are virtually bonded, hence their Euclidean distance can have only slight variations, and that the Euclidean distance between any two non-consecutive amino acid residues cannot fall below a certain threshold. Furthermore, starting from the observation that in α-helices amino acid residues with positions i and $i + 2$ along the chain are closer than in other structures, a contraint on the Euclidean distance between x_i and x_{i+2} has been imposed to specifically simulate all-α conformations.

The global constrained optimization problem described so far has the following formulation:

$$\max F(X) = \max[f(X) + g(X)] \qquad (6)$$

subject to

$$c_1 \leq d(x_i, x_{i+1}) \leq c_2, \qquad \forall i \in \{1, 2, \ldots, n - 1\}, \qquad (7)$$

$$c_3 \leq d(x_i, x_j), \qquad \forall i, j : i > j + 1, \qquad (8)$$

$$c_4 \leq d(x_i, x_{i+2}) \leq c_5, \qquad \forall i \in \{1, 2, \ldots, n - 2\}. \qquad (9)$$

where c_1, c_2, c_3, c_4 and c_5 are real positive constants chosen on the base of experimental observations (see Section 9). The costraints (9) are specifically related to all-α structures.

9 Choice of Model Parameters

The problem (1)-(9) requires the choice of some parameters: the thickness $\bar{D}$ and the related value ϵ in the definition of f (see (2)-(3)), the semiaxis lengths a, b, c in the definition of g (see (4)-(5)) and the constants c_i in the constraints (see (7)-(9)).

The values of $\bar{D}$ and ϵ have been chosen by performing an analysis of a set of 3639 protein structures available in the *PDBSELECT* data collections with *R-factor* < 0.25 and *Resolution* < 2.5 [65]. The thickness of each structure has been evaluated, obtaining the thickness frequency distribution shown in Figure 9.

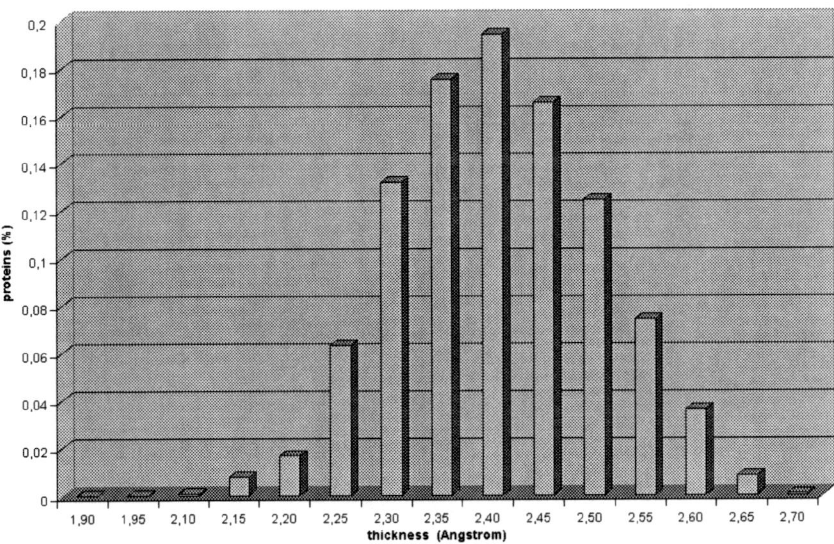

Fig. 7. Frequency distribution of the thickness for a set of 3639 proteins from PDBSELECT.

The thickness mean value is 2.40 Å, with a standard deviation of 0.10 Å; the minimum thickness is 1.91 Å (achieved by only one structure), while the maximum is 2.67 Å. The same analysis has been performed considering all the α-helices (14592 structures) and all the β-sheets (13070 structures) separately. The mean thickness value of the α-helices is 2.65 Å, with a standard deviation of 0.07 Å, a minimum of 2.26 Å and a maximum of 4.58 Å. However, according to the small standard deviation value, more than 98.5% of the α-helices have a thickness ranging between 2.50 Å and 2.90 Å. The frequency distribution of the thickness of the α-helices in the interval $[2.50, 2.90]$ is reported in Figure 9. The previous results agree with the fact that the α-helices have very similar geometries. The mean value of the β-structures is 2.65 Å too, but with a standard deviation of 0.46 Å, a minimum of 2.12 Å and a maximum of 9.75 Å. Taking into account the low variability of the α-helices thickness, in our experiments we focused our attention on α-structures.

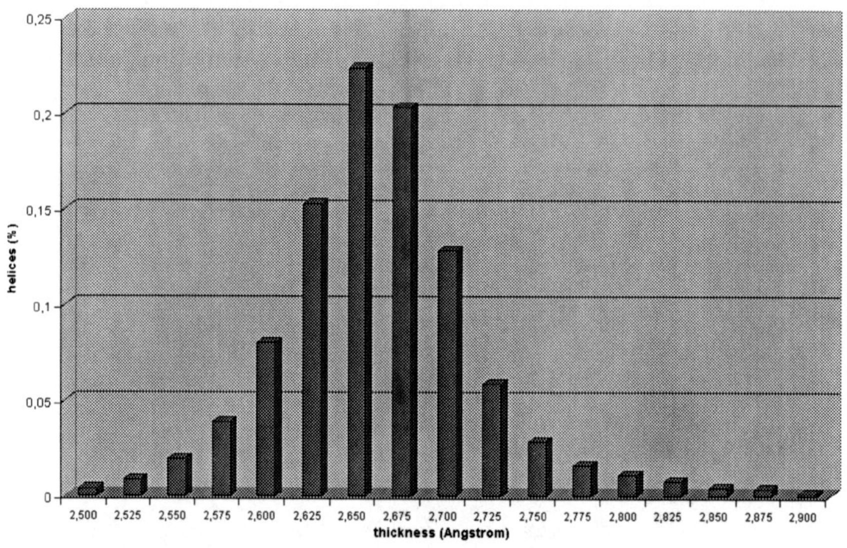

Fig. 8. Frequency distribution of the thickness for a set of 14592 α-helices from PDBSELECT.

A deeper analysis has shown that in the α-helices only few triplets of α-carbons have a three-body radius equal to the thickness. For example, the helix of the crambin (PDB code **1crn**) composed by the amino acid residues $7 \div 17$ has a thickness equal to 2.66 Å, but just the α-carbons 15, 16 and 17 have this three-body radius, while all the other triplets have a three-body radius of at least 2.71 Å.

Since the term f in the objective function (2) counts the number of triplets having a three-body radius close to $\bar{D}$, we made some more studies to find out frequent values of the three-body radius. We first analyzed the so-called perfect helix, that is the stable conformation of the amino acid sequence made only by alanine. In this helix, all the triplets (x_i, x_j, x_k), with constant $i - j$ and $j - k$, have the same radius. All the triplets (x_i, x_{i+h}, x_k), with $h > 0$ and $i + h < k$, and (x_i, x_{k-h}, x_k), with $h > 0$ and $k - h > i$ have the same radius too. The most frequent triplets with the same radius are of the type (x_i, x_{i+1}, x_{i+3}) and (x_i, x_{i+2}, x_{i+3}), but the minimum radius, i.e. the thickness, is achieved by the triplets (x_i, x_{i+1}, x_{i+2}). The corresponding values, reported in Table 9, show that the difference between the minimum radius and the most frequent one amounts to 0.13 Å. We then considered the same type of triplets in the PDBSELECT set and computed the mean radius values, and the corresponding standard deviations, obtaining the results reported in Table 9. In this case, the difference between the mean thickness values (2.65 Å) and the triplet mean values is 0.10 Å for the triplets (x_i, x_{i+1}, x_{i+2}) and 0.23 Å for the triplets (x_i, x_{i+1}, x_{i+3}) and (x_i, x_{i+1}, x_{i+3}).

	perfect helix	PDBSELECT set
(x_i, x_{i+1}, x_{i+2})	2.71	2.75 (0.12)
(x_i, x_{i+1}, x_{i+3})	2.84	2.88 (0.28)
(x_i, x_{i+2}, x_{i+3})	2.84	2.88 (0.28)

Table 1. Three-body radii (Å) of selected triplets of α-carbons in the perfect helix and in a set of 3639 proteins from PDBSELECT. Mean and standard deviation (in brackets) of radius values are reported for the PDBSELECT triplets.

Taking into account the previous analysis, we set $\bar{D} = 2.70$ and $\epsilon = 0.20$, i.e. $[\Delta - \epsilon, \Delta + \epsilon] = [2.50, 2.90]$. This value of $\bar{D}$ is very close to the thickness of the perfect helix (2.71 Å); furthermore, the interval $[2.50, 2.90]$ contains most of the thickness values of the α-helices from PDBSELECT and includes also the most frequent three-body radii of both the perfect helix and the PDBSELECT α-helices.

The semiaxis lengths a, b and c that define the function g have been determined taking into account the volumes of the single amino acids, that are reported in Table 9. For each protein chain, we computed the sum of the volumes of the amino acids, then we increased this sum by 3.8%, to take into account that proteins have cavities [75], and, finally, we set a, b and c in such a way that their products was equal to the cube of the radius s of the sphere with volume equal to the increased sum of amino acid volumes, i.e.

$$a \cdot b \cdot c = s^3, \tag{10}$$

where

$$s = \frac{3}{4\pi} \sqrt[3]{1.038 \cdot \sum_{i=1}^{n} vol_i} \tag{11}$$

and vol_i is the volume of the i-th amino acid in the protein chain. Obviously, the single values of a, b and c are not univocally determined by (10)-(11);

by varying these values, theoretically possible conformations with different shapes can be obtained. Note that, by taking into account the amino acid volumes, we introduce in the model a distinction among the points x_i, that are considered equal in the original tube model.

amino acid	volume	amino acid	volume
ALA	88.6	LEU	166.7
ARG	173.4	LYS	168.7
ASP	111.1	MET	162.9
ASN	114.1	PHE	189.9
CYS	108.5	PRO	112.7
GLU	138.4	SER	89.0
GLN	143.8	THR	116.1
GLY	60.1	TRP	227.8
HIS	153.2	TYR	193.6
ILE	166.7	VAL	140.0

Table 2. The volumes of the 20 amino acids, in $\overset{\circ}{A}{}^3$.

To determine the constants c_1 and c_2, the mean value of the Euclidean distances of all pairs of consecutive α-carbons has been computed for each protein of the PDBSELECT set (the corresponding frequency distribution is shown in Figure 9). However, since the algorithm applied to problem (1)-(9) in our numerical experiments does not change these distances (see Section 10.1), we set c_1 and c_2 both equal to the most frequent mean Euclidean distance, i.e. $c_1 = c_2 = 3.81$ Å.

The remaining constants c_3, c_4 and c_5 have been chosen by observing the perfect helix. In this helix, the Euclidean distance between two α-carbons x_i and x_{i+2} is 5.43 Å, hence we set $c_4 = 5.0$ and $c_5 = 6.0$. Similar observations on the minimum distance between two generic α-carbons led to the choice $c_3 = c_4$. Actually, these choices of the c_i constants have been supported by numerical experiments.

10 Computational Experiments

Computational experiments based on the modified tube model have been carried out to simulate α-helices and an all-α protein, using a Metropolis Monte Carlo Simulated Annealing algorithm to search the conformational space. This algorithm has been implemented in Fortran 77 and in C and the software has been run on a personal computer with a 2 GHz Athlon processor and a 516 MBytes RAM, under the Linux operating system.

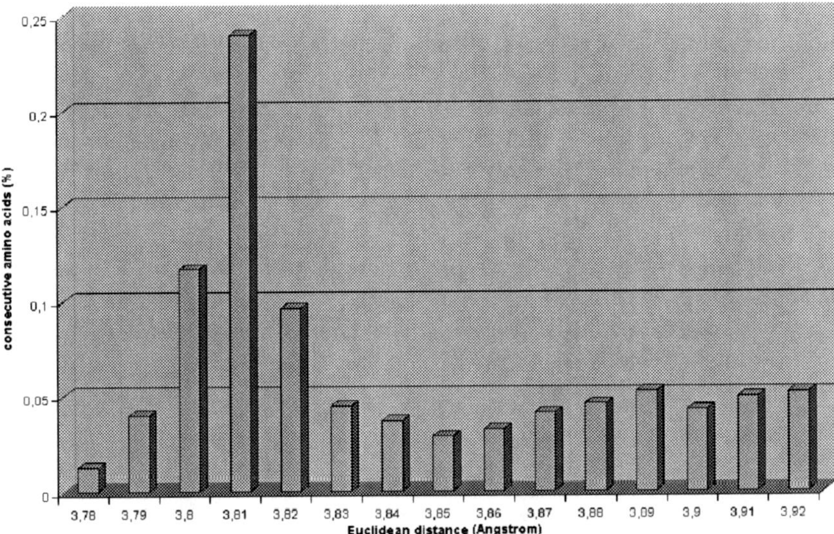

Fig. 9. Frequency distribution of the mean Euclidean distances of the pairs of consecutive amino acids in the proteins of the PDBSELECT set.

A short description of the Simulated Annealing algorithm and a discussion on the results of the computational experiments follow.

10.1 The Metropolis Monte Carlo Simulated Annealing Algorithm

As observed in Section 5, Simulated Annealing (SA) algorithms [40, 52] are based on an analogy with the annealing physical process, in which the temperature of a given system is decreased slowly, in order to obtain a crystalline structure. The structure of a SA algorithm can be described by two nested loops. The inner one generates at each iteration a new candidate approximation to the solution, by applying Monte Carlo perturbations to the previous one. The new approximation is accepted or rejected, by using a random mechanism based on the evaluation of the so-called Metropolis acceptance function, whose value depends on a parameter called temperature. The lower is the temperature, the smaller is the number of accepted approximations. The outer loop controls the decrease of the temperature parameter, i.e. defines the so-called cooling schedule.

From the above description it results that SA algorithms are built up from three basic components: next candidate generation, acceptance strategy and cooling schedule.

To generate the next candidate approximation to the solution, we use operations called *Monte Carlo moves* [88]. In particular, we consider the *pivot*, *multipivot* and *crankshaft* moves. The pivot move randomly selects a pivot point x_i, with $1 < i < n$ and two coordinate axes ξ and η, and then rotates each point x_k, with $i < k \le n$, of a random angle with respect to the axis through x_i and orthogonal to ξ and η. The multipivot move is obtained by performing a sequence of pivot moves. In our case, $n/10$ points x_i, with $1 < i < n$, are randomly selected and used as pivots. Finally, the crankshaft move randomly selects two points x_i and x_j, with $1 \le i < j - 1 < n$, and then rotates the points x_k, with $i < k < j$, of a random angle around the axis passing through x_i and x_j.

The acceptance strategy used in our experiments is based on the well-known *Metropolis acceptance function* [55]. If $X^{(k)}$ is the approximation of the solution at a step k and $\bar{X}$ is a candidate approximation obtained by a Monte Carlo move, then $\bar{X}$ is accepted if

$$A(X^{(k)}, \bar{X}, t^{(k)}) = \min\left\{1, e^{\frac{F(\bar{X}) - F(X^{(k)})}{t^{(k)}}}\right\} > p,$$

where F is the objective function to be maximized (see (1)), $t^{(k)}$ is the temperature value at step k and p is a random number from the uniform distribution in $(0, 1)$. The candidate approximation can be accepted even if it does not increase the value of F, depending on $t^{(k)}$ and p. At high temperatures, many candidate approximations can be accepted, but, as the temperature decreases, the number of candidate approximations decreases, in analogy with the physical process of annealing.

The cooling strategy has an important role in SA. The temperature must be decreased very slowly to avoid trapping into local optima that are far from the global one. This reflects the behaviour of the physical annealing, in which a fast temperature decrease leads to a polycrystalline or amorphous state. In our experiments, a fixed number *nsteps* of Metropolis Monte Carlo iterations is performed at constant temperature and then the temperature value is decreased by a fixed factor $\gamma < 1$. The values of *nsteps* and γ have been experimentally set to $10^3 n$ and 0.99, respectively.

Our algorithm terminates when the value of the objective function F has not been changed for ten outer iterations, or a maximum number of outer iterations, *maxout*, is achieved. We set *maxout* = 300, but this value was never reached in our experiments. A sketch of the whole algorithm is provided in Figure 10.1.

We note that the cost of evaluating the term f in the objective function F (see (1) and (2)-(3)) is usually lower than $O(n^3)$. Indeed, if two points have a Euclidean distance greater than $2(\bar{D} + \epsilon)$, then all the triplets containing these points have a three-body radius greater than $\bar{D} + \epsilon$ (in a circle, a chord is smaller than the diameter) and hence they do not give any contribution to f. Therefore, once the Euclidean distances of all the pairs of points are computed, as required by the constraints (8), the three-body radii are computed only for triplets such that the Euclidean distance of all the pairs in the triplet is not greater than $2(\bar{D} + \epsilon)$.

```
t = t_0
X = random conformation satisfying the constraints
nout = 0

{outer loop}
while ( F(X) not settled down and nout ≤ maxout )
    nout = nout + 1

    {inner loop}
    for k = 1, nsteps
        X^(k) = random MC move on X
        if ( X^(k) satisfies the constraints ) then
            p = uniform random number in (0,1)
            if ( A(X, X^(k), t)) > p ) then
                X = X^(k)
            endif
        endif
    endfor

    t = γ · t
endwhile
```

Fig. 10. Metropolis Monte Carlo Simulated Annealing algorithm.

10.2 Simulation of α-helices

First experiments have been performed with very short amino acid chains and with the objective function of the original tube model, i.e. without considering the compactness term $g(X)$ in the objective function $F(X)$ (see (1)).

Many simulations have been carried out with $n = 10$ amino acids, starting from different initial conformations. All the computed optimal conformations are clock-wise rotated helices with about 3.6 points per helix turn, as in the real α-helices. About 60% of these conformations differ each other by a RMSD value of about 0.5 $\overset{\circ}{\mathrm{A}}$; a maximum RMSD of 2.0 $\overset{\circ}{\mathrm{A}}$ has been observed. The value of the objective function at the solution is always equal to 22 and is due to all the triplets (x_i, x_{i+1}, x_{i+2}), (x_i, x_{i+1}, x_{i+3}) and (x_i, x_{i+2}, x_{i+3}) (8, 7 and 7 triplets, respectively), which have a tree-body radius ranging between $\bar{D} - \epsilon$ and $\bar{D} + \epsilon$, where $\bar{D} = 2.70$ $\overset{\circ}{\mathrm{A}}$ and $\epsilon = 0.20$ $\overset{\circ}{\mathrm{A}}$, as discussed in Section 9. Each simulation was completed in about 7 seconds. An example of computed optimal conformation is shown in Figure 10.2.

Other experiments have been performed by changing the value of $\bar{D}$, but keeping $\epsilon = 0.20$. In this case, the computed conformations are unrealistic helices, with less than 3.6 points per turn if $\bar{D} < 2.70$ and more than 3.6 if

Fig. 11. Two views of a computed optimal conformation with $n = 10$ points $\left(\bar{D} = 2.70 \text{ Å}\right)$.

$\bar{D} > 2.70$. These results support the choice $\bar{D} = 2.70$. Some conformations obtained with different values of $\bar{D}$ are shown in Figure 10.2.

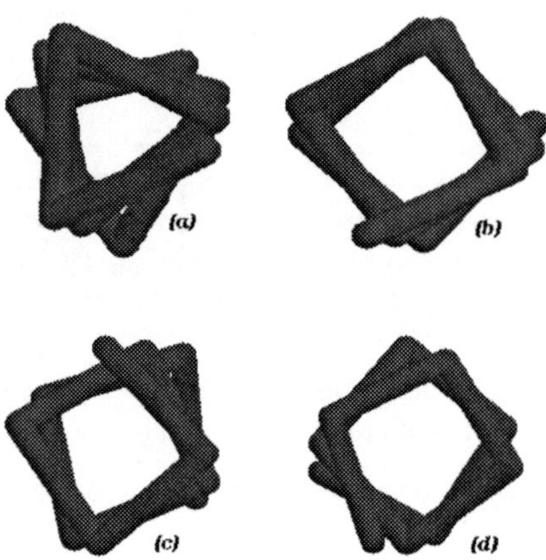

Fig. 12. Conformations obtained with $n = 10$ and different values of $\bar{D}$ ((a) $\bar{D} = 2.60$, (b) $\bar{D} = 2.80$, (c) $\bar{D} = 2.90$, (d) $\bar{D} = 3.20$).

Further experiments with $n > 10$ led to similar results. When $\bar{D} = 2.70$, conformations very close to real α-helices are obtained, while unrealistic helices are generated for $\bar{D} \neq 2.70$. Furthermore, for $n > 30$, single long helices are computed which do not exist in nature, hence the need of introducing a compactness term into the problem objective function.

10.3 Simulation of All-α Proteins

Some experiments have been devoted to generate all-α protein conformations. A globular protein composed of 153 amino acid residues, the sperm whale myoglobin (PDB code 1mbn), has been chosen as reference protein. Obviously, we did not expect to generate conformations very close to the myoglobin one, since the information contained in the considered model is too poor for an accurate fold prediction. On the other hand, we wished to analyze the reliability and accuracy provided by such a simplified model.

The lengths of the ellipsoid semiaxes a, b and c have been computed using the amino acid volumes of the selected protein, as explained in Section 9. According to the whole myoglobin shape, the following lengths have been considered: $a = b = 1.15s$ and $c = 0.76s$, where s is radius of the sphere with volume equal to the sum of the amino acid volumes, increased by 3.8% (see (11)), i.e. $s = 17.32$ Å. A few experiments with different semiaxis lengths have been also performed to analyze the weight of the compactness term $g(X)$ with respect to the thickness term $f(X)$ in the objective function $F(X)$. Sixty simulations have been performed until now, each requiring an execution time of about two hours. Better simulated conformations could be obtained by running a larger number of experiments.

The results obtained so far show that, as a, b and c get closer, the value of the term $f(X)$ at the solution decreases. A minimum value of 300 has been achieved with $a = b = c$. On the other hand, as the difference between two semiaxes increases, and hence the formation of longer helices is allowed, the value of $f(X)$ at the solution usually increases; $f(X) = 360$ has been obtained for $a = b = 1.2s$ and $c = 0.7s$. Conformations with values of $g(X)$ varying between 100 and 153 have been obtained, where larger values of $g(X)$ correspond to smaller values of $f(X)$.

Like the all-α proteins, the computed conformations are globular objects with secondary structures that are very close to real α-helices. For $a = b = 1.15s$ and $c = 0.76s$, i.e. for semiaxis lengths corresponding to the myoglobin shape, we obtained two conformations that have 66.7 and 59.5 identity percentages of secondary structures with respect to the reference protein. If we consider only the α-helices, the identity percentages are 67.8 and 51.6, respectively. This is shown in Figure 10.3. The corresponding three-dimensional representations are given in Figure 10.3. On the other hand, while having a certain similarity, the real protein and the computed conformations can have different numbers of helices, with different lengths and orientations, thus indicating that more information must be included in the model to perform more accurate simulations.

Fig. 13. A comparison of the sperm whale myoglobin with two simulated conformations. First row: amino acid sequence of the sperm whale myoglobin; second row: myoglobin amino acid residues that are contained into α-helices in the original conformation; third and fourth row: myoglobin amino acid residues that are contained into helices in the two simulated conformations. The identity percentage is 66.7 for the conformation called **test1** and 59.5 for the one called **test2**.

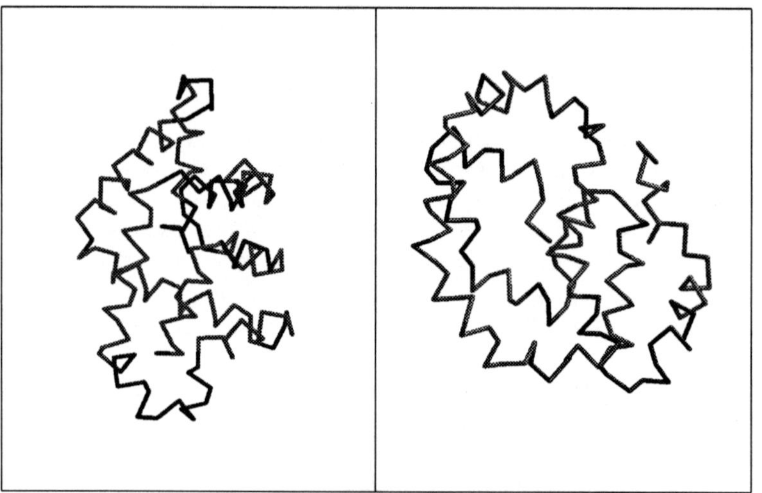

Fig. 14. Conformations obtained from the sperm whale myoglobin protein chain ($a = b = 1.15s$ and $c = 0.76s$). The α-helices are lighter, all the other structures are darker.

11 Conclusions

The great interest in the solution of the protein folding problem strongly pushes the research activity in this area. However, despite the many efforts performed so far, this problem is still considered a big challenge in science.

In this chapter we focused on ab-initio computational methods for protein fold predictions that are potentially able to discover unknown native state conformations. In this context, we analyzed an interesting topological approach, that takes into account geometrical rather than physicochemical protein features. This approach is based on a very simplified model that represents the polymer chain as a non-intersecting tube of nonzero thickness, by explicitly considering only the C_α trace of the protein and describing the amino acid interactions through the use of a suitable metric that measures the "distance" among any three C_α atoms. This model leads to the formulation of a global constrained optimization problem.

To enhance compactness and globularity in the computed conformations, we introduced a modification into the above model, and presented a methodology for choosing the values of characterstic parameters. The results of computational experiments devoted to simulating α-helices and all-α proteins can be considered "promising", especially if we take into account the great simplicity and the relatively low computational cost of the model. Indeed, simulations performed using the sperm whale myoglobin as target protein, generated a conformation with a percentage identity equal to 66.7. Hence, we expect that the model can be significantly improved by adding some physicochemical features to the geometrical ones currently considered. The introduction of the amino acid hydrophobicity into the model and the definition of ad hoc constraints and suitable parameter values for the simulation of β-strands and β-sheets are currently under investigation.

Acknowledgements

We wish to thank Davide Marenduzzo from University of Oxford, Department of Physics, for providing us an implementation of the Metropolis Monte Carlo Simulated Annealing algorithm, that we used as a basis for our implementation, and for some helpful discussions. This work has been partially supported by the MIUR FIRB projects "Large Scale Nonlinear Optimization" (grant no. RBNE01WBBB) and "Identification and functional analysis of gene and molecular modifications of hormone-responsive breast cancer" (grant no. RBNE0157EH_03).

References

1. N.L. Allinger. MM2. A Hydrocarbon Force Field Utilizing V_1 and V_2 Torsional Terms. *Journal of the American Chemical Society*, 99(25): 8127-8134, 1977.

2. N.L. Allinger, Y.H. Yuh, and J.-H. Lii. Molecular Mechanics. The MM3 Force Field for Hydrocarbons. *Journal of the American Chemical Society*, 111(23): 8551-8565, 1989.

3. S.F. Altschul, W. Gish, W. Miller, E.W. Myers, and D.J. Lipman. Basic local alignment search tool. *Journal of Molecular Biology*, 215: 403-410, 1990.

4. S.F. Altschul, T.L. Madden, A.A. Schaffer, J. Zhang, Z. Zhang, W. Miller, and D.J. Lipman. Gapped BLAST and PSI-BLAST: a new generation of protein database search programs, *Nucleic Acids Research*, 25 (17): 3389-402, 1997.

5. C.B. Anfinsen, E. Haber, M. Sela, and F.H. White. The kinetics of formation of native ribonuclease during oxidation of the reduced polypeptide chain. *Proceedings of the National Academy of Sciences*, 47: 1309-1314, 1961.

6. I.P. Androulakis, C.D. Maranas, and C.A. Floudas. αBB: A Global Optimization Method for General Constrained Nonconvex Problems. *Journal of Global Optimization*, 7(4): 337-363, 1995.

7. J.R. Banavar, A. Flammini, D. Marenduzzo, A. Maritan, and A. Trovato. Geometry of Compact Tubes and Protein Structures. *ComPlexUs*, 13: 1-4, 2003.

8. J.R. Banavar, O. Gonzalez, J.H. Maddocks, and A. Maritan, Self-interactions of strands and sheets. *Journal of Statistical Physics*, 110: 35-50, 2003.

9. J.R. Banavar, A. Maritan, C. Micheletti, and F. Seno. *Geometrical aspects of protein folding*, Lectures held at the "Enrico Fermi Summer School", Varenna, Italy, 2001.

10. J.R. Banavar, A. Maritan, C. Micheletti, and A. Trovato. Geometry and Physics of Protein. *Proteins*, 47(3): 315-322, 2002.

11. D. Baker. A surprising simplicity to protein folding. *Nature*, 405: 39-42, 2000.

12. R. Bonneau, J. Tsai, I. Ruczinski, D. Chivian, C. Rohl, C.E.M. Strauss, and D. Baker. Rosetta in CASP4: progress in ab initio protein structure prediction. *Proteins: Structure, Function and Genetics Supplement*, 5: 119-126, 2001.

13. P. Bradley, D. Chivian, J. Meiler, K.M.S. Misura, C.A. Rohl, W.R. Schief, W.J. Wedemeyer, O. Schueler-Furman, P. Murphy, J. Schonbrun, C.E.M. Strauss, and D. Baker. Rosetta Predictions in CASP5: Successes, Failures, and Prospects for Complete Automation. *Proteins: Structure, Function and Genetics Supplement*, 53: 457-468, 2003.

14. C. Caporale, A. Facchiano, L. Bestini, L. Leopardi, G. Chiosi, V. Buonocore, and C. Caruso. Comparing the modelled structures of PR-4 proteins from wheat. *Journal of Molecular Modeling*, 9: 9-15, 2003.

15. CHARMM Home Page, http://www.charmm.org/.

16. EMBnet Home Page, http://www.ch.embnet.org/MD_tutorial/.

17. A.M. Facchiano, P. Stiuso, M.L. Chiusano, M. Caraglia, G. Giuberti, M. Marra, A. Abruzzese, and G. Colonna. Homology modelling of the human eukaryotic initiation factor 5A (eIF-5A). *Protein Engineering*, 14: 881-890, 2001.

18. J.S. Fetrow, A. Giammona, A. Kolinski, and J. Skolnick. The protein folding problem: A Biophysical Enigma. *Current Pharmaceutical Biotecnology*, 3: 329-347, 2002.

19. G.S. Fishman, editor. *Monte Carlo: Concepts, Algorithms, and Applications*. Springer, 1996.

20. C.A. Floudas. *Deterministic global optimization: theory, methods and applications*. Kluwer Academic Publishers, 2000.

21. C.A. Floudas, J.L. Klepeis, and P.M. Pardalos. Global Optimization Approaches in Protein Folding and Peptide Docking. In M. Farach, F.S. Roberts, M. Vingron, and M. Waterman, editors, *Mathematical Support for Molecular Biology*, pages 141–171. DIMACS Series, Volume 47, American Mathematical Society, Providence, RI, 1999.

22. I. Friedberg, T. Kaplan, and H. Margalit. Evaluation of PSI-BLAST alignment accuracy in comparison to structural alignments. *Protein Science*, 9: 2278-2284, 2000.

23. O. Gonzalez and J.H. Maddocks. Global curvature, Thickness and the Ideal Shapes of Knots. *Proceedings of the National Academy of Sciences*, 96: 4769-4773, 1999.

24. T.X. Hoang, M. Cieplak, J. Banavar, and A. Maritan. Prediction of Protein Secondary Structures from Conformational Biases. *Proteins: Structure, Function and Genetics*, 48: 558-565, 2002.

25. D.T. Jones. Protein secondary structure prediction based on position-specific scoring matrices. *Journal of Molecular Biology*, 292: 195-202, 1999.

26. D.T. Jones. Predicting novel protein folds by using FRAGFOLD. *Proteins: Structure, Function and Genetics Supplement*, 5: 127-132, 2001.

27. D.T. Jones. Critically assessing the state-of-art in protein structure prediction. *The Pharmacogenomics Journal*, 1(2): 126-134, 2001.

28. D.T. Jones and L.J. McGuffin. Assembling novel protein folds from supersecondary structural fragments. *Proteins: Structure, Function and Genetics*, 53: 480-485, 2003.

29. JUFO Home Page, `http://www.jens-meiler.de/jufo.html`.

30. W. Kabsch and C. Saunder. Dictionary of protein secondary structure: pattern recognition of hydrogen-bonded and geometrical features. *Biopolymers*, 22: 2577-2637, 1983.

31. M. Karplus and J.N. Kushick. Method for estimating the configurational entropy of macromolecules. *Macromolecules*, 14: 325-332, 1981.

32. S.-Y. Kim, S.J. Lee and J. Lee. Conformational space annealing and an off-lattice frustrated model protein. *Journal of Chemical Physics*, 119: 10274 - 10279, 2003.

33. J.L. Klepeis and C.A. Floudas. Deterministic Global Optimization for Protein Structure Prediction. In C. Caratheodory, N. Hadjisavvas and P.M. Pardalos, editors, *Advances in Convex Analysis and Global Optimization*, pages 31-74, Kluwer, 2001.

34. J.L. Klepeis and C.A. Floudas. ASTRO-FOLD: Ab Initio Secondary and Tertiary Structure Prediction in Protein Folding. In J. van Schijndel, ed-

itor, *European Symposium on Computer Aided Process Engineering*, Volume 12, Elsevier Applied Science, 2002.

35. J.L. Klepeis and C.A. Floudas. Ab initio tertiary structure prediction of proteins. *Journal of Global Optimization*, 25: 113-140, 2003.

36. J.L. Klepeis and C.A. Floudas. ASTRO-FOLD: a combinatorial and global optimization framwork for ab initio prediction of three-dimensional structures of proteins from the amino acid sequence. *Biophysical Journal*, 85: 1-28, 2003.

37. J.L. Klepeis and C.A. Floudas. Prediction of *beta*-Sheet Topology and Disulfide Bridges in Polypeptides. *Journal of Computational Chemistry*, 24: 191-208, 2003.

38. J.L. Klepeis and C.A. Floudas. Analysis and Prediction of Loop Segments in Protein Structures. *Computers & Chemical Engineering*, 29: 423-436, 2005.

39. J.L. Klepeis, Y. Wei, M.H. Hecht and, C.A. Floudas. Ab initio Prediction of the 3-Dimensional Structure of a De novo Designed Protein: A Double Blind Case Study. *Proteins*, 58: 560-570, 2005.

40. S. Kirkpatrick, C.D. Gelatt Jr., and M.P. Vecchi. Optimization by Simulated Annealing. *Science*, 220(4598): 671-680, 1983.

41. P. Koehl and M. Levitt. Improved recognition of native-like protein structures using a family of designed sequences. *Proceedings of the National Academy of Sciences*, 99(2): 691-696, 2002.

42. N. Koga and S. Takada. Roles of Native Topology and Chain-length Scaling in Protein Folding: A Simulation Study with a Go-like Model. *Journal of Molecular Biology*, 313: 171-180, 2001.

43. A. Kolinski, P. Rotkiewicz, B. Ilkowski and J. Skolnick. Protein Folding: Flexible Lattice Models. *Progress of Theoretical Physics*, 138: 292-300, 2000.

44. A. Kolinski, P. Rotkiewicz and J. Skolnick. Structure of proteins: New Approach to Molecular Modeling. *Polish Journal of Chemistry*, 75: 587-599, 2001.

45. A. Kolinski. Protein modeling and structure prediction with a reduced representation. *Acta Biochimica Polonica*, 51(2): 349-371, 2004.

46. M. Kuhn, J. Meiler and D. Baker. Strand-loop-strand motifs: prediction of hairpins and diverging turn in proteins. *Proteins: Structure, Function and Bioinformatics*, 54: 282-288, 2004.

47. J. Lee, H. A. Scheraga and S. Rackovsky. New optimization method for conformational energy calculations on polypeptides: conformational space annealing. *Journal of Computational Chemistry*, 18: 1222-1232, 1997.

48. C. Levinthal. Are there pathways for protein folding? *Chemical Physics*, 65: 44-45, 1968.

49. M. Levitt and A. Hinds. A lattice model for protein structure prediction at low resolution. *Proceedings of the National Academy of Sciences*, 89: 2536-2540, 1992.

50. A. Liwo, M.R. Pincus, R.J. Wawak, S. Rackovsky, and H.A. Scheraga. Calculation of protein backbone geometry from α-carbon coordinates based on peptide-group dipole alignment. *Protein Science*, 2: 1697-1714, 1993.

51. A. Liwo, J. Lee, D.R. Ripoll, J. Pillardy, and H.A. Scheraga, Protein structure prediction by global optimization of a potential energy function. *Proceedings of the National Academy of Sciences*, 96: 5482-5485, 1999.

52. M. Locatelli. Simulated Annealing Algorithms for Continuous Global Optimization. In P.M. Pardalos and H.E. Romeijn, editors, *Handbook of Global Optimization*, Volume 2, pages 179-229. Kluwer Academic Publishers, 2002.

53. C.D. Maranas, L.P. Androulakis, and C.A. Floudas. A Deterministic Global Optimization Approach for the Protein Folding Problem. In P. M. Pardalos, D. Shalloway, and G. Xue, editors, *Global Minimization of Nonconvex Energy Functions: Molecular Conformation and Protein Folding*, pages 133-150. DIMACS Series, Volume 23, American Mathematical Society, Providence, RI, 1996.

54. D. Marenduzzo, A. Flammini, A. Trovato, J.R. Banavar, and A. Maritan. Physics of thick polymers. *Journal of Polymer Science, Part B: Polymer Physics*, 43: 650679, 2005.

55. N. Metropolis, A.W. Rosenbluth, M.N. Rosenbluth. A.H. Teller, and E. Teller. Equation of State Calculations by Fast Computing Machines. *Journal of Chemical Physics*, 21: 1087-1092, 1953.

56. C. Micheletti. Prediction of Folding rates and Transition-State Placement From Native-State Geometry. *Proteins: Structure, Function and Genetics*, 51: 74-84, 2003.

57. F.A. Momany, L.M. Carruthers, R.F. McGuire, and H.A. Scheraga. Intermolecular potentials from crystal data. III. *Journal of Physical Chemistry*, 78: 1595-1620, 1974.

58. F.A. Momany, L.M. Carruthers, and H.A. Scheraga. Intermolecular potentials from crystal data. IV. *Journal of Physical Chemistry*, 78: 1621-1630, 1974.

59. G. Némety, M.S. Pottle, and H.A. Scheraga. Energy Parameters in Polypeptides. 9. *Journal of Physical Chemistry*, 87: 1883-1887, 1983.

60. G. Némety, K.D. Gibson, K.A. Palmer, C.N. Yoon, G. Paterlini, A. Zagari, S. Rumsey, and H.A. Scheraga. Energy Parameters in Polypeptides. 10. *Journal of Physical Chemistry*, 96: 6472-6484, 1992.

61. D.J. Osguthorpe. Ab initio protein folding. *Current Opinion in Structural Biology*, 10: 146-152, 2000.

62. P.M. Pardalos and H.E. Romeijn, editors. *Handbook of Global Optimization*, Volume 2. Kluwer Academic Publishers, 2002.

63. P.M. Pardalos and G. Xue, editors. *Advances in Computational Chemistry and Protein Folding. Journal of Global Optimization*, Special Issue, 4(2), 1994.

64. P.M. Pardalos, D. Shalloway, and G. Xue, editors. *Global Minimization of Nonconvex Energy Functions: Molecular Conformation and Protein Folding*. DIMACS Series, Volume 23, American Mathematical Society, Providence, RI, 1996.

65. PDBSELECT Home page `http://www.cmbi.kun.nl/swift/pdbsel/`.

66. S. Petit-Zeman. Treating protein folding diseases. *Nature*, 2002, available at `http://www.nature.com/horizon/proteinfolding/background/-treating.html`

67. J. Pietzsch. The importance of Protein Folding. *Nature*, 2002, available at `http://www.nature.com/horizon/proteinfolding/background/-importance.html`

68. J. Pietzsch. Protein Folding diseases. *Nature*, 2002, available at `http://www.nature.com/horizon/proteinfolding/background/disease.html`.

69. K.W. Plaxco, K.T. Simons and D. Baker. Contact Order, Transition State Placement and the Refolding Rates of Single Domain Proteins. *Journal of Molecular Biology*, 277: 985-994, 1998.

70. Protein Structure Prediction Center Home Page, `http://predictioncenter.llnl.gov`.

71. PROSPECTOR Home Page, `http://www.bioinformatics.buffalo.edu/new_buffalo/services/threading.html`.

72. D.R. Ripoll, A. Liwo, and H.A. Scheraga. New Developments of the Electrostatically Driven Monte Carlo Method: Test on the Membrane-Bound Portion of Melittin. *Biopolymers*, 46: 117, 1998.

73. D.R. Ripoll and H.A. Scheraga. On the multiple-minima problem in conformational analysis of polypeptides. IV. *Biopolymers*, 30: 165-176, 1990.

74. D.R. Ripoll, M.J. Vàsquez, and H.A. Scheraga. The electrostatically driven Monte Carlo method - Application to conformational analysis of decaglycine. *Biopolymers*, 31: 319-330, 1991.

75. K. Rother, R. Preissner, A. Goede, and C. Frommel. Inhomogeneous molecular density: reference packing densities and distribution of cavities within proteins. *Bioinformatics*, 19(16): 2112-2121, 2003.

76. I. Ruczinski, C. Kooperberg, R. Bonneau and D. Baker. Distributions of Beta Sheets in Proteins With Application to Structure Prediction. *Proteins: Structure, Function and Genetics*, 48: 85-97, 2002.

77. R. Samudrala, Y. Xia, E. Huang, and M. Levitt. Ab initio Protein Structure Prediction Using a Combined Hierarchical Approach. *Proteins: Structure, Function and Genetics Supplement*, 3: 194-198, 1999.

78. J.A. Saunders, K.D. Gibson, and H.A. Scheraga. Ab initio folding of multiple-chain proteins. *Pacific Symposium on Biocomputing*, 7: 601-612, 2002.

79. G. Scapigliati, S. Costantini, G. Colonna, A. Facchiano, F. Buonocore, P. Boss, J.W. Holland, and C.J. Secombes. Modelling of fish interleukin 1 and its receptor. *Developmental and Comparative Immunology*, 28: 429-41, 2004.

80. G. Settanni, A. Cattaneo, and A. Maritan. Role of Native-State Topology in the Stabilitazion of Intracellular Antibodies. *Biophysical Journal*, 81: 2935-2945, 2001.

81. K.T. Simons, C. Kooperberg, E. Huang, and D. Baker. Assembly of Protein Tertiary Structures from Fragments with Similar Local Sequences using Simulated Annealing and Bayesian Scoring Function. *Journal of Molecular Biology*, 268: 209-225, 1997.

82. K.T. Simons, R. Bonneau, I. Ruczinski, and D. Baker. Ab Initio Protein Structure Predictions of CASP III Targets Using ROSETTA. *Proteins: Structure, Function and Genetics Supplement*, 3: 171-176, 1999.

83. J. Skolnick, A. Kolinski, D. Kihara, M. Betancourt, P. Rotkiewicz, and M. Boniecki. Ab initio protein structure prediction via a combination of threading, lattice folding, clustering, and structure refinement. *Proteins: Structure, Function and Genetics Supplement*, 5: 149-156, 2001.

84. J. Skolnick, D. Kihara, H. Lu, and A. Kolinski. TOUCHSTONE: An ab initio protein structure prediction method that uses threading-based tertiary restraints. *Proceedings of the National Academy of Sciences*, 98(18): 10125-10130, 2001.

85. J. Skolnick, D. Kihara, Y. Zhang, H. Lu, and A. Kolinski. Ab initio protein structure prediction on a genomic scale: Application to the Mycoplasma genitalium genome. *Proceedings of the National Academy of Sciences*, 99(9), 5993-5998, 2002.

86. J. Skolnick, Y. Zhang, A.K. Arakaki, A. Kolinski, M. Boniecki, A. Szilágyi, and D. Kihara. TOUCHSTONE: A Unified Approach to Protein Structure Prediction. *Proteins: Structure, Function and Genetics*, 53: 469-479, 2003.

87. J.E. Smith. Genetic Algorithms. In P.M. Pardalos and H.E. Romeijn, editors, *Handbook of Global Optimization*, Volume 2, pages 275-362. Kluwer Academic Publishers, 2002.

88. A.D. Sokal. Monte Carlo methods for the self-avoiding walk. *Nuclear Physics B (Proceedings Supplements)*, 47: 172-179, 1996.

89. R. Srinivasan and G.D. Rose. LINUS: a hierarchic procedure to describe the fold of a protein. *Proteins*, 22: 81-99, 1995.

90. R. Srinivasan and G.D. Rose. A physical basis for protein secondary structure. *Proceedings of the National Academy of Sciences*, 96(25): 14258-14263, 1999.

91. R.H. Swendsen and J.S. Wang. Replica Monte Carlo Simulation of Spin-Glasses. *Physical Review Letters*, 57: 2607-2609, 1986.

92. A. Trovato. *A Geometric Perspective on Protein Structures and Heteropolymer Models*. PhD Thesis, SISSA, Trieste, 2000.

93. S.J. Weiner, P.A. Kollman, D.A. Case, U.C. Singh, C. Ghio, G. Alagona, S. Profeta, and P. Weiner. A New Force Field for Molecular Mechenical Simulation of Nucleic Acids and Proteins. *Journal of the American Chemical Society*, 106: 765-784, 1984.

94. Y. Xia, E. S. Huang, M. Levitt, and R. Samudrala. Ab Initio Construction of Protein Tertiary Structures Using a Hierarchical Approach. *Journal of Molecular Biology*, 300: 171-185, 2000.

95. A. Zemla, C. Venclovas, K. Fidelis, and B. Rost. A Modified Definition of Sov, a Segment-Based Measure for Protein Secondary Structure Prediction Assessment. *Proteins: Structure, Function and Genetics*, 34: 220-223, 1999.

96. Y. Zhang, D. Kihara, and J. Skolnick. Local Energy Landscape Flattering: Parallel Hyperbolic Monte Carlo Sampling of Protein Folding. *Proteins: Structure, Function and Genetics*, 48: 192-201, 2002.

97. Y. Zhang, A. Kolinski, and J. Skolnick. TOUCHSTONE II: A New Approach to Ab Initio Protein Structure Prediction. *Biophysical Journal*, 85: 1145-1164, 2003.

A Topological Characterization of Protein Structure

Bala Krishnamoorthy[1], Scott Provan[2], and Alexander Tropsha[3]

[1] Department of Mathematics
Washington State University
`kbala@wsu.edu`
[2] Department of Statistics and Operations Research
University of North Carolina
`scott_provan@unc.edu`
[3] School of Pharmacy
University of North Carolina
`alex_tropsha@unc.edu`

Summary. We develop an objective characterization of protein structure based entirely on the geometry of its parts. The three-dimensional alpha complex filtration of the protein represented as a union of balls (one per residue) captures all the relevant information about the geometry and topology of the molecule. The neighborhood of a strand of contiguous alpha carbon atoms along the back-bone chain is defined as a "tube" which is a sub-complex of the original complex that has been sub-divided. We then define a retraction for the tube to another complex that is guaranteed to be a 2-manifold with boundary. We capture the topology of the retracted tube by computing the most persistent connected components and holes in the entire filtration. A "motif" for a 3D structure is characterized by the number of persistent 0- and 1-cycles, and the relative persistences of these cycles in the filtration of the "tube" complex. These motifs represent non-random, recurrent, tertiary interactions between parts of the protein back-bone chain that characterize the overall structure of the protein. A basis set of 1300 motifs are identified by analyzing the alpha complex filtrations of several proteins. Any test protein is represented by the number of times each motif from the basis set occurs in it. Preliminary results from the discrimination of protein families using this representation are provided.

Key words: Protein structure, simplicial complexes, homology groups, topological persistence.

Structural Similarity Between Proteins

Understanding the similarities and differences between protein structures is central to the study of connections between the sequence, structure, and the function of the proteins, and also for detecting possible evolutionary relationships. With the number of proteins with known structures currently exceeding 25,000 [21], and rapidly increasing by the day, the need for reliable and automated methods for structural comparison has never been greater. Various techniques for structural comparison have emerged, ranging from those which try to match the geometric coordinates of the back-bone [23], to those which use vector approximations to secondary structure elements [16, 14]. Then there are domain-based methods, which try to classify proteins based on the units of structure (or domains) that they contain. Even though there is no exact definition available, a structural domain is usually considered as a compact and semi-independent unit of a protein, which consists of a small number of contiguous segments of the peptide chain, and forms a structurally "separate" region in the whole three-dimensional structure of the protein. Widely used structural databases such as SCOP [19], and CATH [20] have been constructed using domain-based approaches. On the other hand, one of the most successful automated classifications of proteins uses concepts from knot theory to reproduce the classification provided by the CATH database with a high degree of accuracy.

This chapter is organized as follows. We review the main features of the SCOP database in Section 1. A brief description of the knot theory-based classification follows in Section 2. The drawbacks of these methods which motivated our line of research are outlined. We provide the necessary background material on alpha shapes and homology in Section 3. The definition of the neighborhood of a strand in a protein is given in detail in Section 5. We outline the algorithm used to characterize the topology of these neighborhoods in Section 6. Finally, we describe the salient features of the structural motifs that characterize these neighborhoods in Section 8.

1 The SCOP Database

The Structural Classification of Proteins (SCOP) database is a comprehensive ordering of all proteins of known structure according to their evolutionary and structural relationships. A fundamental unit of classification in this database is the protein domain. A domain is defined as an evolutionary unit observed in nature either in isolation or in more than one context in multi-domain proteins. All Protein domains are hierarchically classified into families, super-families, folds, and classes. The method used to construct this classification is essentially the visual inspection and comparison of structures. Any use of automatic tools in this process is aimed only at making the task manageable. The SCOP database could be considered as containing the most accurate and useful results on protein structure classification. Recent updates of the database [3] reported the introduction of integer identifiers for each node in the hierarchy (called *sunid*), and a new set of concise classification strings (called *sccs*). There is also an initiative [1] to rationalize and integrate the SCOP information with the data about protein families housed by other prominent sequence and structural databases such as InterPro [17], CATH, and others.

The classification in SCOP is done on four hierarchical levels – family, super-family, common fold, and class. These levels embody the evolutionary and structural relationships between the domains. Proteins that have at least 30% sequence identity are classified into the same family. In addition, proteins that have lower (than 30%) sequence identity, but whose functions and structures are *very similar*, are also classified into the same family. Families, whose proteins have low sequence identities, but whose structures, and in many cases, functional features suggest that a common evolutionary origin is possible, are grouped into super-families. In the next level of hierarchy, super-families and families that have some *major* secondary structures in the *same* arrangement with the *same topological* connections are defined to a have common fold. Finally, for the convenience of users, different folds have been grouped into classes. Most folds are assigned to one of the following five structural classes based on their secondary structure composition -

1. all alpha (when the structure is mainly formed by α-helices),
2. all beta (when the structure is mainly formed by β-sheets),
3. alpha and beta (when α-helices and β-strands are largely interspersed),
4. alpha plus beta (when α-helices and β-strands are largely segregated), and
5. multi-domain (for which no homologues are known as of now).

In the latest version of SCOP (2004), the multi-domain class is further subdivided into seven classes, thus giving a total of eleven classes. The CATH database assigns to proteins a unique **Class, Architecture, Topology,** and a **Homological** super-family. The methods used to achieve these assignments are similar to those employed in SCOP.

2 Knot Theory-based Classification

In 2003, Røgen and Fain [22] introduced a novel method of looking at, analyzing, and comparing protein structures that used the concepts from knot theory. The topology of a protein is captured by 30 numbers inspired by Vassiliev knot invariants. A measure for the similarity of protein shapes called the Scaled Gauss Metric (SGM) is created from these 30 numbers. The protein back-bone is analyzed as a curve in 3D space. The primary invariant calculated by the authors is the *writhing number* of the curve. This invariant essentially measures the self-linking of the curve which is the protein back-bone. The first biological applications of this measure were reported in the studies of DNA structure. It is related to the *linking number* and *twisting number* of two curves by the Călugăreanu-Fuller-White formula [24]:

$$Lk = Wr + Tw \tag{1}$$

The formula applies to a narrow closed orientable ribbon in 3D space. Here, Lk is the linking number of the two boundary curves of the ribbon, Wr is the writhing number of the central spine, and Tw is the twisting number of the two boundary curves. For a protein, the back-bone plays the role of the spine, and it is naturally oriented by the residue numbering order. Now imagine projecting the ribbon onto a 2D plane orthogonal to a randomly chosen direction. The curves defining the ribbon will seem to cross each other at certain locations in the plane of projection. Depending on the orientation of the two curve segments and the over-under relationship at each crossing, we assign a $+1$ or a -1 to the crossing. The linking number Lk counts the sum of the signed crossings between the two boundary curves, divided by two. This sum is independent of the direction of projection. The writhing number Wr counts the sum of the signed self-crossings of the ribbon's spine, now averaged over *all* projections. Finally, the twist Tw is a torsion-dependent term that measures how much one boundary curve intertwines with the other.

If we add up the unsigned individual contributions to the writhe, we obtain the (unsigned) average crossing number. A family of structural measures could be constructed using the writhe and the average crossing number as the building blocks. The authors found it sufficient to compute 30 such measures for the purpose of structural classification. Thus each protein is mapped to $\mathbb{R}^{30}$ space. Based on this mapping, the Euclidean distance between two points (or proteins) is defined as the Scaled Gauss Metric (SGM). Unlike the metrics defined by most other methods, SGM is a proper pseudo-metric - it has a zero element, it is symmetric, and most importantly, it satisfies the triangle inequality. The last property enables us to use the SGM to identify meaningful intermediate and marginal similarities, and also to distinguish between various degrees of similarity. Another desirable property of the Gauss metric is that it requires neither structural nor sequential alignment between chains, thus making the pair-wise comparison of proteins almost instantaneous. The authors used SGM to construct an automatic classification procedure for the CATH2.4 database (they essentially clustered the proteins based on the SGM). They could accurately assign more than 95% of the chains into the proper C(class), A(architecture), T(topology), and H(homologous super-family), find all new folds, and detect no false positives.

Erdmann [12] builds on the ideas of Røgen and Fain in using knot theory ideas for studying structural similarity between proteins. Supplementing the knot theory concepts with ideas from geometric convolution, the author proposes a definition of similarity based on atomic motions that preserve local back-bone topology without incurring significant errors. Similarity detection then seeks rigid body motions able to overlay pairs of substructures, each requiring a substructure-preserving motion, without necessarily requiring global structural preservation. This definition has a very broad scope - one could talk about the full rearrangement of one protein into another while preserving the global topology, or about rearrangements of sets of smaller substructures that preserve local topology, but not the global topology, all under the same framework.

The techniques for determining structural similarity based on knot theory concepts prove to be by far the most efficient, and at the same time the most accurate method for assigning protein structure automatically. Further, the ideas presented by Erdmann could be used to develop an efficient residue-wise structural alignment scheme that might also be using the information from the structural classification. Results are awaited on this particular problem. At the same time, there are quite a few open questions that have been created by this work. There are no intuitive interpretations of the Gauss integrals except for the fundamental ones - the writhe and the average crossing number. It is also unclear how these Gauss integrals could be combined. Another question to investigate would be the relative significance of these invariants.

In spite of the amazing success in automatically classifying the CATH database with a high degree of accuracy, the Gauss integrals method has a major drawback. The problem of finding protein domains is not addressed at all. A new structure coming to SGM will not be broken into basic biologically and structurally significant pieces. From this point of view, the most desirable method for determining structural similarity would be the one that identifies protein domains using their geometric and topological properties alone, and would naturally lead to the construction of a pseudo-metric (similar to SGM), based on the definition of these domains, for measuring structural similarity. Developing a (residue-wise) structural alignment of proteins based on such a classification would be the next step. With these aims in mind, we propose the ideas for a novel characterization of tertiary structural units in proteins based on their topological and geometric properties. In the next section, we review the geometrical construction used as the framework for analyzing protein structure, and the relevant topological definitions that will be used in our analysis.

3 Alpha Shapes

An accurate representation of the protein molecule is a collection of balls, one for each atom. The equivalent picture in 2D will be the union of disks. Edelsbrunner et al. analyzed the geometry of a union of disks in 2D as early as 1983 [9]. The results for the union of balls in 3D were presented later by the author in [6]. Let B denote a finite set of balls (solid spheres) in $\mathbb{R}^3$. We specify each ball $b_i = (z_i, r_i)$ by its center $z_i \in \mathbb{R}^3$ and its radius $r_i \in \mathbb{R}$. The weighted distance of a point x from a ball b_i and is defined as the square distance from the center of the ball minus the square of the radius.

$$\pi_i(x) = \| x - z_i \|^2 - r_i^2 \tag{2}$$

The *power Voronoi cell* of a ball b_i under the power distance is the set of points that are at least as close to b_i as to any other ball in B,

$$V_i = \{x \in \mathbb{R}^3 \mid \pi_i(x) \leq \pi_j(x), \forall j\}. \tag{3}$$

All V_i's turn out to be convex polyhedra (see Figure 1). The dual to the power Voronoi diagram will constitute the *weighted Delaunay triangulation* of B, and is the collection of the convex hulls of the centers of those balls whose Voronoi cells have a non-empty common intersection.

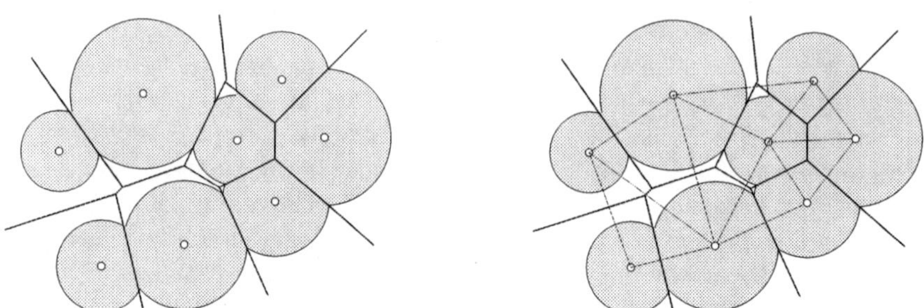

Fig. 1. Power Voronoi diagram of the disks in 2D (left) and the corresponding dual (*weighted*) Delaunay triangulation (right)

Edelsbrunner and Mücke [11] generalized the construction of the Delaunay triangulation given above to consider the dual of the power Voronoi diagram restricted to within the union of the defining balls. The Voronoi cells (3) decompose $\bigcup B$ into convex cells $R_i = \bigcup B \cap V_i = b_i \cap V_i$. The *dual complex* records the non-empty common intersection of these cells,

$$K = \{\sigma_\Lambda \mid \bigcap_{i \in \Lambda} R_i \neq \emptyset\}, \tag{4}$$

where Λ is a subset of the index set, and σ_Λ is the convex hull of the centers of the balls with index in Λ. Equivalently, $\sigma_\Lambda \in K$ is the common intersection of the Voronoi cells that have a non-empty intersection with the union of the balls. The *underlying space* is the set of points contained in the simplices of K, and is denoted by $|K|$. In this context, the authors refer to the underlying space as the *dual shape* of B. The concept is illustrated in 2D in Figure 2. Notable is the special case where the balls have non-empty pair-wise intersections, but have no (non-empty) triple-wise intersections. In this case, K looks like the familiar ball-and-stick diagram of a molecule. Each stick (which originally represents a covalent bond in the molecule) represents the geometric overlap between two balls.

Now consider growing the balls continuously in time and studying how their union changes. We set the weight of each ball b_i as $r_i^2 + t$ at time t and let t go from $-\infty$ to $+\infty$. Each b_i has zero weight at $t = -r_i^2$ and negative weight, and hence imaginary radius, before that time. By construction, the Voronoi cells of the balls remain unchanged. It follows that the dual complexes that

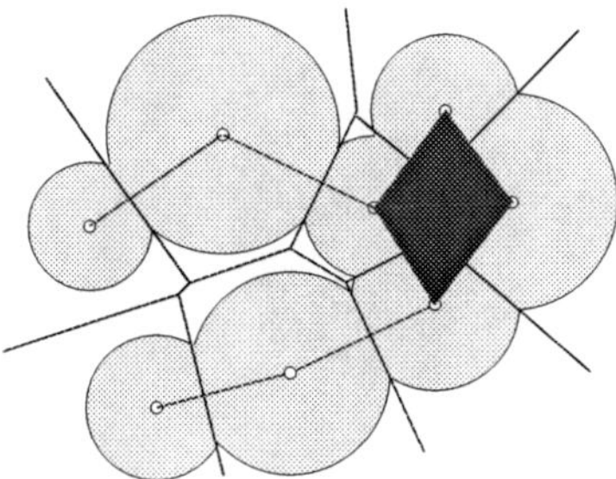

Fig. 2. The dual complex for the union of disks. The nine edges correspond to pairwise intersections and the two triangles correspond to the triple-wise intersections of the clipped Voronoi cells of the balls

arise throughout time are sub-complexes of the same Delaunay triangulation. Also, the dual complexes can only get larger in time. We use the square root $\alpha = \sqrt{t}$, as the index for time varying sets. Under this convention, with $r_i = 0$ (i.e. the ball is originally a point), the radius of the ball b_i at time t is α. Denote by B_α the collection of balls and K_α the dual complex of B_α at time $t \in \mathbb{R}$, indexed by α. We refer to K_α as the α-complex and its underlying space as the α-shape of B. For small enough (large enough negative) time, all radii are imaginary, and $\bigcup B_\alpha = \emptyset$. And for large enough time, the dual complex of B_α is equal to the Delaunay triangulation. We thus obtain a sequence of complexes that begins with the empty complex and ends with the Delaunay triangulation, $\emptyset \subseteq K_\alpha \subseteq K_\beta \subseteq D$, for every $-\infty < \alpha^2 < \beta^2 \leq +\infty$. Since there are only finitely many simplices, there are only finitely many sub-complexes of D that arise as dual complexes during the growth process. We refer to this sequence as a *filtration* of the Delaunay triangulation, $\emptyset = K^1 \subseteq \ldots \subseteq K^m = D$. We illustrate the construction by showing three complexes in the filtration of the union of the disks in the plane in Figure 3. We define a function $j(\alpha^2)$ such that $K_\alpha = K^i$ if $i = j(\alpha^2)$ in order to translate between continuous and discrete rank.

The Delaunay simplices can be sorted in the order in which they enter the dual complex. Define the *birth time* of a simplex $\sigma \in D$ as the minimum time $t = \alpha_\sigma^2$ such that $\sigma \in K_\alpha$ for all $\alpha^2 \geq t$. Thus the difference between two contiguous complexes in the filtration consists of all simplices whose birth-time coincides with the creation of the second complex,

$$K^{i+1} - K^i = \{\sigma \in D \,|\, \alpha_\sigma^2 = j^{-1}(i+1)\} \tag{5}$$

We represent the filtration by sorting the Delaunay simplices by birth time, and in case of a tie by dimension. Remaining ties are broken arbitrarily. Every dual complex K^i is a prefix of this ordering. Due to the tie breaking rule, every such prefix is a complex, even if does not coincide with a dual complex. This property of the ordering will be crucial for the algorithm that we will use to compute the connectivity of K^i.

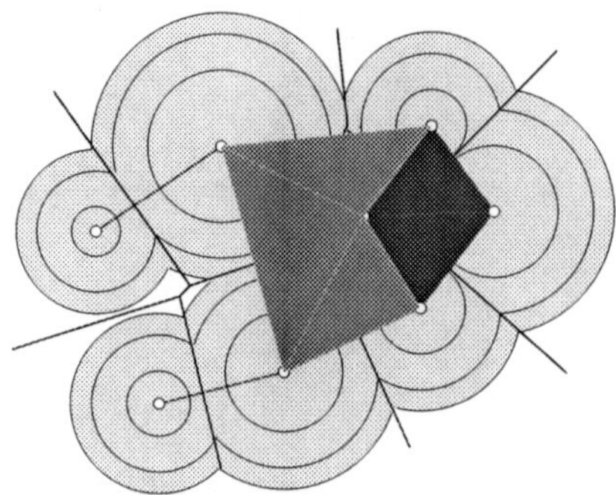

Fig. 3. Three unions of disks and the corresponding dual complexes from the filtration. The first complex consists of only the vertices (as all the balls are disjoint). The second complex is shown in red (same as the one shown in Figure 2). The simplices shown in green get added in the third complex.

4 Homology Groups

We will use homology groups as an algebraic means to study the connectivity of a topological space. The overview of the main concepts presented here follow mainly the treatment given in [8, §IV.2]. Chapter 4 in [13] presents an easy to read discussion of the same subject. Roughly speaking, for any given simplicial complex K, there is one group denoted by $H_p(K)$ in each dimension p with $0 \leq p \leq \dim K$, which measures the number of "independent p-dimensional holes" in K.

We will call a set of k-simplices a k-chain. By definition, the *sum* of two k-chains is the symmetric difference of the two sets.

$$c + d = (c \cup d) - (c \cap d)$$

We define the boundary of a simplex σ as $\partial\sigma = \{\tau \leq \sigma \mid \dim \tau = \dim \sigma - 1\}$. The *boundary* of chain is the sum of the boundaries of its simplices, $\partial c = \sum_{\sigma \in c} \partial\sigma$. Two types of chains are particularly important for us: the ones without boundary and the ones that bound. A k-cycle is a k-chain c with $\partial c = 0$. A k-*boundary* is a k-chain c for which there exists a $(k+1)$-chain d with $\partial d = c$. C_k is the set of k-chains and $(\mathsf{C}_k, +)$ is the group of k-chains. The zero of this chain group is the empty set. Let Z_k and B_k be the set of k-cycles and the set of k-boundaries respectively. Then $(\mathsf{Z}_k, +)$ is a subgroup of $(\mathsf{C}_k, +)$, and $(\mathsf{B}_k, +)$ is a subgroup of $(\mathsf{Z}_k, +)$.

The k-*th homology group* is the quotient of the k-th cycle group divided by the k-th boundary group, $\mathsf{H}_k = \mathsf{Z}_k | \mathsf{B}_k$. The size of H_k is a measure of how many k-cycles are not k-boundaries. If $\mathsf{Z}_k = \mathsf{B}_k$, then H_k is the trivial group consisting of only one element. Two k-cycles are homologous if they belong to the same homology class, $c \sim d$ if $c + d \in \mathsf{B}_k$. Equivalently, $c \sim d$ if there exists $e \in \mathsf{Z}_{k+1}$ with $d = c + \partial e$.

The most useful parameters associated with the homology groups are their ranks, which have intuitive interpretations in terms of the connectivity of the space. Given a subset S of a group G, the subgroup called the linear hull of S ($\lin S$) consists of all $\sum a_i x_i$, with $x_i \in S$ and $a_i \in \{0,1\}$. A *basis* is a minimal subset S that generates the entire group, i.e. $\lin S = G$. The *rank* of G is the cardinality of a basis. If the group is the k-th homology group of a space, $G = H_k$, the rank is known as the *k-th Betti number* of that space, and is denoted by $\beta_k = \rank H_k$. Since $H_k = Z_k | B_k$, we have

$$\rank H_k = \rank Z_k - \rank B_k \tag{6}$$

In general, the 0-th Betti number (β_0) is the number of connected components. Similarly, β_1 gives the number of independent tunnels, and β_2 gives the number of independent (enclosed) voids in the space. For example, consider a torus. There is a single connected component, and hence $\beta_0 = 1$. There are two independent tunnels - one running inside the torus, and the other one is the hole in the middle. Hence $\beta_1 = 2$. There is only one independent closed void, and hence $\beta_2 = 1$. For the 2-sphere, the Betti numbers are $\beta_0 = 1, \beta_1 = 0$, and $\beta_2 = 1$. All higher Betti numbers are zero in both cases.

4.1 Persistent Homology Groups

We will study simplicial complexes for which $\beta_i = 0$ for $i \geq 2$ (details to follow in Section 5). Ideally, we would like to identify the most significant topological features of such a complex - the biggest connected components and the largest holes. At intermediate levels of growth, we would like to identify those features that are persistent – i.e. it takes a long time for them to disappear once they appear. Edelsbrunner, Letscher, and Zomorodian have formalized a notion of topological simplification within the framework of a filtration of the complex [10]. They defined the *persistence* of a non-bounding cycle as a measure of its life-time in the filtration. For each non-bounding cycle, they identify two simplices σ^i and σ^j that respectively *create* and *destroy* the non-bounding cycle in the face of the filtration. Then, the persistence of this feature is defined as $j - i - 1$. For a simplicial complex K^ℓ, the *p-persistent k-th homology group* is defined as

$$H_k^{\ell,p} = Z_k^\ell \,|\, (B_k^{\ell+p} \cap Z_k^\ell). \tag{7}$$

The p-persistent k-th Betti number of K^ℓ is the rank of this homology group - $\beta_k^{\ell,p} = \rank H_k^{\ell,p}$.
To measure the life-time of a non-bounding cycle, we find when its homology class is created and when its class merges with the boundary group. We will defer the description of the method for computing the topological persistences till later. Now we address the main issue facing us – how to define a simplicial complex that captures the geometry of three dimensional structural units of proteins, such that we could use the tools described above to characterize its topology?

5 Definition of Neighborhood

The series of α-complexes in the filtration of the Delaunay triangulation of the protein carries all the information about the geometry of the molecule. Using the filtration as a framework, we can analyze structural units or parts that could be characterized based on their topology, hence leading to the definition of domains (or *structural motifs*). To simplify the treatment, we consider one C_α atom[4] per residue instead of looking at the all-atom model. The local structure in proteins is captured by defining the neighborhood of each C_α atom and each consecutive C_α-C_α edge as the *links* of the respective simplices in the α-complex of the protein at any α-level. We will need certain definitions to achieve this task. The notation used is the same as that given in [7].

Let K be a simplicial complex. The *closure* of a subset $L \subseteq K$ is the smallest sub-complex that contains L.

$$Cl\ L = \{\tau \in K \mid \tau \leq \sigma \in L\} \tag{8}$$

The star of a simplex τ consists of all simplices that contain τ, and the link consists of all faces of simplices in the star that do not intersect τ.

$$St\ \tau = \{\sigma \in K \mid \tau \leq \sigma\} \tag{9}$$

$$Lk\ \tau = \{\sigma \in Cl\ St\ \tau \mid \sigma \cap \tau = \emptyset\} \tag{10}$$

The star is generally not closed, but the link is always a simplicial complex. Given any α-complex K^i, we define the link of each vertex (or C_α atom) and each back-bone edge as follows [13, pg. 111].

$$Lk(v_0) = \{v_1 \mid (v_0v_1) \in K^i\} \cup \{(v_1v_2) \mid (v_0v_1v_2) \in K^i\} \cup \\ \{(v_1v_2v_3) \mid (v_0v_1v_2v_3) \in K^i\} \tag{11}$$

$$Lk(v_0v_1) = \{v_2 \mid (v_0v_1v_2) \in K^i\} \cup \{(v_2v_3) \mid (v_0v_1v_2v_3) \in K^i\} \tag{12}$$

In words, the link of a C_α atom consists of all other C_α atoms that form an edge with it, and all other C_α-C_α edges (not necessarily consecutive) that form a triangle with it, and all other triangles of three C_α atoms that form a tetrahedron with it. The link of a back-bone edge can be interpreted similarly. In Figure 4, We illustrate these definitions in two dimensions.

Naturally, the links defined above will *grow* as the α-complex grows. We can study the connectivity of the links of C_α atoms (and C_α-C_α edges) by finding the homology groups of the links, and observe how the connectivity changes with the growth of the α-complex. We mention here that the ranks of the homology groups of the C_α and C_α-C_α links show specific patterns of variation when we run down an α-helix or a strand in a β-sheet. Such patterns can be used to characterize specific structural domains. The drawback

---

[4] The reader should be careful not to confuse the α used in the context of an alpha carbon atom, with the α used in the context of an α-complex.

with this approach is that the vertex and edge links provide only "local" information. We now describe how we combine a series of back-bone C_α links and C_α-C_α links to effectively capture the neighborhood of a strand, thus providing important non-local information.

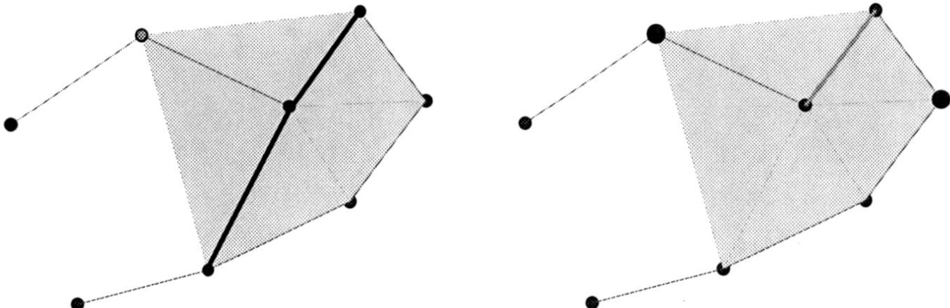

Fig. 4. Alpha-complex in 2D with the back-bone shown in red. Link (shown in black) of a residue shown in green (left figure), and that of a back-bone edge shown in green (right figure).

5.1 Link of a Back-bone Strand

We denote a contiguous strand of back-bone residues and the intermediate edges by $\mathcal{S}$. Formally, a strand of n residues (C_α atoms) and $n-1$ back-bone edges is defined as the sequence of vertices and edges given by $\mathcal{S} = \{v_1, e_1, v_2, \ldots, v_n\}$, where $v_{j+1} = v_j + 1$ and $e_j = (v_j, v_{j+1})$ for $1 \leq j < n$. The *link* (or the *boundary of neighborhood* to be exact) of such a strand in an α-complex K^i is defined as follows.

$$Lk(\mathcal{S}) = \left(\bigcup_{v \in \mathcal{S}} Lk(v) \right) \setminus \left(\bigcup_{v \in \mathcal{S}} St\, v \right) \tag{13}$$

where $Lk(v)$ and $St\,v$ are as defined in (11) and (9) respectively. By construction, the union of the links of all the vertices in $\mathcal{S}$ will include the links of the back-bone edges connecting them. The aim of defining the link of a strand in this way is to capture the non-trivial interactions of the strand with other parts of the protein. This is also the reason why we remove the elements of the star of each vertex in $\mathcal{S}$ from the union in (13). In practice, we are in fact even more careful in removing such "trivial" contacts (or interactions). For the strand $\mathcal{S}$ as defined above, it is natural to expect v_0 to be included in the link of $\mathcal{S}$, as the back-bone edge $e_0 = (v_0, v_1)$ will be part of the α-complex, for sufficiently large α. This observation follows from the fact that the consecutive C_α-C_α edges are among the smallest edges that appear in the Delaunay tessellation of the whole protein. Similarly, one would expect

v_{n+1} also to be included in $Lk(\mathcal{S})$. Hence, we usually modify the link of $\mathcal{S}$ as follows: $Lk(\mathcal{S}) = Lk(\mathcal{S}) \setminus \mathcal{S}'$, where $\mathcal{S}' = \{v_0, e_0, e_{n+1}, v_{n+1}\}$. We illustrate the link of a strand in two dimensions in Figure 5.

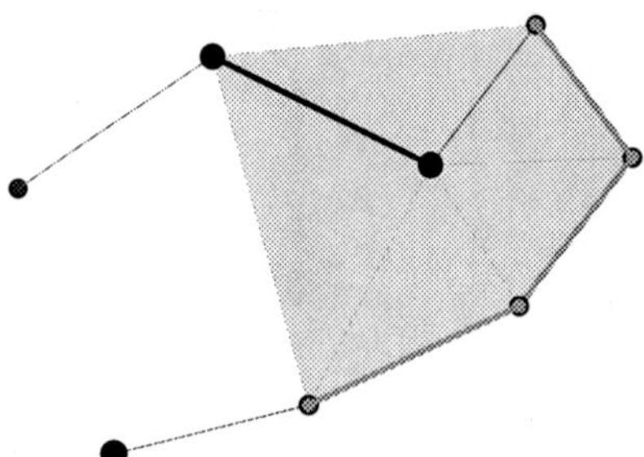

Fig. 5. Link (shown in black) of a back-bone strand (shown in green), consisting of four residues and the three intermediate edges, in the dual complex of the union of disks in 2D. Note that the link shown here is the one before we remove $\mathcal{S}'$ from the union.

The link $Lk(\mathcal{S})$ is defined for each α-complex K^i. By construction, $Lk(\mathcal{S})$ will itself be a simplicial complex. We denote the link of $\mathcal{S}$ defined for K^i by $Lk^i(\mathcal{S})$. One can naturally consider a *filtration* of the link of $\mathcal{S}$ defined for the final Delaunay triangulation D (denoted by $Lk^D(\mathcal{S})$), in the form $\emptyset = Lk^1(\mathcal{S}) \subseteq \cdots \subseteq Lk^m(\mathcal{S}) = Lk^D(\mathcal{S})$. Now we can observe the changes in the connectivity of $Lk^i(\mathcal{S})$ as the complex grows. Specific patterns in the connectivity of $Lk(\mathcal{S})$ as a function of growth (α) could be used to characterize various structural domains.

5.2 From Link to "Tube"

The definition of the link of a strand given in (13) efficiently captures all tertiary interactions made by the strand with other parts of the protein. The topology of the link will indeed be characteristic of these interactions. Nevertheless, there is a drawback with with this definition. Consider a strand that forms little or no contacts with the rest of the protein. The link defined in (13) will possibly be empty in this case. Isolated regions in the protein molecule where the strand merely bends on itself could typically lead to such a situation. The illustration in 2D shown in Figure 5 in fact displays such a strand. The strand of four residues (and the three intermediate back-bone edges) appears to be bending on itself. This information is not provided by the link (shown in black), as the interactions that define the link are located only at the ends of the strand.

We propose the idea of defining the neighborhood of the strand in the form of a "tube" around it. Imagine a tube that has the back-bone chain running through its center. If we thicken the tube uniformly, the parts of the tube around the strand in question would come into contact with the surface of the tube around other parts depending on the interactions between the strand and the other parts of the protein. In the case where the strand does not form any interactions with other parts, the tube around it will also be isolated. At the same time, the tube would touch itself if the strand bends on itself, thus allowing us to characterize domains of this type. In order to implement this idea, we need to modify the representation of the simplicial complex. Consider a residue v' that is present in the link of the strand because it forms an edge $e = (v, v')$ with a vertex v in the strand. Instead of adding v' to the link, we now introduce a new vertex at the mid-point of the edge e, and add the new point to the link of the strand. We can extend this idea to simplices of higher dimensions too. In a way, we are "shrinking" the original link towards the strand to define the "tube" around it. Since the new vertex added is not in $\mathcal{S}$ or $\mathcal{S}'$, the tube might be non-empty even when the link is possibly empty due to $v' \in \mathcal{S}$ or $v' \in \mathcal{S}'$.

Formally, we perform a *barycentric subdivision* of the original alpha complex K^i. As the name suggests, we subdivide every edge in the middle. Every triangle is divided into six smaller triangles by drawing the medians. The division of tetrahedra can be understood in a similar fashion. This construction is used in the classification of closed surfaces [13, Chap. 5]. In general, given a simplicial complex $K \subset \mathbb{R}^n$, a *subdivision* of K is a simplicial complex $K^1 \subset \mathbb{R}^n$ with the property that $|K^1| = |K|$, and given $\sigma_1 \in K^1$, there exists $\sigma \in K$ such that $\sigma_1 \subset \sigma$. Thus, the simplices of K^1 are contained in the simplices of K, but K^1 and K triangulate the same subset of $\mathbb{R}^n$. The barycentric subdivision is one type of a subdivision. The barycentric subdivision of a simplex σ could also be defined as the complex obtained by adding the barycenter (or centroid) of σ as a new vertex, and connecting it to the simplices in the barycentric subdivision of the faces [8, See Exercise II.8]. Since we are working with proteins, consecutive C_α atoms in the chain form the closest interactions. All other interactions, including the tertiary contacts that we are trying to characterize, will appear in the α-complex only after the back-bone edges. Hence, we modify the generic barycentric subdivision so that the back-bone edges are not subdivided. Figure 6 illustrates the proposed barycentric subdivision for proteins in 2D.

Given the barycentrically subdivided complex K^i_{BS} of the α-complex K^i, we apply the definition of the link of strand $\mathcal{S}$ given by (13) on K^i_{BS}. The definitions of vertex and edge links (11) and (12) are also applied on K^i_{BS}. The link that results from this procedure constitutes the "tube" of $\mathcal{S}$. As illustrated in Figure 7, the tube carries all the required information, and will not be empty like the link defined earlier. As we had seen earlier, a filtration of the final tube around $\mathcal{S}$ can be maintained at each index of growth α. We can now study the topological connectivity of the tube around the strand as the complex grows. Patterns observed hence could be used to objectively define tertiary structural domains in proteins.

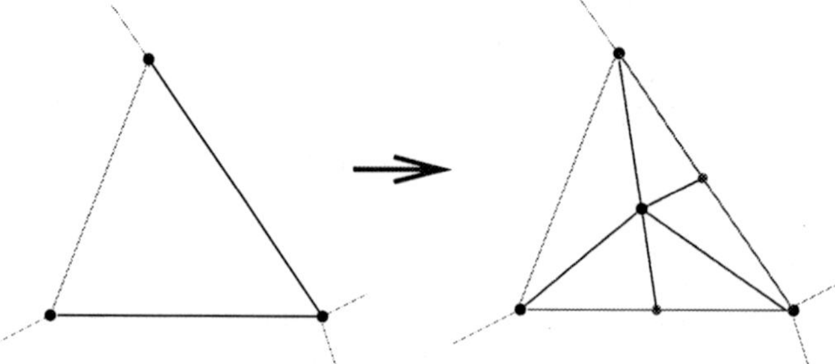

Fig. 6. Barycentric subdivision of a triangle. The edge on the left (in green) is a back-bone edge and hence is not subdivided. One new point is added in the middle of the other two edges, and two new edges are added from this new point to the original vertices in each case. A third point is added at the barycenter of the original triangle, and five new edges connecting this central point to the other points are added. Finally, the five smaller triangles are added in the interior.

5.3 A Retracted Tube

A final step of geometric modification needs to be performed on the tube. Once again, strands that bend on themselves motivate the proposed change. As we have seen, the tube around the strand might touch itself at places. In this process, one or more simplices in the tube complex get identified with certain others. It is desirable if we could actually create a copy of any such simplex, and pull the copy just away from the original simplex, such that tube is not self-intersecting any more. The critical point in performing such a duplication is that we do not desire to alter the (topological) connectivity of the tube in this process. If we could achieve this goal, we would be left with an object that is topologically much nicer to handle than the original tube, but at the same time, it carries the exact connectivity information as before.

We achieve this goal by *retracting* the original tube closer towards the strand S in the following way: every vertex in the tube that forms an edge

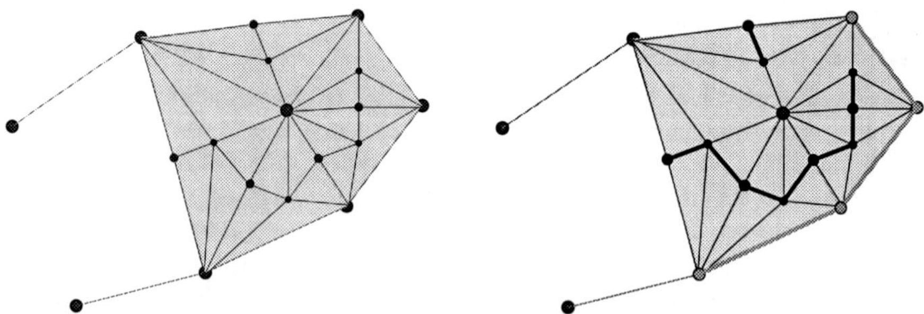

Fig. 7. Barycentric subdivision of the original complex (left) and the "tube" (shown in black) of the strand (in green), now defined on the subdivided complex.

with a residue in $\mathcal{S}$ is retracted half-way towards the residue. This step automatically retracts every triangle in the tube that forms a tetrahedron with a residue in $\mathcal{S}$. The case of an edge in the tube that forms a tetrahedron with a back-bone edge in $\mathcal{S}$ is a little tricky. Retracting the former edge half-way towards the edge in the strand will generate a trapezium. We subdivide this trapezium into two triangles to make sure that the complex is triangulated. By construction, it can be seen that this trapezium will lie in a plane, and hence it can be triangulated by adding either one of its two diagonals. The advantages of this transformation are illustrated in Figure 9.

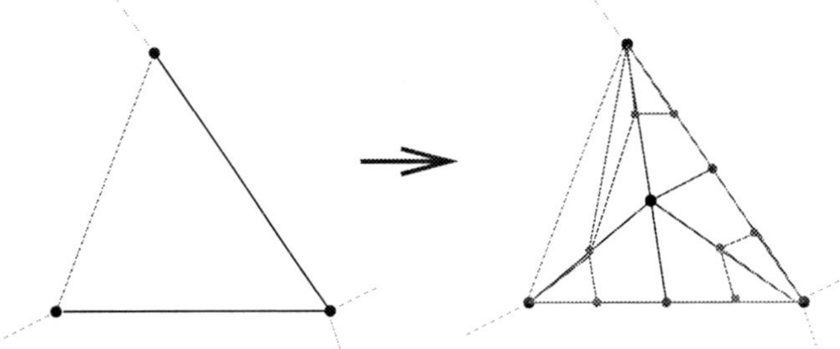

Fig. 8. Barycentric subdivision modified so that the retraction can be defined. Notice that all edges that have one vertex (end-point) as one of the residues are subdivided. Additional edges are added to triangulate any trapezium that gets added, as it happens here near the back-bone edge on the left.

With this additional simplification, the tube will always be a 2-manifold (or a surface) with boundary [15, §22], or a collection of disjoint 2-manifolds with boundary. In addition, the transformation of the tube described above can be shown to be a *strong deformation retraction* [18, §55]. These properties will be important for the method that we will employ in Sections 6 to calculate the ranks of homology groups as well as their topological persistences.

6 Computing the Persistences

We now turn our attention towards the identification of the topologically persistent features of the tube complex. Our task is to pair the positive and negative simplices in the filtration of the tube complex such that each pair represents the life-time of a non-bounding cycle. For 0-cycles, we achieve this pairing while maintaining the UNION-FIND data structure [4, Chap. 22]. When a negative edge comes into the filtration, we pair it with the younger vertex (of the two) if that is unpaired yet. If not, we pair the edge with that vertex in the set to which the younger vertex belongs to, which is the oldest (i,e. has the lowest α-rank). At any point, each set is identified by the oldest unpaired vertex in that set.

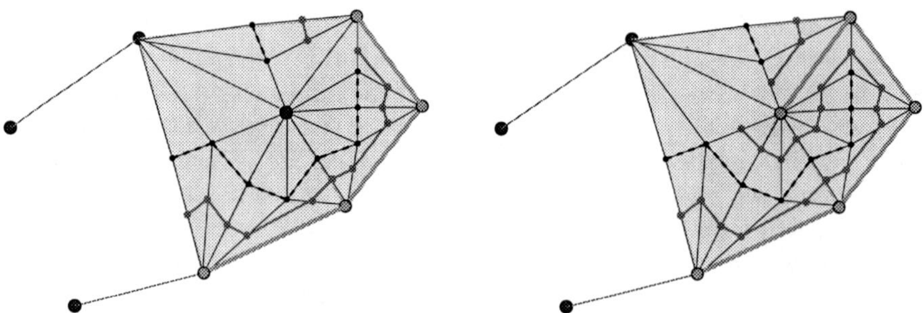

Fig. 9. Retracted "tube" of the strand with four residues (left), and the tube for the strand if we add the residue in the middle of the complex to the strand (right), both shown in magenta. The links defined earlier without retraction will be the same for both the strands (shown in black dotted lines).

The construction of the tube ensures that *all* triangles in the tube complex are negative. Also, the tube complex will be a surface (2-manifold) with boundary, or a set of disjoint surfaces with boundary. Under these conditions, we can make use of the dual relationships between the negative triangles and the positive edges that create the 1-cycles. We maintain a UNION-FIND for β_1's in the following way. The filtration is traversed in the reverse order of time. For each (negative) face, we add a SINGLE SET. Then, for each positive edge that comes in, we do the pairing just as in the case of β_0's, but taking care of the fact that the time scale is reversed now.

Edelsbrunner et al. [10] achieve the pairings in all dimensions simultaneously by performing a cycle search algorithm on a linear array, which acts similar to a hash table [4, Chap. 12]. This algorithm works for any simplicial complex (need not necessarily be a surface with boundary) and has a running time of at most $O(m^3)$, where m is the total number of sub-complexes in the filtration (or the maximum α-rank). The authors do suggest elsewhere, though, that the pairings can be achieved in near-linear time for the simpler case of surfaces, by appropriately modifying the incremental algorithm [5] for calculating Betti numbers. They have also provided the idea of using the dual graph to label faces when running the incremental algorithm. Borrowing these ideas, we achieve the pairings in almost constant time by using weighted merging for the union and path compression for find. Under these conditions, the amortized time per operation is $O(A^{-1}(m))$, where $A^{-1}(m)$ is the inverse of the Ackermann function, which grows very slowly [4, Chap. 22].

Care needs to be taken when treating the positive boundary edges while performing the union-find for the dual graph. We add a dummy dual vertex to represent the external space and assign it the dual rank of zero. A positive boundary edge in the original complex will create a dual edge that connects the dual vertex corresponding to the triangle bounded by the positive edge and the dummy external vertex. For a few of these dual edges, it will happen that the dual vertex (corresponding to the triangle) as well as the external dummy vertex will both have been paired already. We treat such edges as special cases, and record a pairing between the dual edge and the external vertex. We call such a pairing a *forced pairing*. The number of forced pairings will be equal to the β_1 of the fully grown complex under study.

We present the details of our pairing algorithm on the following page, which is essentially a forward and a backward run of a modified incremental algorithm [5]. We assume the sequence of simplices σ^i for $0 \le i < m$ is a filter, and the sequence of sub-complexes $K^i = \{\sigma^j | 0 \le j \le i\}$, for $0 \le i < m$, is the corresponding filtration. Except in the case of forced pairings, each set will have one yet unpaired vertex, which will be the oldest in the set. The set is represented by this vertex, and the rank of the set (denoted by $r(U)$ for set U) will be the rank of this vertex. We maintain two lists of pairings - $\mathcal{P}_0$ and $\mathcal{P}_1$, for β_0 and β_1 pairs respectively. The UNION-FIND data structure supports three operations:

FIND(u): return the representative vertex of the set that contains vertex u.

UNION(u,v): substitute $U \cup V$ for U and V (represented by u and v); $r(U \cup V) = \min\{r(U), r(V)\}$.

ADD(u): add $\{u\}$ as a new singleton set

Once we have the lists of paired simplices, we can calculate the *relative persistence* of the feature represented by each pair (σ^i, σ^j) as $\lambda_{ij} = (j - i - 1)/(m - i_0 - 1)$, where m is the maximum number of α-ranks, and i_0 is the rank at which the first simplex entered the tube complex. This measure evaluates the relative life-time of each feature as compared to the entire life-time of the tube complex. We list the relative persistences of β_0's and β_1's in descending order. Choosing a cut-off value for each of these sets of relative persistences, we will be left with a fixed small number N_λ^0 of λ_{ij}^0's corresponding to the most persistent β_0's, and N_λ^1 of λ_{ij}^1's corresponding to the most persistent β_1's. We present the *persistence signature* of the structural motif represented by the tube complex of the strand $\mathcal{S}$ in question as

$$\text{Sign}(\mathcal{S}) = \{N_{\beta_0}, N_{\beta_1}; \lambda_1^0, \ldots, \lambda_{N_{\beta_0}}^0; \lambda_1^1, \ldots, \lambda_{N_{\beta_1}}^1\} \qquad (14)$$

7 A Basis Set of Motifs

In order to capture all the non-local neighborhoods in a protein, we define the tube complex for a series of strands along the back-bone chain $\mathcal{S}_i$ for $i = 1, 2, \ldots$, each of length $|\mathcal{S}_i| = L$. Here, $\mathcal{S}_i = \{v_i, e_i, v_{i+1}, \ldots, v_{i+L-1}\}$. In other words, we slide a window of contiguous residues (of length L) along the back-bone chain, and study the neighborhood as defined by the tube complex for each strand. The lengths of $L = 8$ and $L = 15$ were chosen. The idea was to capture short-range as well as relatively long-range motifs. As we are going to see, the diversity of the basis motifs is higher for a higher value of L.

list^2 **PAIRING**()
$\mathcal{P}_0 = \mathcal{P}_1 = \emptyset$;
for $i = 0$ to $m - 1$ do
    case σ^i is a vertex u:
        ADD(u); $r(\{u\}) = i$;
    case σ^i is an edge uv:
        $u_r = \text{FIND}(u)$; $v_r = \text{FIND}(v)$;
        if $u_r \neq v_r$
          mark σ^i as *negative*
          UNION(u_r, v_r); $\mathcal{P}_0 = \mathcal{P}_0 \cup (\arg\max\{r(u_r), r(v_r)\}, \sigma^i)$;
        else
          mark σ^i as *positive*
        endif
endfor
ADD(e); $r(\{e\}) = 0$;
for $i = m - 1$ to 0 do
    case σ^i is a triangle (dual vertex u):
        ADD(u); $r(\{u\}) = m - i$;
    case σ^i is a *positive* edge (dual edge uv, dual rank $m - i$):
        $u_r = \text{FIND}(u)$; $v_r = \text{FIND}(v)$;
        if $u_r \neq v_r$
          UNION(u_r, v_r); $\mathcal{P}_1 = \mathcal{P}_1 \cup (\arg\max\{r(u_r), r(v_r)\}, \sigma^i)$;
        else
          $\mathcal{P}_1 = \mathcal{P}_1 \cup (e, \sigma^i)$;
        endif
endfor
return($\mathcal{P}_0, \mathcal{P}_1$);

Algorithm 1: Pairing Algorithm

The particular values were chosen after observing several protein structures for structural units.

For every strand $\mathcal{S}$, we derive the persistence signature (14). The cut-off values for β_0 and β_1 persistences were chosen as $\lambda_0 = 0.43$, $\lambda_1 = 0.37$ for $L = 8$, and $\lambda_0 = 0.42$, $\lambda_1 = 0.35$ for $L = 15$. We initially observed all the relative persistences for each motif in several protein chains. The cut-offs were picked so that most significant topological features would be included in the motif, and at the same time, the total number of motifs to consider would not be too large. A diverse set of 1143 protein chains was selected. In the first run, we identified all possible recurrent motifs. A candidate motif from one of the chains was compared to all the motifs already observed (as maintained in the set of motifs) with the same number of persistent β_0 and β_1 (as denoted by N_{β_0} and N_{β_1} in the signature (14)). If each relative persistence component was within an interval of 0.12 centered at the corresponding component of one of the motifs in the set, the candidate motif was counted as an instance of the particular motif. If not, the candidate motif was added to the set as a new motif. The relative persistences for each motif was averaged over all the instances of the same. Care was also taken to ensure that adjacent similar motifs were not double counted. In several cases, the neighborhood of the strand changes very little when we slide it by one or two residues. Thus we obtain repeated occurrences of the same motif, which in fact should be counted as a single motif that is actually longer than 15 residues. Hence we count a repeated occurrence of the same motif only if we slide the strand by at least 3 residues.

We chose a lower cut-off of 5 for the number of occurrences of a motif in the whole set of protein chains in order to include it the basis set. In case no motif with a particular $(N_{\beta_0}, N_{\beta_1})$ occurred at least 5 times, we grouped all of them with the most frequent one among them (and averaged the relative persistences). After these simplifications, we obtained a basis motif set of 361 motifs for $L = 8$ and 938 motifs for $L = 15$. We discuss the salient features of these structural motifs in the following section.

8 Features of Structural Motifs

Conventionally, one looks at the protein as being made of local units such as alpha helices and strands from beta sheets combined together in 3D arrangements. Typical units of such combinations are distinguished and given names such as a helix bundle, coiled-coil, or alpha-beta barrel. In our analysis, the motifs characterize arrangements in the neighborhood of such a unit (strand). Hence, a 15-residue portion of an alpha helix will form different motifs depending on how the remaining parts of the protein are arranged around it. The values of N_{β_0} and N_{β_1} range from 0 up to 5 for the case of $L = 8$, and from 0 up to 7 for $L = 15$. The two basis sets of motifs cover almost all possible 3D arrangements of strands of the respective lengths in proteins.

Since β_0 measures the number of connected components (section 4), it is straightforward to interpret N_{β_0} can as the number of most prominent interactions that the strand makes with the other parts of the protein. N_{β_1} gives the number of most prominent holes in the filtration of the tube complex, but it is not as easy to see how these persistent holes are created. When there are two or more adjacent dominant contacts (or interactions) of the strand with other parts, space in between two such contacts typically gives rise to a persistent hole in the tube complex. The strand bending on itself usually gives rise to holes - for example, a single turn of an alpha helix creates a hole in the tube complex. At the same time, depending on other tertiary interactions, the holes created due to the strand making helix turns might not be highly persistent. In other words, if we look at an isolated helical strand (which does not interact with any other parts of the protein), the significant holes in the tube complex are the ones due to the helix turns. We illustrate these observations by examining a few motifs in detail.

In Figure 10, we present instances of two $\{2,2\}$ motifs. For the instance of motif 274, the strand appears to make two helix turns and then starts to bend on itself. There is significant interaction with two helical regions and another strand lying around it. On the other hand, there seems to be two prominent interactions made by the strand in the instance of motif 287 with the beta strands on either side of it. The holes in this case are generated due to the strand bending on itself at one end. One would expect the holes in the latter case (of motif 287) to be less persistent than those in the former case, and the β_0 persistences to follow the opposite trend. The sets of persistences show these relationships clearly – for motif 274, they are $\{0.5096\ 0.4572\}\{0.4899\ 0.4060\}$, and for motif 287, they are $\{0.6655\ 0.4490\}\{0.3946\ 0.3681\}$.

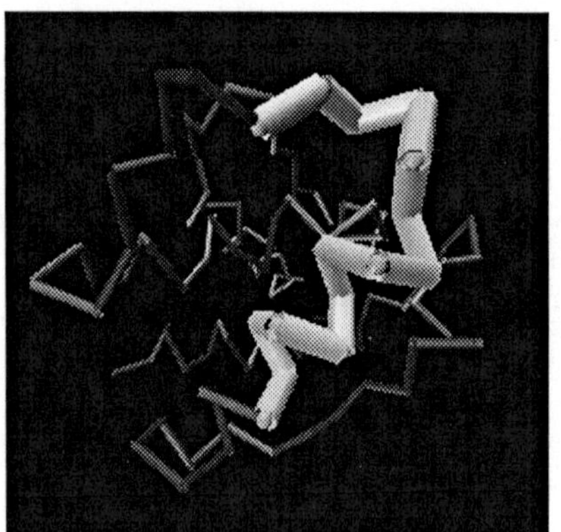 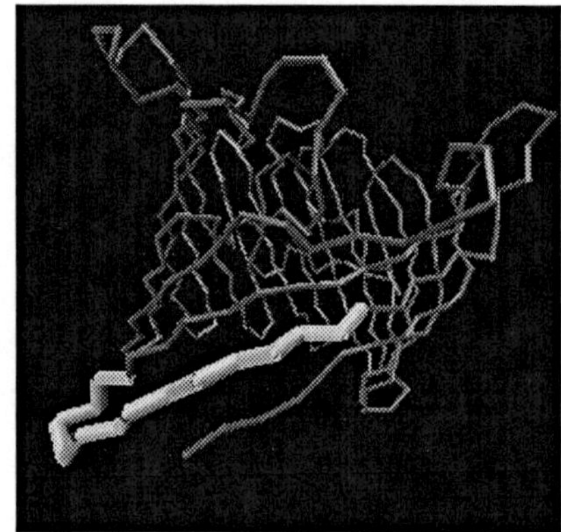

Fig. 10. Instances of $L15$ motifs 274 (in 1BKR) on the left and 287 (in 3PRN) on the right, both with $N_{\beta_0} = 2$, $N_{\beta_1} = 2$.

It typically takes a lot of structure to produce motifs with high N_{β_0} and N_{β_1}. In the same line, strands that have limited interaction with other parts usually give rise to smaller numbers. A straight strand (as opposed to a helical one) lying on the outside of a protein (thus forming limited contact with the rest of the protein) forms motifs with low N_{β_0} and N_{β_1}, as illustrated by the instance of the $L15$ motif 27 shown in Figure 11. The strand interacts with itself and does not produce any other significant contacts, thus producing a $\{1,0\}$ motif. Similarly, the instance of $L8$ motif 162 shown in the figure produces loose interacts with three other parts of the protein, thus providing a $\{3,0\}$ motif.

The instance of the $L15$ motif 891 shown in Figure 12 depicts the strand in the middle of several other portions of the protein, thus forming persistent interactions and holes to give a high-numbered motif ($\{5,3\}$). A very interesting high-numbered $L15$ motif instance is also shown here - that of the $\{7,0\}$ motif 863. The strand appears to lie far from the rest of the protein and seems to forms very little interaction. In fact, the tube complex will not become significant until the original alpha complex (of the entire protein) is grown sufficiently. The long range interactions appear at a later point of time (in the filtration). Since the lifetime of the tube itself is short, we get large relative persistences. One could guess that there is no chance of a hole in the tube here as all the interaction lies to one side of the strand. The end result is an instance of a $\{7,0\}$ motif.

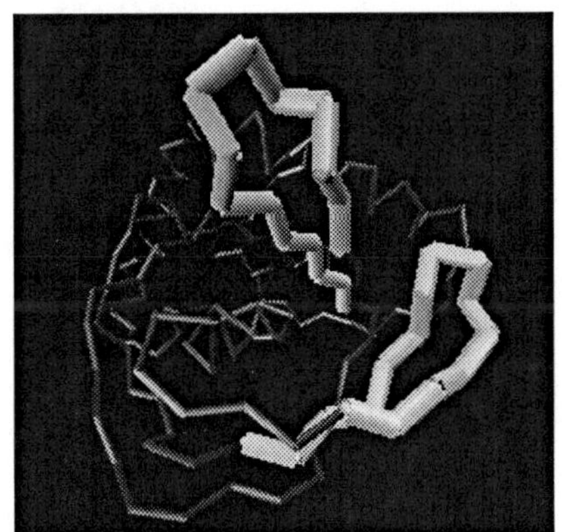
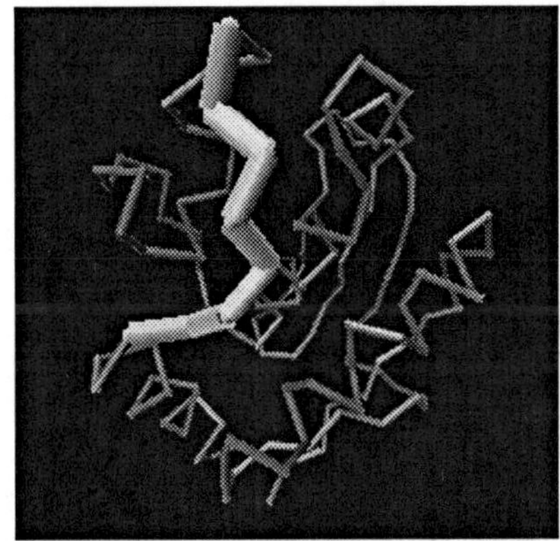

Fig. 11. Instances of $L15$ motif 27 (in 1ICJ) with $N_{\beta_0} = 1$, $N_{\beta_1} = 0$ on the left and $L8$ motif 162 (in 1KUH) with $N_{\beta_0} = 3$, $N_{\beta_1} = 0$ on the right.

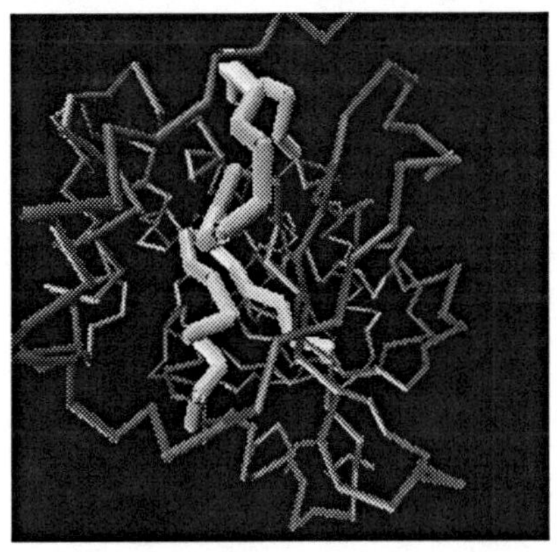
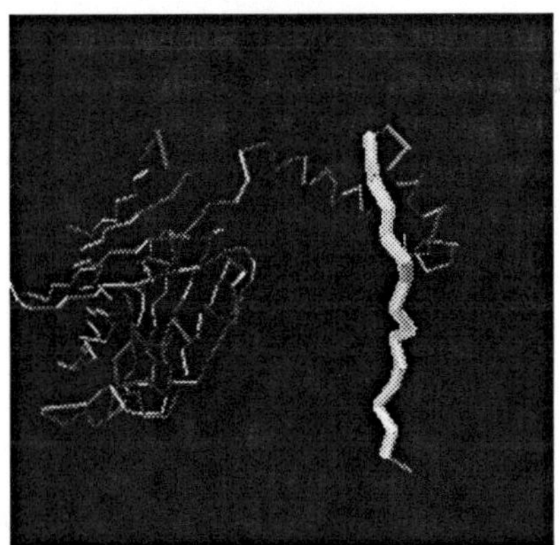

Fig. 12. Instances of $L15$ motif 891 (in 1EDE) with $N_{\beta_0} = 5$, $N_{\beta_1} = 3$ on the left and $L15$ motif 863 (in 1TMY) with $N_{\beta_0} = 7$, $N_{\beta_1} = 0$ on the right.

We have analyzed instances of all the motifs in detail. It is necessary to view the motifs in 3D so that we could rotate the protein and clearly see the different interactions involved. Visualizations are obtained using the software package called VMD. The motifs corresponding to several popular tertiary structural units (such as helix bundle) have been identified. Given the choice of the relative persistence cut-offs and the length of the strand, we believe that these motifs provide a rigorous characterization of parts of proteins using their geometry and topology.

9 The Next Step - Classification

The structural characterization developed by us will prove useful to the biologist only when she could use the same to efficiently classify proteins, similar to the databases such as SCOP or CATH. The ultimate goal is to establish a correspondence between the structural motifs and the function of the proteins. We discuss preliminary progress made in this direction and provide ideas for achieving these goals.

The first step towards developing a classification procedure is the definition of a distance metric between two proteins based on the structural motifs that they consist of. We could collect the number of instances of each $L8$ and each $L15$ motif in a particular protein. One would also expect the overall size of the chain given by the total number of residues to be an important factor. While the counts of the individual motifs are typically single-digit numbers, the number of residues is more than 250 on an average. Hence it makes sense to divide the residue numbers by a factor of 100 and then include it with the individual counts to obtain a 1300-vector (from 938 $L15$ motifs, 361 $L8$ motifs, and the residue number) that represents each protein. We could then calculate the Euclidean distance or the 1-norm between two such vectors. We could naturally think of a way to cluster proteins using such a distance metric.

To get an idea of the accuracy of this method, we considered pairs of proteins from three different families [2] – nuclear receptor ligand-binding domains, serine proteases, and G-proteins. Each pair was distinct due to the function of the protein chains. The protein pairs considered were (2PRG,1A28), (1A0L,1AZZ), and (1EFU,5P21) from the three respective families. It was observed that there were certain motifs that occurred only in one of the families – the $L15$ motif 770 ($\{1,6\}$) occurred only in 1A28, and the $L15$ motif 927 ($\{6,3\}$) occurred only in 2PRG. We tried hierarchically clustering the six chains. The first family was clearly separated from the rest, but the hierarchy was not exact for the other two families. The distance metric used was the 1-norm.

As another preliminary experiment, we took a set containing ten chains from the SCOP class a (all alpha) and ten chains from the SCOP class b (all beta). A cluster analysis was able to successfully group them into two separate hierarchies, with just one chain being mis-classified. A few more similar samples (of 20 chains with 10 each from class a and class b) could be classified with an average accuracy of 80%.

In order to improve the accuracy of the method, it looks essential to add some information about how the individual motifs interact with each other. The order of the motifs along the back-bone chain could also be helpful in making the method more efficient. Defining when two motifs are in contact might not be straightforward. One idea might be to measure the common area of intersection between the corresponding tube complexes before the retraction step. If the area is above certain threshold, we could say that the two motifs are in contact at that particular level of growth. Another piece of information that could prove discriminative might be the joint-occurrences of pairs of motifs in various proteins.

Acknowledgments

The authors would like to thank Prof. Herbert Edelsbrunner from Duke University, and Dr. Afra Zomorodian from Stanford University for providing useful comments on the material presented in this chapter.

References

1. A. Andreeva, D. Howorth, S.E. Brunner, T.J.P. Hubbard, C. Chothia, and A.G. Murzin. SCOP database in 2004: refinements integrate structure and sequence family data. *Nucleic Acids Research*, 32:D226–D229, 2004. Database issue.

2. S.A. Cammer, C.W. Carter, Jr, and A. Tropsha. Identification of sequence-specific tertiary packing motifs in protein structures using Delaunay tessellation. In *Computational Methods for Macromolecules: Challenges and Applications*, volume 24 of *Lecture Notes in Computational Science and Engineering*, pages 477–494, 2000.

3. L. Lo Conte, S.E. Brunner, T.J.P. Hubbard, C. Chothia, and A.G. Murzin. SCOP database in 2002: refinements accommodate structural genomics. *Nucleic Acid Research*, 30(1):264–267, 2002.

4. T.H. Cormen, C.E. Leiserson, and R.L. Rivest. *Introduction to Algorithms*. MIT Press, Cambridge, Massachusetts, 2 edition, 2001.

5. C.J.A. Delfinado and H. Edelsbrunner. An incremental algorithm for betti numbers of simplicial complexes on the 3-sphere. *Computer Aided Geometric Design*, 12:771–784, 1995.

6. H. Edelsbrunner. The union of balls and its dual shape. *Discrete & Computational Geometry*, 13:415–440, 1995.

7. H. Edelsbrunner. *Geometry and Topology for Mesh Generation*. Cambridge University Press, England, 2001.

8. H. Edelsbrunner. CPS 296.1: Bio-Geometric Modeling (Fall 2002) – class notes, 2002. http://cs.duke.edu/education/courses/fall02/cps296.1/.

9. H. Edelsbrunner, D.G. Kirkpatrick, and R. Seidel. On the shape of a set of points in the plane. *IEEE Transactions on Information Theory*, IT-29:551–559, 1983.

10. H. Edelsbrunner, D. Letscher, and A. Zomorodian. Topological persistence and simplification. *Discrete & Computational Geometry*, 28:511–533, 2002.

11. H. Edelsbrunner and E.P. Mücke. Three-dimensional alpha shapes. *ACM Transactions on Graphics*, 13:43–72, 1994.

12. M.A. Erdmann. Protein similarity from knot theory and geometric convolution. Technical Report CMU-CS-03-181, School of Computer Science, Carnegie Mellon University, September 2003.

13. P.J. Giblin. *Graphs, Surfaces and Homology*. Chapman and Hall, London, 2 edition, 1981.

14. H.M. Grindley, P.J. Artimuik, D.W. Rice, and P. Willett. Identification of tertiary structure resemblance in proteins using a maximal common subgraph isomorphism algorithm. *Journal of Molecular Biology*, 229:707–721, 1993.

15. M. Henle. *A Combinatorial Introduction to Topology*. W.H. Freeman and Company, San Francisco, 1979.

16. E.M. Mitchell, P.J. Artymuik, D.W. Rice, and P. Willett. Use of techniques derived from graph theory to compare secondary structure motifs in proteins. *Journal of Molecular Biology*, 212:151–166, 1989.

17. N.J. Mulder, R. Apweiler, T.K. Attwood, A. Bairoch, D. Barrell, and et al. The InterPro database, 2003 brings increased coverage and new features. *Nucleic Acids Research*, 31:315–318, 2003.

18. J.R. Munkres. *Topology*. Prentice Hall, New Jersey, 2 edition, 2000.

19. A.G. Murzin, S.E. Brenner, T. Hubbard, and C. Chothia. SCOP: A structural classification of proteins database for the investigation of sequences and structures. *Journal of Molecular Biology*, 247:536–540, 1995.

20. C.A. Orengo, A.D. Michie, S. Jones, D.T. Jones, M.N. Swindells, and J.M. Thornton. CATH: A hierarchic classification of protein domain structures. *Structure*, 5:1093–1108, 1997.

21. Brookhaven protein data bank. http://www.rcsb.org.

22. P. Røgen and B. Fain. Automatic classification of protein structure by using Gauss integrals. *Proceedings of the National Academy of Sciences*, 100(1):119–124, 2003.

23. W.R. Taylor and C.A. Orengo. Protein structure alignment. *Journal of Molecular Biology*, 208:1–22, 1989.

24. J. White. Self-linking and the Gauss integral in higher dimensions. *American Journal of Mathematics*, 91:693–728, 1969.

## Part V

Applications of Data Mining Techniques to Brain Dynamics Studies

Data Mining in EEG: Application to Epileptic Brain Disorders*

W. Chaovalitwongse[1], P.M. Pardalos[2], L.D. Iasemidis[3],
W. Suharitdamrong[2], D.-S. Shiau[2], L.K. Dance[2], O.A. Prokopyev[4],
V.L. Boginski[5], P.R. Carney[2], and J.C. Sackellares[2]

[1] Rutgers University, USA
   wchaoval@rci.rutgers.edu
[2] University of Florida, USA
[3] Arizona State University, USA
[4] University of Pittsburgh, USA
[5] Florida State University, USA

Summary. Epilepsy is one of the most common brain disorders. At least 40 million people or about 1% of the population worldwide currently suffer from epilepsy. Despite advances in neurology and neuroscience, approximately 25-30% of epileptic patients remain unresponsive to anti-epileptic drug treatment, which is the standard therapy for epilepsy. There is a growing body of evidence and interest in predicting epileptic seizures using intracranial electroencephalogram (EEG), which is a tool for evaluating the physiological state of the brain. Although recent studies in the EEG dynamics have been used to demonstrate seizure predictability, the question of whether the brain's normal and pre-seizure epileptic activities are distinctive or differentiable remains unanswered. In this study, we apply data mining techniques to EEG data in order to verify the classifiability of the brain dynamics. We herein propose a quantitative analysis derived from the chaos theory to investigate the brain dynamics. We employ measures of chaoticity and complexity of EEG signals, including Short-Term Maximum Lyapunov Exponents, Angular Frequency, and Entropy, which were previously shown capable of contemplating dynamical mechanisms of the brain network. Each of these measures can be used to display the state transition toward seizures, in which different states of patients can be classified (normal, pre-seizure, and post-seizure states). In addition, optimization and data mining techniques are herein proposed for the extraction of classifiable features of the brain's normal and pre-seizure epileptic states from spontaneous EEG. We use these features in study of classification of the brain's normal and epileptic activities. A statistical cross validation is implemented to estimate the accuracy of the brain state classifica-

* Research was partially supported by the Medical Research Service of the Department of Veterans Affairs, grants from the Department of Veterans Affairs Research, the NSF grants DBI-980821, EIA-9872509, and NIH grant R01-NS-39687-01A1.

tion. The results of this study indicate that it may be possible to design and develop efficient seizure warning algorithms for diagnostic and therapeutic purposes.

Key words: Chaos Theory, Data Mining, Optimization, Electroencephalogram, Classification, Seizure prediction

1 Introduction

Epilepsy is the second most common serious brain disorder after stroke. Worldwide, at least 40 million people or 1% of population currently suffer from epilepsy. Epilepsy is a chronic condition of diverse etiologies with the common symptom of spontaneous recurrent seizures, which is characterized by intermittent paroxysmal and highly organized rhythmic neuronal discharges in the cerebral cortex. Seizures can temporarily disrupt normal brain functions such as motor control, responsiveness and recall, which typically last from seconds to a few minutes. There is a localized structural change in neuronal circuitry within the cerebrum which produces organized quasi-rhythmic discharges in some types of epilepsy (i.e., focal or partial epilepsy). These discharges then spread from the region of origin (epileptogenic zone) to activate other areas of the cerebral hemisphere. Nonetheless, the mechanism by which these fixed disturbances in local circuitry produce intermittent disturbances of brain function is not well comprehended. The development of the epileptic state can be considered as changes in network circuitry of neurons in the brain. When neuronal networks are activated, they produce changes in voltage potential, which can be captured by an electroencephalogram (EEG), which is one of the most effective tools for evaluating the physiological state of the brain. These changes are reflected by wriggling lines along the time axis in a typical EEG recording. A typical electrode montage for EEG recordings in our study is shown in Figure 1. The 10-sec EEG profiles during the normal (inter-ictal) and pre-seizure (pre-ictal) periods of patient 1 are illustrated in Figures 2 (A) and 2 (B). The EEG onset of a typical epileptic seizure is illustrated in Figure 2 (C). Figure 2 (D) shows the post-seizure (post-ictal) state of a typical epileptic seizure, respectively.

There is a growing body of evidence and interest in predicting epileptic seizures using intracranial EEG. During the past decade, recent studies in the EEG dynamics have been used to demonstrate seizure predictability. Those studies include discoveries previously reported by our group [1, 2, 6, 11, 7, 5, 9, 16], confirmed by several other groups [3, 12, 17, 13, 14], which indicate that it may be possible to predict the onset of epileptic seizures based on analysis of the brain electrical activity through EEG signals. Such analysis was motivated by mathematical models used to analyze multi-dimensional complex systems (e.g., neuronal network in the brain) based on the chaos theory and optimization techniques. The results of those studies demonstrated that seizures can no longer be regarded as a purely stochastic phenomenon but

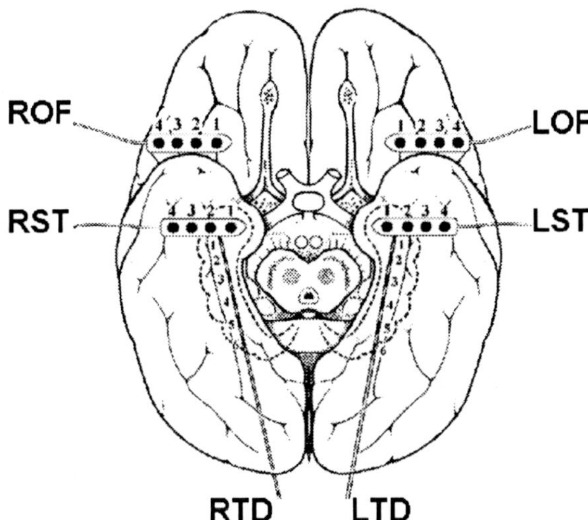

Fig. 1. Inferior transverse views of the brain, illustrating approximate depth and subdural electrode placement for EEG recordings are depicted. Subdural electrode strips are placed over the left orbitofrontal (LOF), right orbitofrontal (ROF), left subtemporal (LST), and right subtemporal (RST) cortex. Depth electrodes are placed in the left temporal depth (LTD) and right temporal depth (RTD) to record hippocampal activity.

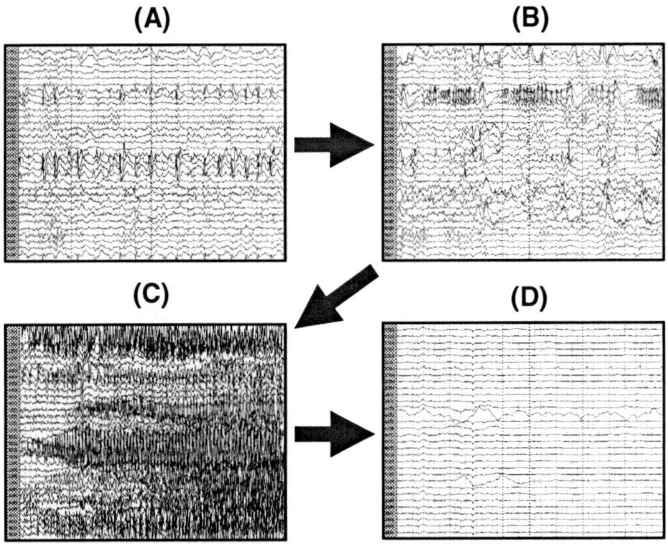

Fig. 2. Twenty-second EEG recordings of (A) Normal Activity (B) Pre-Seizure Activity (C) Seizure Onset Activity (D) Post-Seizure Activity from patient 1 obtained from 32 electrodes. Each horizontal trace represents the voltage recorded from electrode sites listed in the left column (see Figure 1 for anatomical location of electrodes).

they are essentially the reflecting transitions of progressive changes of hidden dynamical patterns in EEG. Such transitions have been shown to be detectable through the quantitative analysis of the brain dynamics [1, 2, 16, 15].

In spite of promising signs of seizure predictability, epilepsy research is still far from being complete. While the previous studies claim that pre-seizure transitions are detectable, the existence of pre-seizure transitions remains to be further investigated with respect to its specificity, that is if it only reflects epileptic activity or it also occurs with other brain activity. In addition, the fundamental question, whether the brain's normal and pre-seizure epileptic activities are distinctive or differentiable, remains unanswered. In order to verify that seizures are predictable, one would have to demonstrate substantial evidence that the brain's normal activity differs from the brain's pre-seizure epileptic activity. This may require a well-trained neurologist who "eyeballs" the direct continuous long-term (over 10 days in duration) EEG recordings over time; however, this laborious task is not guaranteed for the success because of the unrecognizable hidden signatures of the brain's abnormal activity.

In this research, we employ quantitative measures of the brain dynamics, including Short-Term Maximum Lyapunov Exponents, Angular Frequency and Entropy, to contemplate hidden mechanisms of the state transition toward seizures. The study of the brain dynamics is motivated by the chaos theory, in which understanding the brain dynamics is proved capable of providing insights about different states of brain activities reflected from pathological dynamical interactions of the brain network [8, 4, 7]. Especially, these three measures were previously shown capable of contemplating dynamical mechanisms of the brain network [1]. Based on the quantification of the brain dynamics, we herein apply optimization and data mining techniques to the study in the classifiability of the brain's normal and epileptic activities. These techniques are proposed for the extraction of classifiable features of the brain's normal and pre-seizure epileptic states from spontaneous EEG. In fact, the classifiable features are the characteristics of the brain dynamics, which are capable of reflecting the signature of the pre-seizure epileptic activity during the brain's abnormal episodes. We use these features in the study of the classification of the brain's states (normal, pre-seizure, and post-seizure states). A statistical cross validation is implemented to estimate the accuracy of the brain state classification. This framework is an initial proof of concept investigations, which is a necessary first step in differentiating the brain's normal and pre-seizure epileptic activities.

The organization of the succeeding sections of this chapter is as follows. The background including quantification of the brain dynamics and the measures of the brain's chaoticity will be discussed in Section 2. In Section 3, the experimental design, the characteristics of the EEG data, the statistical analysis, the study of multi-parameter in multi-dimensional system, and the statistical cross validation will be described. The results on the statistical evaluation and the performance characteristics of the proposed classification method will be addressed in Section 4. The concluding remarks will be discussed in Section 5.

2 Quantification of Brain Dynamics

Since the brain is a nonstationary system, algorithms used to estimate measures of the brain dynamics should be capable of automatically identifying and appropriately weighing existing transients in the data. In this study, we divide EEG signals into sequential epochs (non-overlapping windows) to properly account for possible nonstationarities in the epileptic EEG. For each epoch of each channel of EEG signals, we quantify the brain dynamics by applying measures of chaos. Measures of chaos, employed to quantify the chaoticity of the attractor, include Lyapunov exponents, Angular Frequency, and entropy. In a chaotic system, orbits that originate from similar initial conditions or nearby points in the phase space diverge exponentially in expansion process. The entropy measures the uncertainty or information about the future state of the system, given information about its previous states in the phase space. The Lyapunov exponents and angular frequency measure the average uncertainty along the local eigenvectors and phase differences of an attractor in the phase space, respectively. In fact, the rate of divergence is an important aspect of the system dynamics and is reflected in the value of Lyapunov exponents and dynamical phase. Next, we will give a short overview of mathematical models used in the estimation of the maximum Lyapunov exponent, the angular frequency, and the entropy.

In the study of the brain dynamics, the initial step in analyzing the dynamical properties of EEG signals is to embed it in a higher dimensional space of dimension p, which enables us to capture the behavior in time of the p variables that are primarily responsible for the dynamics of the EEG. We can now construct p-dimensional vectors $X(t)$, whose components consist of values of the recorded EEG signal $x(t)$ at p points in time separated by a time delay. The construction of the embedding phase space from a data segment $x(t)$ of duration T is performed using the method of delays. The vectors X_i in the phase space are constructed as:

$$X_i = (x(t_i), x(t_i + \tau) \ldots x(t_i + (p-1) * \tau)) \tag{1}$$

where τ is the selected time lag between the components of each vector in the phase space, p is the selected dimension of the embedding phase space, and $t_i \in [1, T - (p-1)\tau]$.

2.1 Estimation of Maximum Lyapunov Exponent

The method for estimation of the Short Term Maximum Lyapunov Exponent (STL_{max}) for nonstationary data (e.g., EEG time series) is previously explained in [4, 7, 18]. In this chapter, we will only give a short description and basic notation of our mathematical models used to estimate STL_{max}. First, let us define the following notation.

- $X(t_i)$ is the point of the fiducial trajectory $\phi_t(X(t_0))$ with $t = t_i$, $X(t_0) = (x(t_0), \ldots, x(t_0 + (p-1)*\tau))$, and $X(t_j)$ is a properly chosen vector adjacent to $X(t_i)$ in the phase space.
- $\delta X_{i,j}(0) = X(t_i) - X(t_j)$ is the displacement vector at t_i, that is, a perturbation of the fiducial orbit at t_i, and $\delta X_{i,j}(\Delta t) = X(t_i + \Delta t) - X(t_j + \Delta t)$ is the evolution of this perturbation after time Δt.
- $t_i = t_0 + (i-1)*\Delta t$ and $t_j = t_0 + (j-1)*\Delta t$, where $i \in [1, N_a]$ and $j \in [1, N]$ with $j \neq i$.
- Δt is the evolution time for $\delta X_{i,j}$, that is, the time one allows $\delta X_{i,j}$ to evolve in the phase space. If the evolution time Δt is given in seconds, then L is in bits per second.
- t_0 is the initial time point of the fiducial trajectory and coincides with the time point of the first data in the data segment of analysis. In the estimation of L, for a complete scan of the attractor, t_0 should move within $[0, \Delta t]$.
- N_a is the number of local L_{max}'s that will be estimated within a duration T data segment. Therefore, if D_t is the sampling period of the time domain data, $T = (N-1)D_t = N_a \Delta t + (p-1)\tau$.

Let L be an estimate of the short term maximum Lyapunov exponent, defined as the average of local Lyapunov exponents in the state space. L can be calculated as follows.

$$L = \frac{1}{N_a \Delta t} \sum_{i=1}^{N_a} \log_2 \frac{|\delta X_{i,j}(\Delta t)|}{|\delta X_{i,j}(0)|} \tag{2}$$

with

$$\delta X_{i,j}(0) = X(t_i) - X(t_j) \tag{3}$$

$$\delta X_{i,j}(\Delta t) = X(t_i + \Delta t) - X(t_j + \Delta t). \tag{4}$$

Per electrode, we computed the STL_{max} profile using the method proposed by Iasemedis et al. [4], which is a modification of the method by Wolf et al. [18]. A modification of Wolf's algorithm is necessary to better estimate of STL_{max} in small epochs that include transients, such as inter-ictal spikes. The modification is primarily in the searching procedure for a replacement vector at each point of a fiducial trajectory. For example, in our analysis of the EEG, we found that the crucial parameter of the L_{max} estimation procedure, in order to distinguish between the pre-ictal, the ictal and the post-ictal stages, is the adaptive estimation in time and phase space of the magnitude bounds of the candidate displacement vector to avoid catastrophic replacements. Results from simulation data of known attractors have shown the improvement in the estimates of L achieved by using the proposed modifications [4].

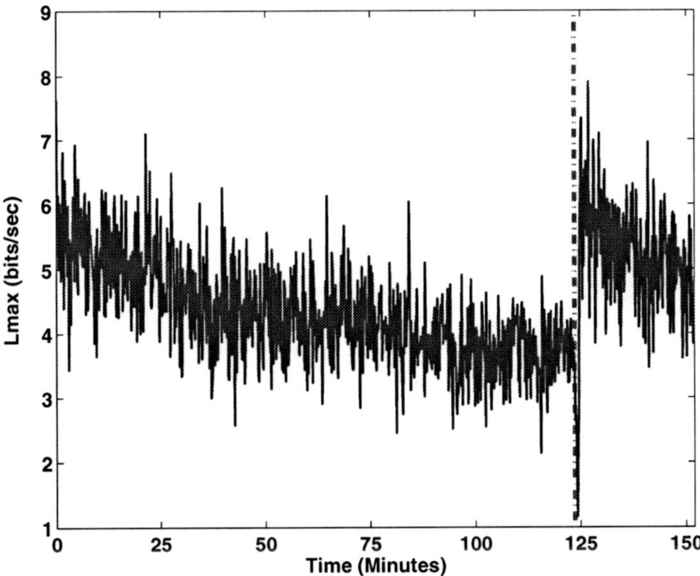

Fig. 3. STL_{max} profile over 2.5 hours estimated from an EEG signal recorded at $RTD-2$ (the epileptogenic hippocampus of patient 1). A seizure started at the vertical dashed line.

An example of a typical STL_{max} profile estimated from an EEG signal recorded at $RTD-2$ (the epileptogenic hippocampus of patient 1) over time is given in Figure 3. The estimation of the L_{max} values was made by dividing the signal into non-overlapping segments of 10.24 sec each, using $p = 7$ and $\tau = 20$ msec for the phase space reconstruction. In the pre-seizure state, one can see a trend of STL_{max} toward lower values over the whole pre-seizure period. This phenomenon can be explained as an attempt of the system toward a new state of less degrees of freedom long before the actual seizure [8].

2.2 Estimation of Dynamical Phase (Angular Frequency)

Similar to the estimation of Lyapunov exponents, the estimation of the angular frequency is motivated by the representation of a state as a vector in the state space. The angular frequency is merely an average uncertainty along the phase differences of an attractor in the phase space.

First, let us define the difference in phase between two evolved states $X(t_i)$ and $X(t_i + \Delta t)$ as $\Delta\Phi_i$ [10]. Then, denoting by $\langle\Delta\Phi\rangle$ the average of the local phase differences $\Delta\Phi_i$ between the vectors in the state space, we have:

$$\Delta\Phi = \frac{1}{N_\alpha} \cdot \sum_{i=1}^{N_\alpha} \Delta\Phi_i \tag{5}$$

where N_α is the total number of phase differences estimated from the evolution of $X(t_i)$ to $X(t_i + \Delta t)$ in the state space, according to:

$$\Delta\Phi_i = \mid \arccos \frac{X(t_i) \cdot X(t_i + \Delta t)}{\parallel X(t_i) \parallel \cdot \parallel X(t_i + \Delta t) \parallel} \mid . \tag{6}$$

Then, the average angular frequency $\bar{\Omega}$ is:

$$\bar{\Omega} = \frac{1}{\Delta t} \cdot \Delta\Phi. \tag{7}$$

If Δt is given in sec, then $\bar{\Omega}$ is given in rad/sec. Thus, while STL_{max} measures the local stability of the state of the system on average, $\bar{\Omega}$ measures how fast a local state of the system changes on average (e.g. dividing $\bar{\Omega}$ by 2π, the rate of the change of the state of the system is expressed in $sec^{-1} = Hz$).

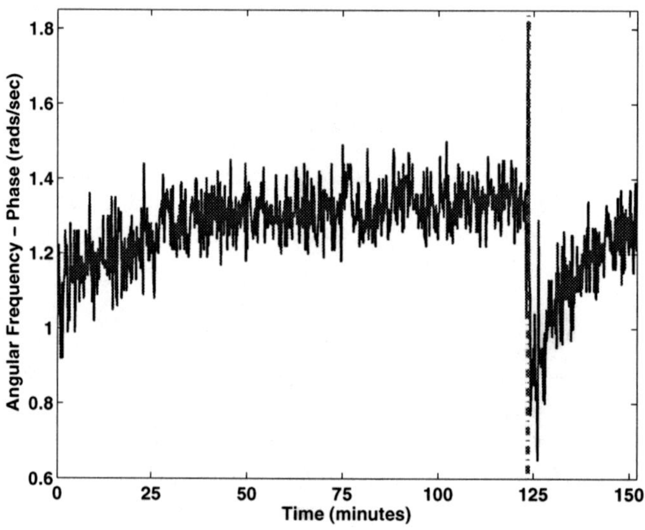

Fig. 4. A typical $\bar{\Omega}$ profile before, during, and after an epileptic seizure, estimated from an EEG signal recorded at electrode $RTD - 2$ (the epileptogenic hippocampus of patient 1). Note that this segment of EEG signals is the same as the one used to estimate the STL_{max} profile in Figure 3, where the seizure occurred at the moment denoted by the vertical dashed line.

An example of a typical $\bar{\Omega}$ profile estimated from an EEG signal recorded at $RTD - 2$ (the epileptogenic hippocampus of patient 1) over time is given in Figure 4. The values are estimated from a 150-minute-long EEG sample recorded from an electrode located in the epileptogenic hippocampus. This segment of EEG signals is the same as the one used to estimate the STL_{max} profile in Figure 3. The EEG sample includes a seizure that occurs at the moment corresponding to 124 minutes of recording. The state space was reconstructed from sequential, non-overlapping EEG data segments of 2048 points (sampling frequency 200 Hz, hence each segment of 10.24 sec in duration) with $p = 7$ and $\tau = 4$, as for the estimation of STL_{max} profiles [10]. The pre-ictal, ictal and postictal states correspond to medium, high and lower values of $\bar{\Omega}$ respectively. The highest $\bar{\Omega}$ values were observed during the ictal period, and higher $\bar{\Omega}$ values were observed during the preictal period than during the postictal period. This pattern roughly corresponds to the typical observation of the decrease in values of the STL_{max} profile. The explanation is that the higher rate of change of the trajectory in the phase space, the lower change in direction (angular frequency) of the trajectory. On the other hand, while the rate of change is low, the higher angle (more diverse directions) the trajectory can change.

2.3 Entropy and Information

One of the key concepts in information theory is that information is conveyed by randomness. Basically, information is defined, in some mathematical sense, as knowledge used to identify structure of the data. It is not difficult to make the connection between randomness and information. The information is related somewhat to the degree of surprise at finding out the answer. In fact, the information can be used to measure the amount of pattern or sequence hidden in the data. On the other hand, entropy can thus be viewed as a self-moment of the probability, in contrast to the ordinary moments. However, the entropy also measures the degree of surprise that one should feel upon learning the results of a measurement. It counts the number of possible states, weighted each with its likelihood. The negative of entropy is sometimes called information.

The entropy can also be considered as the average uncertainty provided by a random variable x. For example, the entropy measures average over quantities that provide a different kind of information than the ordinary moments. In fact, it is a measure of the amount of information required on the average to describe the random variable x.

The entropy $H(p)$ of a probability density p is

$$H(p) = -\int p(x) \log p(x) dx. \tag{8}$$

The entropy of a distribution over a discrete domain is

$$H(p) = -\sum_i p_i \log p_i. \tag{9}$$

The entropy of EEG data describes the extension to which the distribution is concentrated on small sets. If the entropy is low, then the distribution is concentrated at a few values of x. It may, however, be concentrated on several sets from each other. Thus, the variance can be large while the entropy is small. The average entropy explicitly depends on p. The entropy can be written as

$$H(p) = - < \log p(x) > . \tag{10}$$

As we mentioned, the information is referred to as the negative of entropy written as

$$I(p) = -H. \tag{11}$$

For our purposes the distinction between the two is semantic. Entropy is most often used to refer to measures, whereas information typically refers to probabilities. Specifically to the study of the brain dynamics, we use the entropy to measure the information hidden in the EEG data. In Figure 5, a example of the entropy profile from EEG recordings of electrode $RTD - 2$ (the epileptogenic hippocampus of patient 1) over 2-hour interval including one seizure are illustrated.

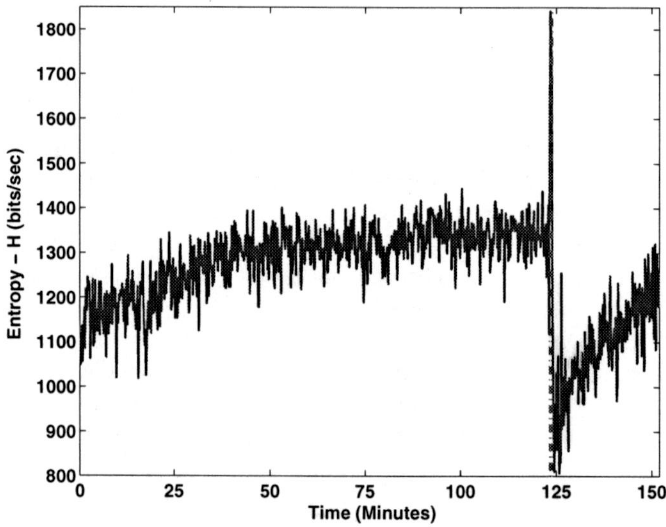

Fig. 5. Plot of the Entropy over time derived from an EEG signal recorded at $RTD - 2$ (the epileptogenic hippocampus of patient 1). Note that this segment of EEG signals is the same as the one used to estimate the STL_{max} profile in Figure 3 and the phase profile in Figure 4, which the seizure occurred at the vertical dashed line.

3 Materials and Methods

3.1 Experimental Design

The present study was undertaken to determine whether it is possible to discriminate different states of the brain's activity (normal, pre-seizure, and post-seizure) by employing quantitative techniques which continuously analyzes the

STL_{max}, phase, and entropy profiles. To do so, we develop a novel statistical classification approach based on optimization and data mining techniques. The primary goal of this study is to develop a novel quantitative approach capable of capturing hidden signatures of the pre-seizure epileptic activity and distinguishing that from the normal activity. These hidden signatures can be considered as a seizure pre-cursor. The results of this study can be extended to the development of a new therapeutic approach, which is used to provide an early seizure warning as well as a seizure susceptibility period detection.

In this framework, we first calculate measures of chaos from EEG signals using the methods described in the previous section. We will then test our hypothesis whether the measures of chaos can be used as features to discriminate different states of the brain dynamics. Our hypothesis can be stated as follow: "The EEG dynamics from the same brain state should be more similar than that from the different brain states". In other word, the characteristics of the brain dynamics during the brain's normal activity should be more similar to each other than the brain's pre-seizure epileptic activity, and vice versa. We also postulate that the brain functional units should execute different activities during the three states of a patient. To classify different brain states, we propose a statistical T-test as a statistical distance between two EEG data. Applying the T-test, a leave-one-out statistical cross-validation will be implemented to evaluate our classification schemes and validate our hypothesis.

3.2 Dataset

In this study, we test the aforementioned hypothesis in the human subject in which long-term (3 to 14 days in duration) multi-channel intracranial EEG recordings were obtained from bilaterally, surgically implanted micro-electrodes in the hippocampus, temporal and frontal lobe cortexes of 3 epileptic patients with medically intractable temporal lobe epilepsy. The recordings were obtained as a part of a pre-surgical clinical evaluation. Each record included a total of 28 to 32 intracranial electrodes (8 subdural and 6 hippocampal depth electrodes for each cerebral hemisphere). These EEG recordings were viewed by two independent electroencephalographers to determine the number and type of recorded seizures, seizure onset and end times, and seizure onset zones. From our EEG recordings, for an individual patient, we divide the EEG data into 3 groups: 1) normal state, 2) pre-seizure state, and 3) post-seizure state. The normal states of EEG data are chosen such that the recordings are far away from seizures (at least 8 hours) and the patient's medical conditions are at a steady state. The pre-seizure epileptic states of EEG data are chosen so that the recordings are within the 30-minute interval before seizures. In addition, we are also interested in the brain recovery period (post-seizure); therefore, the post-seizure states of EEG data are chosen such that the recordings are in the 30-minute interval after seizures.

The characteristics of the patients and the test recordings are outlined in Figure 6. For individual patient, we randomly select 200 samples (epochs) from the group of normal state EEG's. Each epoch is 5 minutes in length. Note that in this analysis we only consider clinical seizures and un-clustered seizures. From the data set, we select 22, 7, and 15 seizures in patients 1, 2, and 3, respectively. For every seizure, we randomly select 3 epochs of pre-seizure EEG data and 3 epochs of post-seizure EEG data. In other words, 66, 21, and 45 epochs of pre-seizure EEG data are selected for patients 1, 2 and 3, respectively. Similarly, 66, 21, and 45 epochs of post-seizure EEG data are selected in patient 1,2, and 3, respectively.

Patient ID	Gender	Onset region	Age (years)	Seizure types	Duration of recordings (days)	Number of seizures	Inter-seizure interval range (hours)	average inter-seizure interval (hours)
1	F	RH	41	CP	9.06	24	0.30 ~ 14.49	3.62
2	F	RH	45	CP, SC	3.63	9	0.52 ~ 47.93	8.69
3	M	RH	29	CP, SC	6.07	19	0.32 ~ 70.70	6.61
Total					18.76 days	52	0.30 ~ 70.70	5.59

Fig. 6. Data characteristics. Onset region: LH, Left Hippacampal; RH, Right Hippacampal; RF, Right Orbitofrontal. Seizure types: CP, Complex Partial; SC, Subclinical

3.3 Statistical Distance

In this section, we propose a similarity measure used to estimate the difference of the EEG data between different groups of the brain states. We employ the T-index as a measure of statistical distance between two epochs of measures of chaos. In this section, we will use the notation of measures of chaos as STL_{max} to simplify the mathematical models. Note that, in practice, we do not only use these equations to calculate the statistical distance of two STL_{max} samples but also use in the case of phase and entropy samples. The T-index at time t between electrode sites i and j is defined as:

$$T_{i,j}(t) = \sqrt{N} \times |E\{STL_{max,i} - STL_{max,j}\}|/\sigma_{i,j}(t) \qquad (12)$$

where $E\{\cdot\}$ is the sample average difference for the $STL_{max,i} - STL_{max,j}$ estimated over a moving window $w_t(\lambda)$ defined as:

$$w_t(\lambda) = \begin{cases} 1 \text{ if } \lambda \in [t - N - 1, t] \\ 0 \text{ if } \lambda \notin [t - N - 1, t], \end{cases}$$

where N is the length of the moving window. Then, $\sigma_{i,j}(t)$ is the sample standard deviation of the STL_{max} differences between electrode sites i and

j within the moving window $w_t(\lambda)$. The thus defined T-index follows a t-distribution with $N-1$ degrees of freedom.

3.4 Cross-Validation

In this research, we propose leave-one-out statistical cross-validation to prove our main hypothesis. In general, cross-validation can be seen as a way of applying partial information about the applicability of alternative classification strategies. In other words, cross-validation is a method for estimating generalization error based on "resampling". The resulting estimates of generalization error are often used for choosing among various decision models (rules). Generally, people refer cross validation to k-fold cross validation. In k-fold cross-validation, the data are divided into k subsets of (approximately) equal size. The decision models are trained k times, in which one of the subsets from training is left out each time, by using only the omitted subset to compute whatever error criterion interests you. If k equals the sample size, this is called "leave-one-out" cross-validation.

In our study, we have already had the decision rule in classifying the brain states. Therefore, we do not have any decision models to train. In fact, we can only apply cross-validation to simply estimate the generalization error of our decision rule and validate our hypothesis. Since we consider the estimate of statistical distance between EEG epochs as our decision rule, we call this technique a "statistical cross-validation". Our decision rule is described as follows.

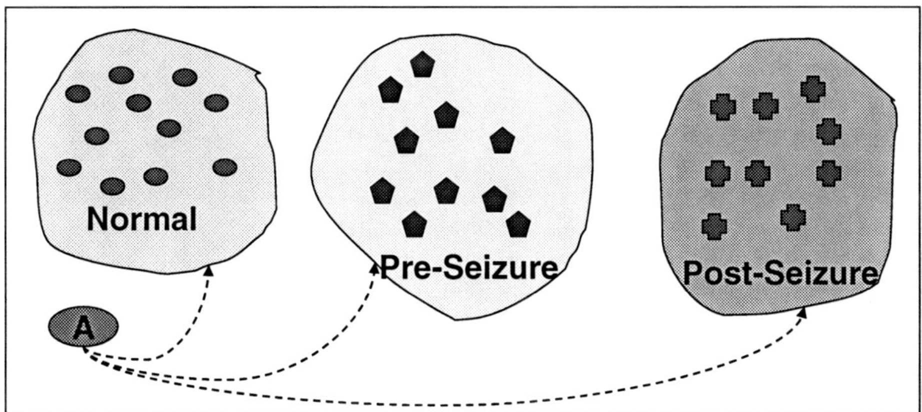

Fig. 7. Cross validation for classification of an unknown-state EEG epoch "A" by calculating the statistical distances between "A" and normal, "A" and pre-seizure, and "A" and post-seizure

Given an unknown-state epoch of EEG signals "A", we calculate the average statistical distance (T-index) between "A" and the groups of normal, pre-seizure, and post-seizure EEG data. Per electrode, we will then obtain 3 T-index values, which are the average mean statistical distances between the epoch "A" and the group of normal, pre-seizure, and post-seizure, respectively (see Figure 7). The EEG epoch "A" will be classified in the group of EEG data (normal, pre-seizure, and post-seizure) that yields the minimum T-index value based on 28–32 electrodes. Since our classifier has 28–32 decision inputs, we proposed different classification schemes which are based on different electrodes and combination of dynamical measures. For all electrodes, the classification is based on: 1) average scheme 2) voting scheme. This study will indicate the dynamical measures that are most useful in classifying different states of the brain. In other words, the proposed framework can be further used to study the feature selection of the brain dynamics. Next, we will discuss the two proposed classification schemes that we employ in this study.

3.5 Classification Schemes

For multi-dimensional systems, such as brain, we have more than one attribute used for decision making or classification. In our case, the attributes are electrode sites and dynamical measures. As we mentioned in the previous section, we apply three measures of chaos derived from 28–32 electrodes, which are considered to be attributes in the classification. In this framework, we treat different classification schemes as combinations of measures of chaos. For instance, one of the classification schemes can be the combination of L_{max} and dynamical phase. Another example can be the combination of L_{max} and entropy. In addition, we still have 26-30 attributes used in decision making. We propose two schemes to incorporate these attributes; namely, averaging scheme and voting scheme. The averaging scheme is one of the most intuitive schemes used to incorporate different attributes (considered as outputs). The voting scheme is a common technique used for construction of reliable decisions in critical systems. For each electrode, an output is separately calculated and is sent to the voter (see Figure 8). Each version for each electrode is usually developed independent of other versions. Based on the 30 electrodes (outputs), the voter then estimates the correct output. The design of the voter is essential to the reliable functioning of the system. Majority voting selects action with maximum number of votes.

4 Results

To evaluate the performance characteristics of our statistical cross-validation technique, we calculate the sensitivity and specificity for each of the proposed classification schemes. In order to select the optimal classification scheme, we propose the receiver operating characteristics (ROC) analysis, which is derived from the detection theory. Basically, the sensitivity and the specificity of each classification scheme for each combination of dynamical measures are

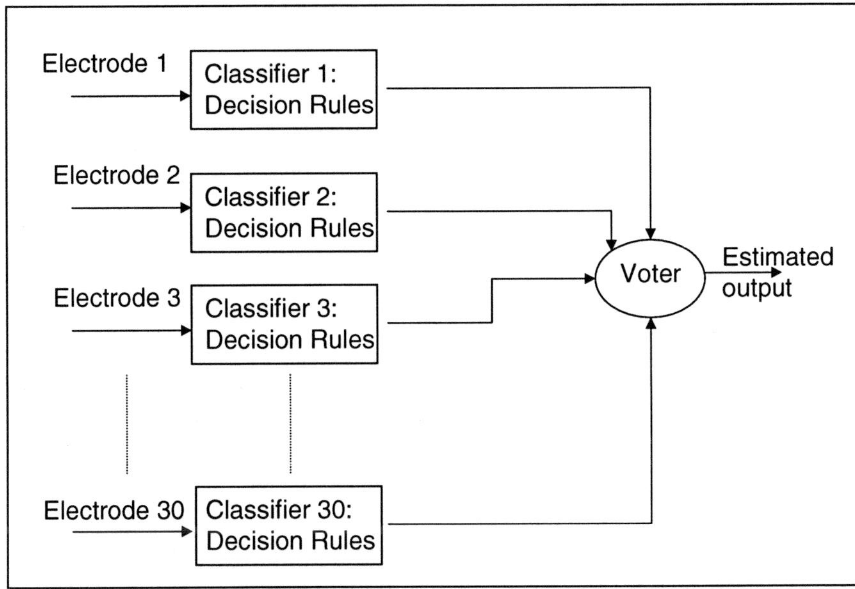

Fig. 8. Voting scheme of 30 electrodes (votes)

calculated. Based on the detection theory, optimal scheme is selected such that it is closest to the ideal scheme (100% sensitivity and 100% specificity). The performance evaluation and the results on each of classification schemes will be discussed in more details later in this section.

4.1 Performance Evaluation of Classification Schemes

To evaluate the classifier, we categorize the classification into two classes, positive and negative. Then we consider four subsets of classification results, 1. True positives (TP): True positive answers of a classifier denoting correct classifications of positive cases; 2. True negatives (TN): True negative answers denoting correct classifications of negative cases; 3. False positives (FP): False positive answers denoting incorrect classifications of negative cases into the positive class; 4. False negatives (FN): False negative answers denoting incorrect classifications of positive cases into the negative class. To better explain the concept of the evaluation of classifiers, let us consider in the case of the detection of abnormal data (see Figure 9). A classification result was considered to be true positive if we classify an abnormal sample as an abnormal sample. A classification result was considered to be true negative if we classify a normal sample as a normal sample. A classification result was considered to be false positive when we classify a normal sample as an abnormal sample. A classification result was considered to be false negative when we classify an abnormal sample as a normal sample.

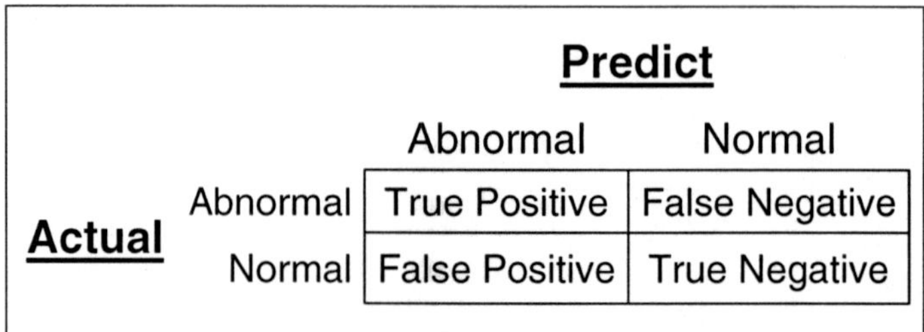

Fig. 9. Evaluation of classification results

In medical community, two classification accuracy measures, sensitivity and specificity, are usually employed. Sensitivity measures the fraction of positive cases that are classified as positive. Specificity measures the fraction of negative cases classified as negative. The sensitivity and specificity are defined as follows:

$$Sensitivity = TP/(TP + FN),$$

$$Specificity = TN/(TN + FP).$$

In fact, sensitivity can be considered as the probability of accurately classifying EEG samples in the positive case. Specificity can be considered as the probability of accurately classifying EEG samples in the negative case. In general, one always wants to increase the sensitivity of classifiers by attempting to increase the correct classifications of positive cases (TP). On the other hand, false positive rate can be considered as 1-Specificity which one wants to minimize. In order to decrease the false positive rate, we should try to decrease the number of incorrect classifications of negative cases into class positive (FP).

4.2 Optimal Classification Scheme

In the algorithm, we have different classification scheme including average and voting schemes for dynamical measures, which need to be optimized. In order to find the optimal scheme, we employed ROC analysis, which is used to indicate an appropriate trade-off that one can achieve between the false positive rate (1-Specificity, plotted on X-axis) that needs to be minimized, and the detection rate (Sensitivity, plotted on Y-axis) that needs to be maximized. For individual patient, the optimal classification scheme can be identified by selecting the scheme whose performance is closest to the ideal point (sensitivity = 1 and specificity = 1); that is, the scheme closest to the top left hand corner on ROC plot will be selected. An example of a ROC plot in patient 1 was illustrated in Figure 10, in which the scheme that incorporates the average of L_{max} and Entropy is the optimal scheme. The performance characteristics of the optimal scheme for individual patient are listed in Table 1.

Figure 11 illustrates the classification results of the optimal scheme for patient 1 (Average Lmax & Entropy). The probabilities of correctly predicting

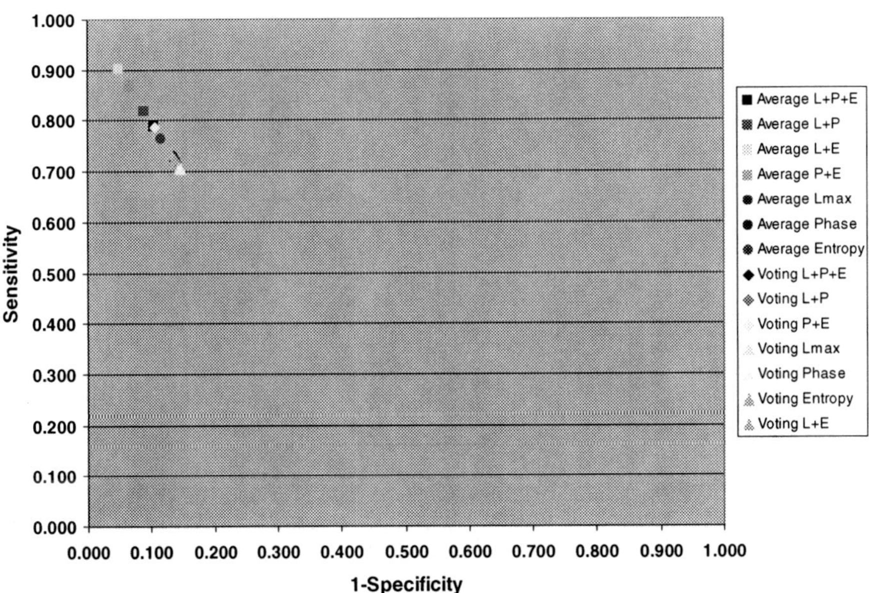

Fig. 10. ROC Plot for all classification schemes in Patient 1

Table 1. Performance characteristics of the optimal classification scheme for individual patient

Patient	Sensitivity	Specificity	Optimal Scheme
1	90.06%	95.03%	Average Lmax & Entropy
2	77.27%	88.64%	Average Lmax & Phase
3	76.21%	88.10%	Average Lmax & Phase

pre-seizure, post-seizure, and normal EEG's are about 90%, 81%, and 94%, respectively. Figure 12 illustrates the classification results of the optimal scheme for patient 2 (Average Lmax & Phase). The probabilities of correctly predicting pre-seizure, post-seizure, and normal EEG's are about 86%, 62%, and 78%, respectively. Figure 13 illustrates the classification results of the optimal scheme for patient 3 (Average Lmax & Phase). The probabilities of correctly predicting pre-seizure, post-seizure, and normal EEG's are about 85%, 74%, and 75%, respectively. Note that in practice classifying pre-seizure and normal EEG's is more meaningful than classifying post-seizure EEG's since the post-seizure EEG's can be easily observed (visualized) after the seizure onset.

The results of this study indicate that we can correctly classify the pre-seizure and normal EEG's with 90% and 83% accuracy, respectively. These

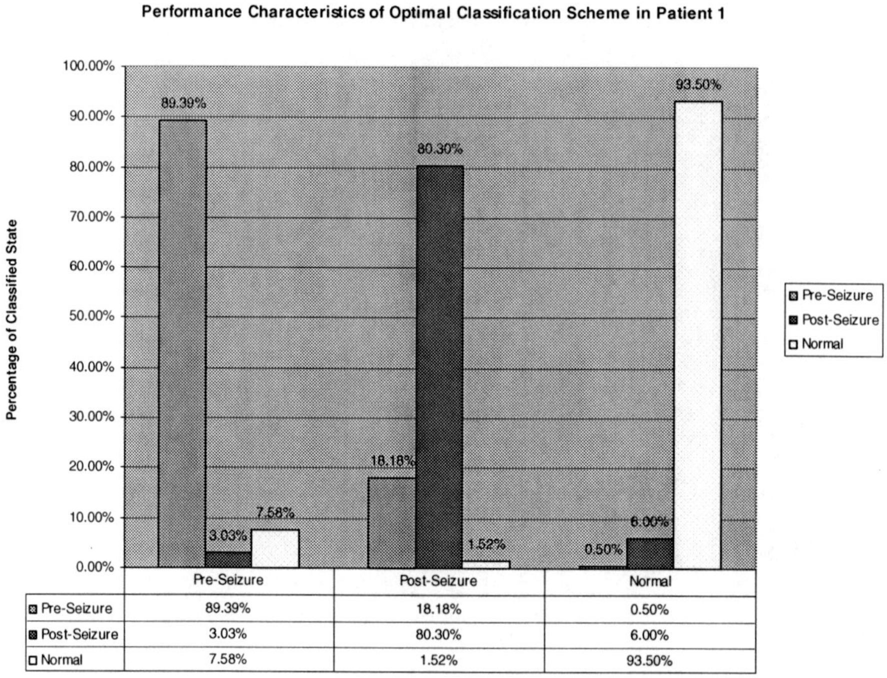

Performance Characteristics of Optimal Classification Scheme in Patient 1

	Pre-Seizure	Post-Seizure	Normal
Pre-Seizure	89.39%	18.18%	0.50%
Post-Seizure	3.03%	80.30%	6.00%
Normal	7.58%	1.52%	93.50%

Fig. 11. Patient 1 Results

results confirm our hypothesis that the pre-seizure and normal EEG's are differentiable. The techniques proposed in this study can be extended to development of the online brain activity monitoring, which is used to detect the brain's abnormal activity and seizure's pre-cursors. From the optimal classification schemes in 3 patients, we note that L_{max} tends to be the most classifiable attribute.

5 Concluding Remarks

Although evidence for the characteristic pre-seizure state transition and seizure predictability were first reported by our group [1, 2, 6, 11, 7, 5, 9, 16] and was confirmed by several other groups [3, 12, 17, 13, 14], further studies are required before a practical seizure prediction algorithm becomes feasible. The open question of whether the brain's normal and pre-seizure epileptic activities are distinctive or differentiable needs to be answered before one can translate the seizure prediction research into novel therapeutic approaches for controlling epileptic seizures.

This research was motivated by the analysis of the brain spatio-temporal dynamics, which was previously shown capable of reflecting the signature of

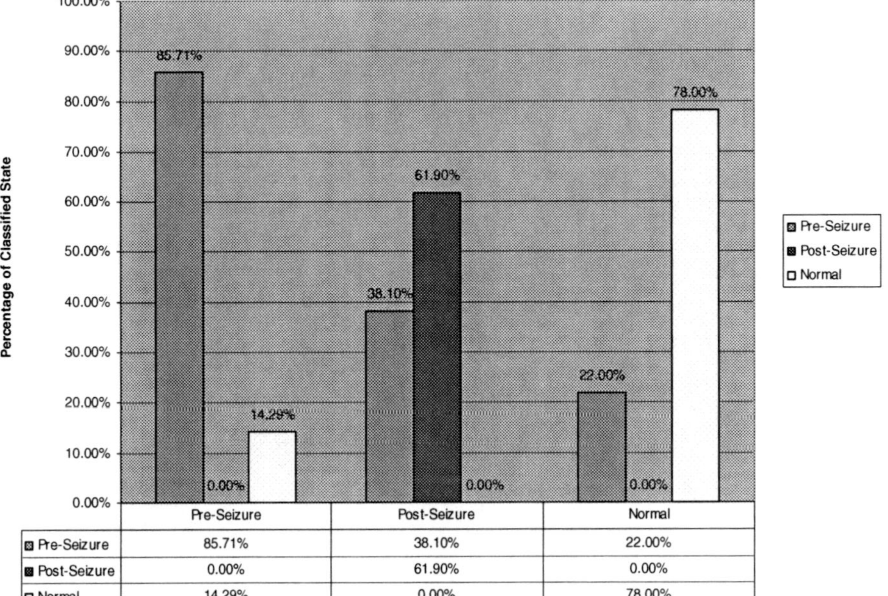

	Pre-Seizure	Post-Seizure	Normal
Pre-Seizure	85.71%	38.10%	22.00%
Post-Seizure	0.00%	61.90%	0.00%
Normal	14.29%	0.00%	78.00%

Fig. 12. Patient 2 Results

the pre-seizure epileptic activity during the brain abnormal episodes [1, 2, 6, 11, 7, 5, 9, 16]. That study of the brain dynamics was successful because understanding the brain dynamics provides insights about different states of brain activities reflected from pathological dynamical interactions of brain network. Based on the study of the brain dynamics, we have employed optimization and data mining techniques to classify dynamical states of an epileptic brain and identify pre-seizure epileptic activities. To our knowledge, this research represents the first attempt to study brain state transitions and develop mathematical models for the classification of the brain's normal and pre-seizure epileptic activities.

The results of this research confirm our hypothesis that it is possible to differentiate and classify the brain's pre-seizure and normal activities based on optimization, data mining, and dynamical system approaches in multichannel intracranial EEG recordings. The reliable classification is conceivable because, for the vast majority of seizures, the spatio-temporal dynamical features of the pre-seizure state sufficiently differs from that of the normal state. It can be easily observed that there are statistical differences in the brain dynamics from different states of the same patient. The experiments in this research can be done in a practical fashion because the herein proposed statistical

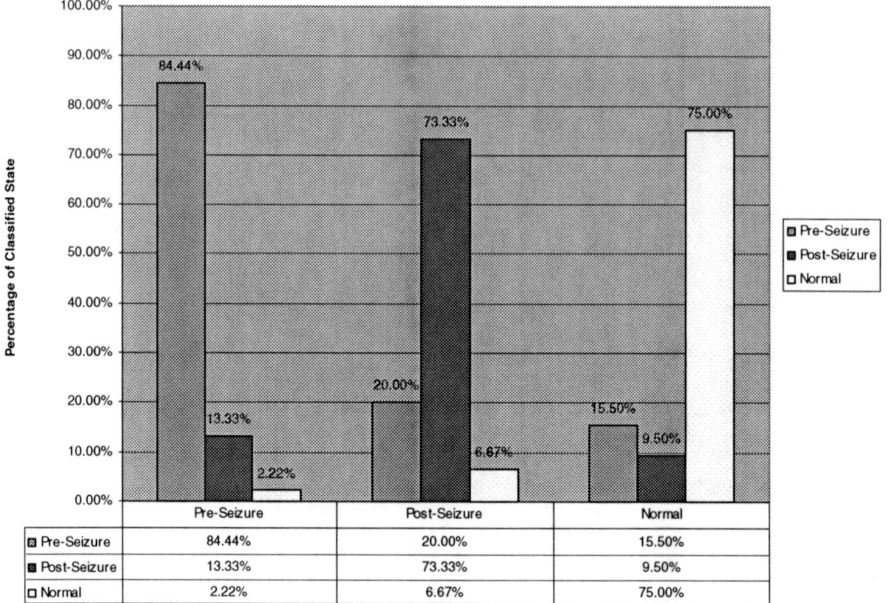

	Pre-Seizure	Post-Seizure	Normal
Pre-Seizure	84.44%	20.00%	15.50%
Post-Seizure	13.33%	73.33%	9.50%
Normal	2.22%	6.67%	75.00%

Fig. 13. Patient 3 Results

cross-validation derived from data mining concepts is fast and efficient and the dynamical measures of brain dynamics derived from the chaos theory can reveal hidden information from EEG signals.

In future, more cases (patients and seizures) will be studied to validate the observation across patients. This pre-clinical research will form a bridge between seizure prediction research and the implementation of seizure prediction/warning devices, which will be a revolutionary approach for handling epileptic seizures, very much similar to the pacemaker. It will also lead to clinical investigations of the effects of medical diagnosis, drug effects, or therapeutic intervention during invasive EEG monitoring of epileptic patients. The future research towards the treatment of human epilepsy and therapeutic intervention of epileptic activities as well as the development of seizure feedback control devices will be feasible. Thus, it represents a necessary first step in the development of implantable biofeedback devices to directly regulate therapeutic pharmacological or physiological intervention to prevent impending seizures or other brain disorders. For example, such an intervention might be achieved by electrical or magnetic stimulation (e.g., vagal nerve stimulation) or by a timely release of an anticonvulsant drug. Another practical application of the proposed approach is to help neurosurgeons to quickly identify the epileptogenic zone without having patients stay in the hospital for the invasive long-time (10-14 days in duration) EEG monitoring. This research has potential to revolutionize the protocol to identify the epileptogenic zone, which can drastically reduce the healthcare cost during the hospital stay for these patients. In addition, this protocol will help physicians to identify epileptogenic zones without the necessity to risk patients' safety by implanting depth electrodes in the brain.

In addition, the results from this study will also contribute to the understanding of the intermittency of other dynamical neurophysiological disorders of the brain (e.g., migraines, panic attacks, sleep disorders, and Parkinsonian tremors). We also expect our research to contribute to the localization of defects (flaws), classification and prediction of spatiotemporal transitions in other high dimensional biological systems.

Acknowledgments

The authors also want to thank Prof. Ioannis P. Androulakis for his fruitful comments and discussion.

References

1. W. Chaovalitwongse, P.M. Pardalos, L.D. Iasemidis, D.-S. Shiau, and J.C. Sackellares. Applications of global optimization and dynamical systems to prediction of epileptic seizures. In P.M. Pardalos, J.C. Sackellares, L.D. Iasemidis, and P.R. Carney, editors, *Quantitative Neuroscience*, pages 1–36. Kluwer, 2003.

2. W. Chaovalitwongse, P.M. Pardalos, and O.A. Prokopyev. A new linearization technique for multi-quadratic 0-1 programming problems. *Operations Research Letters*, 32(6):517–522, 2004.

3. C.E. Elger and K. Lehnertz. Seizure prediction by non-linear time series analysis of brain electrical activity. *European Journal of Neuroscience*, 10:786–789, 1998.

4. L.D. Iasemidis. *On the Dynamics of the Human Brain in Temporal Lobe Epilepsy*. PhD thesis, University of Michigan, Ann Arbor, 1991.

5. L.D. Iasemidis, P.M. Pardalos, J.C. Sackellares, and D.-S. Shiau. Quadratic binary programming and dynamical system approach to determine the predictability of epileptic seizures. *Journal of Combinatorial Optimization*, 5:9–26, 2001.

6. L.D. Iasemidis, J.C. Principe, J.M. Czaplewski, R.L. Gilmore, S.N. Roper, and J.C. Sackellares. Spatiotemporal transition to epileptic seizures: a nonlinear dynamical analysis of scalp and intracranial EEG recordings. In F.L. Silva, J.C. Principe, and L.B. Almeida, editors, *Spatiotemporal Models in Biological and Artificial Systems*, pages 81–88. IOS Press, 1997.

7. L.D. Iasemidis, J.C. Principe, and J.C. Sackellares. Measurement and quantification of spatiotemporal dynamics of human epileptic seizures. In M. Akay, editor, *Nonlinear Biomedical Signal Processing*, pages 294–318. Wiley–IEEE Press, Vol. II, 2000.

8. L.D. Iasemidis and J.C. Sackellares. The evolution with time of the spatial distribution of the largest Lyapunov exponent on the human epileptic cortex. In D.W. Duke and W.S. Pritchard, editors, *Measuring Chaos in the Human Brain*, pages 49–82. World Scientific, 1991.

9. L.D. Iasemidis, D.-S. Shiau, W. Chaovalitwongse, J.C. Sackellares, P.M. Pardalos, P.R. Carney, J.C. Principe, A. Prasad, B. Veeramani, and K. Tsakalis. Adaptive epileptic seizure prediction system. *IEEE Transactions on Biomedical Engineering*, 5(5):616–627, 2003.

10. L.D. Iasemidis, D.-S. Shiau, P.M. Pardalos, and J.C. Sackellares. Phase entrainment and predictability of epileptic seizures. In P.M. Pardalos and J.C. Principe, editors, *Biocomputing*, pages 59–84. Kluwer Academic Publishers, 2001.

11. L.D. Iasemidis, D.-S. Shiau, J.C. Sackellares, and P.M. Pardalos. Transition to epileptic seizures: Optimization. In D.Z. Du, P.M. Pardalos, and J. Wang, editors, *Discrete Mathematical Problems with Medical Applications*, pages 55–74. American Mathematical Society, 2000.

12. K. Lehnertz and C.E. Elger. Can epileptic seizures be predicted? evidence from nonlinear time series analysis of brain electrical activity. *Physical Review Letters*, 80:5019–5022, 1998.

13. B. Litt, R. Esteller, J. Echauz, D.A. Maryann, R. Shor, T. Henry, P. Pennell, C. Epstein, R. Bakay, M. Dichter, and G. Vachtservanos. Epileptic seizures may begin hours in advance of clinical onset: A report of five patients. *Neuron*, 30:51–64, 2001.

14. J. Martinerie, C. Van Adam, and M. Le Van Quyen. Epileptic seizures can be anticipated by non-linear analysis. *Nature Medicine*, 4:1173–1176, 1998.

15. P.M. Pardalos, W. Chaovalitwongse, L.D. Iasemidis, J.C. Sackellares, D.-S. Shiau, P.R. Carney, O.A. Prokopyev, and V.A. Yatsenko. Seizure warning algorithm based on spatiotemporal dynamics of intracranial EEG. *Mathematical Programming*, 101(2):365–385, 2004.

16. P.M. Pardalos, V.A. Yatsenko, J.C. Sackellares, D.-S. Shiau, W. Chaovalitwongse, and L.D. Iasemidis. Analysis of EEG data using optimization, statistics, and dynamical system techniques. *Computational Statistics & Data Analysis*, 44(1–2):391–408, 2003.

17. M. Le Van Quyen, J. Martinerie, M. Baulac, and F. Varela. Anticipating epileptic seizures in real time by non-linear analysis of similarity between EEG recordings. *NeuroReport*, 10:2149–2155, 1999.

18. A. Wolf, J.B. Swift, H.L. Swinney, and J.A. Vastano. Determining Lyapunov exponents from a time series. *Physica D*, 16:285–317, 1985.

Information Flow in Coupled Nonlinear Systems: Application to the Epileptic Human Brain

S. Sabesan[1], K. Narayanan[2], A. Prasad[3], L. D. Iasemidis[2], A. Spanias[1], and K. Tsakalis[1]

[1] Department of Electrical Engineering, Arizona State University, Tempe, AZ 85287 USA
[2] The Harrington Department of Bioengineering, Arizona State University, Tempe, AZ 85287 USA
[3] Department of Physics and Astrophysics, University of Delhi, New Delhi, India 110009

Summary. A recently proposed measure, namely Transfer Entropy (TE), is used to estimate the direction of information flow between coupled linear and nonlinear systems. In this study, we suggest improvements in the selection of parameters for the estimation of TE that significantly enhance its accuracy and robustness in identifying the direction of information flow and quantifying the level of interaction between observed data series from coupled systems. We demonstrate the potential usefulness of the improved method through simulation examples with coupled nonlinear chaotic systems. The statistical significance of the results is shown through the use of surrogate data. The improved TE method is then used for the study of information flow in the epileptic human brain. We illustrate the application of TE to electroencephalographic (EEG) signals for the study of localization of the epileptogenic focus and the dynamics of its interaction with other brain sites in two patients with Temporal Lobe Epilepsy (TLE).

Key words: Nonlinear Dynamics, Coupled Systems, Transfer Entropy, Information Flow, Epilepsy Dynamics, Epileptogenic Focus Localization

1 Introduction

Much of the research in the field of nonlinear dynamics has recently focused on the analysis of coupled nonlinear systems [1,2]. The motivation for this kind of work stems from the fact that the identification of the direction of information flow and estimation of the strength of interaction between coupled, nonlinear, complex systems of unknown structure has potential application to the understanding of their mechanisms of interactions, and the design and implementation of relevant control studies, e.g. control of interacting sets of neurons in the brain. In a recent paper [3], an information theoretic approach to identify the direction of information flow and quantify the strength of coupling between complex systems was suggested. This method is based on the study of transitional rather than static probabilities of the states of the systems under consideration. The proposed measure was termed Transfer Entropy (TE).

We have observed that the direct application of the method as proposed in [3] may not give the expected results, and that tuning of certain involved parameters plays a critical role in obtaining the correct direction of information flow [4,5]. We herein show the significance of the selection of two crucial parameters for the estimation of TE from time series, without any apriori knowledge of the structure of the systems that generate those data sets. The measure of TE and the estimation problems we identified are described in Section 2. The improvements suggested are given in Section 3. In Section 4, results from the application of the method to configurations of coupled Rossler and Lorenz oscillators are shown. In Sections 5 and 6, the application of the method to EEG data in epilepsy is shown. Discussion of these results and conclusions are given in Section 7.

2 Transfer Enotropy

Consider a kth order Markov process [6] described by

$$P(x_{n+1}|x_n, x_{n-1}\ldots, x_{n-k+1}) = P(x_{n+1}|x_n, x_{n-1}\ldots, x_{n-k}) \tag{1}$$

where P is the conditional probability of finding a random process X in state x_{n+1} at time $n + 1$. Eq. (1) implies that the probability of finding a system in a particular state depends only on the past k states $[x_n\ldots\ldots\ldots x_{n-k+1}]$ $\overset{\Delta}{=} x_n^{(k)}$ of the system. The definition given in Eq. (1) can be extended to the case of Markov interdependence of two random processes X and Y. Then, the generalized Markov property

$$P(x_{n+1}|x_n^{(k)}) = P(x_{n+1}|x_n^{(k)}, y_n^{(l)}) \tag{2}$$

where $y_n^{(l)}$ are the past l states of the second random process Y, implies that the state x_{n+1} of the process X depends only on the past k states of the process X and not on the past l states of the process Y. If the process X also depends on the past values of process Y, the divergence from Eq. (2) can be quantified using the Kullback-Leibler measure [7], where $P\left(x_{n+1}|x_n^{(k)}\right)$ is the a priori transition probability and $P(x_{n+1}|x_n^{(k)}, y_n^{(l))})$ is the true underlying transition probability of the system. The Kullback-Leibler measure quantifies the transfer of entropy from the driving process Y to the driven process X, and is given by:

$$TE(Y \rightarrow X) = \sum_{n=1}^{N} P(x_{n+1}, x_n^{(k)}, y_n^{(l)}) \log \frac{P(x_{n+1}|x_n^{(k)}, y_n^{(l)})}{P(x_{n+1}|x_n^{(k)})} \qquad (3)$$

The values of the parameters k and l are the orders of the Markov process for the two coupled processes X and Y respectively. The value of N denotes the total number of points in the state space. The selection of k and l plays a critical role in obtaining reliable values for the Transfer of Entropy. The estimation of TE as suggested in [3], also depends on the neighborhood size (radius r) in the state space that is used for the calculation of the involved joint and conditional probabilities. The value of radius r in the state space defines the maximum norm distance measure between two neighboring state space points. Intuitively, these different radius values correspond to different probability bins that are used in the state space to estimate multidimensional probabilities. The values of radius for which the probabilities are not accurately estimated (typically large r values) may eventually lead to an erroneous estimation of TE.

3 Improved Computation of Transfer Entropy

A. Selection of k:

The value of k used in the calculation of *TE (Y $\rightarrow$ X)* (see Eq. (1)) represents the dependence of the current state x_n of the system on its past k states. One of the reasonable methods to estimate this parameter is the delayed mutual information [8]. The delay at which the mutual information of X reaches its first minimum can be taken as the estimate of the period within which two states of X are dynamically correlated with each other. Then, in units of the sampling period, this delay would be equal to the order k of the Markov process.

B. Selection of l:

The value of l (order of the driving system) is chosen to be 1 in all examples of the systems presented herein. The assumption for this selection of l is that the current state of the driving system is sufficient to produce a considerable change in the dynamics of the driven system within one time step in the future. Larger values of l(delayed influence of Y on X) have been considered and their effect will be shown elsewhere.

C. Selection of radius r:

The multi-dimensional transitional probabilities, involved in the Transfer Entropy defined in Eq.(1), are calculated by joint probabilities using the Bayes'

Rule P(A|B)=P(A,B)/P(B). Therefore, the Transfer Entropy estimate can be rewritten as

$$TE(Y \rightarrow X) = \sum_{n=1}^{N} P(x_{n+1}, x_n^{(k)}, y_n^{(l)}) \log_2 \frac{P(x_{n+1}, x_n^{(k)}, y_n^{(l)}) P(x_n^{(k)})}{P(x_n^{(k)}, y_n^{(l)}) P(x_{n+1}, x_n^{(k)})} \qquad (4)$$

The above multi-dimensional joint probabilities are estimated at resolution r through the generalized correlation integrals $C(r)$ in the state space [9], [10] as:

$$P_r(x_{n+1}, x_n^{(k)}, y_n^{(l)}) \cong \frac{1}{N} \sum_m \Theta \left(r - \left| \begin{matrix} x_{n+1} - x_{m+1} \\ x_n^{(k)} - x_m^{(k)} \\ y_n^{(l)} - y_m^{(l)} \end{matrix} \right| \right) = C(r) \qquad (5)$$

where $\Theta(x > 0) = 1$; $\Theta(x \leq 0) = 0$ and $|.|$ is the maximum distance norm, and $m = [0, \ldots N - 1]$.

The estimation of joint probabilities between two different state space attractors requires simultaneous calculation of distances in both state spaces (see Eq. (3)). In the computation of these multivariate distances, a common value of distance r is desirable. In order to establish a common radius r in the state space for X and Y, the data are first normalized to zero mean and unit variance. The joint probabilities thus obtained in the state space are functions of r. When any of the joint probabilities in log scale is plotted against the corresponding radius r in log scale (see Fig. 2(b)), it is observed that, with increase in the value of the radius, they initially increase (for small values of r) and then saturate (for large values of r). This is similar to what occurs in the calculation of the correlation dimension from the slope in the linear region of $\ln C(r)$ vs. $\ln r$ [9]. We use a similar concept to estimate reliably the values of TE at specific values of r^* within the linear region of the curve $\ln C(r)$ vs. $\ln r$.

4 Simulation Results

A. Simulation examples:

We have applied the thus improved method of TE to several coupled nonlinear systems. Herein, we show the application of the method to coupled Rossler [11] and coupled Lorenz [11] type of oscillators, governed by the following differential Eqs. (6) and (7) respectively, where $\omega_i = 1$, $\alpha_i = 0.38$, $\beta_i = 0.3$, $\gamma_i = 4.5$ are the parameters of the $i-$th Rossler oscillator in Eq. (6), and $\sigma_i = 2.0$, $r_i = 175$, $\beta_i = 8/3$ are the parameters of the $i-$th Lorenz oscillator in Eq. (7), ε_{ji} denotes the strength of the diffusive coupling from oscillator j

to oscillator i (ε_{ii} denotes self-coupling of the i-th oscillator, and it is taken to be zero for all the simulations reported in this study).

$$\frac{dx_i(t)}{dt} = -\omega_i y_i - z_i + \sum_{j=1,i\neq j}^{m} \varepsilon_{j,i} x_j - \varepsilon_{i,i} x_i$$

$$\frac{dy_i(t)}{dt} = \omega_i x_i + \alpha_i y_i \qquad (6)$$

$$\frac{dz_i(t)}{dt} = \beta_i x_i + z_i(x_i - \gamma_i)$$

$$\frac{dx_i(t)}{dt} = -\sigma_i(x_i - y_i) + \sum_{j=1,i\neq j}^{m} \varepsilon_{j,i} x_j - \varepsilon_{i,i} x_i$$

$$\frac{dy_i(t)}{dt} = r_i x_i - y_i - x_i z_i \qquad (7)$$

$$\frac{dz_i(t)}{dt} = -\beta_i z_i + x_i y_i$$

The parameters selected for each type of oscillator place them in the chaotic regime. A total of three different coupled configurations (between any two oscillators) are considered herein: one-directional, bi-directional and no coupling. Results from the application of the method, with and without the suggested improvements, will be shown for each configuration.

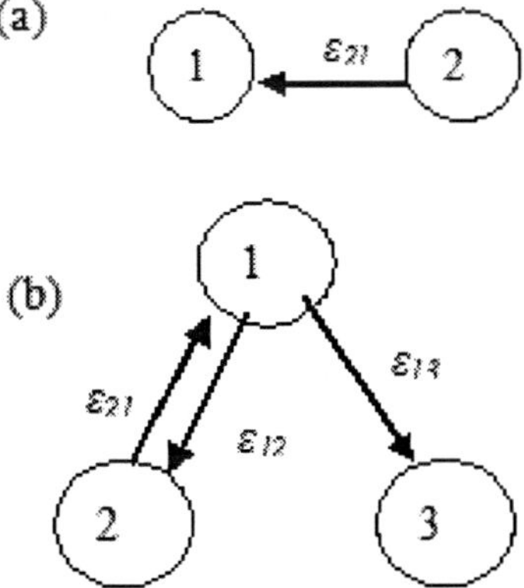

Fig. 1. (a) *Configuration 1:* Two coupled oscillators with $\varepsilon_{12} = 0$, $\varepsilon_{21} = 0.15$; (b) *Configuration 2:* Three coupled oscillators with $\varepsilon_{12} = \varepsilon_{21} = 0.15$, $\varepsilon_{13} = 0.12$, $\varepsilon_{31} = 0.0$, $\varepsilon_{23} = \varepsilon_{32} = 0.0$.

The direction of information flow is from oscillator $2\rightarrow1$ in Fig. 1(a) and $1\rightarrow2$, $2\rightarrow1$ and $1\rightarrow3$ in Fig. 1(b). The data from these model configurations were generated from the previously cited differential equations using an integration step of 0.01 and a 4th order Runge-Kutta integration method. A total of 30,000 points from the x time series of each oscillator were considered for the estimation of each value of the TE after down-sampling the data produced by Runge-Kutta by a factor of 10.

A1. Configuration 1:

Coupled Rossler oscillators: Fig. 2(a) shows the mutual information MI between the two oscillators of Fig. 1(a) for different values of k. The first minimum of MI occurred at k=16. The state spaces were reconstructed from the x time series of both oscillators with embedding dimension $p = k + l$. Fig. 2(b) shows the $\ln C(r)$ vs. $\ln r$. Fig. 3 shows the transfer of entropy TE(r) values without (Fig. 3(a)) and with (Fig. 3(b)) our suggested values for k, l and r. In Fig. 3(a), the TE(r) is illustrated with arbitrary chosen values $k = l$=5. No clear preference between 1→2 and 2→1 direction of information flow can be seen for values of radius in the linear region of $\ln C$vs. $\ln r$. In this configuration, the information flow exists only in the direction of 2→1, whereas Fig. 3(a) incorrectly shows a bi-directional flow for low and high values of r. In Fig. 3(b), the optimal value of k=16 (l=1) is used in the estimation of TE(r). In this figure, the information flow is correctly depicted in the direction of 2→1 for low values of r. The direction is depicted wrongly for large values of r. These observations justify our selection of r^* from the low region of $\ln C(r)$ vs. $\ln r$ (shown at the vertical line in Fig. 2(b)), according to our discussion in Section 3.

Coupled Lorenz oscillators: A system of two Lorenz oscillators in Configuration 1 ($\varepsilon_{12} = 0, \varepsilon_{21} = 2$) shows similar TE results with respect to detection of the information flow (see Fig. 4) as the ones from the Rossler oscillators. The estimated parameters here were k=8 and l=1.

A2. Configuration 2:

Fig. 5 shows the TE of three coupled Rossler oscillators with coupling directions 1→2, 2→1, 2→3 and 3→2, without (see Fig. 5(a)) and with (see Fig. 5(b)) the proposed improvements for k and l.The first minimum of mutual information from the x data series for this model was 15. Therefore k=15 and l=1 were further used in TE estimation. It is evident that, for these values of k and l, we can clearly detect the existing bi-directional information flow between oscillators 1 and 2 (see Fig. 5(b)), as well as the absence of interactions between oscillators 2 and 3 using $r = r^*$ (see Fig. 5(c)).

A3. Configuration 3:

In this configuration, a two Rossler oscillator system (see Fig. 6(a)) is used to illustrate the performance of the proposed measure TE with respect to the quantification of the strength of the directional coupling between the

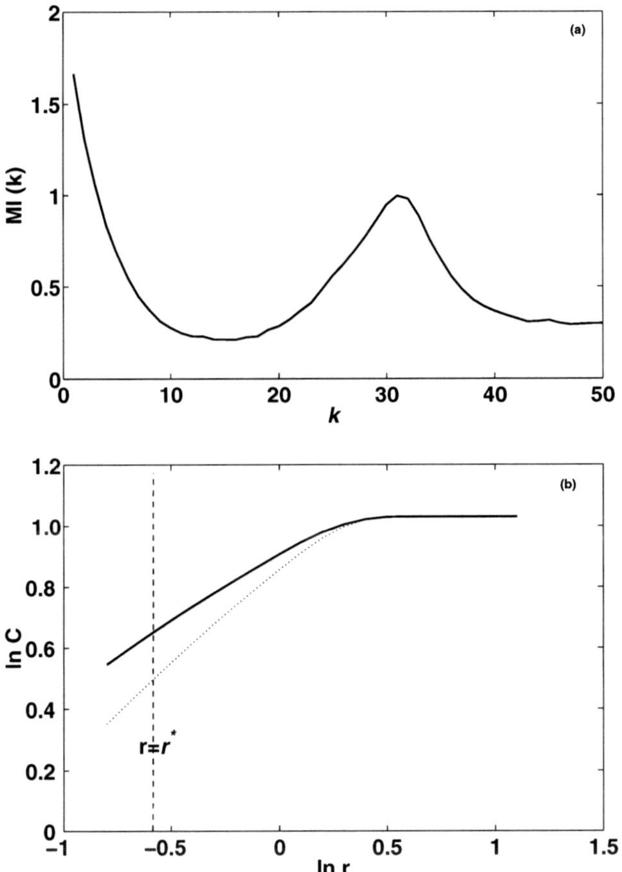

Fig. 2. Configuration 1: Coupled Rossler oscillators. (a) Plot of the mutual information MI(k) vs. k. (b) Plot of ln C vs. ln r , where C and r denote average joint probability (correlation integrals) and radius in the state space respectively, for direction of flow 1→2 (dotted line) and 2→1 (solid line) (k=16, l=1).

oscillators. While the value of the coupling strength ε_{12} was fixed at 0.05, the coupling ε_{21} was let to increase from 0.01 to 0.15.

Fig. 6(b) shows the value of TE as a function of increasing coupling strength ε_{21} at 0.01, 0.05, 0.1 and 0.15, and at a value of r^* chosen as before in the linear region of the respective ln C vs ln r. We can see that TE (2→1) increases for a progressive increase in the values of the coupling strength ε_{21}. In addition, as it should, TE (2→1) $\cong$ TE (1→2) at $\varepsilon_{12} = \varepsilon_{21}$=0.05.

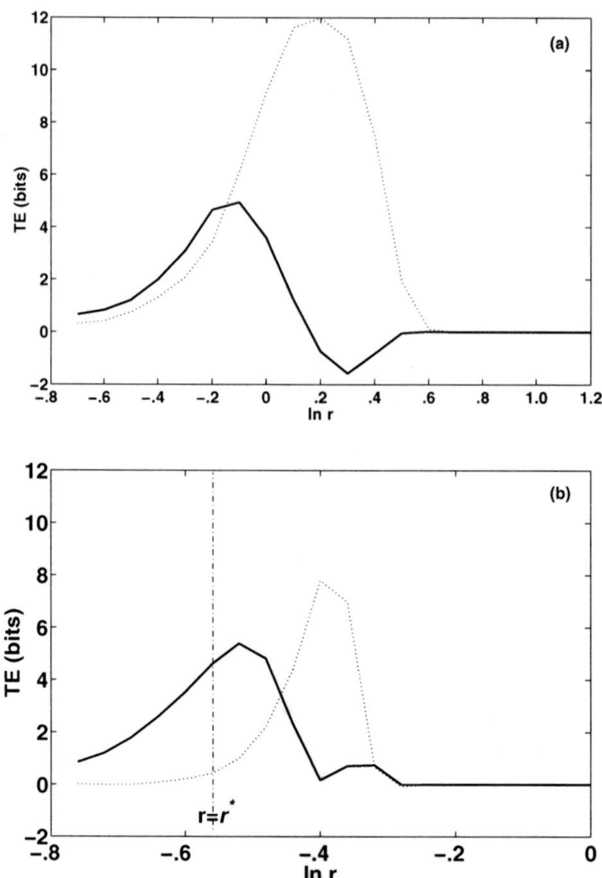

Fig. 3. Configuration 1: Coupled Rossler oscillators.Transfer entropy TE vs. radius r (in ln scale), between Rossler oscillators 1→2 (dotted line) and 2→1 (solid line) for (a) $k = l = 5$ and (b) $k = 16, l = 1$ (values estimated with the proposed methodology). The position of the vertical dotted line in the figures corresponds to the value of the radius r^* within the linear region of ln C(r) vs. ln r (see Fig. 2(b)).

B. Statistical significance of the model results:

The statistical significance of the derived values of TE was tested using the method of surrogate analysis. Since TE calculates the direction of information transfer between systems by quantifying their conditional statistical dependence, a random shuffling, separately applied to the original data series, significantly reduces their statistical inter-dependence. The shuffling was based on generation of white Gaussian noise and reordering of the original data samples according to the order indicated by the thus generated noise values. The

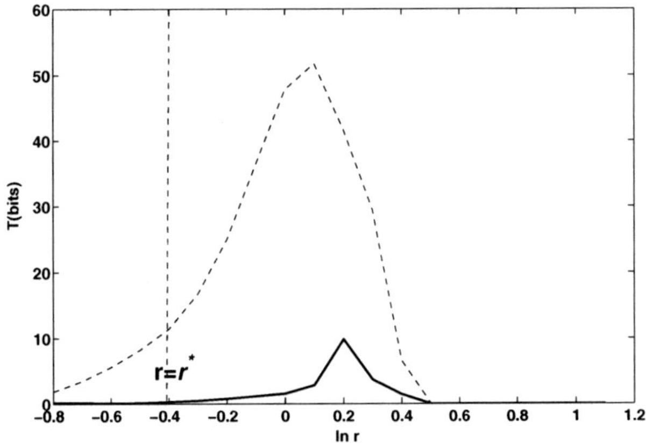

Fig. 4. Configuration 1: Coupled Lorenz oscillators. Transfer entropy TE vs. radius r (in logarithmic scale), between Lorenz oscillators ($\varepsilon_{12} = 0, \varepsilon_{21} = 0.15$) 1→2 (solid line) and 2→1 (dotted line). The position of the vertical dotted line corresponds to the value of the radius r^* within the linear region of the $\ln C(r)$ vs. $\ln r$.

null hypothesis that the obtained values of the directional transfer of entropy are not statistically significant was then tested. Transfer entropy TE_s values from the shuffled datasets of pairs of oscillators were obtained. If the TE_o values from the original time series laid well outside the mean of the TE_s values, the null hypothesis was rejected. A total of 12 surrogate data series were produced for configurations 1 and 2 analyzed in the previous subsection. The TE_s values were calculated at the optimal radius r^*estimated from the original data. The means and standard deviations of TE_s and TE_o for each configuration are shown in Table 1.

In Configuration 1, the null hypothesis is rejected for the direction 2→1, which is the real direction of the information flow, at p ¡ 0.0001 (assuming Gaussian distribution for the TE_s values of the shuffled data sets and using the sample t-test with 11 degrees of freedom , see last column in Table 1). On the contrary, as expected, the 1→2 information flow was not statistically significant (p=0.1188), as the TE_o is no more than 0.49 standard deviations away from the mean of TE_s. Results along these lines were produced for the other configurations and are given in the last rows of Table 1. This statistical analysis confirms that the values of TE, calculated with the suggested improvements, capture the direction of the information flow between the existing subsystems in all different model configurations and types of oscillators investigated herein.

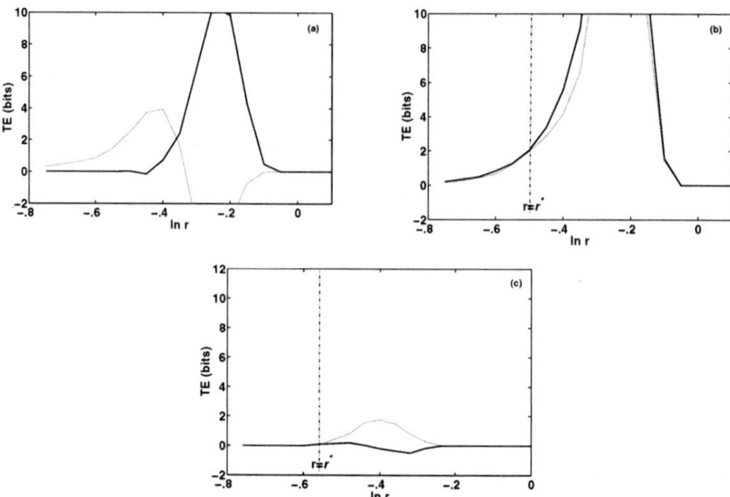

Fig. 5. Configuration 2: Transfer entropy TE vs. radius r (in ln scale, between Rossler oscillators $1{\rightarrow}2$ (dotted line) and $2{\rightarrow}1$ (solid line) (a) without ($k = 5$, $l = 1$), and (b) with improvements ($k = 15$, $l = 1$). (c) Transfer entropy TE between oscillators $3{\rightarrow}2$ (dotted line) and $2{\rightarrow}3$ (solid line). The vertical dotted line in the figures denotes the value of the radius r^* within the linear region of $\ln C(r)$ vs. $\ln r$.

5 Transfer Entropy: Application to EEG in Epilepsy

The need for a reliable method to identify the direction of information flow in physiological systems has assumed an extreme importance in the medical world today. In particular, the functional description of the human brain still poses itself as a perennial problem to the researchers [12-20]. The human brain is a highly complex system with nonlinear characteristics. The identification of direction of information flow in such systems is a formidable task. From a system's perspective, the EEG signals that we obtain are the output of a nonstationary and a multivariable system. Therefore, the importance of the development of methods that incorporate the dynamics of all individual subsystems and quantify their interactions cannot be emphasized enough. Epilepsy is a brain disorder with significant alteration of information flow in the brain intermittently (between seizures) and during seizures.

An epileptogenic focus is electrophysiologically defined as the brain's area that first exhibits the onset of epileptic seizures. Electrical discharges from the focus spread and disrupt the normal operation of the brain. The EEG signals are very helpful in providing evidence for epilepsy but are often not reliable in identifying the epileptogenic focus by visual inspection. The primary objective in presurgical evaluation of patients is to identify the region, which is responsible for generating the patient's habitual seizures. This region

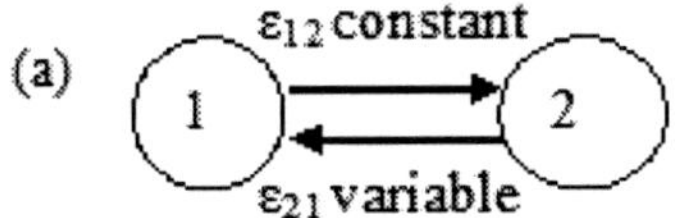

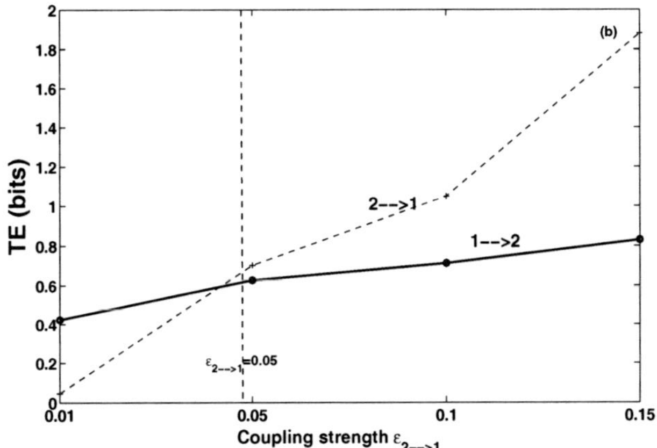

Fig. 6. Configuration 3: (a) Two bi-directionally coupled Rossler Oscillators with increasing coupling in one direction ($\varepsilon_{12} = 0.05$, and $\varepsilon_{21} = 0.01, 0.05, 0.1, 0.15$). (b) Transfer entropy TE at $\ln r^* = 0.6$ versus coupling strength $\varepsilon_{2\rightarrow 1}$ (TE ($1\rightarrow 2$) solid line and TE ($2\rightarrow 1$) dashed line).

(epileptogenic zone) is generally determined visually from long term monitoring of EEG by localization of the earliest electrographic seizure onset. Usually, resection of the focal tissue is sufficient to abolish epileptic seizures in carefully selected unifocal patients. Epileptogenic focus localization, using analytical signal processing and information theoretic techniques on the EEG from epileptic patients, in conjunction with the traditional, visual inspection of EEG records, is very promising in understanding of the underlying dynamics of this disease and a more accurate diagnosis and prognosis. The hypothesis we consider in this respect is that the epileptogenic focus acts as the driver for the brain sites that participate in a seizure activity, especially in the preictal (prior to seizure) periods. In this section, instead of a system of nonlinear oscillators, we analyze the human brain. The individual subsystems of such a system are the brain sites under the recording EEG electrodes. Thus, our hypothesis is that, by taking a driver-driven system approach, one can identify the epileptogenic focus as the electrode or the set of electrodes that drive other critical electrodes in the pre-ictal (before seizures) and ictal (during seizures) state. We apply the method of TE to quantitatively test this hypothesis.

Table 1. Surrogate analyses of Transfer of entropy (* denotes the real direction of coupling

Direction of flow	Transfer Entropy (TE)		TE_o-TE_s (in standard deviation units)	p value of t-test
	Surrogate Data (TE_s) (bits)	Original Data (TE_o) (bits)		
Configuration 1: Coupled Rossler oscillators				
1→2	0.43 ± 0.86	0.01	0.49	.1188
2→1 *	0.30 ± 1.88	4.50	2.23	.0001
Configuration 1: Coupled Lorenz oscillators				
1→2	1.02 ± 1.01	.49	.79	.2383
2→1*	1.08 ± .74	16.72	15.46	.0001
Configuration 2				
1→2*	0.31 ± 0.91	1.98	1.84	.0001
2→1*	0.28 ± 0.97	1.98	1.76	.0001
2→3	0.10 ± 0.90	0.09	0.01	.97
3→2	0.08 ± 0.88	0.05	0.03	.90

A. Transfer Entropy and its application to EEG:

As it was shown in section 3, when TE is applied to standard coupled nonlinear systems, such as the Rossler and Lorenz systems, it requires proper choice of the parameters k (order of the Markov process for the driven system), l (order of the Markov process for the driving system), and r (the multi-dimensional bin size for calculation of the involved probabilities). A straightforward extension of these methods to estimate the involved parameter values from real-life data may not be possible. For example, human epileptogenic EEG data sets do not have unique characteristics in terms of their underlying dynamics and their noise levels. Also, physiological data are nonstationary, and therefore it is difficult to estimate reliably values of the first minimum of mutual information over time [21-23], which may lead to erroneous calculations of the values of TE. From a practical point of view, a statistic that might be used as an optimal solution to this problem is the correlation time constant t_e, which is defined as the time required for the autocorrelation function (AF) to decrease to 1/e of its maximum value [24,25]. AF is an easy metric to compute over time, and has been found to be robust in many simulations. Furthermore, if one looks at the estimation of TE from a state space reconstruction perspective, then a reasonable relation between time delay $\tau = 1$ and embedding dimension $p = k + 1$ is:

$$t_e = [p - 1]\tau = k \tag{8}$$

Finally, in the analysis of EEG signals, we keep l (Markov order of the driving system) to be 1 as in all previous examples investigated herein.

B. Selection of the range of r for EEG:

For physiological data sets such as epileptic EEG (i.e. data of finite length and corrupted with noise), existence of a linear region of $\ln C(r)$ vs. $\ln r$ is very difficult to be found. In fact, the curve $\ln C(r)$ vs. $\ln r$ could be divided into three regions based on the values of $\ln C(r)$ (see also Fig. 2(b)). These three regions generated by the estimation of $\ln C(r)$ from a 10.24 second preictal EEG segment, are shown in Fig. 7.

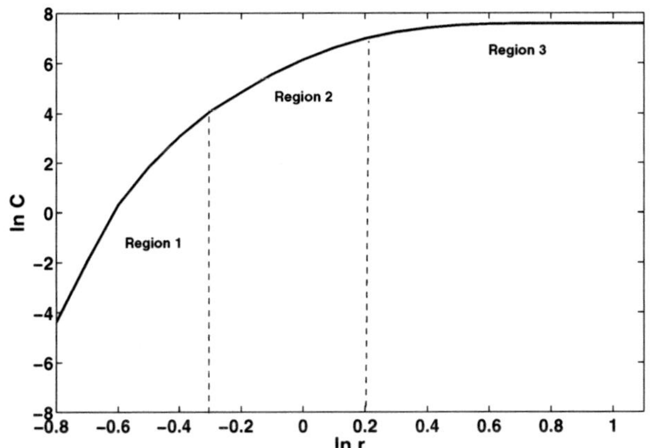

Fig. 7. Plot of $\ln C(r)$ vs. $\ln r$ from a preictal EEG segment ($C(r)$ and r denote average probability and radius respectively) using $k = 16$, $l = 1$.

It is clear that in region 1 (small radius values:$-0.8 \leq \ln r \leq -0.2$), the presence of noise in the data is predominant [17,19] and smears the data into the entire space (high-dimensional). No linear slope can be observed therein. On the other hand, region 3 ($\ln r \geq 0.2$) corresponds to those values of r for which a flat region in $\ln C(r)$ is observed (saturation). In between regions 1 and 3, we observe a quasi-linear region. In the literature of generalized correlation integrals, this region is also called the scaling region or the "linear" region (region 2: $-0.3 \text{i} \ln r \leq 0.2$) [9]. Therefore, a choice of a value for the radius from region 2 will result in a value of TE that is more reliable and sensitive to the direction of information flow. Values of radius from region 2 were used in our study of epileptogenic focus localization and did not differ much from the r^* value we used earlier in the coupled chaotic model configurations ($\ln r^* = -0.3$).

C. Application of Transfer Entropy to Epileptogenic focus localization:

TE was estimated from successive, non-overlapping electrocorticographic (ECoG) and depth EEG segments of 10.24 seconds in duration (2048 points per segment at 200 Hz sampling rate) for a total of 42 minutes (20 minutes pre-ictal, 20 minutes postictal and 2 minutes ictal). Two seizures were analyzed per patient, from two of our patients with temporal lobe epilepsy. The EEG signals were recorded from 6 different areas of the brain with a total of 28 electrodes (see Fig. 8 for the electrode montage used). As TE constitutes a bivariate approach to find the direction of information flow, it is computationally intensive as all pairs of available electrodes are taken into consideration for the analysis. For illustration purposes, we herein consider an analysis with a subset of only 5 out of the 28 available electrodes. The selection of the sites was based on the observed dynamical entrainment of the epileptogenic focus with other cortical sites [12-17].

One value of TE is estimated for every electrode pair per 10.24 seconds of EEG segment. The TE is then calculated for all possible pairs over time. The temporal values of TE are subsequently averaged to obtain one value for the pre-ictal period per electrode pair ($\hat{TE}_{Pre}$) at the relevant radius r^*. For an electrode to be identified as a focal electrode, two conditions were imposed:

a) A brain site i out of N sites is considered to be a probable focus if

$$\hat{TE}_{Pre}(i \rightarrow j) > 1 \text{ bit of information} \qquad (9)$$

where $i \neq j, j = 1, \ldots, N$, and $\hat{TE}(e_{ij})$ is the Transfer Entropy from $i \rightarrow j$.

b) The brain site i that obeys condition (9), and has the largest value of $\hat{TE}_{Pre}(i \rightarrow j)$, for $i = 1, \ldots, N$ and $j = 1, \ldots, N$ is identified as the dominant focal site.

The thus obtained results on the epileptogenic focus localization were compared with the patients' clinical records to test their physiological relevance.

6 EEG Results

A. Epileptogenic Focus Localization / Information Network:

The following 5 electrode sites {RTD2, RTD3, LTD2, RST3, LOF2} are selected for both patients 1 and 2 before the occurrence of a seizure of a focal right temporal lobe origin. These sites were found to be critical sites in the optimal sense described in [14]. TE was calculated for all 10 possible electrode pairs in this selected set of electrodes. The values of $\hat{TE}_{Pro}$ were then calculated at the relevant radius r^*. The optimal value $\ln r^*$ was chosen equal to -0.3.

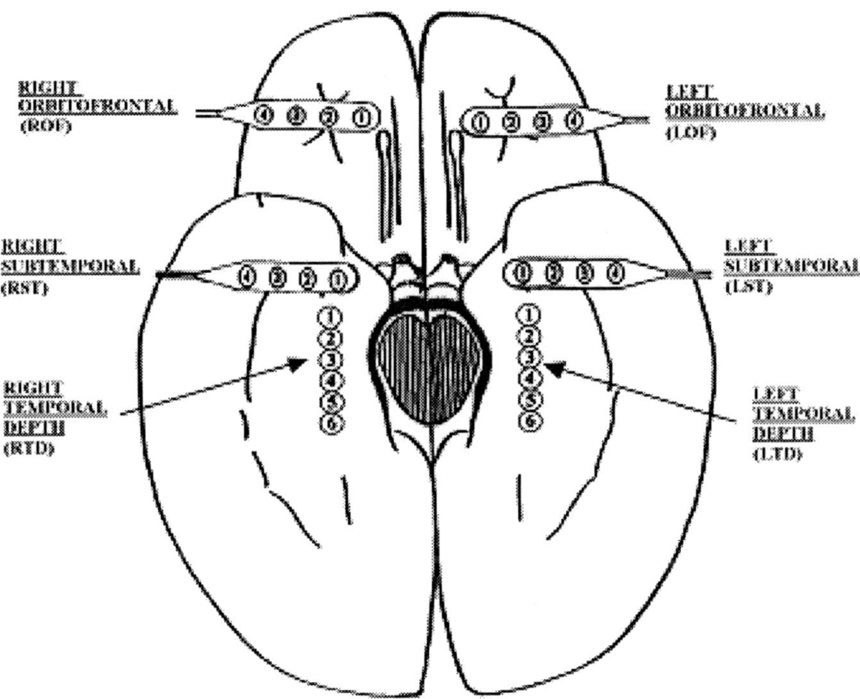

Fig. 8. Schematic diagram of the depth and subdural electrode placement. This view from the inferior aspect of the brain shows the approximate location of depth electrodes, oriented along the anterior-posterior plane in the hippocampi (RTD - right temporal depth, LTD - left temporal depth), and subdural electrodes located beneath the orbitofrontal and subtemporal cortical surfaces (ROF - right orbitofrontal, LOF – left orbitofrontal, RST- right subtemporal, LST- left subtemporal).

1) Patient 1:

Table 2 gives the values of $\hat{TE}_{Pre}$ (preictal $\hat{TE}$ for all possible pairs of electrodes at r^* for patient 1. From Table 2, it is clear that RTD2 drives RTD3 the most among all possible pair-wise interactions, having a $\hat{TE}_{Pre}(RTD2 \rightarrow RTD3) = 7.45$ bits. Therefore, in this particular seizure and according to our aforementioned criteria (a) and (b), RTD2 is the most probable focal electrode. It is easier to visualize Table 2 through an information flow diagram. Such a diagram is shown in Fig. 9. In this figure, the solid arrows mark those interactions where the mean value of TE over a 20 minute preictal period is greater than one bit of information flow in the corresponding directions. The dashed arrows denote all weak interactions, i.e. interactions with less than one bit of information flow in a particular direction.

Similarly, for the preictal period of the second seizure in the same patient 1, the values of $\hat{TE}_{Pre}$ for all possible pairwise interactions are shown in Table 3. The maximum $\hat{TE}_{Pre}$ among all the possible interactions, was again the

Table 2. $\hat{TE}_{Pre}$ values for all pairs of electrodes prior to seizure 1 for patient 1.

		$\hat{TE}_{Pre}$ (bits)				
		Driven				
	Electrodes	RTD2	RTD3	LTD2	RST3	LOF2
	RTD2	0.00	**7.45**	0.27	0.90	0.22
	RTD3	**3.29**	0.00	0.43	0.25	0.25
Driving	LTD2	0.49	0.47	0.00	0.23	0.16
	RST3	**1.19**	**1.71**	0.16	0.00	0.08
	LOF2	0.25	0.35	0.26	0.08	0.00

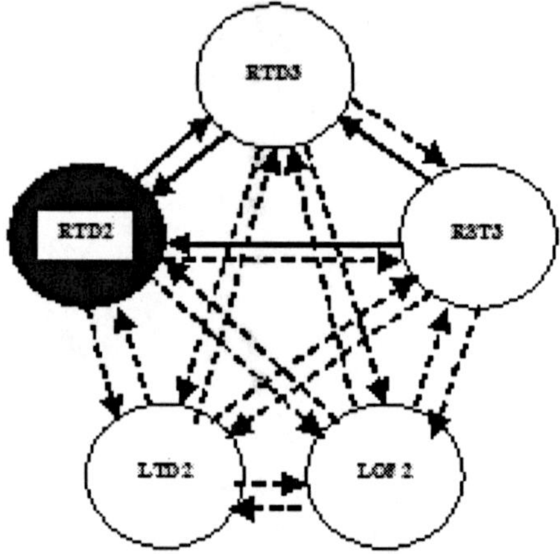

Fig. 9. Transfer Entropy flow diagram of the electrode interactions from Pre-ictal state of seizure 1 in patient 1 (solid arrows indicate stronger interactions, see text for more details).

$RTD2 \rightarrow RTD3$ with a value of 6.31 bits. This leads us to conclude that RTD2 is the driver among all considered electrodes in this seizure too, and thereby the probable focus. An information flow diagram based on Table 3 is shown in Fig. 10. Similar to seizure 1, any transfer of information greater than 1 bit is considered to be significant and is marked with a solid arrow. All other weaker interactions are marked with dashed arrows. Again RTD2 shows up as the most probable focal electrode that drives RTD3 and RST3. In both seizures, these results were in agreement with the identification of the epileptogenic focus by the gold standard used in clinical practice for identification of the epileptogenic focus, that is where the earliest onset of seizure activity occurs.

2) Patient 2:

Similar analysis of two seizures from patient 2 revealed RTD2 as the probable focus in the first seizure ($\hat{TE}_{Pre}(RTD2 \rightarrow RTD3)$ =1.67 bits), whereas RTD3 shows up as the probable focus in the second seizure with a $\hat{TE}_{Pre}(RTD3 \rightarrow$

Table 3. $\hat{TE}_{Pre}$ values for all pairs of electrodes prior to seizure 2 for patient 1.

		$\hat{TE}_{Pre}$ (bits)				
		Driven				
	Electrodes	RTD2	RTD3	LTD2	RST3	LOF2
	RTD2	0.00	**6.31**	0.33	**1.07**	0.13
	RTD3	**3.03**	0.00	0.30	0.76	0.18
Driving	LTD2	0.36	0.43	0.00	0.38	0.16
	RST3	**1.54**	**1.47**	0.24	0.00	0.12
	LOF2	0.15	0.15	0.17	0.07	0.00

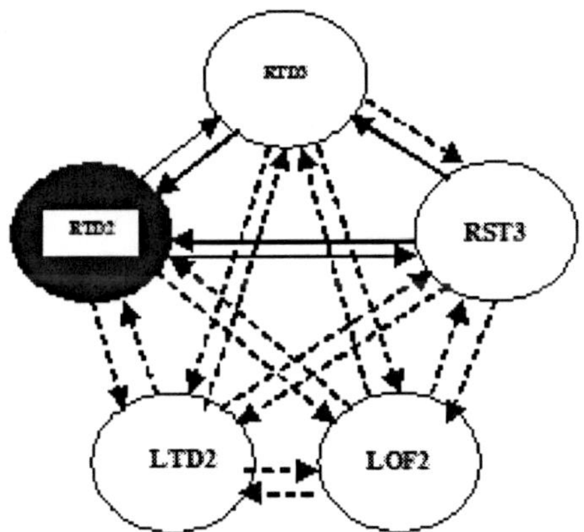

Fig. 10. Transfer Entropy flow diagram of the electrode interactions from Pre-ictal state of seizure 2 in patient 1 (solid arrows indicate stronger interactions, see text for more details).

$RTD2$) value of 1.37 bits. Tables 4 and 5 summarize the $\hat{TE}_{Pre}$ values for these seizures in patient 2. These results are again in agreement with the clinical finding of the epileptogenic focus (RTD) in this patient. The information flow diagrams for both seizures are shown in Figs. 11 and 12.

Table 4. $\hat{TE}_{Pre}$ values for all pairs of electrodes prior to seizure 1 for patient 2

		$\hat{TE}_{Pre}$ (bits)				
		Driven				
	Electrodes	RTD2	RTD3	LTD2	RST3	LOF2
	RTD2	0.00	**1.67**	0.50	0.64	0.29
	RTD3	**1.26**	0.00	0.31	0.94	0.32
Driving	LTD2	0.54	0.23	0.00	0.41	0.33
	RST3	.84	**1.07**	0.48	0.00	0.42
	LOF2	0.27	0.27	0.33	0.25	0.00

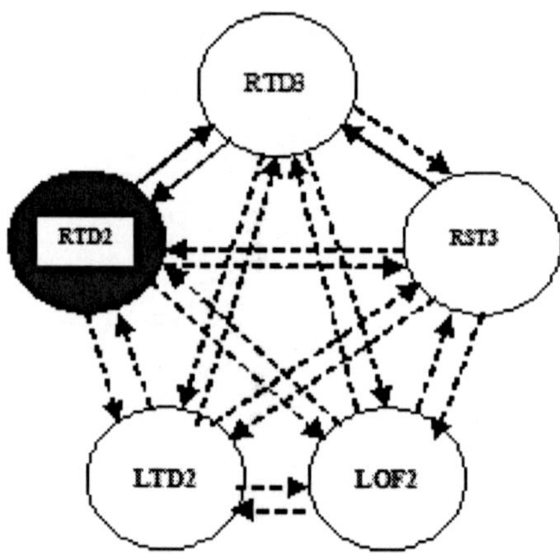

Fig. 11. Transfer Entropy flow diagram of the electrode interactions from Pre-ictal state of seizure 1 in patient 2 (solid arrows indicate stronger interaction, see text for more details).

Table 5. $\hat{TE}_{Pre}$ values for all pairs of electrodes prior to seizure 2 for patient 2

		$\hat{TE}_{Pre}$ (bits)				
		Driven				
	Electrodes	RTD2	RTD3	LTD2	RST3	LOF2
	RTD2	0.00	**1.07**	0.50	0.64	0.29
	RTD3	**1.37**	0.00	0.31	0.94	0.32
Driving	LTD2	0.54	0.23	0.00	0.41	0.33
	RST3	.84	.87	0.48	0.00	0.42
	LOF2	0.27	0.27	0.33	0.25	0.00

B. Epileptogenic Focus Dynamics:

Monitoring of TE over time also may shed light on the underlying dynamics of the information network in the epileptic brain. For example, in Fig. 13, we show the values of the bi-directional TE over the preictal, ictal and postictal period for the most interacting electrode pair RTD2 and RTD3 in patient 1, seizure 1. The hippocampal sites RTD2 and RTD3 are both focal sites in this patient. The vertical line marks the onset of this seizure. The duration of this seizure was 2 minutes. It is noteworthy that the focal electrode RTD2 is preictally driving the other focal electrode RTD3 ($TE(RTD2 \rightarrow RTD3)$ larger than $TE(RTD3 \rightarrow RTD2)$). This driving is postictally reduced to the level of flow in the other direction. In essence, we have a strong uni-directional flow preictally, becoming a strong bi-directional flow about 5 minutes before the seizure onset, and a weak bi-directional flow postictally.

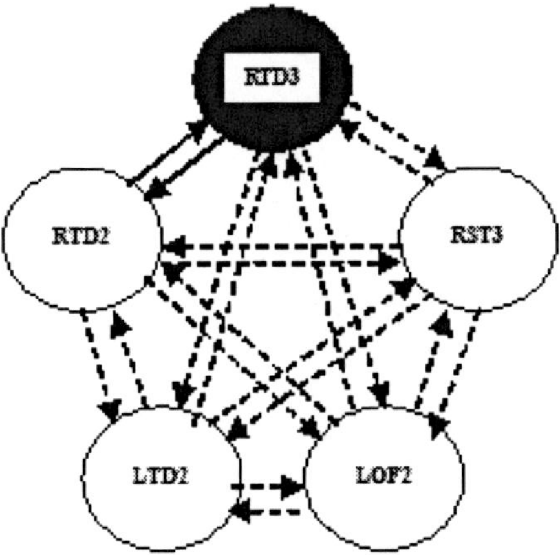

Fig. 12. Transfer Entropy flow diagram of the electrode interactions from Pre-ictal state of seizure 2 in patient 2 (solid arrows indicate stronger interactions, see text for more details).

In summary, the EEG analysis of a limited number of seizures by TE provided focus localization results that are consistent with the clinical localization of the epileptogenic focus in two patients with temporal lobe focal epilepsy. It also appears that the proposed methodology may be a useful tool for the investigation of the dynamical mechanisms involved in the occurrence of epileptic seizures.

7 Conclusion

In this study, we suggested and implemented improvements for the estimation of Transfer Entropy, a measure of the direction and the level of information flow between coupled subsystems. The improvements were shown to be critical in obtaining consistent and reliable results from complex signals generated by systems of coupled, nonlinear (chaotic) oscillators. The application of this methodology to epileptogenic focus localization in two patients with focal temporal lobe epilepsy produced results in agreement with the location of their focus. Accurate localization and further understanding of the dynamics of the epileptogenic focus by the use of TE in more patients could further elucidate the mechanisms of epileptogenesis. We believe that our analytical scheme can have several potential applications in diverse scientific fields, from medicine and biology, to physics and engineering.

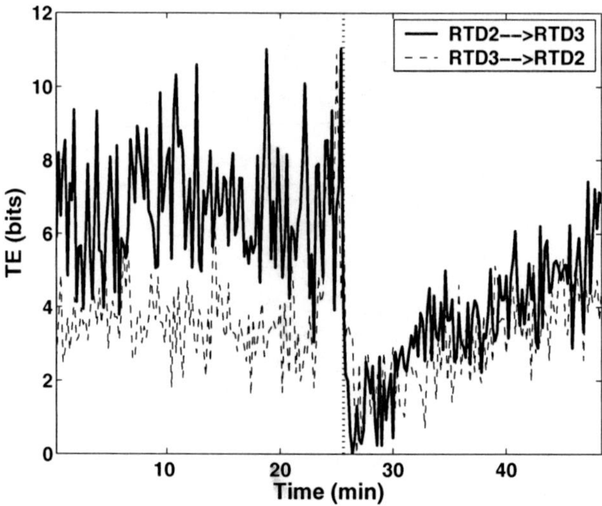

Fig. 13. Long-term dynamics of Information Flow between two focal electrodes before and after seizure 1 of patient 1. The seizure duration is 2 minutes.

References

1. H. Liang, M. Ding, S. L. Bressler. Temporal dynamics of Information Flow in the Cerebral Cortex. *Neurocomputing*, 38: 1429-1435, 2001.
2. J.A. Vastano, H.L. Swinney. Information transport spatiotemporal systems. *Physical Review Letters*, 60: 1773-1776, 1988.
3. T. Schreiber. Measuring information transfer. *Physical Review Letters* , 85: 461-464, 2000.
4. S. Sabesan. *Entropy and Transfer of Entropy-Based Measures: Application to Dynamics of Epileptic Seizures*. M.S. Thesis, Arizona State University, August 2003.
5. S. Sabesan, K. Narayanan, A. Prasad, and L.D. Iasemidis. Improved Measure of Information Flow in Coupled Non-Linear Systems. In *Proceedings of ICMS*, pages 270-274, 2003.
6. A. T. Bharucha-Reid. *Elements of the Theory of Markov Processes and Their Applications*. Dover Pubns, 1997.
7. R.Q. Quiroga, J. Arnhold, K. Lehnertz, and P. Grassberger. Kullback-Leibler and renormalized entropies: Application to electroencephalograms of epilepsy patients. *Physical Review E*, 62(6): 8380-8386, 2000.
8. J.M.Martinerie, A.M.Albano, A.I.Mees, and P.E.Rapp. Mutual information, strange attractors and the optimal estimation of dimension. *Physical Review A*, 45(10): 7058-7064, 1992.
9. P.Grassberger and I. Procaccia. Measuring the strangeness of Strange Attractors. *Physica D*, 9: 189-208, 1983.
10. K. Pawelzik and H.G. Schuster. Generalized dimensions and entropies from a measured time series. *Physical Review A*, 35(1): 481-484, 1987.
11. P. Gaspard and G. Nicolis. What can we learn from homoclinic orbits in chaotic dynamics. *Journal of Statistical Physics*, 31: 499-518, 1983.

12. L.D. Iasemidis, D.S. Shiau, J.C. Sackellares, P.M. Pardalos, and A. Prasad. Dynamical resetting of the human brain at epileptic seizures: Application of nonlinear dynamics and global optimization techniques. *IEEE Transactions on Biomedical Engineering*, 51: 493-506, 2004.

13. L.D. Iasemidis, P.M. Pardalos, D.S. Shiau, W. Chaowolitwongse, K. Narayanan, S. Kumar, P.R. Carney, and J.C. Sackellares. Prediction of human epileptic seizures based on optimization and phase changes of brain electrical activity. *Optimization Methods and Software*, 18: 81-104, 2003.

14. L.D. Iasemidis, D.S. Shiau, W. Chaovalitwongse, J.C. Sackellares, P.M. Pardalos, J.C. Principe, P.R. Carney, A. Prasad, B. Veeramani, and K. Tsakalis. Adaptive epileptic seizure prediction system. *IEEE Transactions on Biomedical Engineering*, 50: 616-627, 2003.

15. L.D. Iasemidis. Epileptic seizure prediction and control. *IEEE Transactions on Biomedical Engineering*, 50: 549-558, 2003.

16. L.D. Iasemidis and J.C. Sackellares. Chaos theory in epilepsy. *The Neuroscientist*, 2: 118-126, 1996.

17. L.D. Iasemidis, J.C. Sackellares, and R.S. Savit. Quantification of hidden time dependencies in the EEG within the framework of nonlinear dynamics. In B.H. Jansen and M.E. Brandt, editors, *Nonlinear Dynamical Analysis of the EEG*, pages 30-47. World Scientific, Singapore, 1993.

18. E. Basar and T.H. Bullock. *Brain Dynamics: Progress and Prospectives*. Springer-Verlag, Heidelberg, Germany, 1989.

19. L.D. Iasemidis, A. Barreto, R.L. Gilmore, B.M. Uthman, S.N. Roper and J.C. Sackellares. Spatio-temporal evolution of dynamical measures precedes onset of mesial temporal lobe seizures. *Epilepsia*, 35S: 133-139, 1994.

20. F. Lopes da Silva. EEG analysis: theory and practice; Computer-assisted EEG diagnosis: pattern recognition techniques. In E. Niedermeyer and F. Lopes da Silva, editors, *Electroencephalography; Basic Principles, Clinical Applications and Related Fields*, 2nd edition, pages 871-919. Wiley, Baltimore, MA, 1987.

21. K. Lehnertz and C.E. Elger. Spatio-temporal dynamics of the primary epileptogenic area in temporal lobe epilepsy characterized by neuronal complexity loss. *Electroencephalography and Clinical Neurophysiology*, 95: 108-117, 1995.

22. T. Schreiber. Determination of noise level of chaotic time series. *Physical Review E*, 48: 13-16, 1993.

23. A.M. Albano, J. Muench, C. Schwartz, A.I. Mees, and P.E. Rapp. Singular value decomposition and Grassberger-Procaccia algorithm. *Physical Review A*, 38(6): 3017-3026, 1988.

24. J.P. Eckmann and D.Ruelle. Ergodic theory of chaos and strange attractors. *Reviews of Modern Physics*, 57: 617-623, 1985.

25. P. Grassberger. Finite sample corrections to entropy and dimension estimates. *Physical Review A*, 128: 369-374, 1988.

Reconstruction of Epileptic Brain Dynamics Using Data Mining Techniques

Panos M. Pardalos[1] and Vitaliy A. Yatsenko[2]

[1] University of Florida, Gainesville, FL 32611-6595, USA
   pardalos@ufl.edu
[2] Institute of Space Research, 40 Glushkov Ave, Kyiv 02022, Ukraine
   vitaliy_yatsenko@yahoo.com

Summary. The existence of complex chaotic, unstable, noisy and nonlinear dynamics in the brain electrical and magnetic activities requires new approaches to the study of brain dynamics. One approach is the combination of certain multichannel global reconstruction concept and data mining techniques. This approach assumes that information about the physiological state comes in the form of nonlinear time series with noise. It also involves a geometric description of the brain dynamics for the purpose of understanding massive amount of experimental data. The novelty in this chapter is in the representation of the brain dynamics by hierarchical and geometrical models. Our approach plays an important role in analyzing and integrating electromagnetic data sets, as well as in discovering properties of the Lyapunov exponents. Further, we discuss the possibility of using our approach to control the Lyapunov exponents, predict the brain characteristics, and "correct" brain dynamics. We represent the Lyapunov exponents by fiber bundle and its functional space. We compare the reconstructed dynamical system with the geometrical model. We discuss an application of this approach to the development novel algorithms for prediction and seizure control through electromagnetic feed-back.

Key words: Data mining, epileptic seizures, Lyapunov exponents, geometric approach.

1 Introduction

Recent years have seen a dramatic increase in the amount of information stored in electronic format. It has been estimated that the amount of information in the world doubles every 20 months and the size and number of databases are increasing even faster. Having concentrated so much attention on the accumulation of data the problem was what to do with this valuable resource? It was recognized that information is crucial for decision making, especially in medicine. For example, prediction of epileptic seizures needs an adequate amount of information for calculation of dynamical and information characteristics.

The term 'Data Mining' (or 'Knowledge Discovery') has been proposed to describe a variety of techniques to identify sets of information or decision-making knowledge in bodies of experimental data, and extracting these in such a way that they can be put to use in areas such as prediction or estimation. Data mining, the extraction of hidden predictive information from large data sets, is a powerful new technology in medicine with great potential to help medical personnel focus on the most important diagnostic information in their data warehouses. Medical data mining tools predict future trends and behaviors, allowing medical doctors to make proactive, knowledge-driven decisions. The automated, prospective analyses offered by data mining move beyond the analyses of past events provided by retrospective tools typical of decision support systems. Data mining tools can answer medical questions that traditionally were too time consuming to resolve. They scour databases for hidden patterns, finding predictive information that experts may miss because it lies outside their expectations. This chapter provides an introduction to the basic dynamical reconstruction and prediction technologies of data mining.

Examples of profitable applications illustrate its relevance to today's medical environment as well as a basic description of how data warehouse architectures can evolve to deliver the value of data mining to end users. Data mining techniques can be implemented rapidly on existing software and hardware platforms to enhance the value of existing information resources, and can be integrated with new products and systems as they are brought on-line. When implemented on high performance client/server or parallel processing computers, data mining tools can analyze massive databases to deliver answers to questions such as, "Which clients are most likely to respond to my next promotional mailing, and why?"

Epilepsy, among the most common disorders of the nervous system, affects approximately 1% of the population [7, 8, 9]. About 25% of patients with epilepsy have seizures that are resistant to medical therapy [10]. Epileptic seizures result from a transient electrical discharge of the brain. These discharges often spread first to the ipsilateral, then the contralateral cerebral cortex, thereby disturbing normal brain function. Clinical effects may include motor, sensory, affective, cognitive, automatic and physical symptomatology. The occurrence of an epileptic seizure appears to be unpredictable and the underlying mechanisms that cause seizures to occur are poorly understood. A recent approach in epilepsy research is to investigate the underlying dynamics that account for the physiological disturbances that occur in the epileptic brain [11, 12, 13].

One approach to understanding the dynamics of epilepsy is to determine the dynamical properties of the electroencephalographic (EEG) signal generated by the epileptic brain. The traditional approach to this problem in time series analysis is to fit a linear model to the data, and determine the optimal order (*dimension*) and the *optimal parameters* of the model. A simple generalization is to fit the best nonlinear dynamical model. However, the results of this approach are usually not very illuminating. Without any prior knowledge, any model that we fit is likely to be *ad hoc*. We are more interested in questions as the following: How nonlinear is the time series? How many degrees of freedom does it have?

The Lyapunov exponents measure quantities which constitute the exponential divergence or convergence of nearly initial points in the phase space of a dynamical model of the human brain. A positive Lyapunov exponent measures the average exponential divergence of two near trajectories, whereas a negative Lyapunov exponent measures exponential convergence of two near trajectories.

Our purpose is to design and investigate algorithms for calculating the local and global Lyapunov exponents such that they would achieve the following three properties:

(1) calculate all local and global Lyapunov exponents of the human brain from observations;

(2) achieve greater accuracy of Lyapunov exponent estimates on a relatively short length of observations;

(3) achieve the accuracy of the estimates that is robust to the human brain as well as measurement noise.

In this chapter, we calculate the global and local Lyapunov exponents, both of which could be important measures for orbital instability.

2 Reconstruction of Global Lyapunov Exponents From Observed Time Series

Assume that the epileptic human brain dynamics can be described by a discrete dynamical system. The Lyapunov exponents for a dynamical system, $f : \mathbb{R}^n \to \mathbb{R}^n$, with the trajectory

$$x_{t+1} = f(x_t), \quad t = 0, 1, 2, \ldots, \tag{1}$$

are measures of the average rate of divergence or convergence of a typical trajectory [1]. For a n-dimensional system as above, there are n exponents which are customarily ranked from largest to smallest:

$$\lambda_1 \geq \lambda_2 \geq \cdots \geq \lambda_n. \tag{2}$$

Associated with each component, $i = 1, 2, \ldots, n$ there are nested subspaces $\Psi^i \subset \mathbb{R}^n$ of dimension $n + 1 - j$ and with the property that

$$\lambda_i = \lim_{t \to \infty} t^{-1} \ln \|(Df')_{x_0} \psi\| \tag{3}$$

for all $\psi \in \Psi^i \setminus \Psi^{i+1}$. It is a consequence of Oseledec's Theorem [3], that the limit in (3) exist for a broad class of functions. Additional properties

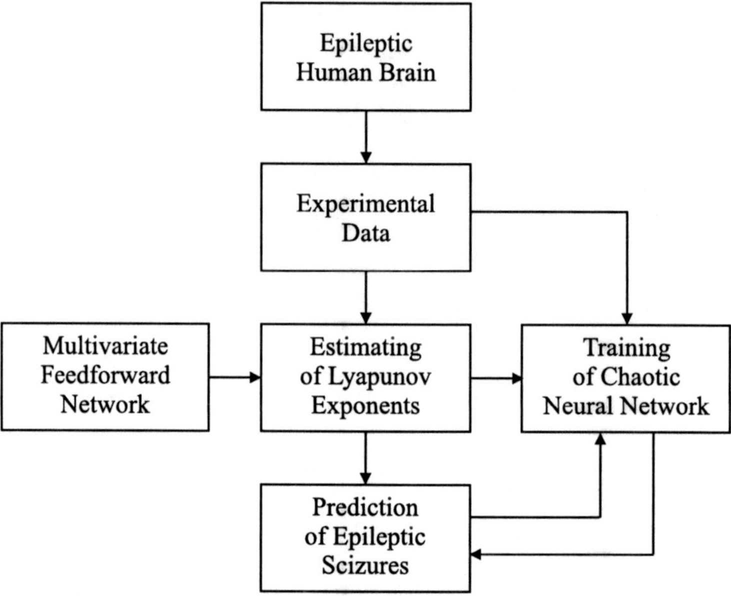

Fig. 1. Functional structure of EEG data analysis adaptive system.

of Lyapunov exponents and a formal definition are given in [1]. Notice that for $i \geq 2$ the subspaces V^i are sets of Lebegue measure zero, and so for almost all $\psi \in \mathbb{R}^n$ the limit in equation (3) equals λ_i. This is the basis for the computational algorithm which is a method for calculating the largest Lyapunov exponents. Since

$$(Df')_{x_t} = (Df)_{x_t}(Df)_{x_{t-1}} \cdots (Df)_{x_0} \tag{4}$$

all of Lyapunov exponents can be calculated by evaluating the Jacobian of the function f along a trajectory $\{x_t\}$.

In [4] the QR decomposition is proposed for extracting the eigenvalues from $(Df')_{x_0}$. The QR decomposition is one of many ways to calculate eigenvalues. One advantage of the QR decomposition is that it performs successive rescaling to keep magnitudes under control. It is also well studied and extremely fast subroutines are available for this type of computation. It is the method that we use here.

An attractor is a set of points toward which the trajectories of f converge. More precisely, Λ is an attractor if there is an open $V \subset \mathbb{R}^n$ with

$$\Lambda = \bigcap_{t \geq 0} f'(\tilde{V}),$$

where $\tilde{V}$ is the closure of V. The attractor Λ is said to be indecomposable if there is no proper subset $A \subset \Lambda$ with $f(A) = A$. An attractor can be chaotic or ordinary (i.e. non-chaotic).

Suppose that there is a function $h : \mathbb{R}^n \to \mathbb{R}$ which generates observations,

$$y_t = h(x_t) \tag{5}$$

For notation purposes let $y_t^m = (y_{t+m-1}, y_{t+m-2}, \ldots, y_t)$. Under general conditions, it is shown in [5] that if the the set $\tilde{V}$ is a compact manifold then for $m \geq 2n + 1$

$$J^m(x) = y_t^m = \left(h(f^{m-1}(x)), h(f^{m-2}(x)), \ldots, h(x)\right) \tag{6}$$

is an embedding of $\tilde{V}$ onto $J^m(\tilde{V})$. Generically, for $m \geq 2n + 1$ there exists a function $g : \mathbb{R}^m \to \mathbb{R}^m$ such that

$$y_{t+1}^m = g(y_t^m) \tag{7}$$

where

$$y_{t+1}^m = (y_{t+m}, y_{t+m-1}, \ldots, y_{t+1}). \tag{8}$$

But notice that

$$y_{t+1}^m = J^m(x_{t+1}) = J^m(f(x_t)). \tag{9}$$

Hence from (7) and (9)

$$J^m(f(x_t)) = g(J^m(x_t)). \tag{10}$$

Under the assumption that J^m is homeomorphism, f is topologically conjugate to g. This implies that certain dynamical properties of f and g are the same. From equation (7) the mapping g may be taken to be

$$g : \begin{pmatrix} y_{t+m-1} \\ y_{t+m-2} \\ \vdots \\ y_t \end{pmatrix} \longrightarrow \begin{pmatrix} V(y_{t+m-1}, y_{t+m-2}, \ldots, y_t) \\ y_{t+m-1} \\ \vdots \\ y_{t+1} \end{pmatrix} \tag{11}$$

and this reduces to estimating

$$y_{t+m} = V(y_{t+m-1}, y_{t+m-2}, \ldots, y_t). \tag{12}$$

In (11), (12) a truncated Taylor series is used to calculate the function n. In [6] the feedforward networks are used to calculate the largest Lyapunov exponent.

The derivative of g is the matrix

$$(Dg)_{y_t^m} = \begin{pmatrix} V_m & V_{m-1} & V_{m_2} & \cdots & V_2 & V_1 \\ 1 & 0 & 0 & \cdots & 0 & 0 \\ 0 & 1 & 0 & \cdots & 0 & 0 \\ \vdots & \vdots & \vdots & \vdots & \vdots & \vdots \\ 0 & 0 & 0 \cdots & & 1 & 0 \end{pmatrix}, \tag{13}$$

where

$$v_m = \frac{\partial v}{\partial y_{t+m-1}}, \quad \ldots, \quad v_1 = \frac{\partial v}{\partial y_t}. \tag{14}$$

Proposition. Assume that M is the manifold of dimension n, $f : M \to M$ and $h : M \to \mathbb{R}$ are (at least) C^2. Define $J^m : M \to \mathbb{R}^m$ by $J^m(x) = \left(h(f^{m-1}(x)), h(f^{m-2}(x)), \ldots, h(x)\right)$. Let $\mu_1(x) \geq \mu_2(x) \geq \cdots \geq \mu_m(x)$ be the eigenvalues of $(Df^m)'_x (DJ^m)_x$, and suppose that

$$\inf_{x \in M} \mu_m(x) > 0, \quad \sup_{x \in M} \mu_1(x) < \infty.$$

There is more then one definition of chaotic attractor in the literature. In practice the presence of a positive Lyapunov exponent is taken as a signal that the attractor is chaotic.

Let $\lambda_1^f \geq \lambda_2^f \geq \cdots \geq \lambda_n^f$ be the Lyapunov exponents of f and $\lambda_1^g \geq \lambda_2^g \geq \cdots \geq \lambda_m^g$ be the Lyapunov exponents of g, where $g : J^m(M) \to J^m(M)$ and $J^m(f(x)) = g(J^m(x))$ on M. Then generically $\lambda_i^f = \lambda_i^g$ for $i = 1, 2, \ldots, n$.

This is the basis of our approach: consider the function g based on the data sequence $\{J^m(x_t)\}$, and calculate the Lyapunov exponents of g. As n increases there is a value between n and $2n+1$ at which the n largest exponents remain constant and the remaining $m - n$ exponents diverge to $-\infty$ as the number of observations increases.

For estimating Lyapunov exponents we use a single layer feedforward network,

$$v_{N,m}(z; \beta, \omega, b) = \sum_{j=1}^{L} \beta_j k \left(\sum_{i=1}^{m} \omega_{ij} z_i + b_i \right), \tag{15}$$

where $z \in \mathbb{R}^m$ is the input, the parameters to be estimated are β, ω, and b; and k is a known hidden unit activation function. Here L is the number of hidden unit weights, and $\omega \in \mathbb{R}^{L \times m}$ and $b \in \mathbb{R}^L$ represent input to hidden unit weights.

For a single layer network, the least squares criterion for a data set of length T is

$$L(\beta, \omega, b) = \sum_{t=0}^{T-m-1} [y_{t+m} - v_{N,m}(y_t^m; \beta, \omega, b)]^2. \tag{16}$$

This is a straightforward multivariate minimization problem.

3 Simulation of Local and Global Lyapunov Exponents by Neural Network

In this section, we use the model of chaotic neural networks with two internal states, $\alpha_i(t)$ and $\beta_i(t)$. Then, defining the $2n$-dimensional state as

$$x(t) = (\alpha_1(t), \ldots, \alpha_n(t), \beta_1(t), \ldots, \beta_n(t)), \tag{17}$$

we can interpret the chaotic neural network as the $2n$-dimensional dynamical system

$$x(t+1) = f(x(t)), \tag{18}$$

where $f : \mathbb{R}^{2n} \to \mathbb{R}^{2n}$ is a nonlinear mapping defined by equations

$$\alpha_i(t+1) = k_m \alpha_i(t) + \sum_{j=1}^{n} \omega_{ij} u_j(t), \tag{19}$$

$$\beta_i(t+1) = k_m \beta_i(t) + \gamma u_i(t) + a, \tag{20}$$

$$\alpha_i(t+1) = f[\alpha_i(t+1) + \beta_i(t+1)], \tag{21}$$

where $u_i(t+1)$ is the output of the chaotic neuron at the discrete time $t+1$; ω_{ij} is the connection weight from the jth chaotic neuron to the ith chaotic neuron; f is the continuous output function.

Considering an infinitesimal derivation $\delta x(t)$ from $x(t)$ in equation (18), we obtain

$$x(t+1) + \delta x(t+1) = f(x(t) + \delta x(t)). \tag{22}$$

By expanding equation (22) to the Taylor series and discarding the higher order terms, the following linear map is obtained:

$$\delta y(t+1) = J_t \delta y(t), \tag{23}$$

where J_t is the Jacobian matrix of $f(t)$.

Let us define the following positive matrix

$$M(x(t), L) = [\{J_t^L\}^T J_t^L]^{\frac{1}{2L}}, \tag{24}$$

where J_t^L is an an L times multiplication of the Jacobian matrix J_t from t to $t+L$, T indicates the transpose of a matrix, and L corresponds to a localization parameter. Namely if L is small, it corresponds to observing the very local dynamics. Oseledec proved that the limit of equation (24) as $L \to \infty$ exists, and the global Lyapunov exponents are defined as the logarithm of eigenvalues of $M(y(0), L)$,

$$\lambda_i = \lim_{L \to \infty} \log \sigma_i(M(y(0), L)). \tag{25}$$

However, due to the ill-defined problem of the above matrices, it is not numerically easy to calculate eigenvalues directly from equation (24). The following orthogonalization scheme is utilized for numerically calculating the Lyapunov spectrum. First, the matrix $J_1 Q_0$ is decomposed as

$$J_1 Q_0 = Q_1 R_1, \tag{26}$$

where Q_1 is an orthogonal matrix, R_1 is an upper triangular matrix with nonnegative diagonal elements and Q_0 is the identity matrix. Next, $J_2 Q_1$ is decomposed as

$$J_2 Q_1 = Q_2 R_2. \tag{27}$$

Generally,

$$J_{j+1} Q_j = Q_{j+1} R_{j+1}. \tag{28}$$

Then, the matrix $[\{J_t^L\}^T J_t^L]$ is decomposed as

$$J_1^T \cdots J_L^T J_L \cdots J_1 = Q_{2L} R_{2L} R_{2L-1} \cdots R_1. \tag{29}$$

Since Q_{2L} is an orthogonal matrix, $[\{J_t^L\}^T J_t^L]$ and $R_{2L} R_{2L-1} \cdots R_1$ have the same growth rates. The global Lyapunov exponents λ_i are estimated by the following equations:

$$\bar{\lambda}_i = \lim_{L \to \infty} \frac{1}{2L} \sum_{j=1}^{2L} \log |r_j^{ii}|, \tag{30}$$

where r_j^{ii} is the ith diagonal element of a matrix R_j, and L is the size of the locality.

The local Lyapunov exponents at time t with L are defined in a similar way:

$$\lambda_i(t, L) = \log \sigma_i(M(y(t), L)), \tag{31}$$

with the same decomposition into successive QR factors as equation (28), it is numerically evaluated by

$$\lambda_i(t, L) = \frac{1}{2L} \sum_{j=1}^{2L} \log |r_j^{ii}|. \tag{32}$$

If the calculated largest Lyapunov exponent λ_1 is at least positive, the dynamic system has orbital instability, which is one of fundamental characteristics of deterministic chaos.

4 Geometrization of Brain Dynamics

During the past decade, there has been growing evidence of the independence of the two properties of instability and predictability of the human brain dynamics. The generic situation of the brain dynamics is instability of the trajectories in the Lyapunov sense. Nowdays such instability is called intrinsic stochasticity, or chaoticity, of the brain dynamics and is a consequence of nonlinearity of the equation of motion.

In order to characterize the dynamical instability, we first examined some properties that chardcterize the EEG signal, including the spectrum of Lyapunov exponents, the energy, geodesics and fiber bundles. We analyzed the parameter spaces as well as related quantities T-index of STLmax, Jacobi–Levi-Civita equation, and the modelling of EEG time series from the experimental point of view.

The approach involves a geometric description of Lyapunov exponents for the purpose of correcting the nonlinear process that provides adaptive dynamic control. The novelty in this section is in the representation of dynamical instability by a Riemannian theory in a way that permits practical applications.

We separate the Lyapunov exponent into a tangent space (fiber bundle) and its functional space. Control involves signal processing, calculation of an information characteristic, measurement of Lyapunov exponents, and feedback to the system. With more information, we can reduce uncertainty by a certain degree.

The actual interest of the Riemannian formulation of dynamics stems from the possibility of studying the instability of brain dynamics through the instability of geodesics of a suitable manifold, a circumstance that has several advantages.

First of all, a powerful mathematical tool exists to investigate the stability or instability of a geodesic flow: the Jacobi–Levi-Civita (JLC) equation for geodesic spread.

The JLC equation describes covariantly how nearby geodesics locally scatter, and it is a familiar concept in both Riemannian geometry and theoretical physics. Moreover, the JLC equation relates the stability or instability of a geodesic flow with curvature properties of the ambient manifold, thus opening a wide and largely unexplored field of investigation, as far as physical systems are concerned, of the connections among geometry, topology, and geodesic instability, hence chaos.

Geometrization of the brain dynamics includes the following stages: 1) Reconstruction of equations of the epileptic brain from experimental data; 2) Realization in local coordinates of a one-parameter group of diffeomorphisms of a manifold M; 3) Estimation of largest Lyapunov exponent; 4) Geometrization of dynamics; 5) Geometric description of dynamical instability; 6) Jacobi–Levi-Civita equation for geodesic spread; 7) Analytical description of the largest Lyapunov exponent.

By transforming the Jacobi–Levi-Civita equation from geodesic spread into a scalar equation for ψ variable, the original complexity of the JLC equation has been considerably reduced. From a tensor equation we have worked out an effective scalar equation formally representing a stochastic oscillator [16].

Our Lyapunov exponent is defined as

$$\lambda = \lim_{t \longrightarrow \infty} \frac{1}{2t} \ln \frac{\psi^2(t) + \dot{\psi}^2(0)}{\psi^2(0) + \dot{\psi}^2(0)}, \tag{33}$$

where $\psi(t)$ is the solution of the equation

$$\frac{d^2\psi}{ds^2} + \Omega(t)\psi = 0, \tag{34}$$

$\Omega(t)$ is a Gaussian stochastic process;

$$\Omega(t) = \langle k_R \rangle_\mu + \mu \frac{1}{\sqrt{N}} \langle \delta^2 K_R \rangle_\mu^{\frac{1}{2}} \eta(t), \tag{35}$$

if the Eisenhart metric is used.

The instability growth rate of ψ measures the instability growth rate of $\|J\|^2$ (geodetic separation field) and thus provides the dynamical instability exponent in our Riemannian framework.

Equation (4) is a scalar equation that, independently of the knowledge of dynamics, provides a measure of the average degree of instability of the dynamics itself through the behavior of $\psi(s)$. The peculiar properties of a given Hamiltonian system enter (4) through the global geometric properties $\langle k_R \rangle_\mu$ and $\langle \delta^2 K_R \rangle_\mu$ of the ambient Riemannian manifold whose geodesics are natural motions and are sufficient to determine the average degree of chaoticity of the dynamics.

5 Conclusion

An approach to global reconstruction of the epileptic brain dynamics using massive data has been considered. It is based on new algorithms for estimation of local and global Lyapunov exponents and a geometric technique. We have shown that Lyapunov exponents can be calculated by estimation of diffeomorphism of the reconstructed system and a chaotic neural network. The performance of the algorithm is very satisfactory in the presence of noise as well as with limited number of observations. We hypothesize three types of changes that the epileptic brain attractors can undergo as a system parameter is varied. The first type leads to the sudden destruction of a chaotic attractor. The second type leads to the sudden widening of a chaotic attractor. In the third type of change two (or more) chaotic attractors merge to form a single chaotic attractor and the merged attractor can-be larger in phase-space extent than the union of the attractors before the change. All three of these types of changes are termed crises and are accompanied by a characteristic temporal behavior of orbits after the crisis.

A geometric approach to the study of the physiological disturbances that occur in human epilepsy was proposed. Under reasonable hypotheses, which obviously restrict the validity of the geometric approach, our results provide the possibility of numerical computation of the state changes in the EEG signals using the largest Lyapunov exponent and the combination of the curvature of the underlying manifold and the geodesics. These geodesics flows may have very specific hidden symmetries, mathematically defined through Killing tensor field.

References

1. A. Wolf, B. Swift, Y. Swinney, and J. Vastano. Determining Lyapunov exponents from a time series. *Physica D*, 16: 285–317, 1985.
2. R. Brocon, P. Bryant, and H. Abarbanel. Computing the Lyapunov exponents of a dynamical system from observed time series, *Physical Review A*, 43: 2787–2806, 1991.
3. V. Oseledec. A multiplicative ergodic theorem Lyapunov characteristic number for a dynamical system from an observed time series, *Transactions of the Moscow Mathematical Society*, 19: 356-362, 1968.
4. J. Eckmann and D. Ruelle. Ergotic theory of strange attractors. *Reviews of Modern Physics*, 57: 617–656, 1985.
5. F. Takens. Detecting strange attractors in turbulence. In D.A. Rand and L.S. Young, editors, *Dynamical Systems and Turbulence*, pages 366–381. Lecture Notes in Mathematics, Vol. 898, Springer, 1981.
6. G. Ramazan and D. Davis. An aldorithm for the n Lyapunov exponents of n-dimensional unknown dynamical system. *Physica D*, 59: 142-157, 1992.
7. M. Brewis, D. Poskanzer, C. Rolland, and H. Miller. *Acta Neurologica Scandinavica*, 42(24): 9–89, 1996.
8. O. Cockerell, I. Eckle, D. Goodridge, J. Sander, and S. Shorvon. Epilepsy in a population of 6000 re-examined: secular trends in first attendance rates, prevalence, and prognosis. *Journal of Neurology, Neurosurgery & Psychiatry*, 58(5): 570-576, 1995.
9. P. Jallon. Epilepsy in developing countries. *Epilepsia*, 38(10): 1143-1151, 1997.
10. C. Elger and K. Lehnertz. Seizure prediction by nonlinear time series analysis of brain electrical activity. *European Journal of Neuroscience*, 10(2): 786-789, 1998.
11. R. Andrzejak, G. Widman, K. Lehnertz, C. Rieke, P. David, and C. Elger. The epileptic process as nonlinear deterministic dynamics in a stochastic environment - An evaluation on mesial temporal lobe epilepsy. *Epilepsy Research*, 44: 129-140, 2001.
12. K. Lehnertz, R. Andrzejak, J. Arnold, G. Widman, W. Burr, P. David, and C. Elger. Possible clinical and research applications of nonlinear EEG analysis in humans. In K. Lehnertz, J. Arnold, P. Grassberger, and C.E. Elger, editors, *Chaos in Brain*, pages 134-155. *World Scientific*, London, 2000.
13. L. Iasemidis, D.-S. Shiau, P.M. Pardalos, and J. Sackellares. Transition to epileptic seizures: Optimization. In D. Du, P.M. Pardalos, and J. Wang, editors, *Discrete Mathematical Problems with Medical Applications*, pages 55-74. DIMACS series, Vol. 55, American Mathematical Society, 2000.
14. H. Abarbanel. *Analysis of Observed Chaotic Data*. Springer-Verlag, New York, 1995.
15. P.M. Pardalos, J.C. Sackellares, and V. Yatsenko. Classical andquantum controlled lattices: self–organization, optimization and biomedical applications. In P. Pardalos and J. Principe, editors, *Biocomputing*, pages 199-224. Kluwer Academic Publishers, 2002.
16. L. Casetti and M. Pettini. Analytic computation of the strong stochasticity threshold in Hamiltonian dynamics using Riemannian geometry. *Physical Review E*, 48: 4320-4332, 1993.

Automated Seizure Prediction Algorithm and its Statistical Assessment: A Report from Ten Patients*

D.-S. Shiau[1], L.D. Iasemidis[2], M.C.K. Yang[3], P.M. Pardalos[4], P.R. Carney[5], L.K. Dance[1], W. Chaovalitwongse[6], and J.C. Sackellares[7]

[1] Department of Neuroscience, University of Florida, Malcolm Randall Department of Veteran's Affairs Medical Center, Gainesville, FL
`{shiau,lkdance}@mbi.ufl.edu`
[2] Department of Biomedical Engineering, Arizona State University, Tempe, AZ
`Leon.Iasemidis@asu.edu`
[3] Department of Statistics, University of Florida, Gainesville, FL
`yang@stat.ufl.edu`
[4] Department of Industrial and Systems Engineering, Computer and Information Science and Engineering, Biomedical Engineering, University of Florida, Gainesville, FL
`pardalos@ufl.edu`
[5] Department of Pediatrics, Neurology, Neuroscience, Biomedical Engineering, University of Florida, Gainesville, FL
`carnepr@peds.ufl.edu`
[6] Department of Industrial and Systems Engineering, Rutgers, The State University of New Jersey
`wchaoval@rci.rutgers.edu`
[7] Department of Neurology, Biomedical Engineering, Neuroscience, Pediatrics and Psychiatry, University of Florida, Malcolm Randall Department of Veteran's Affairs Medical Center, Gainesville, FL
`sackellares@mbi.ufl.edu`

Summary. The ability to predict epileptic seizures well prior to their clinical onset provides promise for new diagnostic applications and novel approaches to seizure control. Several groups of investigators have reported that it may be possible to predict seizures based on the quantitative analysis of EEG signal characteristics. The objective of this chapter is first to report an automated seizure warning algorithm, and second to compare its performance with other, theoretically sound, statistical algorithms. The proposed automated seizure prediction algorithm (ASPA) consists

* This research is supported by the National Institute of Biomedical Imaging and Bioengineering (NIBIB) via a Bioengineering Research Partnership grant for Brain Dynamics (8R01EB002089-03). The facilities used for this research were the Brain Dynamics Laboratories at the University of Florida, Gainesville, FL and at the Arizona State University, Tempe, AZ, USA.

of an optimization method for the selection of critical cortical sites using measures from nonlinear dynamics, and a novel method for the detection of preictal transitions using adaptive transition thresholds according to the current state of dynamical interactions among brain sites. Continuous long-term (mean 210 hours per patient) intracranial EEG recordings obtained from ten patients with intractable epilepsy (total of 130 recorded seizures) were analyzed to test the proposed algorithm. For each patient, the prediction ROC (receiver operating characteristic) curve, generated from ASPA, was compared with the ones from periodic and random prediction schemes. The results showed that the performance of ASPA is significantly superior to each naïve prediction method used (p-value < 0.05). This suggests that the proposed nonlinear dynamical analysis of EEG contains relevant information to prospectively predict an impending seizure, and thus has potential to be useful in clinical applications.

Key words: Epilepsy, Dynamical entrainment, Automated seizure prediction, Naïve prediction schemes, ROC curves.

1 Introduction

Epileptic seizures result from a temporary electrical disturbance of the brain and affect at least 50 million people worldwide, including 1.4 million Americans. For the patients with epilepsy, the occurrences of seizures interfere with their normal life and are sometimes fatal. One of the most disturbing aspects of epilepsy is that the occurrences of seizures appear to be random and unpredictable. Therefore, seizure prediction is listed as one of the most important future directions in epilepsy research [13]. The ability to predict epileptic seizures well prior to their occurrences provides promise for new diagnostic applications and novel approaches to seizure control. Therefore, any methods that can reliably predict or warn of a seizure occurrence would be very clinically significant. An immediate application of such an automatic seizure prediction computer algorithm, incorporated into existing long-term EEG recording equipment, could be used for diagnostic purposes and to enhance patient safety and treatment by alerting the nursing and technical staff of impending seizures.

Several groups of investigators have reported that it may be possible to predict seizures based on analysis of the EEG signal characteristics. For example, Iasemidis and coworkers use STL_{max} to detect preictal state prior to seizure onset [5, 11, 12, 21]; Lehnertz and Elger [3, 14] use the effective correlation dimension to detect a change in dynamics before a seizure; Martinerie and coworkers [16, 18, 19] apply dynamical similarity analysis to show significant difference between preictal and interictal states; Litt and coworkers [15] reported that number of energy bursts starts increasing several hours prior to seizure onset. Most recently, Mormann, Andrzejak and coworkers [17] showed that the period preceding a seizure can be characterized by a decrease in synchronization between different EEG recording sites, with a mean prediction time of 86~102 minutes.

Our previous work indicates that temporal lobe seizures are preceded by a dynamical entrainment of critical cortical sites which can be detected approximately 70 minutes before a seizure and that this entrainment is reset by the seizures [11, 9]. This has led us to use a new model to explain the spontaneous occurrence of epileptic seizures – dynamical ictogenesis theory. Based on this model, we have reported several automated seizure warning systems that detect the preictal transition from selected "critical" electrode sites and provide seizure warnings. Critical sites selection is based on the dynamical entrainment-disentrainment behavior before and after a seizure or an entrainment transition. These algorithms have been previously described in detail [20, 4, 8]. In this chapter, we propose an improved version of our automated seizure warning system using an adaptive transition detection threshold based on the degree of the dynamical entrainment. The details of this algorithm will be described in the Material and Methods section.

Before applying any seizure prediction system for diagnostic and treatment applications, it is necessary to assess its performance and reliability. The most commonly used criterion to assess seizure prediction algorithms is to estimate the sensitivity (probability of a seizure being correctly predicted) and false prediction rate (false predictions per unit time) separately. However, in different studies, the characteristics of the testing EEG recordings (e.g., the duration of the EEG recordings, the seizure frequencies and the number of patients) and the criterion to claim a correct prediction are different. Without objective and standard evaluation methods, it is very difficult to check the progress of a prediction method. Hence, it is important to first compare its performance with the methods that do not use any information from EEG or other brain recordings, as a control method to evaluate a seizure prediction algorithm.

Winterhalder [24] and Aschenbrenner-Scheibe [2] reported studies to assess a seizure prediction method by qualitatively comparing its ROC curve with the ones from non-specified alert systems (periodic and random prediction). In this chapter, we have extended this evaluation approach by (1) testing all compared prediction methods on the same EEG data and (2) quantifying prediction statistics, ROC areas, from ROC curves. The first extension is to make sure that all the compared methods are tested under the same data characteristics (e.g., recording duration, number of seizures, seizure intervals). In this study, all compared prediction methods were tested on long-term continuous (mean 210 hours) multichannel recordings with multiple seizures (average number of seizures = 13) from ten test patients. The second extension is to allow us to compare the performance of prediction methods quantitatively. ROC curves from each of the three prediction methods (periodic, random, and ASPA) are generated for each patient and the ROC areas are estimated as a prediction performance statistic for the evaluation and comparisons. A standard nonparametric statistical test is employed to show the statistical significance of the comparisons between the proposed prediction method ASPA and the two non-EEG based prediction schemes.

The rest of this chapter is organized as follows. Materials and methods are described in Section 2. The results of the study are presented in Section 3. Discussions of the methods and results are given in Section 4, and conclusion remarks of this chapter will be in final Section 5.

2 Materials and Methods

2.1 Recording Procedure and EEG Data Characteristics

Electrographic recordings from bilaterally, surgically implanted macro-electrodes in the hippocampus, temporal and frontal lobe cortexes of epileptic patients with medically intractable temporal lobe, complex, focal epilepsy were analyzed. The recordings were obtained as part of a pre-surgical clinical evaluation. Figure 1 shows our typical electrode montage. The EEG signals were recorded using amplifiers with an input range of ± 0.6 mV, and a frequency range of $0.5{\sim}70$ Hz. Prior to storage, the signals were sampled at 200 Hz using an analog to digital (A/D) converter with 10-bit quantization. The multi-electrode EEG signals (28 to 32 common reference channels) were obtained from long-term (3.18 to 13.45 days) continuous recordings in 10 patients. A total of 130 seizures over 87.53 days were recorded with a mean inter-seizure interval approximately 13.4 hours (see Table 1 for details). In this study, all the EEG recordings were viewed by two independent board-certified electroencephalographers to determine the number and type of recorded seizures, seizure onset and end times, and seizure onset zones.

2.2 Analysis of Nonlinear Dynamics – Short-Term Maximum Lyapunov Exponent (STL_{max})

Short-term maximum Lyapunov exponent was utilized to extract the nonlinear dynamical characteristics (chaoticity) of the EEG signal over time for each recording channel [6, 7, 10]. The rationale is based on the hypothesis that the epileptic brain progresses into and out of the order-disorder states in terms of the theory of phase transitions of nonlinear dynamical systems. The largest Lyapunov exponent (L_{max} or L_1) is defined as the average of local Lyapunov exponents L_{ij} in the state space, that is:

$$L_{\max} = \frac{1}{N_\alpha} \sum_\alpha L_{ij},$$

where N is the total number of the local Lyapunov exponents that are estimated from the evolution of adjacent points (vectors), $Y_i = Y(t_i)$ and $Y_j = Y(t_j)$, in the state space, and

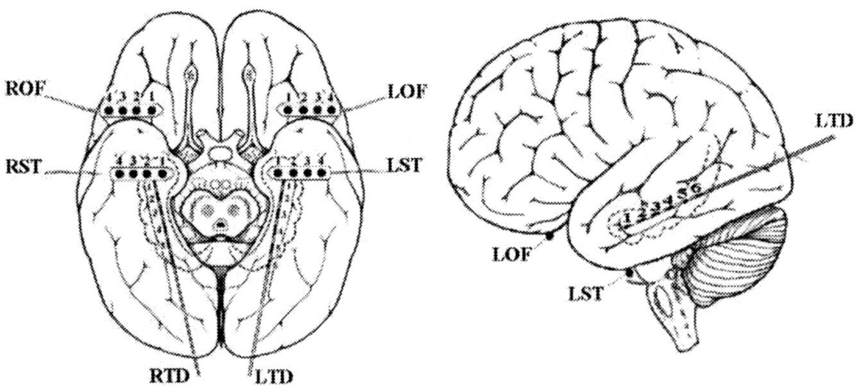

Fig. 1. Diagram of the depth and subdural electrode placement. Electrode strips are placed over the left orbitofrontal (LOF), right orbitofrontal (ROF), left subtemporal (LST), and right subtemporal cortex. Depth electrodes are placed in the left temporal depth (LTD) and right temporal depth (RTD) to record hippocampal EEG activity.

Table 1. Summary of analyzed EEG data and patients

Patient			Duration of EEG recordings (days)	# of seizures	Inter-seizure Interval (hours)	
Gender		Age			Mean	Stand. Dev.
# 1	Female	45	3.63	7	11.58	17.85
# 2	Male	60	11.98	7	20.32	29.43
# 3	Female	41	9.06	18	4.86	3.45
# 4	Male	19	13.45	15	19.55	17.57
# 5	Male	33	12.24	16	17.34	20.88
# 6	Male	38	3.18	8	8.68	6.90
# 7	Male	44	6.24	17	9.09	7.05
# 8	Male	29	6.07	14	10.78	18.87
# 9	Female	37	11.80	17	14.90	16.09
# 10	Male	37	9.88	11	21.44	34.72
Total			**87.53 days (≈ 2100 hrs)**	**130**	**13.39**	**18.29**

$$L_{ij} = \frac{1}{\Delta t} \log_2 \frac{|X(t_i + \Delta t) - X(t_j + \Delta t)|}{|X(t_i) - X(t_j)|}$$

where Δt is the evolution time allowed for the vector difference $\delta_0(x_{ij}) = |Y(t_i) - Y(t_j)|$ to evolve to the new difference $\delta_\kappa(x_k) = |Y(t_i + \Delta t) - Y(t_j + \Delta t)|$, where $\Delta t = k \cdot dt$ and dt is the sampling period of the data $u(t)$. If Δt is given in sec, L_{max} is in bits/sec.

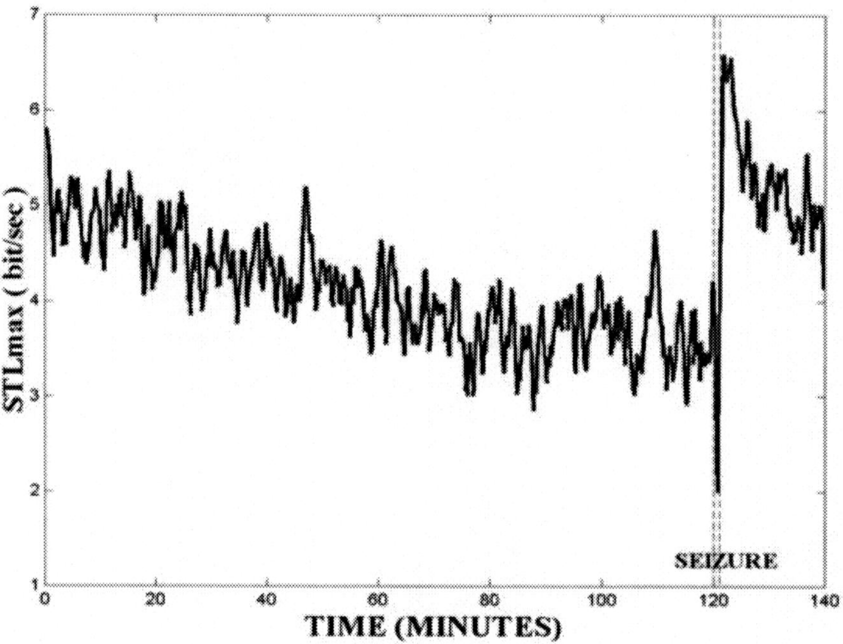

Fig. 2. STLmax profile over time (140 minutes), including a 2.5 minute seizure.

The first step in the STL_{max} analysis is to divide the EEG time series into non-overlapping segments of 10.24 seconds duration (2048 points). Brief segments were used in an attempt to ensure that the signal within each segment was approximately dynamically stationary. By the method described in [6], the STL_{max} values were calculated continuously over time for the entire EEG recordings in each patient. Figure 2 shows a STL_{max} profile in a duration of two hours before and 20 minutes after a seizure. The values over the entire period are positive. This observation has been a consistent finding in all recordings in all patients studied to date. Moreover, the STL_{max} values during the seizure are lower than before and after the onset of the ictal discharge. This indicates that methods can be developed, using sequential calculations of STL_{max}, to detect ictal discharges from the EEG signals.

2.3 Quantification of Dynamical Entrainment

Dynamical entrainment is defined as the convergence of STL_{max} values among the EEG channels within a 10-minute window (approximate 60 STL_{max} values). This convergence is quantified by the average of pair-T statistics over all pairs among the group of channels. We defined this value as T-index. For example, the T-index value for a group of 5 channels is the average of 10 pair-T statistics. The calculation of a pair-T statistic is described as follows:

For channels i and j, if their STL_{max} values in a window W_t of 60 STL_{max} points are

$$L_i^t = \{STL\max_i^t, STL\max_i^{t+1}, \ldots, STL\max_i^{t+59}\}$$
$$L_j^t = \{STL\max_j^t, STL\max_j^{t+1}, \ldots, STL\max_j^{t+59}\},$$

and

$$D_{ij}^t = L_i^t - L_j^t = \{d_{ij}^t, d_{ij}^{t+1}, \ldots d_{ij}^{t+59}\} = \{STL\max_i^t - STL\max_j^t,$$
$$STL\max_i^{t+1} - STL\max_j^{t+1}, \ldots, STL\max_i^{t+59} - STL\max_j^{t+59}\},$$

then, the pair-T statistic at time window W_t between channels i and j is calculated by

$$T_{ij}^t = \frac{\left|\overline{D}_{ij}^t\right|}{\hat{\sigma}_d / \sqrt{60}},$$

where $\overline{D}_{ij}^t$ and $\hat{\sigma}_d$ are the average value and the sample standard deviation of $D_{ij}^t = \{d_{ij}^t, d_{ij}^{t+1}, \ldots d_{ij}^{t+59}\}$.

2.4 Automated STL_{max}-based Seizure Warning algorithm

Based on the STL_{max} and T-index profiles over time, we proposed a new automated seizure warning algorithm that involves the following steps:

(1) Observing the first recording seizure

The algorithm utilizes the spatiotemporal characteristics before and after the first recorded seizure for the selection of the critical groups of channels. In this off-line study, the first seizure time for each patient was given in the algorithm to initiate the seizure prediction procedure. However, in the on-line real-time testing, the system can be activated manually by an EEG technician, a nurse, or a programming engineer when observing the first seizure; or it can be activated automatically by incorporating a reliable seizure detection subroutine with the prediction system.

(2) Selection of anticipated critical groups of channels

One of the most important tasks to accomplish an automated seizure warning algorithm based on the STL_{max} profiles is to identify beforehand the group of electrodes that will participate in the preictal transition of an impending seizure. We defined these groups of electrodes as "critical groups" and they should be detected as early as possible before the impending seizure. We have proposed in the past that the anticipated critical groups of channels can be identified based on the dynamical entrainment/disentrainment (T-index values [20, 4]) in the time windows before and after the first recorded seizure. Here, the selection process chooses groups of electrode channels that exhibit the most disentrainment after the first seizure relative to the entrainment preictally. In other words, ASPA selects the groups of channels which maximize the difference of average T-indices 10 minutes before and after the first recorded seizure, conditional on average T-index is larger after the seizure. This task can be easily accomplished by creating two T-index matrices (one before and one after the first recorded seizure). Therefore, after the first recorded seizure, the algorithm automatically identifies the most critical groups of channels for the prediction of the following seizures. In addition, with more input EEG channels or more complicated selection constraints, a constrained multi-quadratic 0-1 global optimization technique can be employed to reduce the computational time of this selection step.

(3) Monitoring the average T-index profiles among the selected channels

After the groups of critical electrode sites are selected, the average T-index values of these groups are monitored forward in time (i.e., moving W_t 10.24 seconds at a time). Thus, T-index curves over time are generated. It is expected that T-index curves will cross the entrainment threshold line before a seizure. The idea of such a transition is that, the critical electrode sites begin initially disentrained (T-index value larger than an upper threshold, UT), then will gradually converge (T-index value less than a lower threshold, LT) before a seizure.

(4) Detection of entrainment transitions and issue of warnings

The objective of an automated seizure warning algorithm is to prospectively detect a preictal transition in order to warn for an impending seizure. In the proposed algorithm, a warning is declared when an entrainment transition is detected by the T-index curves. An important question is how to determine the entrainment thresholds UT and LT. We herein propose an adaptive scheme to adjust these thresholds according to the dynamical states of the patient. Thus, UT is determined as follows: if the current T-index value is greater than max_{20}, the maximum T-index value in the past 20 minute interval, UT is equal to max_{20}, otherwise, the algorithm continues searching to identify UT. Once UT is identified, LT is equal to UT-D, where D is a preset distance in T-index units. After determining UT and LT, an entrainment transition is detected if an average T-index curve is initially above UT and then gradually drops below LT. Once an entrainment transition is detected, the thresholds are reset and the algorithm will search for a new UT.

Figure 3 shows the workings of the proposed prediction algorithm for two seizures occurring six hours apart from each other. In this example, two groups of three critical channels each were considered and their average T-index values were monitored over time. The algorithm first determined a UT (see star in the Figure 3, $UT \approx 10$) and then issue a warning when the T-index dropped below LT $(=UT\text{-}D=6)$. After this warning, UT is reset, that is algorithm is to determine a new UT. Warnings that occurred within a pre-set time interval of each other were grouped as a single warning. Between the two seizures in Figure 3, the algorithm issued three warnings at approximately 120, 235 and 335 minutes, respectively, where their corresponding UT's were marked by asterisks. After the run, if the prediction horizon is set to one hour, the first two warnings (dashed arrows in the Figure) are considered as false warnings (false positives), and the third warnings (solid arrow) is considered as a true warning.

2.5 Naïve prediction schemes

A natural question that was recently arisen is whether an EEG-based analysis is really necessary for seizure prediction and how prediction results from such an analysis may differ from a pure naïve (non-EEG) statistical analysis [1]. Here, we apply two common non-EEG based prediction schemes for such a comparison: periodic and random predictions. The former is to predict a seizure with a fixed time interval and the latter is to predict according to an exponential distribution with a fixed mean. In the periodic prediction scheme, the algorithm issue a seizure warning at a given time interval T after the first seizure. For each subsequent warning, the process is repeated. The random prediction scheme issues a seizure warning at time interval determined by random number distributed as an exponential distribution with mean λ, $(\exp(\lambda))$. The algorithm first issues a warning at an $\exp(\lambda)$ distributed random time interval after the first seizure. After the first warning, another random time interval is chosen from the same distribution for issuing the next warning. This procedure is repeated after each warning.

2.6 Statistical Evaluation and Comparisons of prediction schemes

To evaluate the prediction accuracy, the prediction horizon, a parameter often referred to as the alert interval [23], is necessary because it is practically impossible to exactly predict the time when an event occurs. A prediction is usually considered true if the event occurs within the prediction horizon. If no event occurs within the window of prediction horizon, the prediction is called a false prediction. The merit of a prediction scheme is then evaluated by its

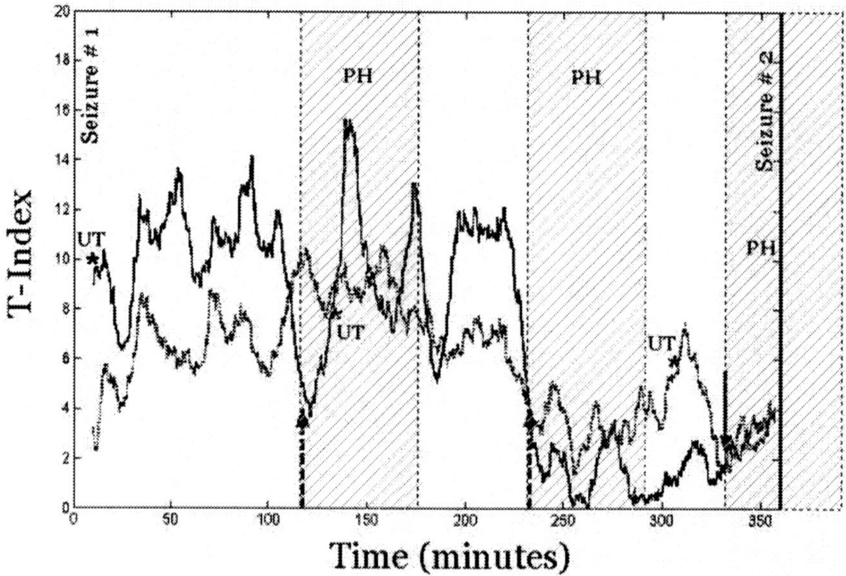

Fig. 3. STL_{max}-based prediction algorithm. Two seizures occurred at t=0 and t=360 minutes. There are only two (from two groups of critical channels) T-index curves shown in this example for better visualization. UT's are automatically determined (marked as "*"). The algorithm issues three warnings at approximately 120, 235 and 335 minutes after the first seizure in the plot. With the prediction horizon (PH) set to one hour, the first two warnings (dashed arrows) are considered false and the third one (solid arrow) is a true warning.

probability of correctly predicting the next seizure (sensitivity) and its false prediction rate (FPR) (specificity). The unit of FPR used here is per hour and thus FPR is estimated as total number of false predictions divided by total number of hours of EEG analyzed. An ideal prediction scheme should have its sensitivity = 1, and FPR = 0.

One can compare any two prediction schemes by their sensitivities at a given FPR, or conversely, compare their FPRs at a given sensitivity. However, in practice it is not always possible to fix the sensitivity or FPR in a sample with a small number of events. Moreover, there is no universal agreement on what is an acceptable FPR or sensitivity. One can always increase the sensitivity at the expense of a higher FPR. A similar situation happens in comparing methods in disease diagnosis where the trade off is between sensitivity, defined as probability of a sick patient being correctly diagnosed, and specificity, defined as the probability of a healthy patient being correctly diagnosed. A common practice in comparing diagnostic methods is to let the sensitivity and the specificity vary together and use their relation, called the receiver operation characteristic (ROC) curve, for comparison purposes.

An ROC curve is estimated by changing a parameter in a given prediction scheme. For example, in the periodic prediction scheme, the period can be changed from 0 to infinity. When the period is 0, the sensitivity is 1. When the period is infinite, the FPR is 0. Thus, the estimated ROC curve is always a closed curve between 0 and 1 and its area can be estimated.

When estimated from real data, the ROC curve may not be smooth and the superiority of one prediction scheme over the other is difficult to establish. Recent literature for ROC comparisons can be found in [25, 22]. Usually, ROC curves are globally parameterized by one value, called the area above (or under) the curve. For seizure prediction ROC curves, since the horizontal axis FPR is not bounded, the area above the curve is the most appropriate measure, that is

$$A = \int_0^\infty [1 - f(x)]dx,$$

where the ROC is expressed as $y = f(x)$ with x and y being the FPR and sensitivity, respectively. Apparently, smaller area A indicates a better prediction performance.

For each patient, three ROC areas were calculated from the three test prediction algorithms. A two-way non-parametric ANOVA test (Friedman's test) was used for overall "algorithm" effects on ROC areas. Wilcoxon signed-rank test was then employed to determine the statistical significance of differences between each pair of algorithms after an overall significance was observed.

3 Results

Seizure prediction results from ten patients (see Table 1) were used to compare the performances of the proposed STL_{max}-based seizure prediction algorithm with the ones from the periodic and random schemes. The ROC curves from each prediction scheme are shown in Figure 4 and their areas A are given in Table 2. The parameter in ROC curves for periodic prediction is the time length used for issue of warnings; for random prediction is the mean value of the exponential-distributed random intervals between two warnings, and for the STL_{max}-based prediction algorithm is the distance D between the two thresholds UT and LT. It is worthwhile to note that, since the random prediction scheme is a random process, each point in ROC curve (i.e., for each λ) is estimated as the mean sensitivity and FPR from 100 Monte Carlo simulations. From Figure 4 and Table 2 it is obvious that the proposed prediction method is better than the other two methods. It exhibits the smallest A across all patients. In particular, the mean ROC area for the STL_{max}-based method is 0.093 over all test patients, whereas the mean areas are 0.152 and 0.155 for periodic and random prediction schemes, respectively.

Friedman's test revealed that there is significant "algorithm" effect (p = 0.0005) on the observed prediction ROC areas. The pairwise comparisons by Wilcoxon sign-rank test showed that the ROC area for the ASPA was significantly less than the ones from the other two prediction schemes (p = 0.002 for both comparisons). As expected, the difference between the two naïve prediction schemes are not significant (p = 0.56). Thus, we can conclude that the information extracted from analyses of EEG, in particular by the proposed prediction method, is useful for the prediction of an impending seizure.

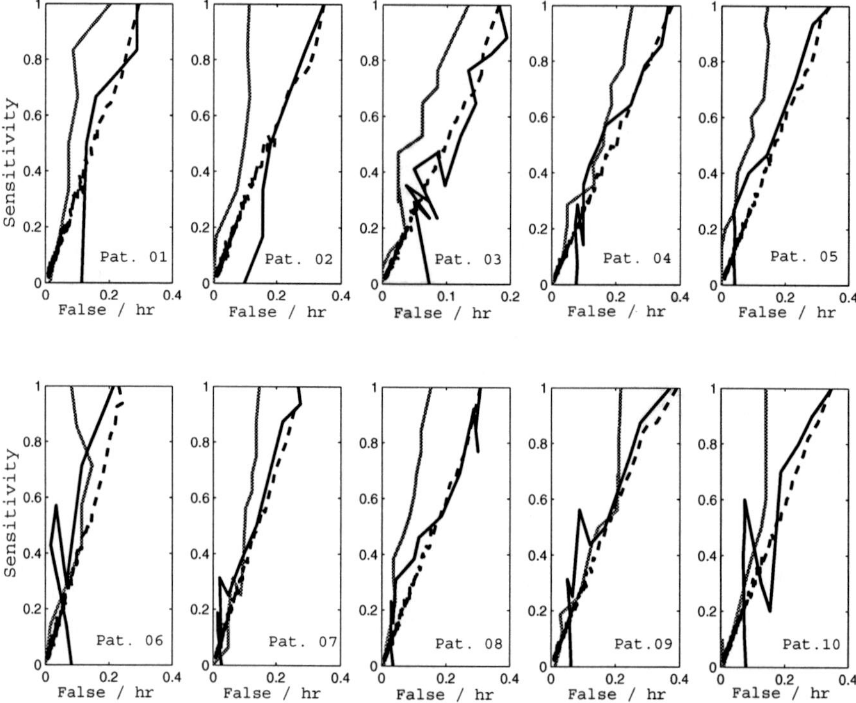

Fig. 4. Estimated ROC curves for ten patients generated by three prediction methods: red line = proposed STL_{max}-based method; black solid line = periodic prediction method, and black dashed line = random prediction method.

Finally, for the proposed ASPA, with at least 80% prediction sensitivity for each test patient, an overall sensitivity of 85.0% and false prediction rate of 0.159 per hour ($\approx$ one false prediction every 6.3 hours) were achieved. The average prediction time for an impending seizure is 63.6 minutes prior to the seizure onset. A summary of these prediction characteristics is given in Table 3.

Table 2. Areas above the ROC curves for the three prediction schemes

Patient	STL_{max}-based	Periodic	Random
1	0.0803	0.1703	0.1535
2	0.0763	0.2058	0.1789
3	0.0545	0.1108	0.0961
4	0.1381	0.1842	0.1884
5	0.0784	0.1486	0.1691
6	0.0857	0.0976	0.1289
7	0.0977	0.1269	0.1252
8	0.0748	0.1534	0.1635
9	0.1406	0.1589	0.1775
10	0.1012	0.1644	0.1710
Mean	**0.0928**	**0.1521**	**0.1552**
(Std)	(0.0277)	(0.0329)	(0.0294)

Table 3. Prediction performance of ASPA with sensitivity larger than 80% per patient

Patient	Sensitivity	False Prediction Rate (per hr)	Prediction Time (mins)	
			Mean	Stand. Dev.
1	5/6 = 83.3%	0.086/hr	63.97	42.54
2	5/6 = 83.3%	0.107/hr	59.09	52.41
3	14/17 = 82.4%	0.097/hr	61.98	35.24
4	12/14 = 85.7%	0.227/hr	56.11	48.42
5	14/15 = 93.3%	0.135/hr	84.16	49.48
6	7/7 = 100%	0.082/hr	67.12	47.56
7	13/16 = 81.3%	0.199/hr	68.86	48.97
8	11/13 = 84.6%	0.121/hr	34.48	49.73
9	13/16 = 81.3%	0.210/hr	74.62	46.93
10	8/10 = 80.0%	0.140/hr	54.12	34.48
Overall	102/120 = 85.0%	0.159/hr	**63.55**	**45.90**

4 Discussion

Ideal EEG characteristics for the assessment of seizure prediction algorithms

For the validation of a seizure prediction algorithm, the ideal EEG test recordings should have: (1) sufficient length with long-term interictal periods, (2) continuous recording without arbitrary or subjective selection of interictal and/or preictal segments, and (3) sufficient number of seizures. These characteristics allow us to have reliable estimations of prediction sensitivity and false prediction rate. Subjective selections of the test interictal or preictal EEG periods make the evaluation questionable. In this study, the proposed prediction algorithm ASPA was tested and evaluated on the continuous long-term EEG recordings from ten patients with multiple seizures. The mean length of the recordings was approximately 210 hours per patient and the mean number of seizures was 13 seizures per patient.

Selection of prediction horizon

Prediction performance depends on the selection of the prediction horizon, that is, the time interval during which a seizure should occur after a warning. Short prediction horizon decreases both prediction sensitivity and specificity (increase of false prediction rate). Different choices of prediction horizon could benefit patients in different clinical prospective. For example, a long prediction horizon could help patients making decisions whether he/she should stay home, ask for medical assistance, drive or take a flight. On the other hand, a more accurate prediction (short prediction horizon) could greatly increase the efficiency of antiepileptic drug or intervention device for the prevention of an impending seizure.

Future prospective

If seizures can be predicted on-line and real-time, major advances in the diagnosis and treatment of epilepsy could be accomplished. For example:

(1) <u>ROUTINE:</u> A reliable seizure prediction system employed in a diagnostic epilepsy-monitoring unit could be used to warn professional staff of an impending seizure or to initiate functional imaging procedures to determine ictal regional cerebral blood flow for a more accurate focus localization,

(2) <u>THERAPEUTIC:</u> A reliable seizure prediction system could be incorporated into digital signal processing chips for use in implantable therapeutic devices. Such devices could be utilized to activate pharmacological or physiological interventions designed to abort impending seizures,

(3) <u>EMERGENCY:</u> Such a prediction scheme could also be used in the ER or ICU with patients in status-epilepticus. Long-term prediction of upcoming seizures may then suggest a change of the medication protocol for the patients to be able to recover.

5 Concluding Remarks

The results of this study confirm the hypothesis that it is possible to predict an impending seizure based on the quantitative analysis of multichannel intracranial EEG recordings. Prediction is possible because the spatiotemporal dynamical features of the preictal transition among the critical groups of cortical sites are robust. This robustness makes it possible to identify electrode sites that will participate in the preictal transition, based on the degree of dynamical entrainment/disentrainment at the first recorded seizure. Further, because the proposed automated seizure prediction algorithm utilized adaptive thresholds based upon the dynamical states of the brain, seizures were predicted during states of alertness and sleep. Thus, brain dynamics were sufficiently distinct to allow seizure prediction independent of the patient's state of alertness.

The prediction performance (quantified by the areas above prediction ROC curves) of the proposed prediction algorithm is superior to the two compared statistical prediction schemes (periodic and random), which indicates that this automated seizure prediction algorithm has potential for clinical applications. These results also provide support for our dynamical ictogenesis theory. However, in order to be beneficial to a variety of clinical applications, such as long-term monitoring procedures for presurgical and diagnostic purposes, intensive care units in patients with frequent uncontrolled seizures, and development of a mechanism to activate implantable therapeutic devices, reducing and understanding the false predictions in the algorithm will be the immediate steps following this study.

References

1. R. Andrzejak, F. Mormann, T. Kreuz, et al. Testing the null hypothesis of the nonexistence of a Preseizure state. *Physical Review E*, 67: 010901-1-4, 2003.

2. R. Aschenbrenner-Scheibe, T. Maiwald, M. Winterhalder, H.U. Voss, J. Timmer, and A. Schulze-Bonhage. How well can epileptic seizures be predicted? An evaluation of a nonlinear method. *Brain*, 126: 1-11, 2003.

3. C.E. Elger and K. Lehnertz. Seizure prediction by non-linear time series analysis of brain electrical activity. *European Journal of Neuroscience*, 10: 786-789, 1998.

4. L.D. Iasemidis, P.M. Pardalos, D.-S. Shiau, W. Chaovalitwongse, M. Narayanan, S. Kumar, P.R. Carney, and J.C. Sackellares. Prediction of Human Epileptic Seizures based on Optimization and Phase Changes of Brain Electrical Activity. *Optimization Methods and Software*, 18(1): 81-104, 2003.

5. L.D. Iasemidis, J.C. Principe, J.M. Czaplewski, R.L. Gilmore, S.N. Roper, and J.C. Sackellares. Spatiotemporal transition to epileptic seizures: A nonlinear dynamical analysis of scalp and intracranial EEG recordings. In F.H. Lopes de Silva, J.C. Principe, L.B. Almeida, editors, *Spatiotemporal Models in Biological and Artificial Systems*, pages 81-88. IOS Press, Amsterdam, 1997.

6. L.D. Iasemidis, J.C. Sackellares, H.P. Zaveri, W.J. Williams. Phase space topography of the electrocorticogram and the Lyapunov exponent in partial seizures. *Brain Topography*, 2: 187-201, 1990.

7. L.D. Iasemidis and J.C. Sackellares. The temporal evolution of the largest Lyapunov exponent on the human epileptic cortex. In D.W. Duck and W.S. Pritchard, editors. *Measuring Chaos in the Human Brain*, pages 49-82. World Scientific, Singapore, 1991.

8. L.D. Iasemidis, D.-S. Shiau, W. Chaovalitwongse, et al. Adaptive epileptic seizure prediction system. *IEEE Transactions in Biomedical Engineering*, 50(5): 616-627, 2003.

9. L.D. Iasemidis, D.-S. Shiau, J.C. Sackellares, P.M. Pardalos, and A. Prasad. Dynamical resetting of the human brain at epileptic seizures: application of nonlinear dynamics and global optimization techniques. *IEEE Transactions on Biomedical Engineering*, 51(3): 493-506, 2004.

10. L.D. Iasemidis. *On the Dynamics of Human Brain in Temporal Lobe Epilepsy*. PhD thesis, University of Michigan, Ann Arbor, 1991.

11. L.D. Iasemidis, P.M. Pardalos, J.C. Sackellares, and D.-S. Shiau. Quadratic binary programming and dynamical system approach to determine the predictability of epileptic seizures. *Journal of Combinatorial Optimization*, 5: 9-26, 2001.

12. L. Iasemidis, D.-S. Shiau, P.M. Pardalos, and J. Sackellares. Transition to epileptic seizures: Optimization. In D. Du, P.M. Pardalos, and J. Wang, editors, *Discrete Mathematical Problems with Medical Applications*, pages 55-74. DIMACS series, Vol. 55, American Mathematical Society, 2000.

13. M.P. Jacobs, G.D. Fischbach,M.R. Davis, M.A. Dichter, R. Dingledine, D.H. Lowenstein, M.J. Morrell, J.L. Noebels, M.A. Rogawski, S.S. Spencer, and W.H. Theodore. Future directions for epilepsy research. *Neurology*, 57: 1536-42, 2001.

14. K. Lehnertz and C.E. Elger. Can epileptic seizures be predicted? Evidence from nonlinear time series analysis of brain electrical activity. *Physical Review Letters*, 80: 5019-5022, 1998.

15. B. Litt, R. Esteller, J. Echauz, M. D'Alessandro, R. Short, T. Henry, P. Pennell, C. Epstein, R. Bakay, M. Dichter, and G. Vachtsevanos. Epileptic seizures may begin hours in advance of clinical onset: A report of five patients. *Neuron*, 30: 51-64, 2001.

16. J. Martinerie, C. Adam, M. Le Van Quyen, M. Baulac, S. Clemenceau, B. Renault, and F.J. Varela. Epileptic seizures can be anticipated by non-linear analysis. *Nature Medicine*, 4: 1173-1176, 1998.

17. F. Mormann, R.G. Andrzejak, T. Kreuz, C. Rieke, P. David, C.E. Elger, and K. Lehnertz. Automated detection of a preseizure state based on a decrease in synchronization in intracranial electroencephalogram recordings from epilepsy patients. *Physical Review E*, 67: 021912-1-10, 2003.

18. M.L.V. Quyen, J. Martinerie, V. Navarro, P. Boon, M. DHave, C. Adam, B. Renault, F. Varela, and M. Baulac. Anticipation of epileptic seizures from standard EEG recordings. *Lancet*, 357: 183-188, 2001.

19. M.L.V. Quyen, J. Martineric, M. Baulac, and F. Varela. Anticipating epileptic seizures in real time by non-linear analysis of similarity between EEG recordings. *NeuroReport*, 10: 2149-2155, 1999.

20. J.C. Sackellares, L.D. Iasemidis, P.M. Pardalos, W. Chaovalitwongse, D.S. Shiau, S.N. Roper, R.L. Gilmore, and J.C. Principe. Performance Characteristics of an Automated Seizure Warning Algorithm (ASWA) Utilizing Dynamical Measures of the EEG Signal and Global Optimization Techniques. *Epilepsia*, 42(S7): 40, 2001.

21. J.C. Sackellares, L.D. Iasemidis, R.L. Gilmore, and S.N. Roper. Epilepsy - when chaos fails. In K. Lehnertz, J. Arnhold, P. Grassberger, and C.E. Elger, editors, *Chaos in the Brain?*, pages 112-133. World Scientific, Singapore, 2000.

22. A.Y. Toledano. Three methods for analyzing correlated ROC curves: A comparison in real data sets from multi-reader, multi-case studies with a factorial design. *Statistics in Medicine*, 22: 2919-2933, 2003.

23. D. Vere-Jones. Forecasting earthquakes and earthquake risk. *International Journal of Forecasting*, 11: 530-538, 1995.

24. M. Winterhalder, T. Maiwald, H.U. Voss, R. Aschenbrenner-Scheibe, J. Timmer, and A. Schulze-Bonhage. The seizure prediction characteristic: a general framework to assess and compare seizure prediction methods. *Epilepsy and Behavior*, 4: 318-325, 2003.

25. D.D. Zhang, X.H. Zhou, D.H. Freeman, J.L. Freeman. A non-parametric method for comparison of partial areas under ROC curves and its application to large health care data sets. *Statistics In Medicine*, 21: 701-715, 2002.

Seizure Predictability in an Experimental Model of Epilepsy*

S.P. Nair[1], D.-S. Shiau[2], L.D. Iasemidis[3], W.M. Norman[4], P.M. Pardalos[5], J.C. Sackellares[6], and P.R. Carney[7]

[1] Department of Biomedical Engineering, University of Florida, Malcolm Randall Department of Veteran's Affairs Medical Center, Gainesville, FL
`spnair@mbi.ufl.edu`
[2] Department of Neuroscience, University of Florida, Malcolm Randall Department of Veteran's Affairs Medical Center, Gainesville, FL
`shiau@mbi.ufl.edu`
[3] Department of Biomedical Engineering, Arizona State University, Tempe, AZ
`Leon.Iasemidis@asu.edu`
[4] Department of Pediatrics, University of Florida, Gainesville, FL
`normanw@mail.vetmed.ufl.edu`
[5] Department of Industrial and Systems Engineering, Computer and Information Science and Engineering, Biomedical Engineering, University of Florida, Gainesville, FL
`pardalos@ufl.edu`
[6] Department of Neurology, Biomedical Engineering, Neuroscience, Pediatrics and Psychiatry, University of Florida, Malcolm Randall Department of Veteran's Affairs Medical Center, Gainesville, FL
`sackellares@mbi.ufl.edu`
[7] Department of Pediatrics, Neurology, Neuroscience, Biomedical Engineering, University of Florida, Gainesville, FL
`carnepr@peds.ufl.edu`

Summary. We have previously reported preictal spatiotemporal transitions in human mesial temporal lobe epilepsy (MTLE) using short term Lyapunov exponent (STL_{max}) and average angular frequency ($\overline{\Omega}$). These results have prompted us to apply the quantitative nonlinear methods to a limbic epilepsy rat (CLE), as this model has several important features of human MTLE. The present study tests the hypothesis that preictal dynamical changes similar to those seen in human MTLE exist in the CLE model. Forty-two, 2-hr epoch data sets from 4 CLE rats (mean seizure duration 74±20 sec) are analyzed, each containing a focal onset seizure and intracranial data beginning 1 hr before the seizure onset. Three nonlinear measures, correlation integral, short-term largest Lyapunov exponent and average angular frequency are used in the current study. Data analyses show multiple transient drops

* This study was supported by NIH grant RO1EB002089, Children's Miracle Network, University of Florida Division of Sponsored Research, and Department of Veterans Affairs.

in STL_{max} values during the preictal period followed by a significant drop during the ictal period. Average angular frequency values demonstrate transient peaks during the preictal period followed by a significant peak during the ictal period. Convergence among electrode sites is also observed in both STL_{max} and $\overline{\Omega}$ values before seizure onset. Results suggest that dynamical changes precede and accompany seizures in rat CLE. Thus, it may be possible to use the rat CLE model as a tool to refine and test real-time seizure prediction, and closed-loop intervention techniques.

Key words: Epilepsy, Hippocampus, Temporal Lobe Epilepsy, Seizure prediction, Nonlinear dynamics, Limbic epilepsy model

1 Introduction

Over the last two decades, a prime focus in epilepsy research has been the application of quantitative measures on continuous EEG recordings obtained from patients with epilepsy for the purpose of seizure detection and prediction. Advancing this area of research to a laboratory setting using animal models is a practical way to initiate and evaluate future control strategies for seizures. Several animal models that reflect various mechanisms underlying human epilepsy have been developed in recent years. It has been shown that status epilepticus (SE), in addition to structural brain damage and lasting neurological deficits, can cause a condition of chronic epilepsy. The rat CLE model originally described by Lothman et al. [49, 50], in which, spontaneous seizures develop following SE induced by a period of continuous hippocampal electrical stimulation, is widely accepted as a desirable model of human temporal lobe epilepsy. It has the important features of spontaneity, chronicity, hippocampal histopathology and temporal distribution of seizures associated with human TLE [5, 61, 62]. Hence we have chosen this model for our current study. Given the similarities with human epilepsy, one would also expect there to be state changes in this model, which can be quantified using algorithms that have been applied to human epilepsy.

By applying nonlinear time series analysis and surrogate time series techniques, several studies have demonstrated the existence of nonlinear components in human EEG [13, 14, 60]. These observations have led to the development of nonlinear techniques to extract information in the EEG signal that traditional signal processing techniques may fail to reveal. Our group, using methods based on the short-term maximum Lyapunov exponent, discovered that seizures are preceded and accompanied by dynamical changes and that these dynamical changes can be used to predict seizures several minutes prior to the actual onset [33, 35, 36, 37].

To gain insight into the rat CLE model as a biological test paradigm for preclinical seizure prediction and control therapies, we conducted nonlinear quantitative analysis of long-term EEG data from spontaneously seizing

Sprague-Dawley rats. The principal goal of this study is to determine whether the system properties of the CLE model, as reflected by the EEG, are comparable to those observed in patients with intractable mesial temporal epilepsy. These properties include: (1) preictal EEG characterized by transient changes in dynamical measures, followed by a progressive convergence (entrainment) of these measures at specific anatomical areas in the brain; (2) ictal EEG characterized by changes in the dynamical values representing a transition from a complex state to a less complex state; (3) postictal EEG characterized by a divergence (disentrainment) of dynamical values at specific anatomical brain areas. The nonlinear measures we have employed do not assume any particular model nor do they rely on a priori understanding of underlying neuronal mechanisms responsible for the data. However, it is important to realize that the brain is not an autonomous system and the EEG signal is not a stationary time series and thus, only approximations of the dynamical measures of the states of the system within a short interval of time can be defined.

The chapter is organized as follows. We start with a brief explanation of the measures used in this study. They are, power spectrum, correlation integral, short-term Lyapunov exponent (STL_{max}) and average angular frequency ($\overline{\Omega}$). Applications of these measures on EEG time series is then demonstrated and the observations are discussed. We further create and quantify spatial maps of multiple STL_{max} and $\overline{\Omega}$ time series to demonstrate the spatiotemporal dynamics of the EEG signal that precede and accompany a seizure.

2 Experimental Setup and Data Characteristics

2.1 Animal Preparation

Experiments were performed on two month old (250 g) male Harlan Sprague Dawley rats ($n = 4$) weighing 210-265 g. Protocols and procedures were approved by the University of Florida Institutional Animal Care and Use Committee. Four 0.8mm stainless steel screws (small parts) were placed in the skull to anchor the acrylic headset (*Figure 1*). Two were located 2mm rostral to bregma and 2mm laterally to either side of the midline. One was 3mm caudal to bregma and 2mm lateral to the midline. One of these served as a screw ground electrode. The last, which served as a screw reference electrode, was located 2mm caudal to lambda and 2mm to the right of midline. Holes were drilled to permit insertion of 2 bipolar twist electrodes for electrical stimulation and recording (AP: -5.3; left and right lateral: +/-4.9mm; vertical: -5 mm to dura) and 2 monopolar recording (AP: 3.2mm, lateral: 1mm left, vertical: -2.5mm; AP: 1mm, lateral: 3mm right, vertical: -2.5mm) electrodes. Electrode pins were collected into a plastic strip connector and the entire headset was glued into place using cranioplast cement (Plastics One, Inc.). Rats were allowed to recover for a week after surgery before further procedures were performed.

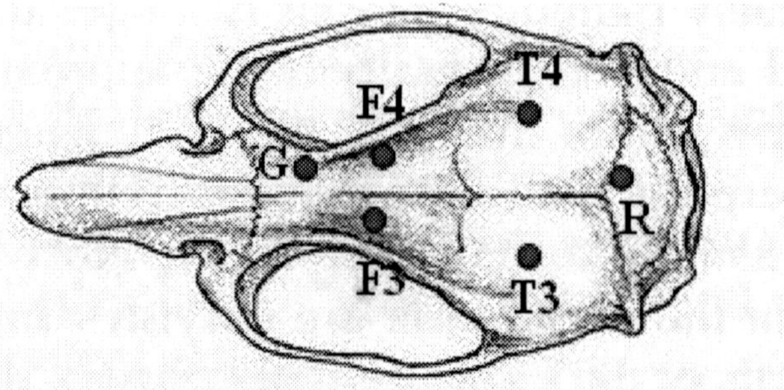

Fig. 1. Electrode placement on the rat head: electrodes are designated arbitrarily to indicate relative positions on the brain. Odd numbers indicate left hemisphere and even numbers indicate right hemisphere. F: frontal; T: temporal; R: reference screw electrode; G: ground screw electrode.

Rats were electrically stimulated to induce seizures 1 wk after surgery. During stimulation and EEG acquisition, rats were housed in specially made chambers [8]. Baseline EEG data was collected using the 2 frontal monopolar electrodes and 1 branch of each hippocampal, bipolar electrode. After recording baseline data, the cable was changed so that an electrical stimulus could be administered to one hippocampal site through a pair of bipolar twist electrodes. The stimulus consisted of a series of 10 sec trains (spaced 2 seconds apart) of 1 msec, biphasic square pulses at 50 Hz, at an intensity of 300-400 μA, for 50-70 minutes [7]. During the stimulus, a normal response was to display 'wet dog shakes' and increased exploratory activity. After approximately 20-30 min, convulsive seizures (up to 1 min duration) were usually observed about every 10 min. At the end of the stimulus period, the EEG trace was observed for evidence of slow waves in all 4 monopolar traces. If this was not the case, the stimulus was re-applied for 10 minute intervals on another 1-3 occasions until continual slow waves appeared after the stimulus was terminated. Rarely ($<$10%), unresponsive rats were noted. The lack of response was attributed to inaccurate placement of the stimulating electrode.

With successful seizure induction, the EEG continued to demonstrate $<$5 Hz activity for 12-24 hrs and intermittent and spontaneous electrographic seizures (30 s - 1 min duration) for 2-4 hrs following an electrical stimulation session. Rats were observed for 12-24 hrs after stimulation for seizure activity. Once their behavior stabilized, they were returned to their home room for 6 weeks while spontaneous seizures developed.

2.2 Data Acquisition

Each animal was connected through a 6-channel electrical commutator and shielded cable to the EEG recording system, which consists of an analog amplifier (Grass Telefactor-Model 10), a 12 bit A/D converter (National Instruments, Inc), and recording software (HARMONIE 5.2, Stellate Inc. Montreal), which is synchronized to a video unit for time-locked monitoring behavioral changes. Each channel is sampled at a uniform rate of 200 Hz and filtered using analog high and low pass filters at cutoff frequencies of 0.1 Hz and 70 Hz, respectively. The recording system uses a 4 channel referential montage and is set to a continuous mode so that prolonged data sets containing ictal as well as interictal data can be collected for analysis. The recording electrodes were named according to their relative positions on the rat brain. The saved EEG and video data is then transferred to a 1.4 TB RAID server for future off-line review and analysis. EEG data pre-processing included removal of baseline wander using a Butterworth filter.

2.3 Classification of Seizures and Data Selection

The test data sets consisted of 42 epochs, each containing a seizure and 1 hr continuous intracranial EEG preceding and following the ictal event, obtained from four stimulated rats, hereafter referred to as rats A, B, C and D (Table 1). Stimulation was done on the left hippocampus in rats A and C and on the right in rats B and D. The 'focal' electrode referred to in later sections refers to the electrode. Seizure 'onset' was defined electrographically as the first sustained change in the EEG clearly different from the background activity. Seizure 'offset' was defined as the time at which the rhythmic activity dies out and postictal spike and wave discharges appear.

Table 1. Seizure characteristics of four rats obtained from qualitative (visual) analysis of EEG

Rat ID	Number of Seizures Analyzed	Seizure Duration (seconds)		Inter-Seizure Interval (hours)		
		Mean	SD	Range	Mean	SD
A	5	52.8	3	20~218	71.5	97.7
B	8	96.6	20.8	2.5~99	42.2	41
C	21	80	8.6	1.5~164	16.8	40
D	8	65	11	12.5~100	43.6	34.2
Overall	42	77.1	12.03	1.5~218	33.25	48.7

3 Signal Processing Methods

3.1 Spectral Analysis

Power spectral analysis has been traditionally used to categorize electrophysiological signals. Any time series can be decomposed into the sum of its sine wave components using Fourier transforms. To get a better understanding of the temporal resolution along with the frequency resolution a spectrogram method was used. The spectrogram is the squared magnitude of the windowed short-time Fourier transform (STFT). It considers the squared modulus of the STFT to obtain a spectral energy density of the locally windowed signal $x(u)h^*(u - t)$:

$$S_x(t, f) = \left| \int_{-\infty}^{\infty} x(u)h^*(u - t)e^{-j2\pi fu} du \right|^2$$

where $h(t)$ is a short time analysis window located around $t = 0$ and $f = 0$. Thus, we can interpret the spectrogram as a measure of the energy of the signal contained in the time-frequency domain centered on the point (t, f).

3.2 Phase Space Reconstruction of the EEG Signal

The EEG, being the output of a multidimensional system, has both spatial and temporal statistical properties. Components of the brain (neurons) are densely interconnected and there exists an inherent relation between EEG recorded from one site and the activity at other sites. This makes the EEG a multivariable time series. A well-established technique for visualizing the dynamical behavior of a multidimensional (multivariable) system is to generate a state space portrait of the system. A state space portrait is created by treating each time-dependent variable of the system as a component of a vector in a multidimensional space. Each vector in the state space represents an instantaneous state of the system. These time-dependent vectors are plotted sequentially in the state space to represent the evolution of the state of the system over time. For many systems, this graphical mapping creates an object confined over time to a sub-region of the phase space. Such sub-regions of the phase space are called "attractors." The geometrical properties of these attractors provide information about the global state of the system.

When the variables of a system are related over time, which is a salient characteristic of a dynamical system, proper analysis of a single observable can provide information about all variables related to this observable. The state space reconstruction of the EEG signal can be done using the method of delays described by Takens [70]. Figure 2b shows the phase space reconstruction of an ictal EEG segment recorded from a single electrode located on the hippocampus (Figure 2a). According to Takens [70], the embedding dimension p should be at least equal to $(2D + 1)$ in order to correctly embed an attractor in the phase space. The measure most often used to estimate D is the phase space correlation dimension (ν). Methods for calculating the correlation dimension from experimental data described in [56] were employed in our work to approximate D of the epileptic attractor.

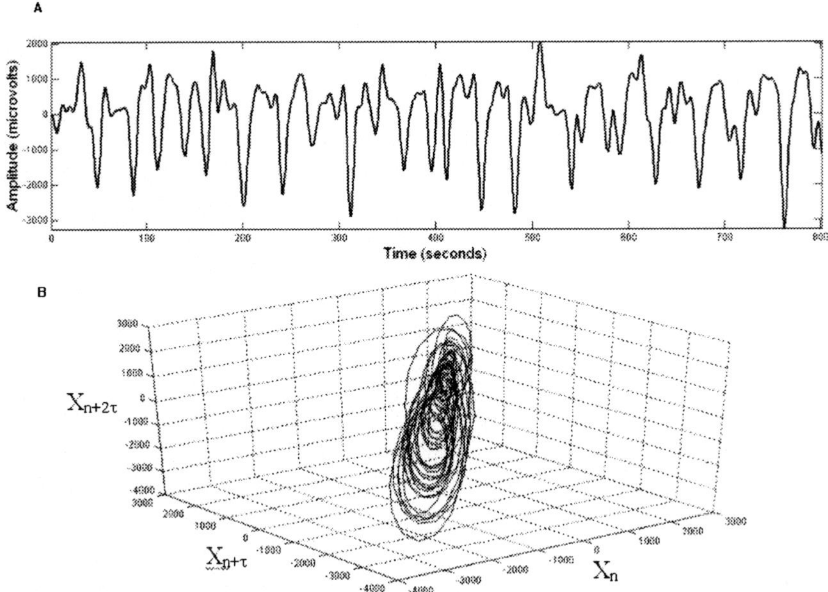

Fig. 2. (a) Ictal segment of raw EEG data from a hippocampal electrode. (b) The reconstructed EEG segment in phase space.

3.3 Test for Nonlinearity – Correlation Integral

The instantaneous state of a dynamical system is characterized by a point in phase space. A sequence of such states subsequent in time defines the phase space trajectory. If the system is governed by deterministic laws, then after a while, it will arrive at a permanent state regime. This fact is reflected by the convergence of ensembles of phase space trajectories towards an invariant subset of phase space, also called the attractor of the system. The correlation integral, simply put, is a measure of the spatial organization of a point in a chaotic attractor. In other words, it is the probability of two vectors in state space being closer than a specified distance r, thus the output of the correlation integral depends on the given radius of the phase space neighborhood.

Two hour epochs of EEG, one hour before and after the seizure from each of the 4 rats were analyzed in order to test for signal nonlinearities. The EEG segment was first divided into non-overlapping segments of 10.24 sec duration. To demonstrate the presence of nonlinearity, the original signal was compared to surrogate datasets generated from the original signal. We used the randomization technique described by Theiler [72] to generate the surrogate datasets. In this procedure, the fast Fourier transform (FFT) of the original time series is calculated. A phase randomization technique was then applied to eliminate the nonlinear components while the linear properties are preserved. Next, an inverse FFT is computed to obtain the surrogate time series. A total of 10 surrogate datasets were obtained by repeating the procedure. All data sets were transformed to produce distributions with zero mean and unit variance.

The correlation integral is estimated after embedding the time series in a multi-dimensional state space using the method of delays [70]. For this study we used an embedding dimension $p = 5$ and a time delay $\tau=3$ (equivalent to 15 ms). The correlation integral $C(r)$ is the probability that two vectors selected at random lie within a distance r of each other. The correlation integral of a time series $x_1, x_2, ..., x_N$ as defined by Grassberger and Procaccia [26] is as follows:

$$C(r) = (1/N_p) \sum_{i=1}^{K} \sum_{j=i+1+w}^{K} h(r - d_{ij})$$

where $K = N - (p - 1)\tau$ is the number of p-dimensional vectors, $v_t = (x_t, x_{(t-\tau)}, \ldots x_{(t-(p-1)\tau)})$, $N_p = K(K - 1 - w)/2$ is the number of distinct pairs of vectors, and h is the Heaviside or step function $h(\lambda) = \begin{cases} 1 & \text{if } \lambda \geq 0 \\ 0 & \text{if } \lambda < 0. \end{cases}$
$d_{ij} = \max_{0 \leq m \leq p-1} |x_{i-m\tau} - x_{j-m\tau}|$ was used for computational speed [26]. We chose the Theiler correction $w = 50$ (equivalent to 250ms) to avoid autocorrelation effects on the computation of the correlation integral [70]. The correlation integral profile from each of the 4 electrodes was generated and compared to the ones generated from the corresponding surrogate datasets. Statistical significance was defined as: $S = (\log_{10} C(r) - \bar{x})/\sigma$, where $\bar{x}$ is the mean and σ is the standard deviation of the logarithm of the correlation integrals of the 10 surrogate datasets, for each 10.24 second segment A difference was considered statistically significant if $S > 5$. The parameters $p = 5$, $\tau=3$ and $r=0.1$ were chosen in order to optimize the significance of non-linearity in as many EEG segments as possible.

3.4 Estimation of Short-Term Largest Lyapunov Exponent (STL_{max})

The Lyapunov exponent is a measure of chaoticity of a signal. It is estimated by examining, for each point of phase space, how quickly nearby trajectories that begin at this point diverge (or converge) over time. For the application on non-stationary EEG time series with spike transients, we have adopted short term largest Lyapunov exponent modified from Wolf's algorithm [74]. The algorithm has been described in detail elsewhere [29]. Mathematically, the largest Lyapunov exponent is defined as the average of local Lyapunov exponents L_{ij} in the state space as follows:

$$L_{max} = \tfrac{1}{N} \sum L_{ij}$$

where N is the total number of the local Lyapunov exponents that are estimated from the evolution of adjacent points (vectors) in the state space according to:

$$L_{ij} = \frac{1}{\Delta t} \cdot \log_2 \frac{|X(t_i + \Delta t) - X(t_j + \Delta t)|}{|X(t_i) - X(t_j)|}$$

where Δt is the evolution time allowed for the vector distance $\delta_{x_k}^{(\lambda)} = |X(t_i) - X(t_j)|$ to evolve to the new distance $\delta_{x_k}^{(\lambda)} = |X(t_i + \Delta t) - X(t_j + \Delta t)|$, where $\lambda = 0, \ldots N - 1$ and $\Delta t = k \times dt$ with dt the sampling period of the original time series ($dt = 5$ milliseconds in our data).

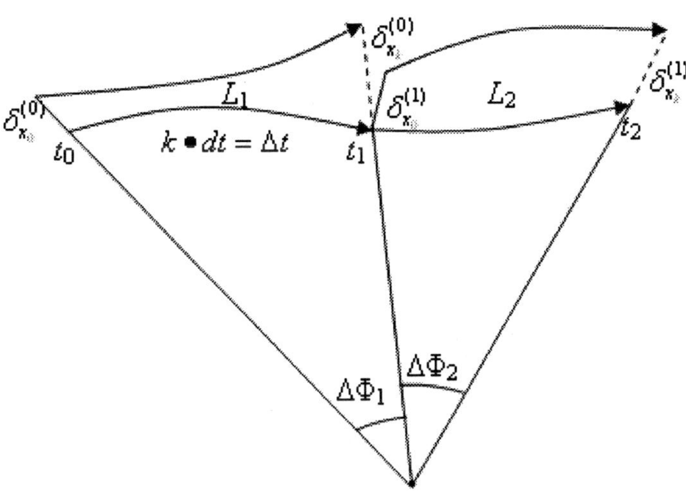

Fig. 3. Calculation of average angular frequency $(\overline{\Omega})$ and L_{max}. $\Delta\Phi_i$ represents the local phase difference between 2 evolved states in the state space.

For the estimation of STL_{max} (measured in bits/sec), a state space reconstruction for each EEG channel time series was performed using the method of delays [70] from sequential, non-overlapping data segments of 10.24 seconds in duration (2048 data points) with embedding parameters of $p = 7$ and $\tau = 3$; where p is the embedding dimension and τ is the time delay between two coordinates in the state space vector. We chose these parameters based on results from earlier experiments with training sets of EEG. An illustration of STL_{max} estimation is given in Figure 3.

3.5 Estimation of Average Angular Frequency ($\overline{\Omega}$)

In addition to capturing the local stability of the system, we were also interested in measuring the rate at which the local state of the system changed over time, and we have used a measure termed as average angular frequency [37]. After the state space reconstruction, the first step of this algorithm is to estimate the difference in phase between two evolved states $X(t_i)$ and $X(t_i + \Delta t)$ is defined as $\Delta\Phi_i$. The average of the local phase differences $\Delta\Phi_i$ between the vectors in the state space is then given by:

$$\Delta\Phi = \frac{1}{N} \cdot \sum_{i=1}^{N} \Delta\Phi_i$$

where N is the total number of phase differences estimated from the evolution of $X(t_i)$ to $X(t_i + \Delta t)$ in the state space:

$$\Delta\Phi_i = \left| \arccos\left(\frac{X(t_i) \cdot X(t_i + \Delta t)}{\|X(t_i)\| \cdot \|X(t_i + \Delta t)\|} \right) \right|$$

Then, the average angular frequency, defined by

$$\overline{\Omega} = \frac{1}{\Delta t} \Delta\Phi$$

measures how fast the local state of a system changes on average (e.g. dividing $\overline{\Omega}$ by 2π, the rate of the change of the state of the system is expressed in $\sec^{-1}$(Hz). For estimating the average angular frequency $\overline{\Omega}$ (rad/sec), the state space reconstruction is done in the same manner as that for estimating STL_{max}.

4 Quantification of Spatiotemporal Dynamics

4.1 Statistical T-index

We will now examine the spatiotemporal dynamical changes in the EEG that precede and accompany seizures by quantifying the observed progressive locking and unlocking of dynamical measures ($STL_{max}/\overline{\Omega}$) over time and space. We use the T-index from the statistical paired-T statistic to measure the degree of convergence/divergence of $STL_{max}/\overline{\Omega}$ between electrode sites. The T-index for each electrode pair was calculated in each 10 minute epoch (60 values of $STL_{max}/\overline{\Omega}$) by dividing the mean difference of the measure considered (i.e. STL_{max} or $\overline{\Omega}$) between the two electrode sites by its standard deviation. The T-index at time t between electrode sites i and j is defined as:

$$T_{ij}(t) = \frac{|\overline{D}_{ij}^{t}|}{\hat{\sigma}_{ij}^{t}/\sqrt{N}}$$

where $|\overline{D}_{ij}^{t}|$ denotes the absolute value of the average of all pairwise differences $D_{ij}^{t} = \{STL\max_{i}^{t} - STL\max_{j}^{t} \,|\, t \in w(t)\}$ within a moving window $w(t)$ defined as:

$$w(t) = \left[\frac{t}{10.24s} - N + 1, \frac{t}{10.24s}\right]$$

where N is the length (# of STL_{max} points) of the moving window, and $\hat{\sigma}_{ij}^{t}$ is the sample standard deviation of D_{ij}^{t} within $w(t)$. Asymptotically, $T_{ij}(t)$ index follows a t-distribution with $N - 1$ degrees of freedom.

The spatiotemporal behavior of the system is quantified by the T-index profiles calculated from multiple electrode sites over time. An electrode pair is said to be dynamically entrained if the T-index values calculated from the pair falls below a critical threshold $T_c = 2.662$. We have chosen this critical value from the T-distribution with a significance level $\alpha = 0.01$, based on earlier human studies. We have defined an 'entrainment transition' as the drop in T-index values from a preset upper threshold to T_c. A postictal 'disentrainment' is defined as a rise in T-index values from the T_c to a value greater than the upper threshold within a 20 minute window after the seizure.

5 Results

5.1 Frequency Evolution

Figure 4 shows seizure #1 of rat A and its corresponding power spectra represented in the time-frequency domain. In the EEG, the onset of the seizure is at second 72 and is accompanied by a clear spike and wave discharge. In the time-frequency plot, this is correlated with a sudden increase in power in the 0-10 Hz range. About 12 seconds into the seizure, high amplitude rhythmic activity starts, and this is correlated with an increase in spectral power at higher frequencies up to 25 Hz. This activity progressively slows down to about 1 Hz towards the end of the seizure. The postictal state is characterized by spikes and slow wave activity, and this is correlated with localized power distribution in the 0-7 Hz range. This frequency evolution is very similar to the pattern seen in depth EEG recordings from human patients [63].

5.2 Nonlinearities in the EEG Time Series

Figure 5(a) shows the correlation integral estimates calculated from the EEG signal obtained from a 'focal' electrode and 10 surrogate datasets created from the original time series, plotted as a function of time. Visual inspection reveals that the correlation integrals for the original EEG signal are uniformly higher than the correlation integrals estimated from any of the 10 surrogates. The sudden increase in the correlation integrals of the original time series and the surrogates during the postictal period is due to the increased autocorrelation

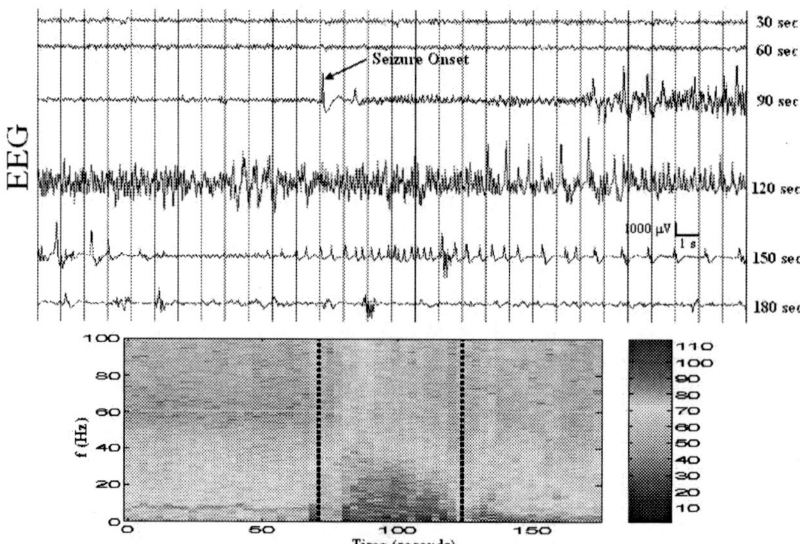

Fig. 4. Three minutes of EEG data from channel T3-R showing seizure #1 from Rat A and corresponding power spectra. The vertical dashed lines represent the seizure onset and offset.

in the EEG during this period. Figure 5(b) shows the statistical significance S of nonlinearity in the same EEG epoch. 79% of the total segments (10.24 sec in duration) in the EEG epoch showed statistically significant nonlinearity ($S > 5$).

5.3 Dynamical Changes in System Chaoticity

Temporal STL_{max} Profiles

Figure 6 shows STL_{max} profiles of a 'critical' electrode pair (includes the 'focal' electrode), for a 10 minute epoch containing a seizure. The STL_{max} curves generally show a drop during a seizure reaching its minimum value during the ictal period followed by a gradual rise to a value greater than the average preictal value before returning to the normal values. The 'focal' electrodes have a consistently lower value of STL_{max} when compared to the other electrodes. This may account for the observation of a less significant drop in the STL_{max} values in the 'focal' electrode when compared to the other electrodes. During the seizure the STL_{max} values of the electrodes are very close to each other but gradually drift apart during the postictal state.

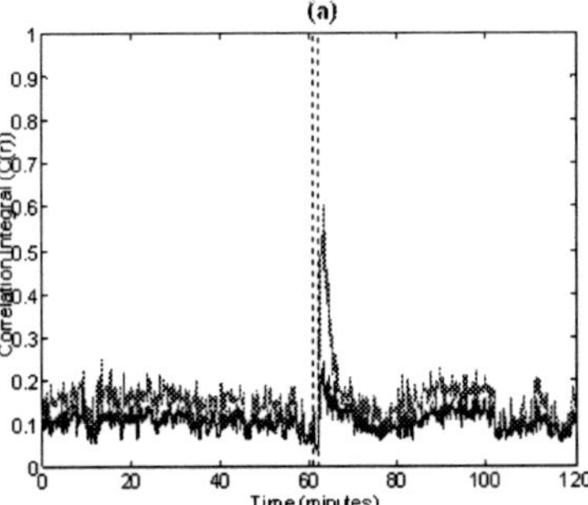
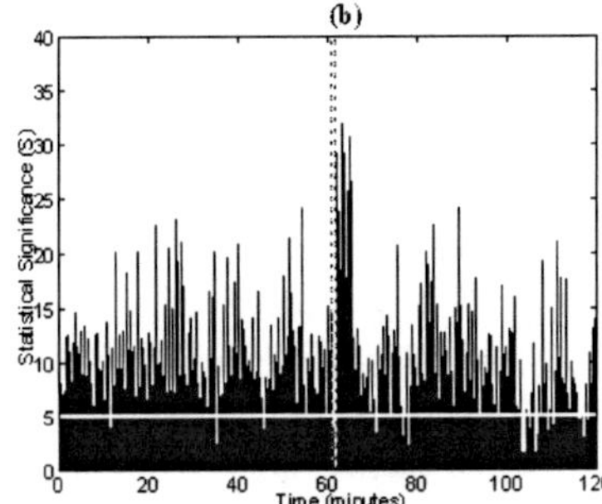

Fig. 5. (a) The values of correlation integral $C(r)$ of the recording from the stimulated left hippocampal electrode(solid red line) and values of 10 surrogate datasets (blue dotted lines), as a function of time. (b) The statistical significance S of nonlinearity in the EEG signal recorded from the same hippocampal electrode. The white horizontal line represents a threshold equal to 5 SD. Statistically significant nonlinearities are present in 79% of the total segments (10.24 sec in duration) in the entire EEG epoch. Vertical dashed lines represent the seizure onset and offset.

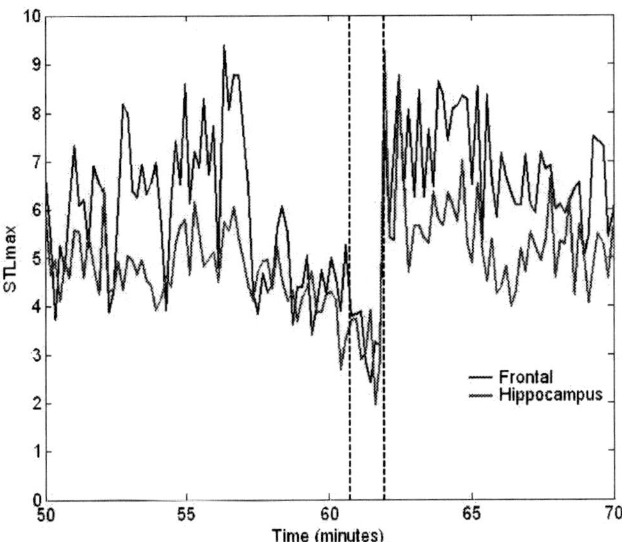

Fig. 6. Sample STL_{max} profile for a 10 minute epoch including a grade 5 seizure. The dashed line represents the 'focal' electrode (T3-R in this case) and the bold line represents the contralateral hippocampal electrode (T4-R). Seizure onset and offset are indicated by vertical arrows at the bottom of the plot. Note that the sudden drop in STL_{max} values during the seizure is clearly visible for T4-R while in T3-R the drop is more gradual.

Temporal $\overline{\Omega}$ Profiles

The $\overline{\Omega}$ profiles follow a general trend opposite to that seen in the case of STL_{max}, with a peak during the seizure and sharp drop during the postictal stage. This pattern roughly corresponds to the typical observation of higher frequencies in the original EEG signal during the ictal period and is consistent with findings from time-frequency analysis of the seizure epochs (Figure 4). Figure 7 shows $\overline{\Omega}$ profiles of a pair of critical electrode sites (including the 'focal' electrode) 5 minutes before and after a seizure. The hippocampal electrode has a consistently higher value of $\overline{\Omega}$ compared to the frontal electrode except at the seizure when the $\overline{\Omega}$ values calculated from frontal electrodes reach the same level as those from hippocampal electrodes, sometimes exceeding it during certain seizures. A gradual convergence in $\overline{\Omega}$ values from the frontal and hippocampal electrode can be seen prior to the seizure, followed by a certain degree of divergence. In the case of Rat C, even though $\overline{\Omega}$ values showed a characteristic peak at the seizure, there was no preictal entrainment transition and postictal disentrainment in a significant fraction of the seizures.

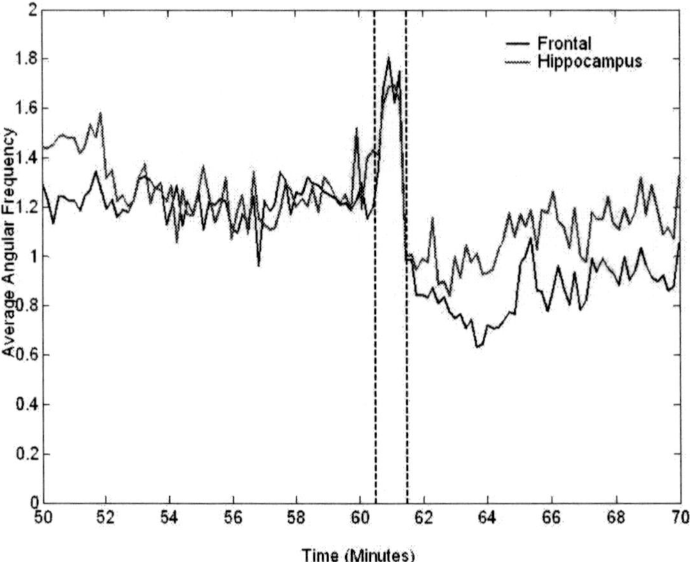

Fig. 7. Sample $\overline{\Omega}$ profile for a 10 minute epoch including a grade 5 seizure. The dashed line represents the 'focal' electrode (T4-R in this case) and the bold line represents the ipsilateral frontal electrode (F4-R). Seizure onset and offset are indicated by vertical arrows at the bottom of the plot. Note that the 'focal' electrode peaks first, followed by the frontal electrode, indicating that the seizure progresses from the 'focal' site to the frontal lobe.

5.4 Spatiotemporal Profiles of STL_{max} and $\overline{\Omega}$

Figures 8 and 9 show examples of the average T-index curves calculated from STL_{max} and $\overline{\Omega}$ values of a pair of electrodes including the 'focal' electrode over a period of 2 hours, from each of the 4 rats. This pattern shows the entrainment of STL_{max} and $\overline{\Omega}$ values among the 'critical' electrode sites before a seizure and the disentrainment after the seizure. From a dynamical perspective, this represents an increase in spatio-temporal interactions between the brain sites during the preictal period and a postictal desynchronization of these brain sites following a seizure. We can see that, except for Rat C, all T-index curves show a transition towards a critical value. We have termed this transition as 'entrainment transition'. A sudden overshoot during the postictal stage followed by significant disentrainment indicates the resetting feature of the seizure (the seizure restores the pre-seizure entrainment to a more normal state). Note that the degree of disentrainment is significantly less in Rat C. Moreover the critical electrode sites are dynamically entrained for almost the entire epoch in this rat. This pattern was consistent in a majority of seizures analyzed from Rat C. A possible explanation for this observation is provided in the discussion section. A summary of test results for preictal transition and postictal resetting is given in Table 2.

Table 2. Summary of test results obtained from non-linear analysis on 42 seizures from 4 rats

Rat ID	# Seizures	# Seizures with preictal transition		# Seizures with postictal disentrainment	
		STLmax	$\overline{\Omega}$	**STLmax**	$\overline{\Omega}$
A	5	5	5	5	5
B	8	6	6	7	8
C	21	14	18	8	15
D	8	7	6	8	8
Overall	42	32(76.2%)	35(83.3%)	28(66.7%)	35(83.3%)

6 Discussion

The results of this study point to three major findings. First, the onset of the seizure represents a temporal transition of the system from a chaotic state to a more ordered state (less chaotic) as revealed by the behavior of non-linear dynamical measures over the preictal, ictal and postictal periods. Second, spatiotemporal dynamical analysis with multiple electrode sites reveals a preictal entrainment and postictal disentrainment, which we have termed as 'resetting', in a large subset of the seizures analyzed. This finding is particularly

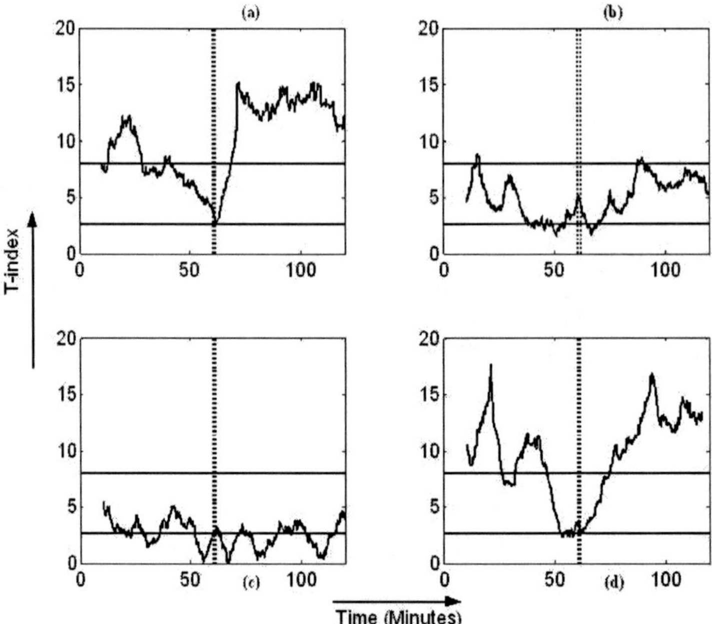

Fig. 8. T-index profiles calculated from STLmax values of a pair of electrodes from each rat. Figures a, b, c and d correspond to Rats A, B, C and D. Each pair includes the 'focal' electrode and a frontal electrode. Vertical dotted lines represent seizure onset and offset. The horizontal dashed line represents critical entrainment threshold.

relevant, because it suggests the ability to identify a pre-seizure state. The ability to predict an impending seizure in this model ahead of its clinical or electrographic onset would be extremely useful in designing new diagnostic and therapeutic applications that could trigger interventions well before the occurrence of a seizure.

6.1 Nonlinearity and Temporal Dynamics in the CLE Model

The EEG is a complex signal whose statistical properties depend on both time and space. The presence of highly significant nonlinearities in electrographic signals supports the concept that the epileptogenic brain in this model of epilepsy is a nonlinear system. Several investigations have employed analytic techniques with the objective to study dynamical characteristics associated with epilepsy. Preliminary results obtained from non-linear EEG analysis in the CLE model show dynamical patterns similar to those previously observed in human TLE. In all 4 animals analyzed, a reduction in EEG chaoticity

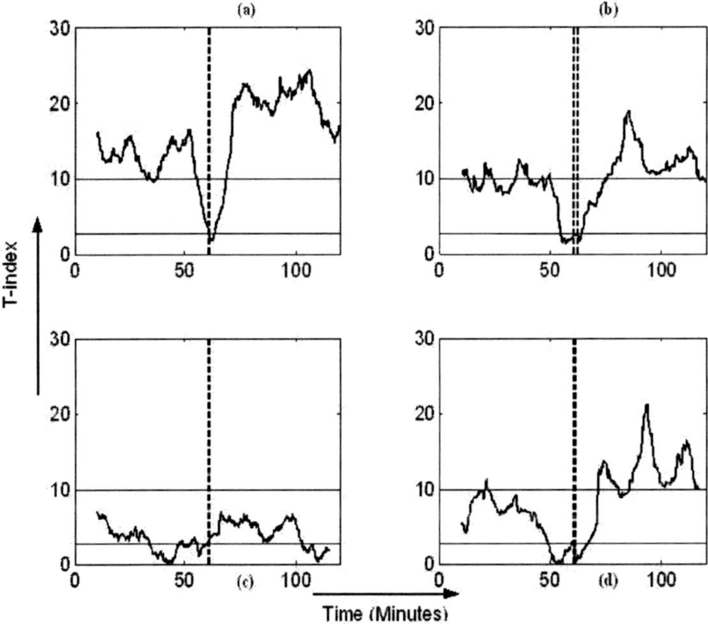

Fig. 9. T-index profiles calculated from $\overline{\Omega}$ values of a pair of electrodes from each rat. Figures (a), (b), (c) and (d) correspond to Rats A, B, C and D. Each pair includes the 'focal' electrode and a frontal electrode. Vertical dotted lines represent seizure onset and offset. The horizontal dashed line represents critical entrainment threshold.

was observed during the ictal period after several transient decreases during the preictal period. Nonlinear methods have also been said to be useful in identifying the seizure foci, for detecting and localizing ictal onset and for studying spatial spread of ictal discharges in human epilepsy. This finding seems to extend to the CLE model as well, since dynamical measures obtained from EEG recorded from the stimulated site of the hippocampus show a clear difference in their mean level from the remaining areas. We must, however, point out that the assumption here is that the site of stimulation is indeed the seizure focus in this model which could of course be not the case.

6.2 Spatiotemporal Dynamics, Seizure Prediction and Control

Perhaps the most exciting discovery to emerge from dynamical analysis of the EEG in temporal lobe epilepsy is that seizures are preceded by dynamical changes in the signal occurring several minutes before the seizure. The analysis of spatiotemporal dynamics of long-term EEG recordings in human patients has revealed that seizures are preceded by dynamical changes well in advance of the actual seizure onset. In 3 out of 4 rats, the preictal state was characterized by a gradual entrainment transition among 'critical' electrode sites followed by a high degree of disentrainment during the postictal period, which is similar to the pattern seen in humans [29, 33]. This observation provides further proof that seizures in this model represent the formation of self-organizing spatiotemporal patterns. This preictal entrainment transition and postictal disentrainment pattern was however, not observed in a major fraction of seizures from one rat (Rat C). One possible explanation for this behavior can be given on the basis of the resetting feature (or lack thereof) of the seizure in this particular animal. Note that the value of the mean inter-seizure interval for this rat could be misleading because of the fact that seizures in this rat occurred in two clusters significantly apart in time from each other. The inter-seizure interval within these clusters was in the order of 2-4 hours, suggesting that the seizures in this rat did not completely reset the brain. This could account for the high frequency of seizures observed. In this regard, we can assume that the brain dynamics in this rat was 'abnormal' and not representative of the general dynamical behavior seen in this model. We have used the term 'abnormal' to convey that the animal does not exhibit normal seizure dynamical properties.

Preliminary results suggest that nonlinear quantitative analysis of multiple regions of brain structures may be useful in detecting alterations in the behavior of the underlying network before actual seizure manifestation. The spatio-temporal patterns seen before the seizures demonstrate that it may be possible to anticipate the seizure by several minutes by automated non-linear analysis. A seizure warning system utilizing such a model could be used to activate pharmacological or physiological interventions designed to prevent an impending seizure. Several fundamental questions remain unresolved in the field of seizure control. One of the long term goals of this research is to find answers to fundamental questions in the field of seizure control such as where in the brain the stimulus should be delivered and what type of stimulation would be most effective.

6.3 Comparison with Human Studies

The results presented in this study reveal a number of similarities in both linear and nonlinear dynamical properties between the CLE model and human TLE. The analysis of spatial and temporal dynamical patterns of long-term intracranial EEG recordings, recorded for clinical purposes in patients with medically intractable temporal lobe epilepsy has demonstrated preictal transitions characterized by progressive convergence (entrainment) of dynamical measures (e.g. maximum Lyapunov exponent) at specific anatomical areas. These dynamical changes have been attributed to spatial interactions or synchronization between the underlying nonlinear components (neurons) of the brain. We have demonstrated that the dynamical properties of the preictal,

ictal and interictal states in the CLE model are distinctly different and can be defined quantitatively.

In humans, seizures represent a dynamical state of increased spatiotemporal order, and also act as a resetting mechanism to return brain dynamics back to a more normal state. Qualitative comparison of dynamical patterns in the CLE model with those obtained from previous studies in humans indicates that the epileptic brain in this model behaves in a similar fashion. Typical dynamical studies involving human intracranial EEG recordings have employed 28-32 channels. Using smaller micro-electrode arrays in future experiments could help overcome one of the major limitations of this study, i.e. the number of recording electrodes. It could be possible to sample from a wider range of brain areas and gain a better understanding of seizure progression and subtle dynamic interactions between nearby sites. Identifying and characterizing the preictal transition process in this model is extremely important in order to develop a seizure anticipation method that can be used in conjunction with an intervention scheme. Additional investigations into the underlying neurobiology in the CLE model could help us understand the basic mechanisms responsible for these dynamical changes.

7 Summary

Preliminary results from quantitative EEG analysis indicate that it may be possible to use the post-status chronic limbic epilepsy model as a tool to refine and test real-time seizure detection and prediction algorithms. Similarities in spatiotemporal dynamical properties between the CLE model and humans, as revealed from quantitative non-linear EEG analysis, may reflect a similarity in underlying network properties and its role in the genesis and expression of these seizures. We plan to use this model to investigate new therapeutic approaches for controlling epileptic seizures such as administering a single fixed dose of anticonvulsant or timed electrical stimuli, based on specific state changes of the epileptic brain. Results from this study could be used for designing future experiments in which the spatiotemporal measures discussed, can be utilized as control parameters for a closed-loop seizure control system. This would be a critical step in developing implantable biofeedback sensors that can regulate drug delivery or electrical stimulation as a means of preventing or aborting seizures. However, further investigation and statistical verification need to be done before extending any validation results from this animal model to the case of humans and clinical applications.

References

1. R. Agarwal and J. Gotman. Adaptive segmentation of electroencephalographic data using a nonlinear energy operator. *Proceedings of the 1999 IEEE International Symposium on Circuits and Systems*, Orlando, FL, 1999.
2. T. Babb and I. Najm. Hippocampal Sclerosis: Pathology, Electrophysiology, and Mechanisms of Epileptogenesis. In E. Wyllie, editor, *The Treatment of Epilepsy: Principles and Practice*, pages 105-114. Williams and Wilkins, 2001.
3. S.M. Bawin, A.R. Sheppard, M.D. Mahoney, M. Abu-Assal, and W.R. Adey. Comparison between the effects of extra cellular direct and sinusoidal currents on excitability in hippocampal slices. *Brain Research*, 362: 350-354, 1986.
4. E.H. Bertram, E.W. Lothman, and N.J. Lenn. The hippocampus in experimental chronic epilepsy: a morphometric analysis. *Annals of Neurology*, 27: 43-48, 1990.
5. E.H. Bertram and J. Cornett. The ontogeny of seizures in a rat model of limbic epilepsy: evidence for a kindling process in the development of chronic spontaneous seizures. *Brain Research*, 625: 295-300, 1993.
6. E.H. Bertram and J. Cornett. The evolution of a rat model of chronic spontaneous limbic seizures. *Brain Research*, 661: 157-162, 1994.
7. E.H. Bertram. Functional anatomy of spontaneous seizures in a rat model of limbic epilepsy. *Epilepsia*, 38: 95-105, 1997.
8. E.H. Bertram, J.M. Williamson, J.F. Cornett, S. Spradlin, and Z.F. Chen. Design and construction of a long term continuous video-EEG monitoring unit for simultaneous recording of multiple small animals. *Brain Research Protocols*, 2: 85-97, 1997.
9. M. Bikson, J. Lian, P.J. Hahn, W.C. Stacey, C. Sciortino, and D.M. Durand. Suppression of epileptiform activity by high frequency sinusoidal fields in rat hippocampal slices. *The Journal of Physiology*, 531: 181-191, 2001.
10. T.R. Browne and G.L. Holmes. Epilepsy. *New England Journal of Medicine*, 344: 1145-1151, 2001.
11. P.R. Carney, D.S. Shiau, P.M. Pardalos, L.D. Iasemidis, W. Chaovalitwongse, and J.C. Sackellares. Nonlinear neurodynamical features in an animal model of generalized epilepsy. In P.M. Pardalos, J.C. Sackellares, P.R. Carney, and L.D. Iasemidis, editors, *Quantitative Neuroscience*, pages 37-51. Kluwer Academic Publishers, Boston, 2004.
12. P.R. Carney, S.P. Nair, L.D. Iasemidis, D.S. Shiau, P.M. Pardalos, D. Shenk, W. Norman, and J.C. Sackellares. Quantitative analysis of EEG in the rat limbic epilepsy model. *Neurology*, 62 (7, Suppl. 5): A282-A283, 2004.
13. M.C. Casdagli, L.D. Iasemidis, J.C. Sackellares, S.N. Roper, R.L. Gilmore, R.S. Savit. Characterizing nonlinearity in invasive EEG recordings from temporal lobe epilepsy. *Physica D*, 99: 381-399, 1996.
14. M.C. Casdagli, L.D. Iasemidis, R.S. Savit, R.L. Gilmore, S.N. Roper, and J.C. Sackellares. Non-linearity in invasive EEG recordings from patients with temporal lobe epilepsy. *Electroencephalography and Clinical Neurophysiology*, 102: 98-105, 1997.

15. M. D'Alessandro, R. Esteller, G. Vachtsevanos, A. Hinson, J. Echauz, and B. Litt. Epileptic seizure prediction using hybrid feature selection over multiple intracranial EEG electrode contacts: a report of four patients. *IEEE Transactions on Biomedical Engineering*, 50(5): 603-615, 2003.

16. H. Degn, A.V. Holden, and L.F. Olsen. *Chaos in Biological Systems*. Kluwer Academic/Plenum Publishers, 1986.

17. M.J. Denslow, T. Eid, F. Du, R. Schwarcz, E.W. Lothman, and O. Steward. Disruption of inhibition in area CA1 of the hippocampus in a rat model of temporal lobe epilepsy. *Journal of Neurophysiology*, 86(5): 2231-2245, 2001.

18. D.M. Durand. Electric field effects in hyperexcitable neural tissue: A review. *Radiation Protection Dosimetry*, 106: 325-331, 2003.

19. H.G. Eder, D.B. Jones, R.S. Fisher. Local perfusion of diazepam attenuates interictal and ictal events in the bicuculline model of epilepsy in rats. *Epilepsia*, 38: 516-521, 1997.

20. H.G. Eder, A. Stein, and R.S. Fisher. Interictal and ictal activity in the rat cobalt/pilocarpine model of epilepsy decreased by local perfusion of diazepam. *Epilepsy Research*, 29: 17-24, 1997.

21. E.E. Fanselow, A.P. Reid, and M.A. Nicolelis. Reduction of pentylenetetrazole-induced seizure activity in awake rats by seizure-triggered trigeminal nerve stimulation. *Journal of Neuroscience*, 20: 8160-8168, 2000.

22. N.B. Fountain, J. Bear, E.H. Bertram, and E.W. Lothman. Responses of deep entorhinal cortex are epileptiform in an electrogenic rat model of chronic temporal lobe epilepsy. *Journal of Neurophysiology*, 80: 230-240, 1998.

23. L. Glass, A.L. Goldberger, M. Courtemanche, and A. Shrier. Nonlinear dynamics, chaos and complex cardiac arrhythmias. *Proceedings of the Royal Society of London, Series A, Mathematical and Physical Sciences*, A413, pp. 9-26, 1987.

24. P. Gloor. Neurobiological substrates of ictal behavioral changes. *Advances in Neurology*, 55: 1-34, 1991.

25. P. Grassberger and I. Procaccia. Measuring the strangeness of strange attractors. *Physica D*, 9: 189-208, 1983.

26. P. Grassberger. An optimal box-assisted algorithm for fractal dimensions. *Physics Letters A*, 148: 63-68, 1990.

27. M. Gruenthal. Electroencephalographic and histological characteristics of a model of limbic status epilepticus permitting direct control over seizure duration. *Epilepsy Research*, 29: 221-232, 1998.

28. W.A. Hauser. Incidence and prevalence. In J. Engel Jr. and T.A. Pedley, editors, *Epilepsy: A Comprehensive Textbook*, pages 47-57. Lippincott-Raven, Philadelphia, 1997.

29. L.D. Iasemidis, J.C. Sackellares, H.P. Zaveri, and W.J. Williams. Phase space topography of the electrocorticogram and the Lyapunov exponent in partial seizures. *Brain Topography*, 2: 187-201, 1990.

30. L.D. Iasemidis and J.C. Sackellares. The temporal evolution of the largest Lyapunov exponent on the human epileptic cortex. In D.W. Duke and W.S. Pritchard, editors, *Measuring Chaos in the Human Brain*, pages 49-82. World Scientific, Singapore, 1991.

31. L.D. Iasemidis. *On the Dynamics of the Human Brain in Temporal Lobe Epilepsy.* PhD Thesis, University of Michigan, Ann Arbor, 1991.

32. L.D. Iasemidis, L.D. Olson, R.S. Savit, and J.C. Sackellares. Time dependencies in the occurrence of epileptic seizures. *Epilepsy Research,* 17: 81-94, 1994.

33. L.D. Iasemidis, J.C. Principe, J.M. Czaplewski, et al. Spatiotemporal transition to epileptic seizures: A nonlinear dynamical analysis of scalp and intracranial EEG recordings. In F.H. Lopes de Silva, J.C. Principe, and L.B. Almeida, editors, *Spatiotemporal Models in Biological and Artificial Systems,* pages 81-88. IOS Press, Amsterdam, 1997.

34. L.D. Iasemidis, J.C. Principe, J.C. Sackellares. Measurement and quantification of spatiotemporal dynamics of human epilepogenic seizures. In: M. Akay, editor, *Nonlinear Signal Processing in Medicine.* IEEE Press, 1999.

35. L.D. Iasemidis, P.M. Pardalos, J.C. Sackellares, and D.S. Shiau. Quadratic binary programming and dynamical system approach to determine the predictability of epileptic seizures. *Journal of Combinatorial Optimization,* 5: 9-26, 2001.

36. L.D. Iasemidis, D.S. Shiau, W. Chaowolitwongse, J.C. Sackellares, P.M. Pardalos, J.C. Principe, P.R. Carney, A. Prasad, B. Veeramani, and K. Tsakalis. Adaptive epileptic seizure prediction system. *IEEE Transactions on Biomedical Engineering,* 50(5): 616-627, 2003.

37. L.D. Iasemidis, P.M. Pardalos, D.S. Shiau, W. Chaovalitwongse, M. Narayanan, S. Kumar, P.R. Carney, and J.C. Sackellares. Prediction of human epileptic seizures based on optimization and phase changes of brain electrical activity. *Optimization Methods and Software,* 18: 81-104, 2003.

38. B.H. Jansen. Is it and so what? A critical review of EEG-chaos. In D.W. Duke and W.S. Pritchard, editors, *Measuring Chaos in the Human Brain.* World Scientific, Singapore, 1991.

39. K.Y. Jung, J.M. Kim, and D.W. Kim. Nonlinear dynamic characteristics of electroencephalography in a high-dose pilocarpine-induced status epilepticus model. *Epilepsy Research,* 54: 179-188, 2003.

40. J.F. Kaiser. On a simple algorithm to calculate the "energy" of a signal. *ICASSP,* 381-384, 1990.

41. J.A.S. Kelso, A.J. Mandell, M.F. Shlesinger. *Dynamic patterns in complex systems.* World Scientific, Singapore, 1988.

42. M. Le Van Quyen, J. Martinerie, V. Navarro, M. Baulac, and F.J. Varela. Characterizing neurodynamic changes before seizures. *Journal of Clinical Neurophysiology,* 18: 191-208, 2001.

43. M. Le Van Quyen, J. Martinerie, V. Navarro, P. Boon, M. D'Have, C. Adam, B. Renault, F. Varela, and M. Baulac. Anticipation of epileptic seizures from standard EEG recordings. *Lancet,* 357: 183-188, 2001.

44. K. Lehnertz and C.E. Elger. Can epileptic seizures be predicted? Evidence from nonlinear time series analysis of brain electrical activity. *Physical Review Letters,* 80: 5019-5022, 1998.

45. K. Lehnertz, R.G. Andrzejak J. Arnhold, T. Kreuz, F. Mormann, C. Rieke, G. Widman, and C.E. Elger. Nonlinear EEG analysis in epilepsy: Its possible use for interictal focus localization, seizure anticipation, and prevention. *Journal of Clinical Neurophysiology,* 18: 209-222, 2001.

46. B. Litt, R. Esteller, J. Echauz, M. D'Alessandro, R. Short, T. Henry, P. Pennell, C. Epstein, R. Bakay, M. Dichter, and G. Vachtsevanos. Epileptic seizures may begin hours in advance of clinical onset: a report of five patients. *Neuron*, 30: 51-64, 2001.

47. W. Loscher. Animal models of intractable epilepsy. *Progress in Neurobiology*, 53: 239-258, 1997.

48. W. Loscher. Animal models of epilepsy for the development of antiepileptogenic and disease-modifying drugs. A comparison of the pharmacology of kindling and post-status epilepticus models of temporal lobe epilepsy. *Epilepsy Research*, 50: 105-123, 2002.

49. E.W. Lothman, E.H. Bertram, J. Kapur, and J.L. Stringer. Recurrent spontaneous hippocampal seizures in the rat as a chronic sequela to limbic status epilepticus. *Epilepsy Research*, 6: 110-118, 1990.

50. E.W. Lothman, E.H. Bertram, and J.L. Stringer. Functional anatomy of hippocampal seizures. *Progress in Neurobiology*, 37: 1-82, 1991.

51. P.S. Mangan and E.W. Lothman. Profound disturbances of pre- and postsynaptic GABA(B)-receptor-mediated processes in region CA1 in a chronic model of temporal lobe epilepsy. *Journal of Neurophysiology*, 76: 1282-1296, 1996.

52. R. Manuca and R. Savit. Stationarity and nonstationarity in time series analysis. *Physica D*, 99: 134-161, 1996.

53. M. Markus, S.C. Muller, G. Nicolis. *From chemical to biological organization*. Springer-Verlag, Berlin, New York, 1988.

54. J. Martinerie, C. Adam, M. Le Van Quyen, M. Baulac, S. Clemenceau, B. Renault, and F.J. Varela. Epileptic seizures can be anticipated by nonlinear analysis. *Nature Medicine*, 4: 1173-1176, 1998.

55. R.M. May. Simple mathematical models with very complicated dynamics. *Nature*, 261: 459-467, 1976.

56. G. Mayer-Kress. *Dimension and entropies in chaotic systems*, Springer-Verlag, Berlin, 1986.

57. S.P. Nair, D.S. Shiau, W.M. Norman, D. Shenk, W. Suharitdamrong, L.D. Iasemidis, P.M. Pardalos, J.C. Sackellares, and P.R. Carney. Dynamical changes in the rat chronic limbic epilepsy model. *Epilepsia*, 45(S7): 211-212, 2004.

58. M. Palus. Testing for nonlinearity using redundancies: quantitative and qualitative aspects. *Physica D*, 80: 186-205, 1995.

59. M. Palus. Nonlinearity in normal human EEG: cycles, temporal asymmetry, nonstationarity and randomness, not chaos. *Biological Cybernetics*, 75: 389-396, 1996.

60. J.P. Pijn, J. Van Neerven, A. Noest, and F.H. Lopes da Silva. Chaos or noise in EEG signals; dependence on state and brain site. *Electroencephalography and Clinical Neurophysiology*, 79: 371-381, 1991.

61. M. Quigg, E.H. Bertram, T. Jackson, and E. Laws. Volumetric magnetic resonance imaging evidence of bilateral hippocampal atrophy in mesial temporal lobe epilepsy. *Epilepsia*, 38: 588-594, 1997.

62. M. Quigg, M. Staume, M. Menaker, and E.H. Bertram. Temporal distribution of partial seizures: Comparison of an animal model with human partial epilepsy. *Annals of Neurology*, 43: 748-755, 1998.

63. R.Q. Quiroga, H. Garcia, and A. Rabinowicz. Frequeny evolution during tonic-clonic seizures. *Electromyography and Clinical Neurophysiology*, 42: 323-331, 2002.

64. R.J. Racine. Modification of seizure activity by electrical stimulation. II. Motor seizure. *Electroencephalography and Clinical Neurophysiology*, 32: 281-294, 1972.

65. K.A. Richardson, B.J. Gluckman, S.L. Weinstein, C.E. Glosch, J.B. Moon, R.P. Gwinn, K. Gale, and S.J. Schiff. In vivo modulation of hippocampal epileptiform activity with radial electric fields. *Epilepsia* 44: 768-777, 2003.

66. J.C.Sackellares, L.D. Iasemidis, H.P. Zaveri, and W.J. Williams. Measurement of chaos to localize seizure onset. *Epilepsia*, 30(5): 663, 1989.

67. Y. Salant, I. Gathe, and O. Henriksen. Prediction of epileptic seizures from two-channel EEG. *Medical and Biological Engineering and Computing*, 36: 549-556, 1998.

68. J.P. Stables, E.H. Bertram, H.S. White, D.A. Coulter, M.A. Dichter, M.P. Jacobs, W. Loscher, D.H. Lowenstein, S.L. Moshe, J.L. Noebels, and M. Davis. Models for epilepsy and epileptogenesis: Report from the NIH workshop, Bethesda, Maryland. *Epilepsia*, 43: 1410-1420, 2002.

69. S. Sunderam, I. Osorio, M.G. Frei, and J.F. Watkins. Stochastic modeling and prediction of experimental seizures in Sprague-Dawley rats. *Journal of Clinical Neurophysiology*, 18: 275-282, 2001.

70. F. Takens. Detecting strange attractors in turbulence. In D.A. Rand and L.S. Young, editors, *Dynamical systems and turbulence: lecture notes in mathematics*, pages 366-381. Springer-Verlag, 1981.

71. J. Theiler. Spurious dimension from correlation algorithms applied to limited time-series data. *Physical Review A*, 34: 2427-2433, 1986.

72. J. Theiler, B. Galdrikian, S. Eubank, and J.D. Farmer. Using surrogate data to detect non-linearity in time series. In M.C. Casdagli and S. Eubank, editors, *Nonlinear Modeling and Forecasting*, pages 163-188. Addison-Wesley, Reading, MA, 1991.

73. L. Vercueil, A. Benazzouz, C. Deransart, K. Bressand, C. Marescaux, A. Depaulis, and A.L. Benabid. High-frequency stimulation of the subthalamic nucleus suppresses absence seizures in the rat: comparison with neurotoxic lesions. *Epilepsy Research*, 31: 39-46, 1998.

74. A. Wolf, J.B. Swift, H.L. Swinney, and J.A. Vastanao. Determining Lyapunov exponents from a time series. *Physica D*, 16: 285-317, 1985.

75. J. Zabara. Inhibition of experimental seizures in canines by repetitive vagal stimulation. *Epilepsia*, 33: 1005-1012, 1992.

76. H. Zaveri, W.J. Williams, and J.C. Sackellares. Energy based detectors of seizures. Presented at 15th Annual International Conference on Engineering and Medicine in Biology, 1993.

77. H.P. Zaveri, W.J. Williams, J.C. Sackellares, A. Beydoun, R.B. Duckrow, and S.S. Spencer. Measuring the coherence of intracranial electroencephalograms. *Clinical Neurophysiology*, 110: 1717-1725, 1999.

Network-Based Techniques in EEG Data Analysis and Epileptic Brain Modeling

Oleg A. Prokopyev[1], Vladimir L. Boginski[2], Wanpracha Chaovalitwongse[3], Panos M. Pardalos[4,7], J. Chris Sackellares[5,6,7], and Paul R. Carney[6,8]

[1] Department of Industrial Engineering, University of Pittsburgh, USA
  prokopyev@engr.pitt.edu
[2] Department of Industrial Engineering, Florida State University, USA
  boginski@eng.fsu.edu
[3] Department of Industrial and Systems Engineering, Rutgers, The State University of New Jersey, USA
  wchaoval@rci.rutgers.edu
[4] Department of Industrial and Systems Engineering, University of Florida, USA
  pardalos@ufl.edu
[5] Department of Neuroscience, University of Florida, USA
  sackellares@mbi.ufl.edu
[6] Department of Neurology, University of Florida, USA
[7] Department of Biomedical Engineering, University of Florida, USA
[8] Department of Pediatrics, University of Florida, USA

Summary. We discuss a novel approach of modeling the behavior of the epileptic human brain, which utilizes network-based techniques in combination with statistical preprocessing of the electroencephalographic (EEG) data obtained from the electrodes located in different parts of the brain. In the constructed graphs, the vertices represent the "functional units" of the brain, where electrodes are located. Studying dynamical changes of the properties of these graphs provides valuable information about the patterns characterizing the behavior of the brain prior to, during, and after an epileptic seizure.

Key words: Graph theory, data analysis, EEG data, brain, epilepsy.

1 Introduction

Human brain is one of the most complex systems ever studied by scientists. Enormous number of neurons and the dynamic nature of connections between them makes the analysis of brain function especially challenging. One of the most important directions in studying the brain is treating disorders of the central nervous system. For instance, *epilepsy* is a common form of such disorders, which affects approximately 1% of the human population. Essentially, epileptic seizures represent excessive and hypersynchronous activity of the neurons in the cerebral cortex.

During the last several years, significant progress in the field of epileptic seizures prediction has been made. The advances are associated with the extensive use of *electroencephalograms* (*EEG*) which can be treated as a quantitative representation of the brain function. Rapid development of computational equipment has made possible to store and process huge amounts of EEG data obtained from recording devices. The availability of these massive datasets gives a rise to another problem - utilizing mathematical tools and data mining techniques for extracting useful information from EEG data. Is it possible to construct a "simple" mathematical model based on EEG data that would reflect the behavior of the epileptic brain?

In this chapter, we make an attempt to create such a model using *graph-theoretical approach*. A *graph* (*network*) - a set of vertices (dots) and edges (links) - is a structure that can be easily understood and visualized. The methodology of representing massive datasets arising in diverse areas as graphs are widely discussed in the literature nowadays [1, 2, 3, 4, 5, 8, 19]. In many cases, studying the structure of such graph may give a non-trivial information about the properties of the real-life system it represents.

In the case of the human brain and EEG data, we apply a relatively simple network-based approach. We represent the electrodes used for obtaining the EEG readings, which are located in different parts of the brain, as the vertices of the constructed graph. The data received from every single electrode is essentially a time series reflecting the change of the EEG signal over time. Later in the chapter we will discuss the quantitative measure characterizing statistical relationships between the recordings of every pair of electrodes - so called T-index. The values of the T-index T_{ij} measured for all pairs of electrodes i and j enable us to establish certain rules of placing edges connecting different pairs of vertices i and j depending on the corresponding values of T_{ij}. Using this technique, we develop several graph-based mathematical models and study the dynamics of the structural properties of these graphs. As we will see, these models can provide useful information about the behavior of the brain prior to, during, and after an epileptic seizure.

2 Graph Theory Basics

In this section we give a brief introduction to definitions and notations from the graph theory used later in the chapter [7].

A *graph* is a pair $G = (V, E)$, where V is any set, called the *vertex set*, and the *edge set* E is any subset of the set of all 2-element subsets of V. The elements of V are called *vertices* (or *nodes*, or *points*) of the graph G, the elements of E are called its *edges* (or *arcs*). The number of vertices of a graph $G = (V, E)$ is denoted usually as $|G|$, or $|V|$ and its number of edges is denoted by $||G||$, or $|E|$.

Graphs with a number of edges roughly quadratic in their number of vertices are usually called dense. The value $||G||/\binom{|G|}{2}$, which represents the ratio of the actual number of edges in the graph and the maximum possible number of edges, is called the *edge density* of the graph.

A graph having a weight, or a number (which are usually taken to be positive), associated with each edge is called *weighted*.

Two vertices x, y of G are *adjacent*, or *neighbors*, if (x, y) is and edge of G, i.e. $(x, y) \in E$. If all the vertices of G are pairwise adjacent (i.e., the graph contains all possible edges), then G is *complete*. Any complete subgraph of G is called a *clique*. A clique of a maximum size is called a *maximum clique*.

The *degree* $d_G(v) = d(v)$ of a vertex v is the number of edges at v, i.e. the number of neighbors of v. The number

$$d(G) = \frac{1}{|V|} \sum_{v \in V} d(v)$$

is the *average degree* of G.

A *path* is a non-empty graph $P = (V, E)$ of the form

$$V = x_0, x_1, \ldots, x_k \quad E = x_0 x_1, x_1 x_2, \ldots, x_{k-1} x_k,$$

where x_i are all distinct . We can refer to a path by the sequence of its vertices, that is $P = x_0 x_1 \ldots x_k$. The vertices x_o and x_k are linked by P and are called *ends*. The number of edges of a path is its *length*. If $P = x_0 \ldots x_{k-1}$ is a path then the graph $C = P + x_{k-1} x_0$ is called a *cycle*. As with paths we can denote cycle by its (cyclic) sequence of vertices; the above cycle C can be written as $x_0 x_1 \ldots x_{k-1} x_0$.

A graph G is called *connected* if any two of its vertices are linked by a path in G. A maximal connected subgraph of G is called a *component* of G.

A *spanning tree* of a graph is a subset of $|V| - 1$ edges which form a tree. Every connected graph contains a *spanning tree* and any minimal connected spanning subgraph is a tree. A minimum-weight tree in a weighted graph which contains all of the graph's vertices is called a *minimum spanning tree*.

3 Statistical Preprocessing of EEG Data

3.1 Datasets

The datasets consisting of continuous long-term (3 to 12 days) multichannel intracranial EEG recordings that had been acquired from 4 patients with medically intractable temporal lobe epilepsy. Each record included a total of 28 to 32 intracranial electrodes (8 subdural and 6 hippocampal depth electrodes for each cerebral hemisphere). A diagram of electrode locations is provided in Figure 1.

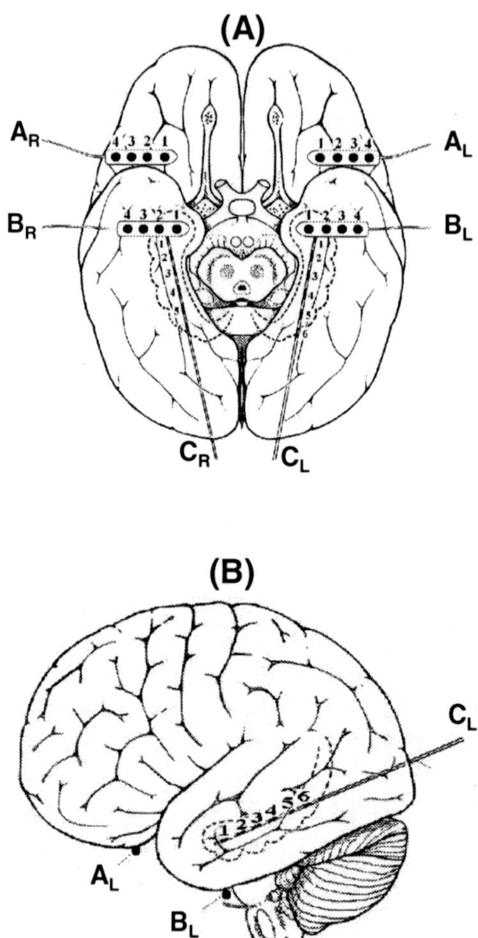

Fig. 1. (A) Inferior transverse and (B) lateral views of the brain, illustrating approximate depth and subdural electrode placement for EEG recordings are depicted. Subdural electrode strips are placed over the left orbitofrontal (A_L), right orbitofrontal (A_R), left subtemporal (B_L), and right subtemporal (B_R) cortex. Depth electrodes are placed in the left temporal depth (C_L) and right temporal depth (C_R) to record hippocampal activity.

3.2 STL_{max} and T-statistics

In this subsection we give a brief introduction to nonlinear measures and statistics used to analyze EEG data (for more information see [11, 13, 15]).

Since the brain is a nonstationary system, algorithms used to estimate measures of the brain dynamics should be capable of automatically identifying and appropriately weighing existing transients in the data. In a chaotic system, orbits originating from similar initial conditions (nearby points in the state space) diverge exponentially (expansion process). The rate of divergence is an important aspect of the system dynamics and is reflected in the value of Lyapunov exponents. The method used for estimation of the short time largest Lyapunov exponent STL_{max}, an estimate of L_{max} for nonstationary data, is explained in detail in [10, 12, 18].

By splitting the EEG time series recorded from each electrode into a sequence of non-overlapping segments, each 10.24 sec in duration, and estimating STL_{max} for each of these segments, profiles of STL_{max} over time are generated.

Having estimated the STL_{max} temporal profiles at an individual cortical site, and as the brain proceeds towards the ictal state, the temporal evolution of the stability of each cortical site is quantified. The spatial dynamics of this transition are captured by consideration of the relations of the STL_{max} between different cortical sites. For example, if a similar transition occurs at different cortical sites, the STL_{max} of the involved sites are expected to converge to similar values prior to the transition. Such participating sites are called "critical sites", and such a convergence "dynamical entrainment". More specifically, in order for the dynamical entrainment to have a statistical content, we allow a period over which the difference of the means of the STL_{max} values at two sites is estimated. We use periods of 10 minutes (i.e. moving windows including approximately 60 STL_{max} values over time at each electrode site) to test the dynamical entrainment at the 0.01 statistical significance level. We employ the T-index (from the well-known paired T-statistics for comparisons of means) as a measure of distance between the mean values of pairs of STL_{max} profiles over time. The T-index at time t between electrode sites i and j is defined as:

$$T_{i,j}(t) = \sqrt{N} \times |E\{STL_{max,i} - STL_{max,j}\}|/\sigma_{i,j}(t) \tag{1}$$

where $E\{\cdot\}$ is the sample average difference for the $STL_{max,i} - STL_{max,j}$ estimated over a moving window $w_t(\lambda)$ defined as:

$$w_t(\lambda) = \begin{cases} 1 \text{ if } \lambda \in [t - N - 1, t] \\ 0 \text{ if } \lambda \notin [t - N - 1, t], \end{cases}$$

where N is the length of the moving window. Then, $\sigma_{i,j}(t)$ is the sample standard deviation of the STL_{max} differences between electrode sites i and j within the moving window $w_t(\lambda)$. The T-index follows a t-distribution with

N-1 degrees of freedom. For the estimation of the $T_{i,j}(t)$ indices in our data we used $N = 60$ (i.e., average of 60 differences of STL_{max} exponents between sites i and j per moving window of approximately 10 minute duration). Therefore, a two-sided t-test with $N - 1(= 59)$ degrees of freedom, at a statistical significance level α should be used to test the null hypothesis, H_o: "brain sites i and j acquire identical STL_{max} values at time t". In this experiment, we set the probability of a type I error $\alpha = 0.01$ (i.e., the probability of falsely rejecting H_o if H_o is true, is 1%). For the T-index to pass this test, the $T_{i,j}(t)$ value should be within the interval $[0, 2.662]$. We will refer to the upper bound of this interval as $T_{critical}$.

4 Graph Structure of the Epileptic Brain

In studying real-life complex systems it is very important to construct an appropriate mathematical model describing this system, using its certain characteristic properties. Network-based approach is one of the most promising techniques in this area [6, 9, 19]. In many cases this approach may significantly simplify the analysis of a system, and provide a new insight into its structural properties.

4.1 Key Idea of the Model

If we model the brain (with epilepsy) by a graph (where nodes are "functional units" of the system and edges are connections between them) we need to answer the following questions: what properties the model has, i.e. what the properties of this graph are; how the properties of the graph change prior to, during, and after epileptic seizures. We try to answer this question using the following idea – we study the system of the electrodes as a weighted graph where nodes are electrodes and weights of the edges between nodes are values of the corresponding T-index. More specifically, we consider three types of graphs constructed using this principle:

- *GRAPH-I* is a complete graph, i.e., it has all possible edges,
- *GRAPH-II* is obtained from the complete graph by removing all the edges (i, j) for which the corresponding value of T_{ij} is *greater* than $T_{critical}$,
- *GRAPH-III* is obtained from the complete graph by removing all the edges (i, j) for which the corresponding value of T_{ij} is *less* than $T_{critical}$ *10 minutes after the seizure point* and *greater* than $T_{critical}$ *at the seizure point.*

Interpretation of the Considered Graph Models

Before proceeding with the further discussion, we need to give a conceptual interpretation of the ideas lying behind introducing the aforementioned graphs.

- *GRAPH-I* contains all the edges connecting the considered brain sites, and it is considered in order to reflect the general distribution of the values of T-indices between each pair of vertices (i.e., the weights of the corresponding edges).

- *GRAPH-II* contains only the edges connecting the brain sites (electrodes) that are statistically *entrained* at a certain time, which means that they exhibit a similar behavior. Recall that a pair of electrodes is considered to be entrained if the value of the corresponding T-index between them is less than $T_{critical}$, that is why we remove all the edges with the weights greater than $T_{critical}$. The main point of our interest is studying the *evolution* of the properties of this graph over time. As we will see in the next subsections, this analysis can help in revealing the *dynamical patterns* underlying the functioning of the brain during preictal, ictal, postictal, and interictal states. Therefore, this graph can be used as a basis for the mathematical model describing some characteristics of the epileptic brain.

- *GRAPH-III* is constructed to reflect the connections only between those electrodes that are entrained during the seizure, but are not entrained 10 minutes after the seizure. The motivation for introducing this graph is the existence of "resetting" of the brain after the seizure [14, 16, 17], which is essentially the divergence of the profiles of the STL_{max} time series. As it was indicated above, this divergence is characterized by the values of T-index greater than $T_{critical}$.

4.2 Properties of the Graphs

In this subsection, we investigate the properties of the considered graph models and give an intuitive explanation of the observed results. As we will see, there are specific tendencies in the evolution of the properties of the considered graphs prior to, during, and after epileptic seizures, which indicates that the proposed models capture certain trends in the behavior of the epileptic brain.

Edge Density

Recall that *GRAPH-II* was introduced to reflect the connections between brain sites that are statistically entrained at a certain time moment. Figure 2 illustrates the typical evolution of the number of edges in *GRAPH-II* over time. As it was indicated above, edge density of the graph is proportional to the number of edges in a graph. It is easy to notice that the number of edges in *GRAPH-II* dramatically increases at seizure points (represented by dashed vertical lines), and it decreases immediately after seizures. It means that the global structure of the graph significantly changes during the seizure and after the seizure, i.e. the density of increases during ictal state and decreases in postictal state, which supports the idea that the epileptic brain (and *GRAPH-II* as the model of the brain) experiences a "phase transition" during the seizure.

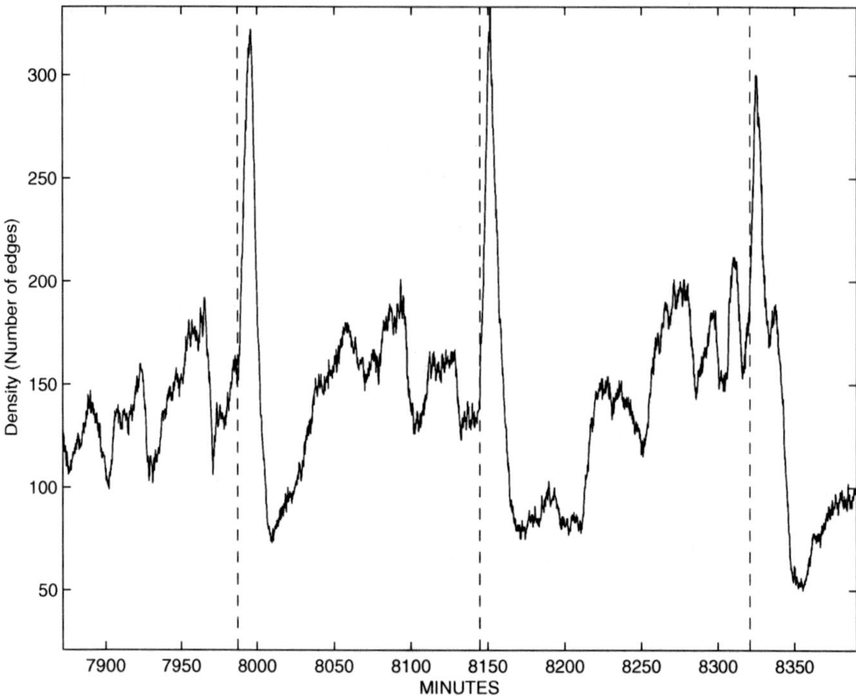

Fig. 2. Number of edges in *GRAPH-II*

Connectivity

Another important property of *GRAPH-II* that we are interested in is its *connectivity*. We need to check if this graph is connected prior to, during, and after epileptic seizures, and if not, find the size of its largest connected component. Clearly, this information will also be helpful in the analysis of the structural properties of the brain. If *GRAPH-II* is connected (i.e., the size of the largest connected component is equal to the number of vertices in the graph), then all the functional units of the brain are "linked" with each other by a path, and in this case the brain can be treated as an "integrated" system, however, if the size of the largest connected component in *GRAPH-II* is significantly smaller than the total number of the vertices, it means that the brain becomes "separated" into smaller disjoint subsystems.

The size of the largest connected component of the *GRAPH-II* is presented in Figure 3. One can see that *GRAPH-II* is connected during the interictal period (i.e., the brain is a connected system), however, it becomes disconnected after the seizure (during the postical state): the size of the largest connected component significantly decreases. This fact is not surprising and can be intuitively explained, since after the seizure the brain needs some time to "reset" [14, 16, 17] and restore the connections between the functional units.

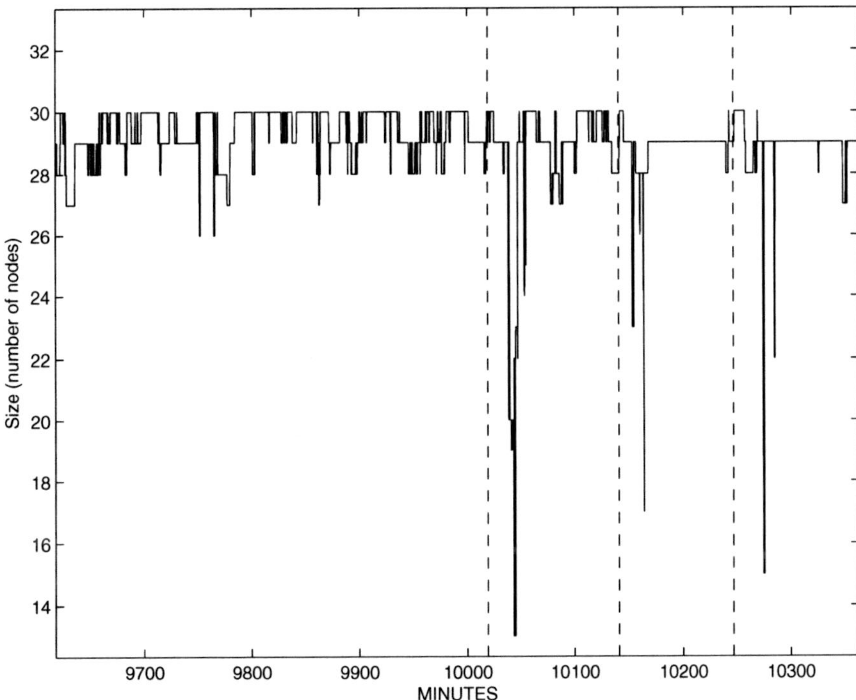

Fig. 3. The size of the largest connected component in *GRAPH-II*. Number of nodes in the graph is 30.

Minimum Spanning Tree

The next subject of our discussion is the analysis of *minimum spanning trees* of *GRAPH-I*, which was defined as the graph with all possible edges, where each edge (i, j) has the weight equal to the value of T-index T_{ij} corresponding to brain sites i and j. The definition of *Minimum Spanning Tree* was given in Section 2. Studying minimum spanning trees in *GRAPH-I* is motivated by the hypothesis that the seizure signal in the brain propagates to all functional units *according to the minimum spanning tree,* i.e. along the edges with small values of T_{ij}. This hypothesis is partially supported by the behavior of the average T-index of the edges corresponding to the Minimum Spanning Tree of *GRAPH-I*, which is shown in Figure 4.

However, this hypothesis cannot be verified using the considered data, since the values of average T-indices are calculated over a 10-minute interval, whereas the the seizure signal propagates in a fraction of a second. Therefore, in order to check if the seizure signal actually spreads along the minimum spanning tree, one needs to introduce other nonlinear measures to reflect the behavior of the brain over short time intervals.

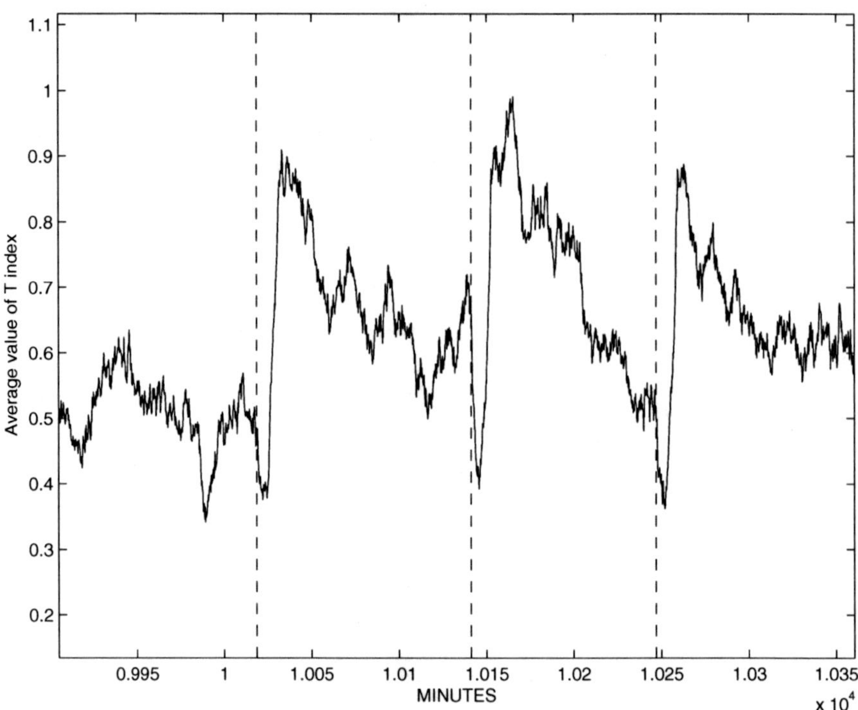

Fig. 4. Average value of T index of the edges in Minimum Spanning Tree of *GRAPH-I*.

Also, note that the average value of the T index in the Minimum Spanning Tree is less than $T_{critical}$, which also supports the above statement about the connectivity of the system.

Degrees of the Vertices

Another important issue that we analyze here is the *degrees of the vertices* in *GRAPH-II*. Recall that the degree of a vertex is defined simply as the number of edges emanating from it.

We look at the behavior of the average degree of the vertices in *GRAPH-II* over time. Clearly, this plot is very similar to the behavior of the edge density of *GRAPH-II* (see Figure 5).

We are also particularly interested in *high-degree vertices*, i.e., the functional units of the brain that are at a certain time moment connected (entrained) with many other brain sites. Interestingly enough, the vertex with a maximum degree in *GRAPH-II* usually corresponds to the electrode which is located in RTD (right temporal depth) or RST (right subtemporal cortex), in other words, the vertex with the maximum degree is located *near the epileptogenic focus*.

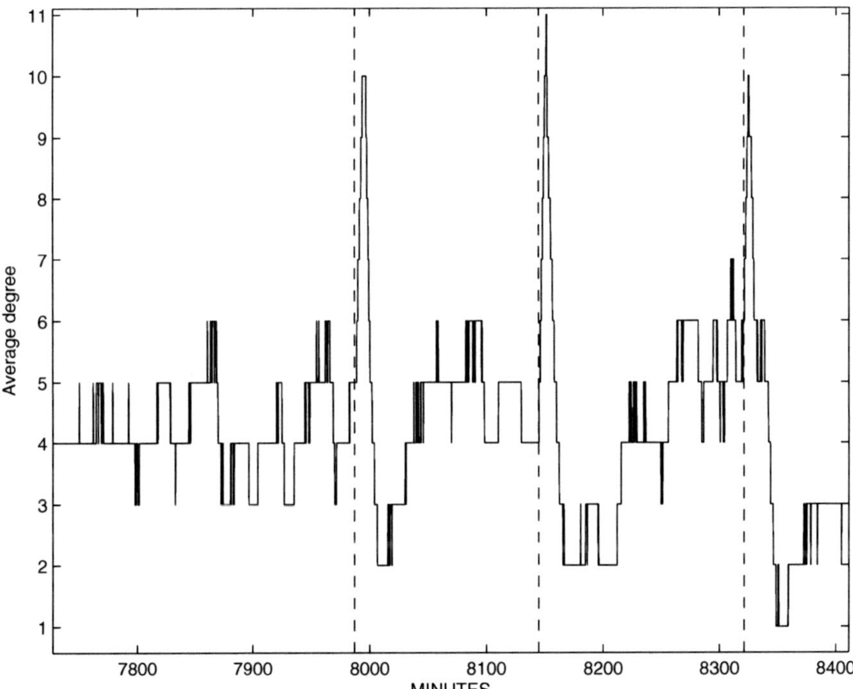

Fig. 5. Average degree of the nodes of *GRAPH-II*.

Maximum Cliques

In the previous works in the field of epileptic seizure prediction, a quadratic 0–1 programming approach based on EEG data was introduced [13]. In fact, this approach utilizes the same preprocessing technique (i.e., calculating the values of T-indices for all pairs of electrode sites) as we apply in this chapter. In this subsection, we will briefly describe this quadratic programming technique and relate it to the graph models introduced above.

The main idea of the considered quadratic programming approach is to construct a model that would select a certain number of so-called "critical" electrode sites, i.e., those that are the most entrained during the seizure. According to Section 3, such group of electrode sites should produce a minimal sum of T-indices calculated for all pairs of electrodes within this group. If the number of critical sites is set equal to k, and the total number of electrode sites is n, then the problem of selecting the optimal group of sites can be formulated as the following quadratic 0-1 problem [13]:

$$\min \quad x^T A x \tag{2}$$

$$\text{s.t.} \quad \sum_{i=1}^{n} x_i \; = k. \tag{3}$$

$$x_i \in \{0, 1\} \; \forall i \in \{1, ..., n\} \tag{4}$$

In this setup, the vector $x = (x_1, x_2, \ldots, x_n)$ consists of the components equal to either 1 (if the corresponding site is included into the group of critical sites) or 0 (otherwise), and the elements of the matrix $A = [a_{ij}]_{i,j=1,...,n}$ are the values of T_{ij}'s at the seizure point.

However, as it was shown in the previous studies, one can observe the "resetting" of the brain after seizures' onset [17, 14, 16], that is, the divergence of STL_{max} profiles after a seizure. Therefore, to ensure that the optimal group of critical sites shows this divergence, one can reformulate this optimization problem by adding one more quadratic constraint:

$$x^T B x \geq T_{critical} \, k \, (k-1), \tag{5}$$

where the matrix $B = [b_{ij}]_{i,j=1,...,n}$ is the T-index matrix of brain sites i and j within 10 minute windows after the onset of a seizure.

This problem is then solved using standard techniques, and the group of k critical sites is found. It should be pointed out that the number of critical sites k is *predetermined*, i.e., it is defined empirically, based on practical observations. Also, note that in terms of *GRAPH-I* model this problem represents finding a subgraph of *GRAPH-I* of a fixed size, satisfying the properties specified above.

Now, recall that we introduced *GRAPH-III* using the same principles as in the formulation of the above optimization problem, that is, we considered the connections only between the pairs of sites i, j satisfying both of the two conditions: $T_{ij} < T_{critical}$ at the seizure point, and $T_{ij} > T_{critical}$ 10 minutes after the seizure point, which are exactly the conditions that the critical sites must satisfy. A natural way of detecting such a groups of sites is to find *cliques* in *GRAPH-III*. Since a clique is a subgraph where all vertices are interconnected, it means that all pairs of electrode sites in a clique would satisfy the aforementioned conditions. Therefore, it is clear that the size of the *maximum clique* in *GRAPH-III* would represent the *upper bound* on the number of selected critical sites, i.e., the maximum value of the parameter k in the optimization problem described above.

Computational results indicate that the maximum clique sizes for different instances of *GRAPH-III* are close to the actual values of k empirically selected in the quadratic programming model, which shows that these approaches are consistent with each other.

5 Graph as a Macroscopic Model of the Epileptic Brain

Based on the results obtained in the sections above, we now can formulate the graph model which describes the behavior of the epileptic brain at the macroscopic level. The main idea of this model is to use the properties of *GRAPH-I*, *GRAPH-II*, and *GRAPH-III* as a characterization of the behavior of the brain prior to, during, and after epileptic seizures. According to this graph model, the graphs reflecting the behavior of the epileptic brain demonstrate the following properties:

- Increase and decrease of the edge density and the average degree of the vertices during and after the seizures respectively;
- The graph is connected during the interictal state, however, it becomes disconnected right after the seizures (during the postictal state);
- The vertex with the maximum degree corresponds to the epileptogenic focus.

Moreover, one of the advantages of the considered graph model is the possibility to detect special formations in these graphs, such as cliques and minimum spanning trees, which can be used for further studying of various properties of the epileptic brain.

6 Concluding Remarks and Directions of Future Research

In this chapter, we have made the initial attempt to analyze EEG data and model the epileptic brain using network-based approaches. Despite the fact that the size of the constructed graphs is rather small, we were able to determine specific patterns in the behavior of the epileptic brain based on the information obtained from statistical analysis of EEG data. Clearly, this model can be made more accurate by considering more electrodes corresponding to smaller functional units.

Among the directions of future research in this field, one can mention the possibility of developing *directed* graph models based on the analysis of EEG data. Such models would take into account the natural "asymmetry" of the brain, where certain functional units control the other ones. Also, one could apply a similar approach to studying the patterns underlying the brain function of the patients with other types of disorders, such as Parkinson's disease, or sleep disorder. Therefore, the methodology introduced in this chapter can be generalized and applied in practice.

Acknowledgement

This work was funded, in part, by The National Institutes of Health, Epilepsy Foundation of America, UF Division of Sponsored Research and Children's Miracle Network. The authors also want to thanks Shands Health System EEG Lab.

References

1. J. Abello, P.M. Pardalos, and M.G.C. Resende. *Handbook of Massive Data Sets*, Kluwer Academic Publishers, 2002.

2. J. Abello, P.M. Pardalos, and M.G.C. Resende. On maximum clique problems in very large graphs, In *DIMACS Series*, 50: 119-130, 1999.

3. V. Boginski, S. Butenko, and P.M. Pardalos. Modeling and Optimization in Massive Graphs. In P.M. Pardalos and H. Wolkowicz, editors, *Novel Approaches to Hard Discrete Optimization*, pages 17-39. Fields Institute Communications Series, American Mathematical Society, 2003.

4. V. Boginski, S. Butenko, and P.M. Pardalos. On Structural Properties of the Market Graph. In A. Nagurney, editor, *Innovations in Financial and Economic Networks*, pages 29-45. Edward Elgar Publishing, 2003.

5. A. Broder, R. Kumar, F. Maghoul, P. Raghavan, S. Rajagopalan, R. Stata, A. Tomkins, and J. Wiener. Graph structure in the Web. *Computer Networks*, 33: 309–320, 2000.

6. C. Cherniak, Z. Mokhtarzada, U. Nodelman. Optimal-Wiring Models of Neuroanatomy. In G. A. Ascoli, editor, *Computational Neuroanatomy: Principles and Methods*, pages 71-82. Humana Press, Totowa, NJ, 2002.

7. R. Diestel. *Graph Theory*. Springer-Verlag, New York, 1997.

8. B. Hayes. Graph Theory in Practice. *American Scientist*, 88: 9-13 (Part I), 104-109 (Part II), 2000.

9. C.C. Hilgetag, R. Kötter, K.E. Stephen, and O. Sporns. Computational Methods for the Analysis of Brain Connectivity, In G. A. Ascoli, editor, *Computational Neuroanatomy*, Humana Press, 2002.

10. L.D. Iasemidis, J.C. Sackellares. The evolution with time of the spatial distribution of the largest Lyapunov exponent on the human epileptic cortex. In D.W. Duke and W.S. Pritchard, editors, *Measuring Chaos in the Human Brain*, pages 49-82. World Scientific, Singapore, 1991.

11. L.D. Iasemidis, J.C. Principe, J.M. Czaplewski, R.L. Gilmore, S.N. Roper, and J.C. Sackellares. Spatiotemporal transition to epileptic seizures: a nonlinear dynamical analysis of scalp and intracranial EEG recordings. In F.L. Silva, J.C. Principe, and L.B. Almeida, editors, *Spatiotemporal Models in Biological and Artificial Systems*, pages 81-88. IOS Press, Amsterdam, 1997.

12. L.D. Iasemidis, J.C. Principe, and J.C. Sackellares. Measurement and quantification of spatiotemporal dynamics of human epileptic seizures. In M. Akay, editor, *Nonlinear Biomedical Signal Processing*, vol. II, pages 294-318. IEEE Press, 2000.

13. L.D. Iasemidis, P.M. Pardalos, J.C. Sackellares, and D.-S. Shiau. Quadratic binary programming and dynamical system approach to determine the predictability of epileptic seizures. *Journal of Combinatorial Optimization*, 5: 9-26, 2001.

14. L.D. Iasemidis, D.-S. Shiau, J.C. Sackellares, P.M. Pardalos, and A. Prasad. Dynamical resetting of the human brain at epileptic seizures: application of nonlinear dynamics and global optimization tecniques. *IEEE Transactions on Biomedical Engineering*, 51(3): 493-506, 2004.

15. P.M. Pardalos, W. Chaovalitwongse, L.D. Iasemidis, J.C. Sackellares, D.-S. Shiau, P.R. Carney, O.A. Prokopyev, and V.A. Yatsenko. Seizure Warn-

ing Algorithm Based on Spatiotemporal Dynamics of Intracranial EEG. *Mathematical Programming*, 101(2): 365-385, 2004.

16. J.C. Sackellares, L.D. Iasemidis, R.L. Gilmore, S.N. Roper. Epileptic seizures as neural resetting mechanisms. *Epilepsia*, 38(S3): 189, 1997.

17. D.S. Shiau, Q. Luo, S.L. Gilmore, S.N. Roper, P.M. Pardalos, J.C. Sackellares, L.D. Iasemidis. Epileptic seizures resetting revisited. *Epilepsia*, 41(S7): 208-209, 2000.

18. A. Wolf, J.B. Swift, H.L. Swinney, and J.A. Vastano. Determining Lyapunov exponents from a time series. *Physica D*, 16: 285-317, 1985.

19. Networks in Biology, *Science*, 301(5641): 1797–1980, 2003.

Index

Printed in the United States
72759LV00001B/25-39